Excel

MATHEMATICS
STUDY GUIDE
YEARS 9–10

Get the Results You Want!

Allyn Jones

Reprinted 2017, 2019, 2021, 2022, 2024, 2025

ISBN 978 1 74125 479 2

Pascal Press
PO Box 250
Glebe NSW 2037
www.pascalpress.com.au

Publisher: Vivienne Joannou
Project Editors: Leanne Poll and Karen Enkelaar
Edited by Karen Enkelaar
Indexed by Puddingburn Publishing Services
Answers checked by Peter Little
Typeset by Nikki M Group Pty Ltd
Page design by Larissa Petryca
Cover by DiZign Pty Ltd
Printed by Vivar Printing/Green Giant Press

Students
All care has been taken in compiling this study guide, but please check with your teacher about the exact requirements of the course as these can change from year to year.

Table of Contents

How to use this book

- This book covers **the topics in the Year 9 and 10 Australian Curriculum (Mathematics)**, as well as the New South Wales Mathematics Stage 5.1 and Stage 5.2 syllabuses. It is best to work through it bit by bit over the whole two years. However, you can also use it for revision before tests or examinations. You will also find it useful for finding out about particular topics.
- Try to work through the book in **chapter order** (Chapter 1 first, then Chapter 2, Chapter 3, and so on). Your teacher may cover topics in a different order. If you are not sure which chapter of the book relates to your classroom work, then ask your teacher for assistance.
- If you want to practise a particular topic, then look up that topic in the Table of contents (the pages before this one) or the index (at the back of the book).

Chapter by Chapter ...

- If you want to work through a whole chapter start at the contents page. Here you will find **an overview** of the topics covered in that chapter.

Keywords

Coefficient	Literal
Elimination	Quadratic
Equation	Simultaneous
Exact	Solution
Factorise	Subject
Formula	Substitution
Formulae	Surd
Inequality	Variable

Keywords from Chapter 4: Equations.

- You should also study the **keywords** at the start of the chapter. Look out for these important terms as you work through the chapter.

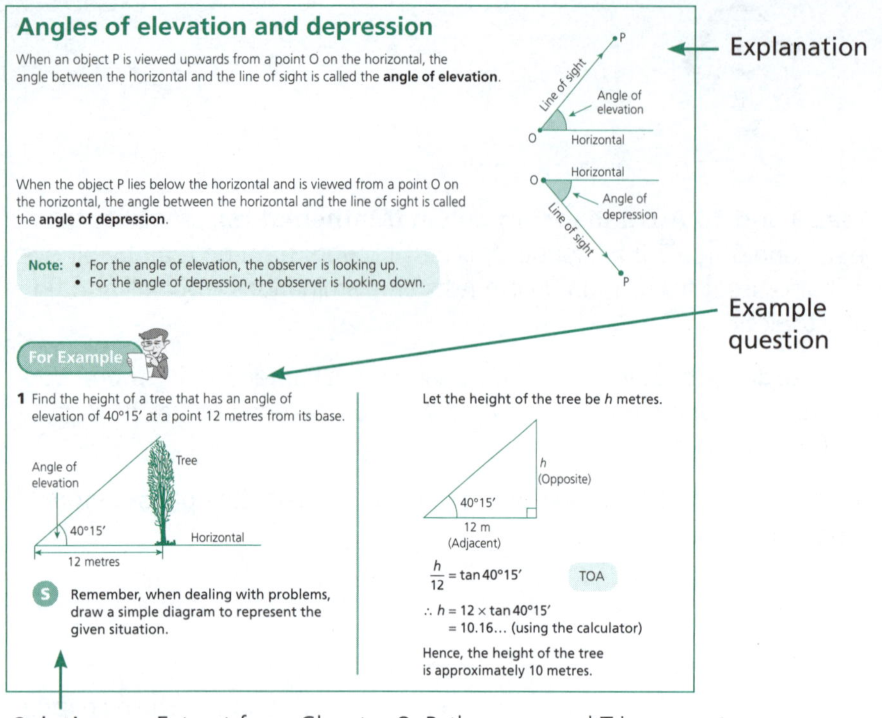

Angles of elevation and depression

When an object P is viewed upwards from a point O on the horizontal, the angle between the horizontal and the line of sight is called the **angle of elevation**.

When the object P lies below the horizontal and is viewed from a point O on the horizontal, the angle between the horizontal and the line of sight is called the **angle of depression**.

Note:
- For the angle of elevation, the observer is looking up.
- For the angle of depression, the observer is looking down.

For Example

1 Find the height of a tree that has an angle of elevation of 40°15′ at a point 12 metres from its base.

Remember, when dealing with problems, draw a simple diagram to represent the given situation.

Let the height of the tree be h metres.

$\frac{h}{12} = \tan 40°15'$ TOA

$\therefore h = 12 \times \tan 40°15'$
$= 10.16\ldots$ (using the calculator)

Hence, the height of the tree is approximately 10 metres.

Explanation

Example question

Solution

Extract from Chapter 8: Pythagoras and Trigonometry

- Each concept is explained using both an **explanation** and an **example question**. Read the explanation and then work through the example question. The worked solution to the **example question** follows and is signalled by the S solution icon. Make sure you understand how the answer was found. If you are not sure, read through the topic again or ask your teacher for help.
- Use an exercise book to make notes as you go along. This will help you remember what you have learned.

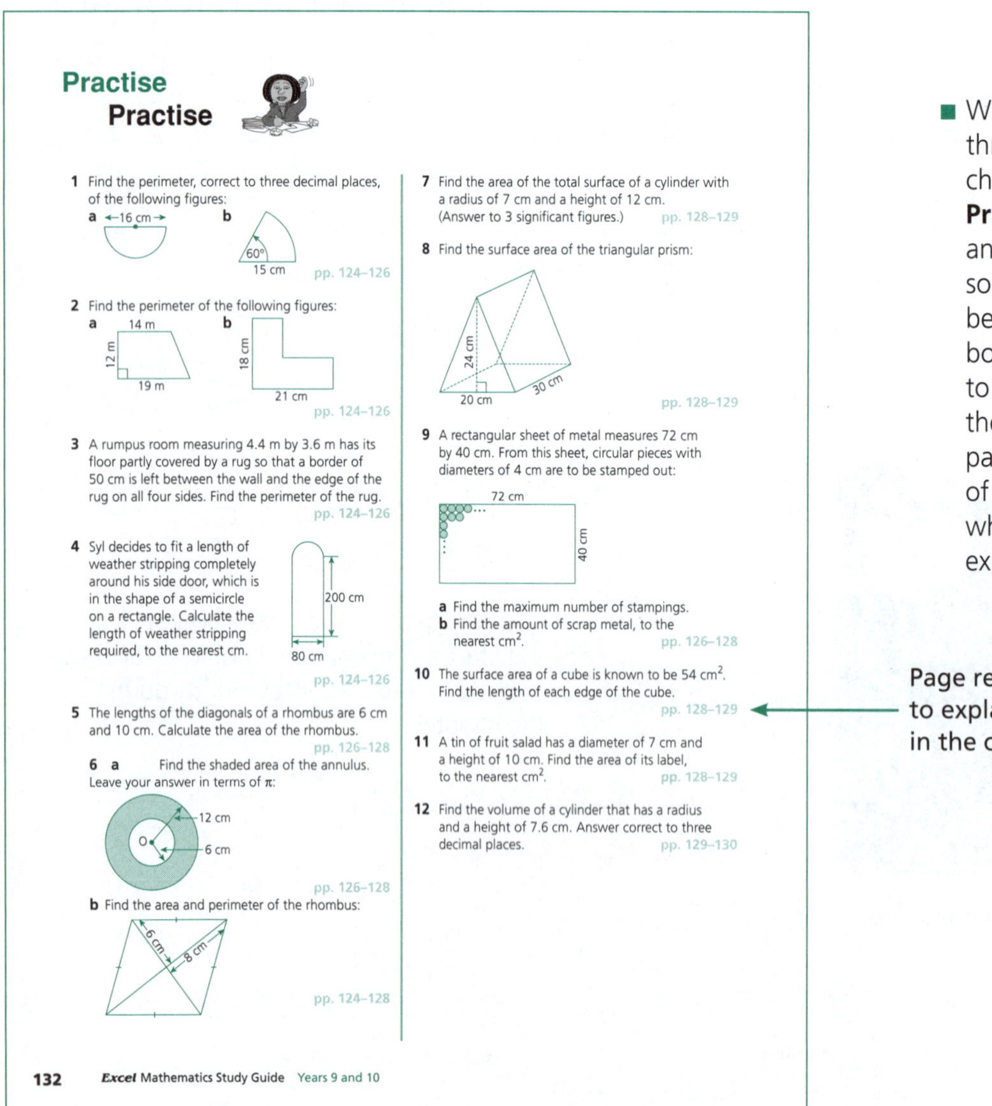

Practise Practise

1 Find the perimeter, correct to three decimal places, of the following figures:
a ← 16 cm → b 60° 15 cm pp. 124–126

2 Find the perimeter of the following figures:
a 14 m, 12 m, 19 m b 18 cm, 21 cm pp. 124–126

3 A rumpus room measuring 4.4 m by 3.6 m has its floor partly covered by a rug so that a border of 50 cm is left between the wall and the edge of the rug on all four sides. Find the perimeter of the rug. pp. 124–126

4 Syl decides to fit a length of weather stripping completely around his side door, which is in the shape of a semicircle on a rectangle. Calculate the length of weather stripping required, to the nearest cm. 200 cm, 80 cm pp. 124–126

5 The lengths of the diagonals of a rhombus are 6 cm and 10 cm. Calculate the area of the rhombus. pp. 126–128

6 a Find the shaded area of the annulus. Leave your answer in terms of π: 12 cm, 6 cm, O pp. 126–128

b Find the area and perimeter of the rhombus: 6 cm, 8 cm pp. 124–128

7 Find the area of the total surface of a cylinder with a radius of 7 cm and a height of 12 cm. (Answer to 3 significant figures.) pp. 128–129

8 Find the surface area of the triangular prism: 24 cm, 20 cm, 30 cm pp. 128–129

9 A rectangular sheet of metal measures 72 cm by 40 cm. From this sheet, circular pieces with diameters of 4 cm are to be stamped out: 72 cm, 40 cm
a Find the maximum number of stampings.
b Find the amount of scrap metal, to the nearest cm^2. pp. 126–128

10 The surface area of a cube is known to be 54 cm^2. Find the length of each edge of the cube. pp. 128–129

11 A tin of fruit salad has a diameter of 7 cm and a height of 10 cm. Find the area of its label, to the nearest cm^2. pp. 128–129

12 Find the volume of a cylinder that has a radius and a height of 7.6 cm. Answer correct to three decimal places. pp. 129–130

132 *Excel* Mathematics Study Guide Years 9 and 10

- When you have worked through all topics in the chapter, try the **Practise Practise** section. Quick answers and worked solutions for this section may be found at the back of the book, with page references to answers and solutions at the end of the section. The page reference on the right of each question tells you where each concept is explained in the chapter.

Page reference to explanations in the chapter

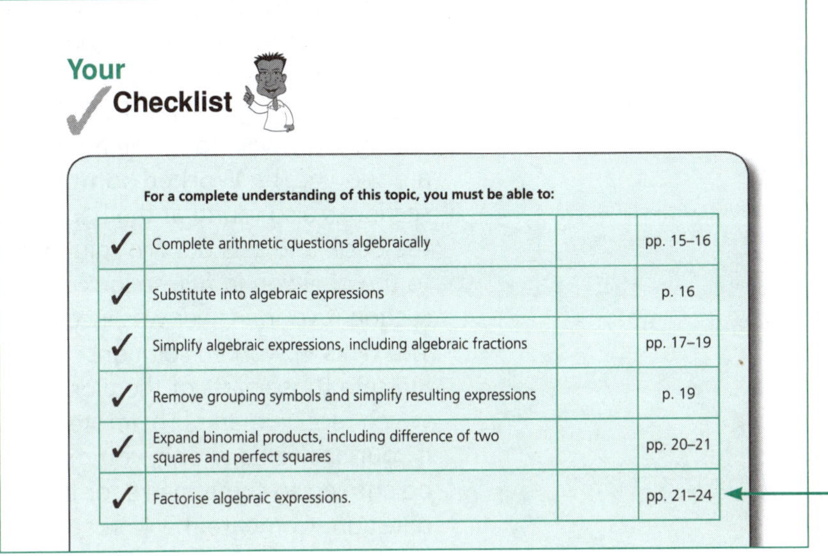

Your Checklist

For a complete understanding of this topic, you must be able to:

✓	Complete arithmetic questions algebraically		pp. 15–16
✓	Substitute into algebraic expressions		p. 16
✓	Simplify algebraic expressions, including algebraic fractions		pp. 17–19
✓	Remove grouping symbols and simplify resulting expressions		p. 19
✓	Expand binomial products, including difference of two squares and perfect squares		pp. 20–21
✓	Factorise algebraic expressions.		pp. 21–24

- Go through the **Checklist** near the end of the chapter to ensure that you understand the key concepts of the chapter. Put a tick next to each checklist item that you have covered. You need to be able to tick every item to be able to do the tests. If you have missed any items, go through the chapter to ensure that you cover the relevant topics. The page reference next to each item tells you where the concept is explained.

Page reference

Testing times ...

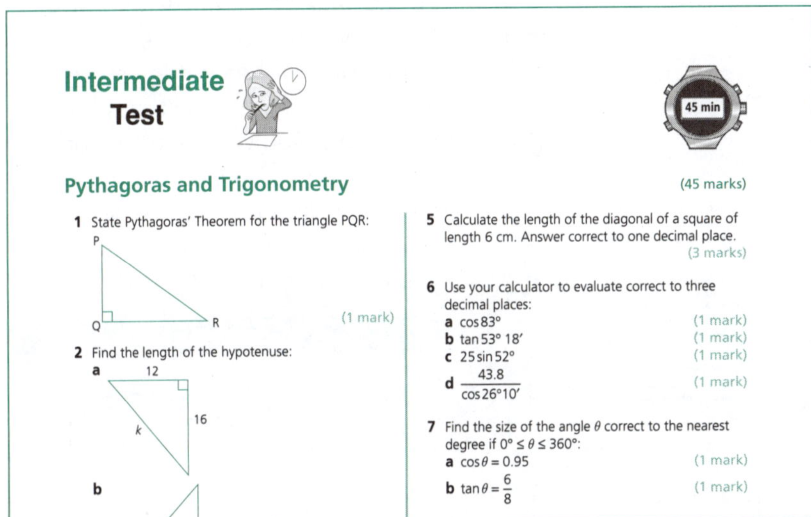

Intermediate Test

45 min

Pythagoras and Trigonometry (45 marks)

1 State Pythagoras' Theorem for the triangle PQR: (1 mark)

2 Find the length of the hypotenuse:
a
b

5 Calculate the length of the diagonal of a square of length 6 cm. Answer correct to one decimal place. (3 marks)

6 Use your calculator to evaluate correct to three decimal places:
a $\cos 83°$ (1 mark)
b $\tan 53° 18'$ (1 mark)
c $25 \sin 52°$ (1 mark)
d $\frac{43.8}{\cos 26°10'}$ (1 mark)

7 Find the size of the angle θ correct to the nearest degree if $0° \le \theta \le 360°$:
a $\cos\theta = 0.95$ (1 mark)
b $\tan\theta = \frac{6}{8}$ (1 mark)

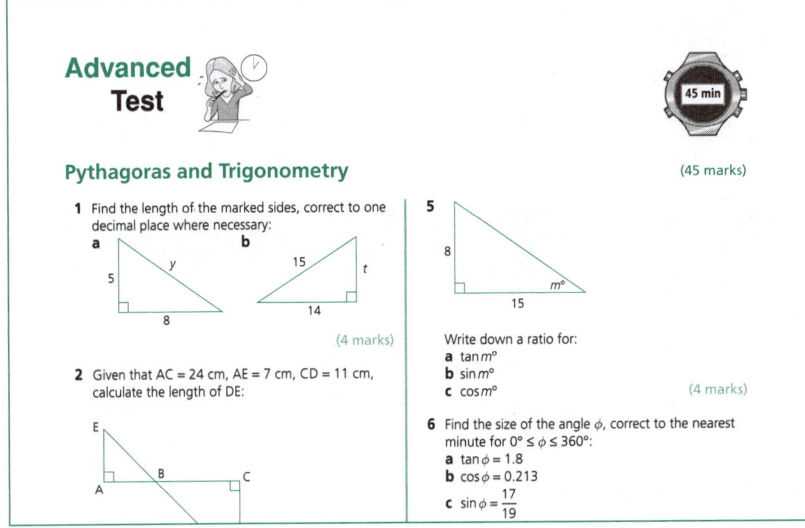

Advanced Test

45 min

Pythagoras and Trigonometry (45 marks)

1 Find the length of the marked sides, correct to one decimal place where necessary:
a
b
(4 marks)

2 Given that AC = 24 cm, AE = 7 cm, CD = 11 cm, calculate the length of DE:

5
Write down a ratio for:
a $\tan m°$
b $\sin m°$
c $\cos m°$
(4 marks)

6 Find the size of the angle ϕ, correct to the nearest minute for $0° \le \phi \le 360°$:
a $\tan\phi = 1.8$
b $\cos\phi = 0.213$
c $\sin\phi = \frac{17}{19}$

- Next, attempt both **Tests**. The tests are of two levels of difficulty: the intermediate tests are of average difficulty and the advanced tests are above average. Start with the intermediate test. If you find it easy, it will still be good practice before you attempt the advanced test. Complete all the questions you can—even if you find them difficult. Because there are worked solutions to each question at the back of the book, you will always be able to find out how to obtain each answer.
- The tests are designed to **prepare you for your school tests or examinations** so try to keep to the time on the stopwatch at the start of the test. Avoid looking at your textbook or notes while doing the test—this will help you develop your examination technique.
- **Marks** are allocated for each question. These are similar to the marks you will be trying for in your school tests and exams. Spend more time on the questions that are worth the most marks. For example, if there are 20 marks in total, and 20 minutes have been allocated for completion of the test, then spend about one minute on each mark.

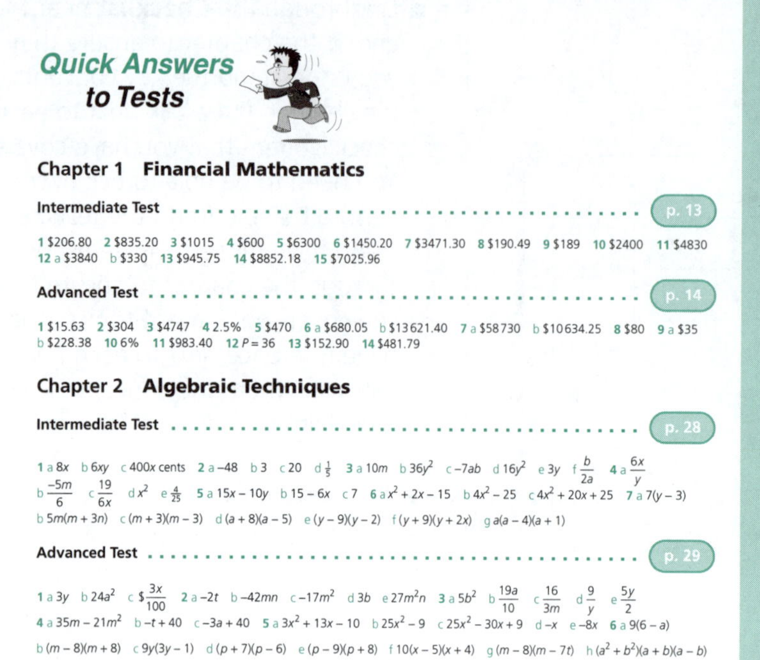
Quick Answers to Tests

Chapter 1 Financial Mathematics

Intermediate Test .... p. 13

1 \$206.80 2 \$835.20 3 \$1015 4 \$600 5 \$6300 6 \$1450.20 7 \$3471.30 8 \$190.49 9 \$189 10 \$2400 11 \$4830 12 a \$3840 b \$330 13 \$945.75 14 \$8852.18 15 \$7025.96

Advanced Test .... p. 14

1 \$15.63 2 \$304 3 \$4747 4 2.5% 5 \$470 6 a \$680.05 b \$13 621.40 7 a \$58 730 b \$10 634.25 8 \$80 9 a \$35 b \$228.38 10 6% 11 \$983.40 12 $P = 36$ 13 \$152.90 14 \$481.79

Chapter 2 Algebraic Techniques

Intermediate Test .... p. 28

1 a $8x$ b $6xy$ c $400x$ cents 2 a -48 b 3 c 20 d $\frac{1}{5}$ 3 a $10m$ b $36y^2$ c $-7ab$ d $16y^2$ e $3y$ f $\frac{b}{2a}$ 4 a $\frac{6x}{y}$ b $\frac{-5m}{6}$ c $\frac{19}{6x}$ d x^2 e $\frac{4}{25}$ 5 a $15x - 10y$ b $15 - 6x$ c 7 6 a $x^2 + 2x - 15$ b $4x^2 - 25$ c $4x^2 + 20x + 25$ 7 a $7(y - 3)$ b $5m(m + 3n)$ c $(m + 3)(m - 3)$ d $(a + 8)(a - 5)$ e $(y - 9)(y - 2)$ f $(y + 9)(y + 2x)$ g $a(a - 4)(a + 1)$

Advanced Test .... p. 29

1 a $3y$ b $24a^2$ c $\$\frac{3x}{100}$ 2 a $-2t$ b $-42mn$ c $-17m^2$ d $3b$ e $27m^2n$ 3 a $5b^2$ b $\frac{19a}{10}$ c $\frac{16}{3m}$ d $\frac{9}{y}$ e $\frac{5y}{2}$ 4 a $35m - 21m^2$ b $-t + 40$ c $-3a + 40$ 5 a $3x^2 + 13x - 10$ b $25x^2 - 9$ c $25x^2 - 30x + 9$ d $-x$ e $-8x$ 6 a $9(6 - a)$ b $(m - 8)(m + 8)$ c $9y(3y - 1)$ d $(p + 7)(p - 6)$ e $(p - 9)(p + 8)$ f $10(x - 5)(x + 4)$ g $(m - 8)(m - 7t)$ h $(a^2 + b^2)(a + b)(a - b)$

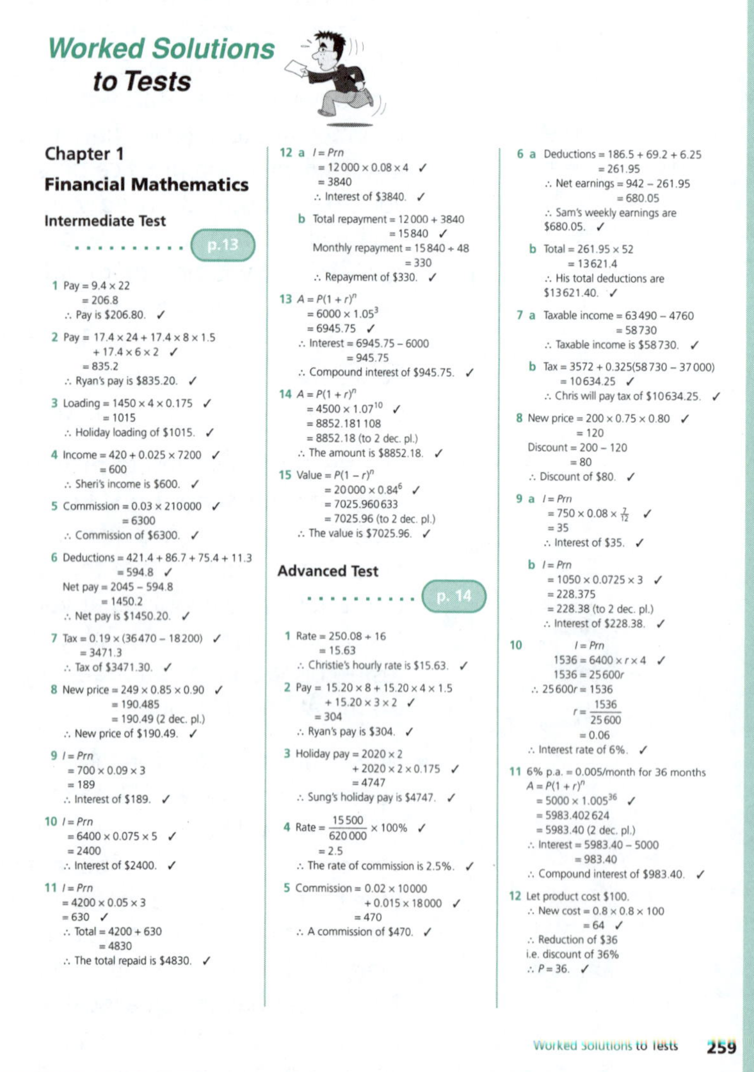
Worked Solutions to Tests

Chapter 1 Financial Mathematics

Intermediate Test p.13

1 Pay = 9.4 × 22
= 206.8
∴ Pay is \$206.80. ✓

2 Pay = 17.4 × 24 + 17.4 × 8 × 1.5 + 17.4 × 6 × 2 ✓
= 835.2
∴ Ryan's pay is \$835.20. ✓

3 Loading = 1450 × 4 × 0.175 ✓
= 1015
∴ Holiday loading of \$1015. ✓

4 Income = 420 + 0.025 × 7200 ✓
= 600
∴ Sheri's income is \$600. ✓

5 Commission = 0.03 × 210 000 ✓
= 6300
∴ Commission of \$6300. ✓

6 Deductions = 421.4 + 86.7 + 75.4 + 11.3
= 594.8 ✓
Net pay = 2045 − 594.8
= 1450.2
∴ Net pay is \$1450.20. ✓

7 Tax = 0.19 × (36 470 − 18 200) ✓
= 3471.3
∴ Tax of \$3471.30. ✓

8 New price = 249 × 0.85 × 0.90 ✓
= 190.485
= 190.49 (2 dec. pl.)
∴ New price of \$190.49. ✓

9 $I = Prn$
= 700 × 0.09 × 3
= 189
∴ Interest of \$189. ✓

10 $I = Prn$
= 6400 × 0.075 × 5 ✓
= 2400
∴ Interest of \$2400. ✓

11 $I = Prn$
= 4200 × 0.05 × 3
= 630 ✓
∴ Total = 4200 + 630
= 4830
∴ The total repaid is \$4830. ✓

12 a $I = Prn$
= 12 000 × 0.08 × 4 ✓
= 3840
∴ Interest of \$3840. ✓

b Total repayment = 12 000 + 3840
= 15 840 ✓
Monthly repayment = 15 840 ÷ 48
= 330
∴ Repayment of \$330. ✓

13 $A = P(1 + r)^n$
$= 6000 \times 1.05^3$
= 6945.75 ✓
∴ Interest = 6945.75 − 6000
= 945.75
∴ Compound interest of \$945.75. ✓

14 $A = P(1 + r)^n$
$= 4500 \times 1.07^{10}$ ✓
= 8852.181 108
= 8852.18 (to 2 dec. pl.)
∴ The amount is \$8852.18. ✓

15 Value = $P(1 - r)^n$
$= 20\,000 \times 0.84^6$ ✓
= 7025.960 633
= 7025.96 (to 2 dec. pl.)
∴ The value is \$7025.96. ✓

Advanced Test .... p. 14

1 Rate = 250.08 ÷ 16
= 15.63
∴ Christie's hourly rate is \$15.63. ✓

2 Pay = 15.20 × 8 + 15.20 × 4 × 1.5 + 15.20 × 3 × 2 ✓
= 304
∴ Ryan's pay is \$304. ✓

3 Holiday pay = 2020 × 2 + 2020 × 2 × 0.175 ✓
= 4747
∴ Sung's holiday pay is \$4747. ✓

4 Rate = $\frac{15\,500}{620\,000} \times 100\%$ ✓
= 2.5
∴ The rate of commission is 2.5%. ✓

5 Commission = 0.02 × 10 000 + 0.015 × 18 000 ✓
= 470
∴ A commission of \$470. ✓

6 a Deductions = 186.5 + 69.2 + 6.25
= 261.95
∴ Net earnings = 942 − 261.95
= 680.05
∴ Sam's weekly earnings are \$680.05. ✓

b Total = 261.95 × 52
= 13 621.4
∴ His total deductions are \$13 621.40. ✓

7 a Taxable income = 63 490 − 4760
= 58 730
∴ Taxable income is \$58 730. ✓

b Tax = 3572 + 0.325(58 730 − 37 000)
= 10 634.25 ✓
∴ Chris will pay tax of \$10 634.25. ✓

8 New price = 200 × 0.75 × 0.80 ✓
= 120
Discount = 200 − 120
= 80
∴ Discount of \$80. ✓

9 a $I = Prn$
$= 750 \times 0.08 \times \frac{7}{12}$ ✓
= 35
∴ Interest of \$35. ✓

b $I = Prn$
= 1050 × 0.0725 × 3 ✓
= 228.375
= 228.38 (to 2 dec. pl.)
∴ Interest of \$228.38. ✓

10 $I = Prn$
$1536 = 6400 \times r \times 4$ ✓
$1536 = 25\,600r$
∴ $25\,600r = 1536$
$r = \frac{1536}{25\,600}$
= 0.06
∴ Interest rate of 6%. ✓

11 6% p.a. = 0.005/month for 36 months
$A = P(1 + r)^n$
$= 5000 \times 1.005^{36}$ ✓
= 5983.402 624
= 5983.40 (2 dec. pl.)
∴ Interest = 5983.40 − 5000
= 983.40
∴ Compound interest of \$983.40. ✓

12 Let product cost \$100.
∴ New cost = 0.8 × 0.8 × 100
= 64 ✓
∴ Reduction of \$36
i.e. discount of 36%
∴ $P = 36$. ✓

Worked Solutions to Tests 259

- **Mark your work** by referring to the answers at the back of the book. The **Quick answers** section allows you to see straight away if you have the right answer—only the answers are shown. The **Worked solutions** section is also found at the back of the book and sets out the solutions to the question in full, so look at this section if your answer was incorrect. The **ticks** in worked solutions indicate those parts of the working which receive marks. Therefore, even if your answer is wrong, you may be entitled to some marks for the question. Compare the worked solution to your own working to find out if you are entitled to any marks for the question.
- Write your own score in the **Your Feeedback** space at the end of the test. From this you will be able to work out your percentage score. If your score for the test was less than 50%, you will need to work through the chapter again. If you are still having trouble, ask your teacher to explain those topics that you can't understand.

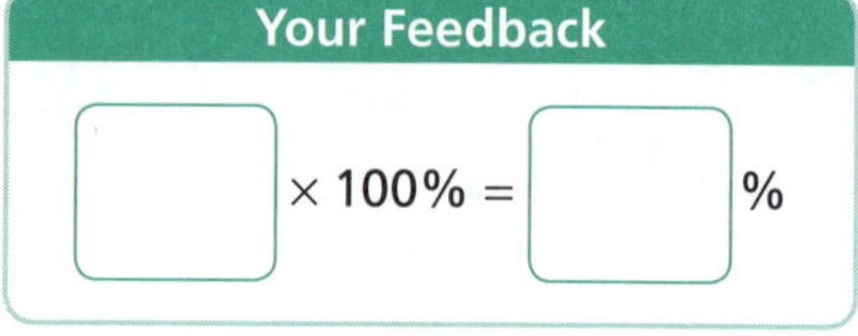

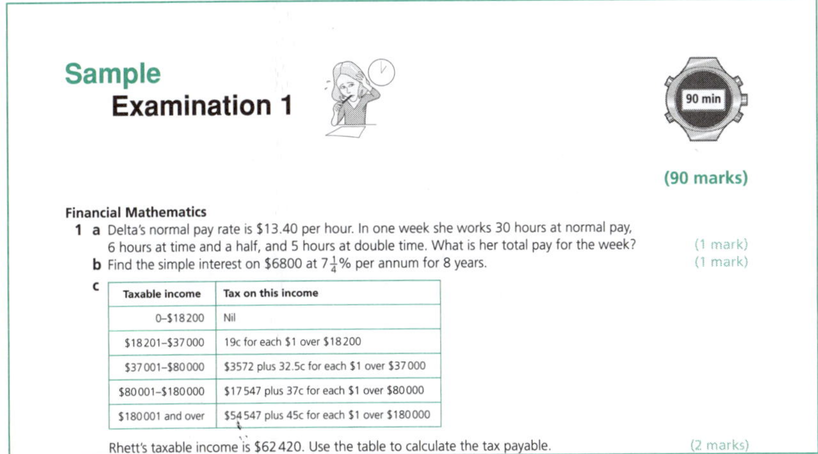

Sample Examination 1

90 min

(90 marks)

Financial Mathematics

1 a Delta's normal pay rate is \$13.40 per hour. In one week she works 30 hours at normal pay, 6 hours at time and a half, and 5 hours at double time. What is her total pay for the week? (1 mark)

b Find the simple interest on \$6800 at $7\frac{1}{4}\%$ per annum for 8 years. (1 mark)

c

Taxable income	Tax on this income
0–\$18 200	Nil
\$18 201–\$37 000	19c for each \$1 over \$18 200
\$37 001–\$80 000	\$3572 plus 32.5c for each \$1 over \$37 000
\$80 001–\$180 000	\$17 547 plus 37c for each \$1 over \$80 000
\$180 001 and over	\$54 547 plus 45c for each \$1 over \$180 000

Rhett's taxable income is \$62 420. Use the table to calculate the tax payable. (2 marks)

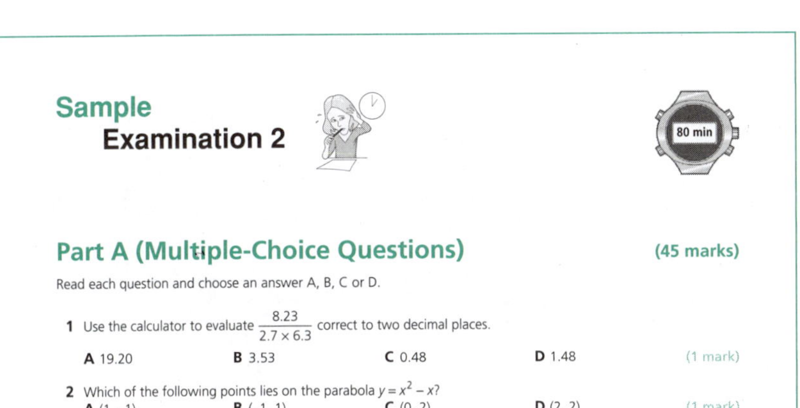

Sample Examination 2

80 min

Part A (Multiple-Choice Questions) (45 marks)

Read each question and choose an answer A, B, C or D.

1 Use the calculator to evaluate $\frac{8.23}{2.7 \times 6.3}$ correct to two decimal places.

A 19.20 **B** 3.53 **C** 0.48 **D** 1.48 (1 mark)

2 Which of the following points lies on the parabola $y = x^2 - x$?

A (1, –1) **B** (–1, 1) **C** (0, 2) **D** (2, 2) (1 mark)

- When you have completed all of the chapters you can try the **Sample Examination Papers**. The Sample Examination Papers cover the entire course's work, so you should only attempt the papers if you have covered all the topics in this book. (Ask your teacher if you are not sure.) The Sample Examination Papers will provide you with good practice for your final examination. Set aside the time allowed for each paper and complete them under exam conditions.

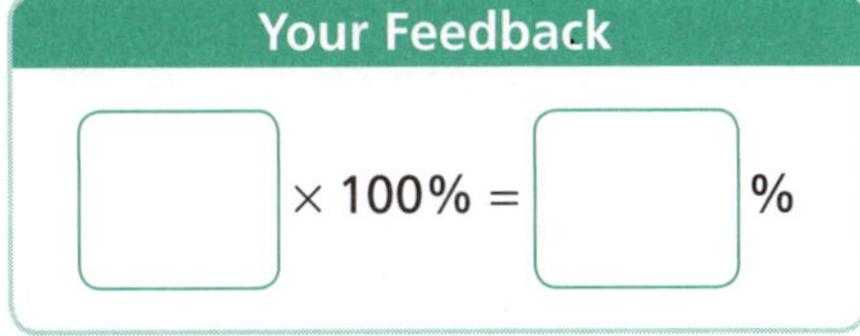

- Write your own score in the box at the end of the exam paper. From this you will be able to work out your percentage score.

Some general advice on succeeding in Mathematics

- Mathematics is a subject that **builds on knowledge**. Often you need to master one topic in order to be able to understand a later topic. This is why you should work through the book from beginning to end.
- Many students fail to **read questions carefully**. This accounts for a large number of the mistakes in exams. Although your time is limited, you must still take enough time to read each question carefully. If necessary, read the question a second (or even a third) time. Don't start your answer until you are sure you understand the question. For the longer questions, take a moment or two to **plan** your answer.
- Take care with your working. Write all answers down the page and avoid skipping any steps, as this often leads to errors. Your teachers like to see all your working, so it is wise to get into good habits. For questions that require a lot of working (such as equations), **write down each step of the solution on a separate line, lining up any equal signs** (=). This helps your teachers follow your working and will also help you check your work.
- Set out all of your working because in Mathematics you may get some marks for your working, even if your answer is wrong.
- Always work through the solutions to any questions you answer incorrectly. This helps you learn. Remember: if you have attempted the questions, you are entitled to look at the answers—this is not cheating. If you don't get feedback, you won't be able to improve!
- You cannot expect to score 100% all the time. Remember that this book has been designed to help you identify your strengths as well as your weaknesses, so it is OK to make mistakes. The key to success is to learn from those mistakes.

Chapter 1
Financial Mathematics

Salaries

A salary is expressed as a yearly amount that is usually paid monthly, fortnightly or weekly.

Note: 1 year = 52 weeks
1 year = 26 fortnights
1 fortnight = 2 weeks

Keywords

Commission
Compound interest
Deduction
Depreciation
Holiday loading
Investment
Per annum (p.a.)
Principal
Simple interest
Superannuation
Taxation

1 An architect earns a salary of $62 062 each year. What is her fortnightly pay?

S Fortnightly pay = 62 062 ÷ 26
= 2387

∴ Her pay is $2387.

2 Costa earns $3208 per month. What is his annual salary?

S Annual salary = 3208 × 12
= 38 496

∴ His salary is $38 496.

Wages and overtime

Wage earners are usually paid at a normal hourly rate, and overtime rates are paid for hours worked in excess of the normal working day. The rate for the overtime hours is called 'time and a half' ($1\frac{1}{2}$ × normal rate) or 'double time' (2 × normal rate).

1 Jan works as a sales assistant in a dress shop and she earns $20.20 per hour. Find her wage in a week where she worked 36 normal hours.

S Wage = 36 × 20.20
= 727.20

∴ Jan receives $727.20.

2 Paul earns $548.80 in a week in which he works 28 normal hours. Find his hourly rate.

S Hourly rate = 548.80 ÷ 28
= 19.60

∴ Paul is paid at the rate of $19.60 per hour.

3 Calculate a worker's wage during a week where he worked 38 normal hours, 6 hours at time and a half and $2\frac{1}{2}$ hours at double time. He is paid at the rate of $23.50 per hour.

Note: to find the wage, multiply the hourly rate by the total number of paid hours.

S Total number of paid hours
$= 38 + (6 \times 1.5) + \left(2\frac{1}{2} \times 2\right)$
$= 38 + 9 + 5$
$= 52$

∴ Wage = 52 × 23.50
= 1222

∴ Wage is $1222 per week.

Other forms of income

Workers can receive additional forms of payment such as *holiday pay* and *commission*.

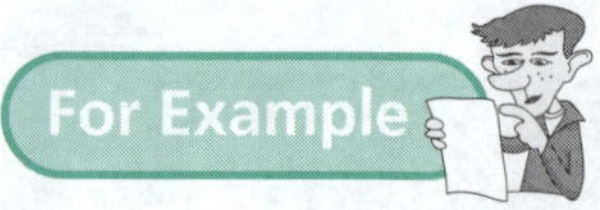

1 A real estate agent charges commission as follows for selling a house:

- $2500 on the first $85 000
- 2% on the remaining amount.

The real estate agent sold a house for $675 000. Calculate his commission.

S Remaining amount
$= 675\,000 - 85\,000$
$= 590\,000$

Agent's commission
$= 2500 + (2\% \text{ of } 590\,000)$
$= 2500 + \left(\frac{2}{100} \times 590\,000\right)$
$= 2500 + 11\,800$
$= 14\,300$

∴ The agent receives $14 300 commission for selling the house.

2 Maddison is paid $1260 per fortnight. She is paid 4 weeks' holiday loading of $17\frac{1}{2}\%$. Find her total holiday pay.

S Weekly income = $630

Holiday pay $= 630 \times 4 + 630 \times 4 \times 0.175$
$= 2961$

∴ Maddison is paid $2961 during her holidays.

Deductions from income

A worker's *net pay* is the *gross pay* minus *deductions*.

1 Mrs Green is paid $25.20 per hour. Find her net income for a week when she worked 36 hours normal time, 6 hours at 'time and a half' and she had the following deductions:

Taxation	$223.80
Health fund	$18.20
Superannuation	$15.00
Union fees	$9.15

S **Note:**
Net wage = Gross wage − Deductions

Hours to be paid $= 36 + (6 \times 1.5)$
$= 36 + 9$
$= 45$

Gross wage $= 45 \times 25.20$
$= 1134$

Total deductions
$= 223.80 + 18.20 + 15.00 + 9.15$
$= 266.15$

Net wage $= 1134 - 266.15$
$= 867.85$

∴ Mrs Green's net wage for that week is $867.85.

2 Kay receives an annual salary of $57 730.40. From this amount Kay's employer deducts $11 215.36 tax over the year. Her weekly pay slip has the following deductions: health fund $34.40, superannuation $86.50, union fees $6.70, and credit union $98.50.

a Find Kay's total annual deductions.

S **Total annual deductions**
= 11 215.36 + (52 × [34.40 + 86.50 + 6.70 + 98.50])
= 11 215.36 + 11 757.20
= 22 972.56

∴ Kay has total annual deductions of $22 972.56.

b Calculate Kay's annual net income.

S **Annual net income**
= 57 730.40 – 22 972.56
= 34 757.84

∴ Kay's annual net income is $34 757.84.

Income tax

A tax table provided by the Australian Taxation Office (ATO) is used by workers to calculate the tax payable on their taxable incomes. It contains *tax brackets*, which represent different ranges of incomes.

The table below allows taxpayers to work out their tax payable for the financial year:

Taxable income	Tax on this income
$0–$18 200	Nil
$18 201–$37 000	19c for each $1 over $18 200
$37 001–$80 000	$3572 plus 32.5c for each $1 over $37 000
$80 001–$180 000	$17 547 plus 37c for each $1 over $80 000
$180 001 and over	$54 547 plus 45c for each $1 over $180 000

1 Paul's taxable income is $43 500. Use the above table to calculate the amount of tax he must pay.

S **Tax payable = 3572 + 0.325 × 6500**
= 5684.5

∴ Paul pays $5684.50.

2 Susan's gross income is $89 750. Her total deductions are $3450.

a Find her taxable income.

S **Taxable income = Gross income – Deductions**
= 89 750 – 3450
= 86 300

∴ Taxable income is $86 300.

b Calculate the amount of tax she must pay.

S **Tax payable = 17 547 + 0.37 × 6300**
= 19 878

∴ Susan pays $19 878.

c If she pays $385 per week in tax, how much refund will she receive for the year?

S **Annual tax paid = 385 × 52**
= 20 020

Refund = 20 020 – 19 878
= 142

∴ Susan receives $142 back from the tax office.

Budgeting

Budgeting involves the managing of an individual's income. Good budgeting requires a balance between income and expenses.

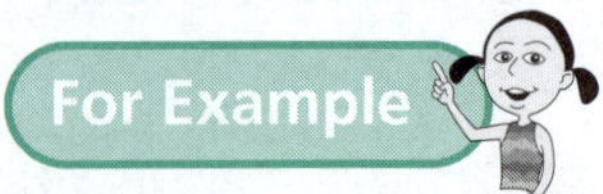

1 John is on unemployment benefits and receives an income of $138.50 per week. To supplement his income he mows lawns and receives a further $30 per week. He decides to budget his weekly income in the following way: rent $53, food $58, entertainment $13.50, gas and electricity $9.20, car expenses $18.50, and saving the balance.

a Make out a weekly budget table for John, showing income and expenses.

S **A budget table usually contains income on the left and expenses on the right.**

Total expenses
= 53.00 + 58.00 + 13.50 + 9.20 + 18.50
= 152.20

∴ **$152.20**

Total income
= 138.50 + 30.00
= 168.50

∴ **$168.50**

Balance
= Total income − Total expenses
= 168.50 − 152.20
= 16.30

∴ **$16.30**

Therefore, John's budget account should be:

Income		**Expenses**	
Unemployment benefit Mowing lawns	$138.50 $30.00	Rent Food Entertainment Gas and electricity Car expenses Balance	$53.00 $58.00 $13.50 $9.20 $18.50 $16.30
Total	$168.50	Total	$168.50

b Balance the account and calculate his weekly savings.

S **Since John saves the balance of his weekly account, he saves $16.30 per week.**

Simple interest

Interest earned when investing at a simple (or flat) interest rate is found by the formula:

$I = Prn$

where I = interest charged ($)
P = principal (amount invested) ($)
r = percentage interest rate per period, expressed as a decimal
n = number of periods

When borrowing money, the same formula can be used to find the extra money paid for the loan (the interest charged on the loan).

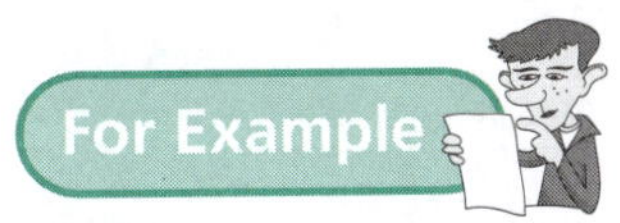

1 Calculate the amount of interest earned if \$5000 is invested for 3 years at $8\frac{1}{2}\%$ p.a.

S
$$I = Prn \qquad \text{where } P = 5000$$
$$= 5000 \times 0.085 \times 3 \qquad r = 0.085$$
$$= 1275 \qquad n = 3$$
$\therefore$ \$1275

2 Tom borrows \$3500 from a bank to buy a car. He has to pay it back, plus 8% p.a. simple interest over 36 months. How much interest does the bank charge Tom?

S
$$I = Prn \qquad \text{where } P = 3500$$
$$= 3500 \times 0.08 \times 3 \qquad r = 0.08$$
$$= 840 \qquad n = 3$$
$\therefore$ \$840

Therefore, the bank charges Tom \$840 interest over 36 months.

3 Determine the interest rate if \$2300 invested at simple interest for 2 years earns \$402.50 interest.

S In this example you are asked to calculate r, given $P = 2300$, $n = 2$ and $I = 402.50$.

From $I = Prn$ we rearrange the formula to get:

$$r = \frac{I}{Pn}$$
$$= \frac{402.50}{2300 \times 2}$$
$$= 0.0875$$

$\therefore$ The interest rate is 8.75% p.a.

Loan repayments

A loan repayment is found by first subtracting the **deposit** from the **cash price** to find the **balance owing**. Interest is then calculated over the length of the loan and added to the balance owing to find the **total repayments**. This is then divided by the number of repayments to find the size of each repayment.

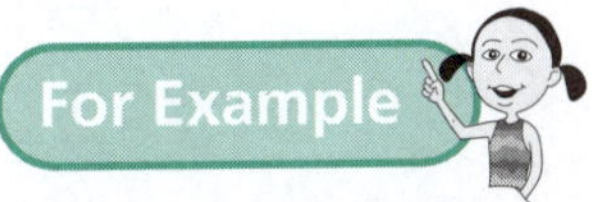

1 Susan borrows \$5400 from a bank to buy a car. She has to pay it back, plus $9\frac{1}{2}\%$ p.a. simple interest over a 5-year period.

a How much extra money does Susan have to pay the bank?

S The extra amount she has to repay equals the interest charged by the bank.

$$I = Prn \qquad P = 5400$$
$$= 5400 \times 0.095 \times 5 \qquad r = 0.095$$
$$= 2565 \qquad n = 5$$

Therefore, Susan has to repay \$2565 extra to the bank.

b How much does she have to repay altogether?

S The total amount to be repaid
= Amount borrowed + Interest
= 5400 + 2565
= 7965

$\therefore$ \$7965

c If Susan repays the loan in monthly repayments over the 5-year period, how much does she pay per month?

Note: 5 years = 60 months

S Monthly repayment
= Total amount to be repaid ÷ Number of months
= 7965 ÷ 60
= 132.75

Therefore, Susan pays an instalment of \$132.75 per month for 5 years to pay back the loan.

Time payment

Time payment is used to buy goods without paying the cost immediately. Interest is charged on the balance owing after a deposit (or trade-in) is taken off the cost price. Repayments are usually weekly or monthly, spread over a period of time.

For Example

1 A lounge suite worth $1850 is bought under a time-payment agreement. The terms required are a deposit of $110 and payment of 24 equal monthly instalments of $96.

a Find the total cost of the lounge suite.

S Total price = Deposit + Instalments
= 110 + 24 × 96
= 110 + 2304
= 2414

The total cost of the lounge suite is $2414.

Note: the extra amount paid is the interest charged.

b How much do you save by paying cash?

S Interest = Total price − Cash price
= 2414 − 1850
= 564

Therefore, if you pay cash you will save $564.

2 A car can be obtained for $25 840 cash, or through a time-payment agreement involving a deposit of 15% of the cash price and 36 monthly instalments of $875.50. Find the total cost of the car.

S Deposit = 15% of 25 840
= 0.15 × 25 840
= 3876

Total price = Deposit + Instalments
= 3876 + 36 × 875.50
= 35 394

The total cost of the car is $35 394.

Successive discounts

Sometimes retailers may offer discounts on items that have already been discounted in price. If an item is discounted in price by 15% it means the purchaser pays 85% of the original price. A second discount would not be added to the original percentage discount, but instead would be applied to the discounted price.

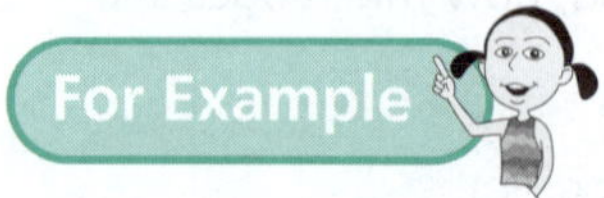

1 A car is discounted by 20%, but after a hailstorm damaged the car it dropped in price by another 30%. If the car was originally priced at $28 000 what is the new price?

S New price = 0.80 × 0.70 × 28 000
= 15 680

∴ The car's new price is $15 680.

2 All power tools have been dropped in price by 15%. If a further $7\frac{1}{2}$% discount is given for cash, how much will be saved when purchasing a drill originally priced at $96?

S New price = 0.85 × 0.925 × 96
= 75.48

∴ Savings = 96 − 75.48
= 20.52

∴ Savings of $20.52.

Compound interest

Compound interest is more realistic than simple interest and assumes depositors will gain interest on their interest as their money lies untouched in their account, rather than withdrawing interest on the day it is earned.

For example, $1000 gaining compound interest of 12% per annum over 3 years can be calculated the following way:

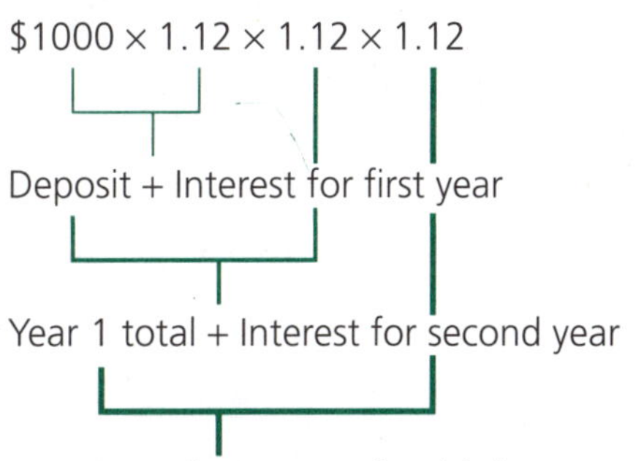

Remember: to increase $1000 by 12% you could find 12% of $1000 and add to $1000, but it is easier to simply multiply: $1000 × 1.12

That is, total after 3 years $= 1000 \times 1.12 \times 1.12 \times 1.12$
$= 1000 \times 1.12^3$
$= 1404.93$ (to the nearest cent)

Compound interest $= 1404.93 - 1000$ ← $1000 is the original deposit.
$= 404.93$ ∴ $404.93

We can use the compound interest formula:

$A = P(1 + r)^n$

where A = accumulated balance
P = principal (amount invested)
r = percentage interest rate per period, expressed as a decimal
n = number of periods

For Example

1 Terry invested $42 000 in a savings account that attracted an 8.5% interest rate compounded annually.

Find out how much money Terry has in the account after 4 years.

S $A = P(1 + r)^n$
$= 42\,000\,(1 + 0.085)^4$
$= 42\,000 \times 1.085^4$
$= 58\,206.07$ (to the nearest cent)

Terry has $58 206.07 in his savings account.

2 Mercia works as a manager of a clothing factory. Her pay conditions involve an increase of 6% every year. If this year she has a salary of $48 320, what will her salary be in 7 years' time?

S $A = P(1 + r)^n$
$= 48\,320(1 + 0.06)^7$
$= 48\,320 \times 1.06^7$
$= 72\,655.41$ (to the nearest cent)

Mercia will be paid a salary of $72 655.41.

3 Megan can choose between two accounts: the first offering simple interest at a rate of 12% and the second compound interest at a rate of 11%. If she had $400 to invest for 3 years, which account should she use to gain the greater amount of interest?

S Simple interest $= 400 \times 0.12 \times 3$
$= 144 \quad \therefore \$144$

For compound interest,
$A = 400(1.11)^3$
$= 547.05$ (to 2 dec. pl.) $\therefore \$547.05$

Compound interest $= \$547.05 - \$400$
$= \$147.05$

Megan would gain more interest by using the compound interest account.

4 Noel deposits $1000 in an account that offers interest at 6% per annum compounded monthly. He invests the money for 2 years. Find his balance after 2 years.

S As the interest is determined monthly, there will be 24 payments over the 2 years. Also, the rate is 6% per annum, that is:

$\frac{6}{12} = 0.5\%$ per month

Note: $0.5\% = 0.005$

$A = P(1 + r)^n$
$= 1000(1 + 0.005)^{24}$
$= 1000(1.005)^{24}$
$= 1127.16$

Noel will have $1127.16 in his account after 2 years.

Depreciation

Whereas compound interest is calculating values that are increasing, depreciation is the opposite—that is, calculating values that are *decreasing*.

We can use this formula for depreciation:

$A = P(1 - r)^n$

where A = final value
P = initial value
r = percentage interest rate per period, expressed as a decimal
n = number of periods

1 Lloyd buys a second-hand car for $14 400. If it depreciates in value at 13% per annum, find its value after 4 years.

S $A = P(1 - r)^n$
$= 14\,400(1 - 0.13)^4$
$= 14\,400(0.87)^4$
$= 8249.73$ (to the nearest cent)

Lloyd's car is valued at $8249.73.

2 Christine purchases a computer valued at $2100. If it depreciates at 20% per annum, how much will it depreciate in its third year?

S $A = P(1 - r)^n$

Value after 2 years:
$2100(1 - 0.2)^2 = 2100(0.8)^2$
$= 1344 \quad \therefore \$1344$

Value after 3 years:
$2100(1 - 0.2)^3 = 2100(0.8)^3$
$= 1075.20 \quad \therefore \$1075.20$

Amount of depreciation in third year:
$= 1344 - 1075.20$
$= 268.80$

The computer depreciates $268.80 in its third year.

Practise Practise

1 An engineer earns a salary of $77 200 a year. What is her weekly pay? p. 1

2 Susan worked $6\frac{1}{2}$ hours per day for 5 days. Find her wage if she receives $16.20 per hour. p. 1

3 Julie earns $52 390 per year. How much does she earn per fortnight? p. 1

4 James earns $739.10 for a 38-hour week. Find his hourly rate. p. 1

5 Peter earns $26.20 per hour. How much would he earn for 5 hours' work at time and a half? p. 1

6 Joanne works as a cashier and earns $14.40 per hour. Last week, Joanne worked 10 hours at the normal rate and 4 hours at time and a half. How much did Joanne earn? p. 1

7 Calculate John's wage if he earns $17.56 per hour and worked 35 normal hours, 6 hours at time and a half plus 3 hours at double time. p. 1

8 Jerry gets $36 for delivering 600 papers. He receives an extra $15 per day for delivering in wet weather. How much does he earn if he delivers 600 papers on a wet day? p. 2

9 Paul works as a postman on a yearly salary of $43 500. From this, Paul's employer deducts $8600 tax over the year.
- **a** What is Paul's yearly net salary?
- **b** If the tax is deducted in equal fortnightly amounts, how much does Paul receive per fortnight? p. 2

10 A boy has a part-time job at McDonald's. If he is paid $9.20 per hour:
- **a** How much would he earn for 8 hours' normal work?
- **b** How much would he earn for 5 normal hours and 2 hours at time and a half?
- **c** How many normal hours does he have to work to earn $138? p. 1

11 Amanda works as a sales assistant at a shoe shop. She is paid $18.75 per hour before 5.00 pm and time and a half after 5.00 pm. On Friday she worked from 2.00 pm until 9.00 pm without rest. How much did Amanda earn that day? p. 1

12 Loukia works from 10.00 am to 2.00 pm on a Saturday. She is paid $18 per hour for morning work and double time for afternoon work. How much does Loukia earn on Saturday? p. 1

13 Complete the following pay sheet for the Nopayne Insurance Company: p. 1

Name of employee	Hourly rate	Number of hours		Tax	Total net pay
		Normal	Time & $\frac{1}{2}$		
N. Logarithm	$18.93	35	–	$114.32	
L. Surd	$16.15	30	6	$125.10	
S. Pye	$17.10	32	7	$141.50	

14 Eleni worked 6 hours a day for 5 days and 4 hours on Saturday. Hours worked on Saturday are regarded as overtime and she is paid time and a half. What are her wages for the week if she is paid at $18.55 per hour? p. 1

15 John earned $774.00 in a week, during which he worked 36 hours at normal rate and 6 hours at time and a half. Find his hourly rate. p. 1

16 A salesperson earns $345 per week, plus 6% commission on all sales. Calculate his wage if he sold $4850 worth of goods. p. 2

17 A real estate agent charges commission as follows for selling a house:
- $2900 on the first $90 000
- 3% on the remaining amount.

The agent sold a house for $528 500.
- **a** Calculate the commission paid to the real estate agent.
- **b** What amount did the owner receive after deducting the agent's fee plus 1% stamp duty on the selling price of the house? p. 2

18 Don is paid $1040 per week. He receives 4 weeks' holiday loading of 17.5%. Find his total:
- **a** holiday loading
- **b** holiday pay. p. 2

19 Mr J. Brown is paid $17.62 per hour. Find his net pay for a week when he worked 35 hours at normal pay, 6 hours at time and a half, 5 hours double time and had the following deductions on his pay slip:

Taxation	$175.64
Health fund	$21.08
Insurance	$15.00
Union fees	$12.35.

p. 2

20 Mr Todd earns $1683.84 per week.
a Find his annual salary.
b Calculate his annual net income if he pays $24 385 in tax over the year.
c What is his fortnightly pay after tax has been deducted? p. 2

21 A person earned $42 580 taxable income last year. Find the tax payable on this income if the tax is $2850 plus 30 cents for each dollar over $25 000. p. 3

22 The following table allows taxpayers to work out their tax payable for the financial year:

Taxable income	Tax on this income
$0–$18 200	Nil
$18 201–$37 000	19c for each $1 over $18 200
$37 001–$80 000	$3572 plus 32.5c for each $1 over $37 000
$80 001–$180 000	$17 547 plus 37c for each $1 over $80 000
$180 001 and over	$54 547 plus 45c for each $1 over $180 000

Use the table to calculate the tax payable on the following taxable incomes:
a $4973 **b** $81 235
c $41 583 **d** $193 540 p. 3

23 Alan's gross income is $41 520. His total deductions are $1870.
a Find his taxable income.
b Calculate the amount of tax he must pay. (Use the table in Question 22.)
c If Alan pays $190 per fortnight in tax, how much refund will he receive for the year? p. 3

24 A university student receives a scholarship income of $152.00 per week. He works at a petrol station part time to earn a further $76 per week. He decides to budget his weekly income in the following way: rent $45, food $58, travelling expenses $12, clothes $17, entertainment $43.50, saving the balance.
a Make out a weekly budget table showing income and expenses.
b Balance the account and calculate his weekly savings.
c With his savings he wants to buy a phone costing $472.50. How many weeks of saving will he need? p. 4

25 A council charges rates at 0.642 cents in the dollar on the value of a property. How much will the council charge for a property valued at $430 000? p. 4

26 To pay off his housing loan of $360 000, Mr Green must pay $2700 per month for 12 years. What is the total amount he will pay? p. 5

27 Find the simple interest on:
a $1800 for 5 years at 7% p.a.
b $450 for 3 years at $8\frac{1}{2}$% p.a.
c $1200 for 18 months at 9% p.a. p. 5

28 $5200 is invested for 2 years at 12% p.a. How much simple interest is earned? p. 5

29 The Great Building Society offers the following interest rates for investments:
- 9% p.a. for a 1-year term
- 11% p.a. for a 2-year term
- $12\frac{1}{2}$% p.a. for a 3-year term.

If Jill invests $1200 for a 2-year term, how much simple interest will she earn on her investment? p. 5

30 The cash price for a sofa was marked at $1080. The Cox family bought it on time payment. They paid no deposit and $49 per month for 2 years. What was the total amount they paid? p. 5

31 Vicky borrows $20 000. She has to pay it back, plus 9% p.a. simple interest over 4 years.
a How much must she repay altogether?
b What is her monthly repayment? p. 5

32 Mr Smith borrowed $280 000 to buy a house. The terms of his loan are $1400 per month for 21 years.
a Calculate the total amount Mr Smith will pay.
b Calculate the total interest he will pay over the 21 years. p. 5

33 A bed valued at $1495 is bought on terms of $150 deposit and $11.50 per week for 4 years.
a What does the bed cost if bought on terms?
b How much interest was charged? p. 6

34 The following terms are offered by four different retailers for the purchase of goods with a cash price of \$2400. Which one is the cheapest buy?

A No deposit and \$250 per month for a year.
B \$1500 deposit and \$100 per month for a year.
C \$1000 deposit and \$150 per month for a year.
D 20% deposit on the cash price and \$210 per month for a year. p. 6

35 The cash price for a bike is \$240. Paul bought it on term payment over a 1-year period. A deposit of \$60 was paid and 12% p.a. interest was charged on the balance owing. Calculate the:

a actual interest charged
b total amount to be repaid
c amount of each monthly instalment. p. 6

36 A television (cash price \$670) is bought for a deposit of 20% of the cash price and monthly repayments to be made over $1\frac{1}{2}$ years. Interest of 12% is charged on the balance owing. Calculate the amount of each payment. p. 6

37 A furniture store held a '20% off everything' weekend sale. Shareholders in the store receive a further 10% discount. Find the price paid by a shareholder for a television originally priced at \$1450. p. 6

38 Which is the larger discount?

A Successive discounts of 15% and 25%.
B One discount of 35%. p. 6

39 Topside steak usually costs \$11.90/kg. It is discounted by 15%. Jake gains a further 10% discount when buying more than 5 kg of any product. Find the cost for Jake of 7.5 kg of topside steak. p. 6

40 Find the compound interest on:

a \$6400 at 6% p.a. over 3 years
b \$4900 at $8\frac{1}{2}$ p.a. over 5 years
c \$2600 at 6% p.a. over 2 years, compounded monthly
d \$14 000 at 7% p.a. over 4 years, compounded 6-monthly. p. 7

41 If \$9400 is invested earning 8% p.a. compounded quarterly, how much is in the account after 10 years? p. 7

42 The town of Singleton had a population in 1980 of 11 240 and it increased at the rate of 3% per year. Find the population in 2005. p. 7

43 A computer is purchased for \$1600. If the value decreases by 15% each year, what is its value after 3 years? p. 8

44 A motorbike is purchased in 2014 for \$16 300. If it depreciates by 18% per annum, find its value in 2019. p. 8

45 The number of motor vehicle fatalities in NSW this year is 340. The goal of the road safety authority is to reduce this number by 5% per year. What number of fatalities is projected in 10 years' time? p. 8

Go to p. 217 for **Quick Answers** or to pp. 226–227 for **Worked Solutions**

For a complete understanding of this topic, you must be able to:

✓	Calculate weekly, fortnightly, monthly and yearly incomes		p. 1
✓	Calculate income earned in casual (part-time) jobs using special rates such as time and a half and double time		p. 1
✓	Calculate commission		p. 2
✓	Calculate holiday loadings		p. 2
✓	Calculate net earnings considering deductions such as taxation and superannuation		pp. 2–3
✓	Calculate a budget by comparing earnings and expenditure		p. 4
✓	Calculate simple interest and apply to problems related to investing money		pp. 4–5
✓	Calculate and compare the cost of purchasing goods using loans and buying on terms		pp. 5–6
✓	Calculate the results of successive discounts		p. 6
✓	Use the formula for compound interest and apply to problems related to investing money		pp. 7–8
✓	Calculate depreciation by modifying the compound interest formula.		p. 8

Now you are ready to do the tests!

Intermediate Test

Financial Mathematics

(30 marks)

1 Lee works at a fast-food franchise, earning \$9.40 per hour. How much will he receive if he works 22 hours? (1 mark)

2 Ryan works as a baker's helper and his hourly rate is \$17.40. In one week he works 24 hours at normal pay, 8 hours at time and a half and 6 hours at double time. What is Ryan's total pay for the week? (2 marks)

3 Jeremy is paid \$1450 per week and receives 4 weeks' holiday loading at $17\frac{1}{2}\%$. Find his total holiday loading. (2 marks)

4 Sheri's weekly pay is \$420 plus $2\frac{1}{2}\%$ commission of her sales. This week her sales total \$7200. What is her income for this week? (2 marks)

5 Conan Real Estate charge 3% commission on the sale of blocks of land in a new subdivision. How much will be charged on a block sold for \$210 000? (2 marks)

6 Veronica is paid \$2045 per fortnight. The following deductions are made: taxation \$421.40, health fund \$86.70, superannuation \$75.40, union fees \$11.30. Find Veronica's net fortnightly pay. (2 marks)

7 Brian's taxable income is \$36 470. Use the table to calculate the tax payable: (2 marks)

Taxable income	Tax on this income
\$0–\$18 200	Nil
\$18 201–\$37 000	19c for each \$1 over \$18 200
\$37 001–\$80 000	\$3572 plus 32.5c for each \$1 over \$37 000
\$80 001–\$180 000	\$17 547 plus 37c for each \$1 over \$80 000
\$180 001 and over	\$54 547 plus 45c for each \$1 over \$180 000

8 A heater was priced at \$249. It was discounted by 15% and then a further 10% discount was applied during a weekend sale. What was the new price of the heater? (2 marks)

9 Find the simple interest on \$700 at 9% p.a. over 3 years. (1 mark)

10 Find the amount of simple interest on \$6400 at $7\frac{1}{2}\%$ p.a. over 5 years. (2 marks)

11 Guy borrows \$4200 from his father to buy a secondhand motorbike. He was charged simple interest of 5% per annum over a period of 3 years. How much will be repaid to his father? (2 marks)

12 Natasha borrows \$12 000 from a credit union. She is charged a flat rate (simple) interest rate of 8% p.a. over a period of 4 years.

a How much interest is Natasha charged by the credit union? (2 marks)

b If Natasha repays the loan in monthly repayments, how much will she pay each month? (2 marks)

13 Find the amount of interest on \$6000 at 5% p.a. compounded annually for 3 years. (2 marks)

14 What is the compounded amount if \$4500 is invested for 10 years at 7% p.a. compound interest? (2 marks)

15 Find the value of a car in 6 years' time if it is purchased for \$20 000 and loses its value at a rate of 16% per annum. (2 marks)

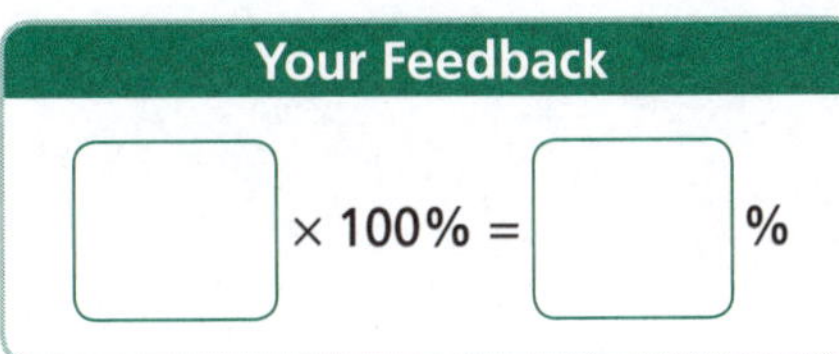

QA PAGE 221

WS PAGE 259

Advanced Test

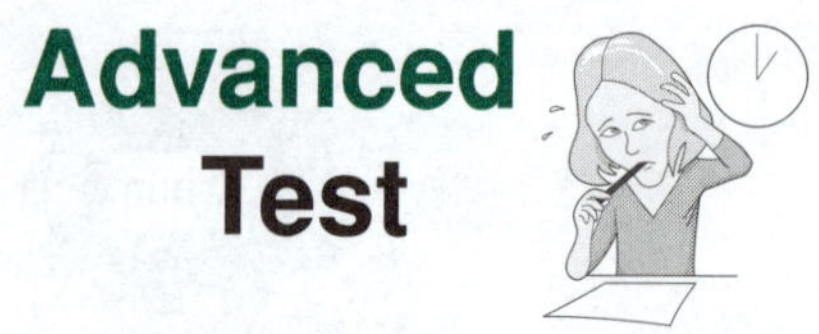

Financial Mathematics

(30 marks)

1 If Christie is paid $250.08 for working 16 hours, find her hourly rate. (1 mark)

2 Ryan is paid normal pay for the first 8 hours' work each day, time and a half for the next 4 hours, and double time for any work after that. If his hourly rate is $15.20, what is his pay for working 15 hours on a particular day? (2 marks)

3 Sung earns $2020 per fortnight. Her holiday loading is $17\frac{1}{2}$% of her normal pay for 4 weeks. Find her total holiday pay. (2 marks)

4 The commission earned by a real estate agency on the sale of a house for $620 000 was $15 500. What is the rate of commission charged by the agency? (2 marks)

5 Keis Car Auctions charge commission on the sale of cars in the following way:

- 2% on first $10 000
- 1.5% on value above $10 000.

What commission is charged on a car sold for $28 000? (2 marks)

6 Sam's gross earnings for one week are $942. The following deductions are made: taxation $186.50, superannuation $69.20, union fees $6.25. Find her:

a net weekly earnings

b total annual deductions. (2 marks)

7 Use the taxation table from Question 7 in the Intermediate Test. Chris' gross income is $63 490 and she has deductions totalling $4760. Find:

a Chris' taxable income (1 mark)

b the amount she will pay in tax. (2 marks)

8 Sam decided to buy a bracelet originally priced at $200. The shop owner reduced the price by 25% and then by another 20% when Sam offered cash. What was the total discount on the bracelet? (2 marks)

9 Find the amount of simple interest on:

a $750 at 8% p.a. over 7 months (2 marks)

b $1050 at $7\frac{1}{4}$% p.a. over 3 years. (2 marks)

10 Find the interest rate if $6400 is invested at simple interest for 4 years and earns $1536. (2 marks)

11 Find the amount of compound interest earned on $5000 at 6% p.a. compounded monthly for 3 years. (2 marks)

12 Sheridan found that successive discounts of 20% and 20% is the same as a single discount of P%. Find the value of P. (2 marks)

13 Find the difference between simple and compound interest on a deposit of $4000 at 6% p.a. over a period of 5 years. (2 marks)

14 A laptop computer is purchased for $2490. If it decreases in value at an annual rate of 28%, what is its value after 5 years? (2 marks)

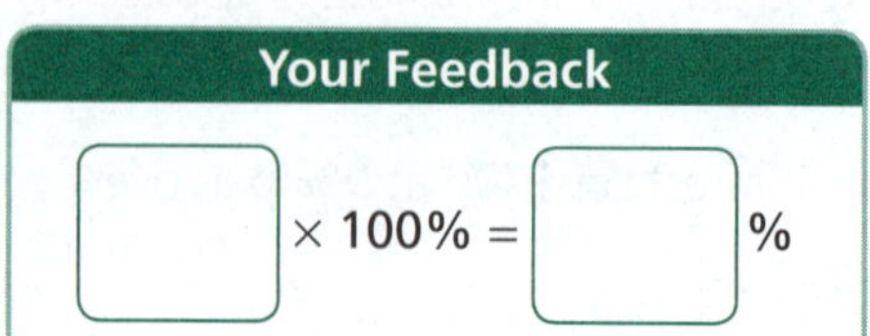

QA PAGE 221

WS PAGE 259

Chapter 2
Algebraic Techniques

Generalised arithmetic

To find the number of days in p hours, it may be helpful to replace temporarily the pronumerals with numbers: alter the question to ask how many days in 48 hours? The answer is $\frac{48}{24} = 2$, so the answer to the original question is $\frac{p}{24}$.

Keywords

Algebraic	Factor
Binomial	Factorise
Coefficient	Monic
Common	Quadratic
Expand	Simplify
Expansion	Substitution
Expression	Trinomial

For Example

1 The sum of x and y is: $x + y$

Sum: add
Product: multiply
Difference: subtract
Quotient: divide

2 The average of a, b and c is: $\frac{a+b+c}{3}$

$$\text{Average} = \text{Mean} = \frac{\text{Sum of scores}}{\text{No. of scores}}$$

3 The number 4 more than c is: $c + 4$

4 The next three consecutive whole numbers after x are: $x + 1$, $x + 2$ and $x + 3$.

Note: in generalised arithmetic, it can be helpful to substitute numbers for the pronumerals.

5 If y is odd, find the next three consecutive odd numbers.

S $y + 2$, $y + 4$, $y + 6$

All odd, and even, numbers are separated by two.

6 Convert:

a \$$y$ to cents

S $100 \times y = 100y$

\$$y = 100y$ cents

Try \$7
$\therefore 7 \times 100$
i.e. \$7 = 700c

b p litres to mL

S $p \times 1000 = 1000p$

p litres $= 1000p$ mL

Try 8 litres.
$\therefore 8 \times 1000 = 8000$ mL

c y minutes to hours

S $y \div 60 = \frac{y}{60}$

y minutes $= \frac{y}{60}$ hours

Try 120 minutes.
$\therefore \frac{120}{60} = 2$ hours

7 Find the area of the rectangle:

3 cm

(2x + 4) cm

S $A = (2x + 4) \times 3$
$= 3(2x + 4)$

Area is $3(2x + 4)$ cm^2.

8 Find the change from $5 if *y* cakes are purchased at *k* cents each.

S Cost $= y \times k$ or $k \times y$
$= yk$ cents or ky cents.

Change $= 500 - ky$
Change is $(500 - ky)$ cents.

Substitution into algebraic expressions

The pronumerals in algebraic expressions are replaced with numbers.

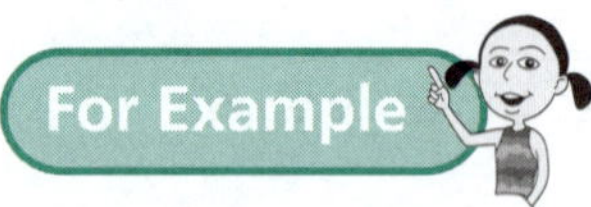

1 If $a = 3$, $b = 4$ and $c = -5$, evaluate:

a $ab + c$

S $ab + c = 3 \times 4 + (-5)$
$= 12 - 5$
$= 7$

b $b - c$

S $b - c = 4 - (-5)$
$= 4 + 5$
$= 9$

c $\dfrac{bc - 1}{a}$

S $\dfrac{bc - 1}{a} = \dfrac{4 \times -5 - 1}{3}$
$= \dfrac{-20 - 1}{3}$
$= \dfrac{-21}{3}$
$= -7$

d $c(a - b)$

S $c(a - b) = -5(3 - 4)$
$= -5(-1)$
$= 5$

e $b^2 + c^2$

S $b^2 + c^2 = (4)^2 + (-5)^2$
$= 16 + 25$
$= 41$

f $2c^2 - (2c)^2$

S $2c^2 - (2c)^2 = 2(-5)^2 - (2 \times -5)^2$
$= 2(25) - (-10)^2$
$= 50 - 100$
$= -50$

Simplifying algebraic expressions

Like terms can be added or subtracted, but **unlike terms** cannot.

For Example

Simplify the following expressions:

1 $3x + 5x + 12x$

S $3x + 5x + 12x = 20x$

2 $4x - 2y + 3x + 4y$

S $4x - 2y + 3x + 4y = 7x + 2y$

3 $12xy - 3yx$

S $12xy - 3yx = 12xy - 3xy$
$= 9xy$

Remember: $ab = ba$

4 $4a \times (-2b)$

S $4a \times (-2b) = -8ab$

5 $(-6y)^2$

S $(-6y)^2 = -6y \times -6y$
$= 36y^2$

6 $cd \div c$

S
$$cd \div c = \frac{cd}{c} = \frac{{}^{1}\cancel{c} \times d}{\cancel{c}_{1}} = \frac{d}{1} = d$$

7 $12ab \div 3a$

S
$$12ab \div 3a = \frac{12ab}{3a} = \frac{{}^{4}\cancel{12} \times \cancel{a}^{1} \times b}{{}_{1}\cancel{3} \times \cancel{a}_{1}} = \frac{4 \times b}{1} = 4b$$

8 $5pq \div p^2q$

S
$$5pq \div p^2q = \frac{5pq}{p^2q} = \frac{5 \times \cancel{p}^{1} \times \cancel{q}^{1}}{{}_{1}\cancel{p} \times p \times \cancel{q}_{1}} = \frac{5 \times 1}{1 \times p} = \frac{5}{p}$$

9 $\dfrac{12a + 3a}{5}$

S
$$\frac{12a + 3a}{5} = \frac{15a}{5} = \frac{{}^{3}\cancel{15} \times a}{\cancel{5}_{1}} = 3a$$

Simple algebraic fractions

Addition and subtraction

Find a common denominator and then add or subtract the numerators.

For Example

Add or subtract the following fractions:

1 $\frac{5y}{2} + \frac{y}{3}$

S $\frac{5y}{2} + \frac{y}{3} = \frac{15y}{6} + \frac{2y}{6}$

$= \frac{17y}{6}$

- Lowest common denominator of 2 and 3 is 6
- 2 times 3 is 6
  ∴ $5y$ times $3 = 15y$ and so on …

2 $\frac{3x}{4} - \frac{x}{2}$

S $\frac{3x}{4} - \frac{x}{2} = \frac{3x}{4} - \frac{2x}{4}$

$= \frac{x}{4}$

3 $\frac{4}{x} - \frac{3}{2x}$

S $\frac{4}{x} - \frac{3}{2x} = \frac{8}{2x} - \frac{3}{2x}$

$= \frac{5}{2x}$

4 $\frac{5}{2y} + \frac{8}{5y}$

S $\frac{5}{2y} + \frac{8}{5y} = \frac{25}{10y} + \frac{16}{10y}$

$= \frac{41}{10y}$

Multiplication and division

To multiply: cancel and then multiply numerators and denominators.
To divide: find the **reciprocal** of the second fraction (that is, turn it upside down) and then multiply.

For Example

Multiply or divide the following fractions:

1 $\frac{x}{3} \times \frac{9}{2x}$

S $\frac{x}{3} \times \frac{9}{2x} = \frac{{}^{1}\cancel{x}}{\cancel{3}_{1}} \times \frac{\cancel{9}^{3}}{2\cancel{x}_{1}}$

$= \frac{3}{2}$

$= 1\frac{1}{2}$

2 $\dfrac{4y}{7}\times\dfrac{21}{6y}$

S $$\frac{4y}{7}\times\frac{21}{6y}=\frac{{}^{2}\cancel{4}\,\cancel{y}^{1}}{\cancel{7}_{1}}\times\frac{\cancel{21}^{\cancel{3}^{1}}}{{}_{1}\cancel{3}\,\cancel{6}\,\cancel{y}_{1}}$$
$$=\frac{2}{1}$$
$$=2$$

3 $\dfrac{x+2}{6}\times\dfrac{18}{x+2}$

S $$\frac{x+2}{6}\times\frac{18}{x+2}=\frac{{}^{1}\cancel{x+2}}{\cancel{6}_{1}}\times\frac{\cancel{18}^{3}}{\cancel{x+2}_{1}}$$
$$=3$$

4 $\dfrac{ab}{4}\div\dfrac{a}{6}$

S $$\frac{ab}{4}\div\frac{a}{6}=\frac{{}^{1}\cancel{a}b}{{}_{2}\cancel{4}}\times\frac{\cancel{6}^{3}}{\cancel{a}_{1}}$$
$$=\frac{3b}{2}$$

5 $\dfrac{4x}{5ab}\div\dfrac{12}{10b}$

S $$\frac{4x}{5ab}\div\frac{12}{10b}=\frac{{}^{1}\cancel{4}x}{{}_{1}\cancel{5}a\cancel{b}_{1}}\times\frac{{}^{2}\cancel{10}\,\cancel{b}^{1}}{\cancel{12}_{3}}$$
$$=\frac{2x}{3a}$$

Removing grouping symbols

The term outside the grouping symbols multiplies the contents of the grouping symbols.

For Example

1 Expand and simplify:

a $3(2x+5y)$

S $3(2x+5y)=3\times 2x+3\times 5y$
$=6x+15y$

b $a(2a-7)$

S $a(2a-7)=a\times 2a-a\times 7$
$=2a^2-7a$

c $-(3-4y)$

S $-(3-4y)=-1(3-4y)$
$=-1\times 3-(-1)\times 4y$
$=-3+4y$

Note: the – sign before the grouping symbols has the effect of negating the contents of grouping symbols.

d $5(2x-4)-3(5-x)$

S $5(2x-4)-3(5-x)=10x-20-15+3x$
$=13x-35$

e $x(x+3)-2(x+3)$

S $x(x+3)-2(x+3)=x^2+3x-2x-6$
$=x^2+x-6$

2 Find the difference between $7x^2-4x$ and $3x+5x^2$.

S $7x^2-4x-(3x+5x^2)=7x^2-4x-3x-5x^2$
$=2x^2-7x$

Binomial products

A binomial expression has two terms, for example $2x + 1$, so a binomial product is the result of multiplying two binomial expressions.

Each term in the first binomial expression multiplies each term in the second expression.

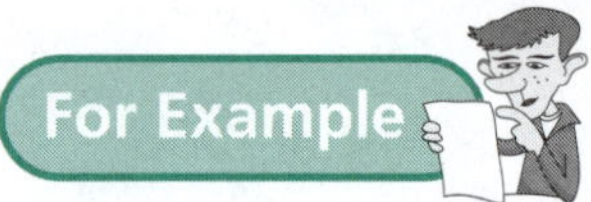

1 Expand and simplify:

a $(x + 2)(x + 4)$

S $(x + 2)(x + 4) = x(x + 4) + 2(x + 4)$
$= x^2 + 4x + 2x + 8$
$= x^2 + 6x + 8$

b $(x - 2y)(x - 3y)$

S $(x - 2y)(x - 3y) = x(x - 3y) - 2y(x - 3y)$
$= x^2 - 3xy - 2yx + 6y^2$
$= x^2 - 5xy + 6y^2$

c $(a - b)(a + b)$

S $(a - b)(a + b) = a(a + b) - b(a + b)$
$= a^2 + ab - ba - b^2$
$= a^2 - b^2$

This method can be shortened of course, or other methods used, such as:

- The **Robin Hood method** (with arrows …).

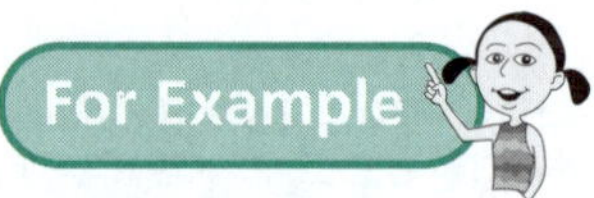

1 Expand and simplify:
$(x + 4)(x - 3)$

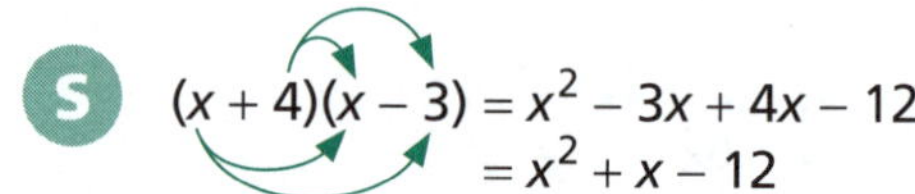

S $(x + 4)(x - 3) = x^2 - 3x + 4x - 12$
$= x^2 + x - 12$

- The **FOIL method** (First, Outside, Inside, Last).

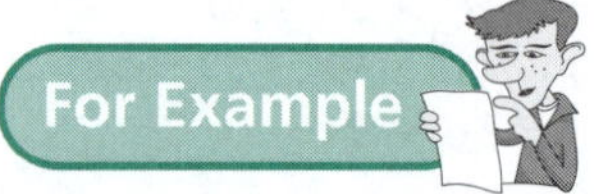

1 Expand and simplify:
$(3x + 1)(2x + 5)$

S $(3x + 1)(2x + 5)$
$= (3x)\,(2x) + (3x)\,(5) + (1)\,(2x) + (1)\,(5)$
↑ first ↑ outside ↑ inside ↑ last
$= 6x^2 + 15x + 2x + 5$
$= 6x^2 + 17x + 5$

Special products

The following results are very important and must be known for success in Year 10:

$$(a + b)^2 = a^2 + 2ab + b^2$$
$$(a - b)^2 = a^2 - 2ab + b^2$$
$$(a - b)(a + b) = a^2 - b^2$$

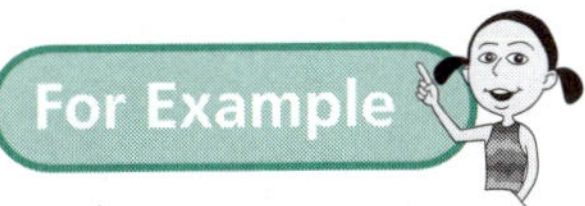

1 Expand and simplify:

a $(x + 5)^2$

S $(x + 5)^2 = (x)^2 + 2(x)(5) + (5)^2$
$= x^2 + 10x + 25$

b $(a - 2)^2$

S $(a - 2)^2 = (a)^2 - 2(a)(2) + (2)^2$
$= a^2 - 4a + 4$

c $(3x - 4)^2$

S $(3x - 4)^2 = (3x)^2 - 2(3x)(4) + (4)^2$
$= 9x^2 - 24x + 16$

d $(d - 2)(d + 2)$

S $(d - 2)(d + 2) = (d)^2 - (2)^2$
$= d^2 - 4$

e $(5y + 3)(5y - 3)$

S $(5y + 3)(5y - 3) = (5y)^2 - (3)^2$
$= 25y^2 - 9$

f $(x^2 - y^2)(x^2 + y^2)$

S $(x^2 - y^2)(x^2 + y^2) = (x^2)^2 - (y^2)^2$
$= x^4 - y^4$

Common factors

We look for the highest, or largest, factor common to the terms in the expression—this is the opposite to expanding.

Factorise:

1 $4x - 6$

S $4x - 6 = 2(2x - 3)$

Check by expanding.

2 $xy - 3x$

S $xy - 3x = x(y - 3)$

3 $12ab - 14a$

S $12ab - 14a = 2a(6b - 7)$

You must take both 2 and *a* as common factors.

4 $3x^2 - 6x - 3$

S $3x^2 - 6x - 3 = 3(x^2 - 2x - 1)$

5 $-12x - 4$

S $-12x - 4 = -4(3x + 1)$

6 $9a^2 + 12a^2b$

S $9a^2 + 12a^2b = 3a^2(3 + 4b)$

7 $4x(x + y) + 3(x + y)$

S $4x(x + y) + 3(x + y) = (x + y)[4x + 3]$
$= (x + y)(4x + 3)$

Factorising by grouping in pairs

We can factorise four-term expressions by grouping in pairs.

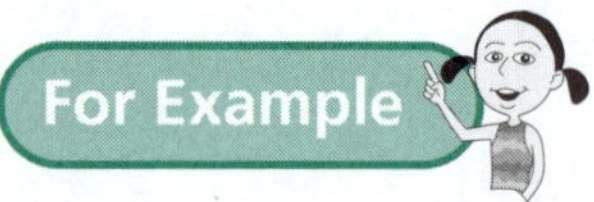

Factorise:

1 $ax + bx + ay + by$

S $ax + bx + ay + by = x(a + b) + y(a + b)$
$= (a + b)[x + y]$
$= (a + b)(x + y)$

2 $xy + y^2 - x - y$

S $xy + y^2 - x - y = y(x + y) - 1(x + y)$
$= (x + y)(y - 1)$

3 $p^2 + mq + pq + mp$

S $p^2 + mq + pq + mp = p^2 + pq + mq + mp$
$= p(p + q) + m(p + q)$
$= (p + q)(p + m)$

Difference of two squares

We can reverse an earlier rule: $a^2 - b^2 = (a - b)(a + b)$

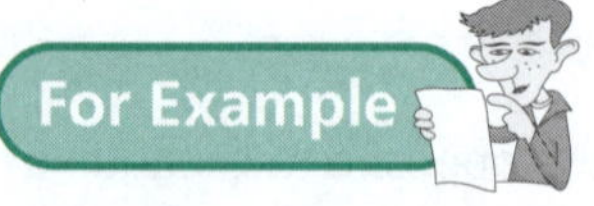

Factorise:

1 $c^2 - 9$

S $c^2 - 9 = (c)^2 - (3)^2$
$= (c - 3)(c + 3)$

Note: it does not matter if your answer has this order: $(c + 3)(c - 3)$.

The order of factors is not important, that is, $2 \times 3 = 3 \times 2$.

2 $49 - 4a^2$

S $49 - 4a^2 = (7)^2 - (2a)^2$
$= (7 - 2a)(7 + 2a)$

3 $x^4 - 1$

S $x^4 - 1 = (x^2)^2 - (1)^2$
$= (x^2 - 1)(x^2 + 1)$
$= (x - 1)(x + 1)(x^2 + 1)$

The monic quadratic trinomial

An expression with three terms is called a **trinomial**.

A trinomial with a highest power of 2 is called a **quadratic**.

If the *coefficient* (number in front) of the term with the power of 2 is 1, the quadratic trinomial is **monic**,
$\therefore x^2 + 5x + 6$ is a monic quadratic trinomial.

1 Which of the following is a quadratic trinomial?

A $2x - 1$

B $x^2 + 3x - 1$

C $2x^3 - 3x + 1$

D 2

S $x^2 + 3x - 1$ has 2 as its highest power, and so is a quadratic trinomial.

The correct answer is B.

2 Which of the following is a monic quadratic trinomial?

A $5x^2 + 4x - 1$

B $2x^2 + x - 1$

C $1 + 6x - x^2$

D $x^2 + 2x - 5$

S $x^2 + 2x - 5$ is monic as there is only one x^2 in the expression.

The correct answer is D.

Factorising monic quadratic trinomials

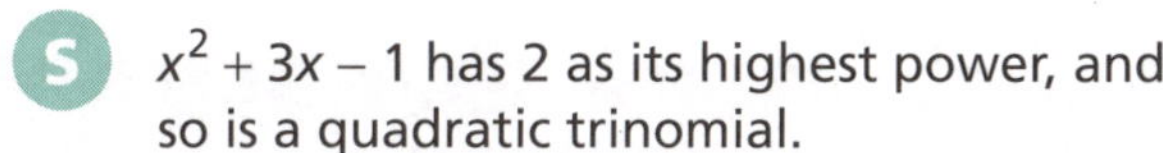

As $\quad (x + a)(x + b) = x^2 + bx + ax + ab$
$\quad = x^2 + (a + b)x + ab$

then, factorising, $x^2 + (a + b)x + ab = (x + a)(x + b)$.

Hence, to factorise $x^2 + 5x + 6$ we are looking for two numbers that add together to give 5 (that is, $a + b = 5$) and multiply together to give 6 (that is, $ab = 6$).

Factorise:

1 $x^2 + 5x + 4$

S $x^2 + 5x + 4$
(that is, $a + b = 5$, $ab = 4$)
$= (x + 4)(x + 1)$

2 $x^2 - 5x + 6$

S $x^2 - 5x + 6$
(that is, $a + b = -5$, $ab = 6$)
$= (x - 3)(x - 2)$

3 $x^2 - 3x - 4$

S $x^2 - 3x - 4$
(that is, $a + b = -3$, $ab = -4$)
$= (x - 4)(x + 1)$

4 $x^2 + 5x - 14$

S $x^2 + 5x - 14$
(that is, $a + b = 5$, $ab = -14$)
$= (x + 7)(x - 2)$

Other helpful rules to remember are:

- For $x^2 + 5x + 4$
  ↑ ↑ positive here, we say 'both'
  'this sign', that is, **both positive**.
- For $x^2 - 3x - 4$
  ↑ ↑ negative here, we say 'the bigger number is'
  'this sign', that is, **bigger one is negative**.

Factorise:

1 $x^2 + 7x + 12$

S $x^2 + 7x + 12 = (x + 4)(x + 3)$
↑
both positive

2 $x^2 - 8x + 15$

S $x^2 - 8x + 15 = (x - 5)(x - 3)$
↑
both negative

3 $x^2 - 4x - 12$

S $x^2 - 4x - 12 = (x - 6)(x + 2)$

The 'bigger number' is negative.

4 $x^2 + x - 20$

S $x^2 + x - 20 = (x + 5)(x - 4)$

The 'bigger number' is positive.

Practise Practise

1 Find the:
- **a** sum of x and y
- **b** difference between p and q (if $p > q$)
- **c** product of 7 and y
- **d** quotient of a and b p. 15

2 Convert:
- **a** \$$p$ to cents
- **b** w weeks to days
- **c** k litres to millilitres
- **d** x cm to millimetres
- **e** m grams to kilograms
- **f** y months to years p. 15

3 Find the average of:
- **a** x, y and z
- **b** x and $x + 1$
- **c** a, b, c, d p. 15

4 Write the next two consecutive whole numbers after x. p. 15

5 Write the next three consecutive even numbers after x, if x is even. p. 15

6 Write the next three consecutive even numbers after x, if x is odd. p. 15

7 Find the area of a rectangle with length $(x + 3)$ cm and width 4 cm. p. 16

8 If p golf balls are purchased at q cents each, what is the total cost? p. 16

9 A jug contains m millilitres of juice. If n millilitres is poured out of the jug, what amount remains? p. 16

10 If $a = -2$, $b = 3$ and $c = -5$, evaluate:
- **a** ab
- **b** $a + c$
- **c** $bc - ac$
- **d** $a^2 - c^2$
- **e** $2c - b$
- **f** $a^2 - 2b$
- **g** $(b - a) \div c$
- **h** $2c^2$
- **i** $\dfrac{ab + 1}{c}$
- **j** $(c - a)^2$ p. 16

11 Simplify:
- **a** $3x + 2x + x$
- **b** $4xy + 2yx$
- **c** $5a + 2b + 3a$
- **d** $7a + 12a$
- **e** $4x^2 + 3x + 2x^2 + x$
- **f** $7a + 11a + a$
- **g** $-10y + 10y$
- **h** $-3a + 7a$
- **i** $-4y + 10y$
- **j** $-6ab + 7ba$ p. 17

12 Simplify:
- **a** $5y - 2y$
- **b** $3q - q$
- **c** $4a - 3a$
- **d** $10xy - 3yx$
- **e** $11y^2 - y^2$
- **f** $6a - 8a$
- **g** $-4y - 2y$
- **h** $-6p + 7p$
- **i** $4a^2 - 12a^2$
- **j** $5m - 4m - 7m$ p. 17

13 Simplify:
- **a** $4a + 7b + 3a + 2b$
- **b** $6y + 3z - 2y - z$
- **c** $6ab - 3a + 7ab - 2a$
- **d** $11 - 4x + 7x - 2$
- **e** $3ab - 2ba + a + b$
- **f** $4cd - 3d + 7cd - cd$
- **g** $-4p + 7q - 2p - 8q$
- **h** $3m + 7n - 5m - 2n$
- **i** $10g - 5k - 3k - 11g$
- **j** $6a - 10b - 3a + 4b$ p. 17

14 Simplify:
- **a** $4a \times 3$
- **b** $5y \times 6$
- **c** $3r \times 2s$
- **d** $4m \times m$
- **e** $-4a \times 5$
- **f** $6y \times (-3)$
- **g** $5p \times (-2p)$
- **h** $k \times k \times k$
- **i** $a \times b \times c \times d$
- **j** $(3b)^2$ p. 17

15 Simplify:
- **a** $20w \div 5$
- **b** $14y \div 2$
- **c** $6a \div 3a$
- **d** $-10y \div 5$
- **e** $-12g \div -4$
- **f** $\dfrac{10ab}{5}$
- **g** $\dfrac{12p}{2p}$
- **h** $\dfrac{4xy}{2y}$
- **i** $\dfrac{-35m}{7m}$
- **j** $\dfrac{-24w}{-3}$ p. 17

16 Simplify:
- **a** $4 \times y + 2y \times 3$
- **b** $6 \times a \times b - 3a \times 4$
- **c** $12mn - 4m \times 2n$
- **d** $\dfrac{10t - 4t}{3t}$ p. 17

17 Simplify:
- **a** $\dfrac{x}{2} + \dfrac{x}{3}$
- **b** $\dfrac{2y}{3} + \dfrac{y}{5}$
- **c** $\dfrac{5a}{3} - \dfrac{a}{4}$
- **d** $\dfrac{6y}{5} - \dfrac{y}{2}$
- **e** $\dfrac{a}{b} + \dfrac{c}{d}$
- **f** $\dfrac{2}{m} + \dfrac{3}{m}$
- **g** $\dfrac{3}{x} + \dfrac{5}{2x}$
- **h** $\dfrac{4y}{3} - \dfrac{2y}{5} - \dfrac{y}{2}$ p. 18

18 Simplify:

a $\frac{x}{2} \times \frac{3}{x}$ **b** $\frac{4y}{3} \times \frac{3}{2y}$

c $\frac{5a}{2} \times \frac{2}{3a}$ **d** $\frac{6xy}{5} \times \frac{15}{8x}$

e $\frac{10ab}{7b} \times \frac{21b}{5a}$

pp. 18–19

19 Simplify:

a $\frac{x}{2} \div \frac{x}{4}$ **b** $\frac{2a}{3} \div \frac{4a}{9}$

c $\frac{4y}{5} \div \frac{3y}{10}$ **d** $\frac{6p}{5} \div \frac{2p}{15}$

e $\frac{7ab}{5xy} \div \frac{21b}{15y}$

pp. 18–19

20 Expand:

a $2(a + b)$ **b** $3(y - 6)$
c $5(2a - 1)$ **d** $4(y - 3)$
e $2(3p - 5q)$ **f** $7(4a + 7b)$
g $-2(m - 1)$ **h** $-3(p - 4)$
i $-6(2r + 5)$ **j** $-(4g - 1)$
k $-(2x + 3)$ **l** $-(6 - 4y)$
m $-(x + 7)$ **n** $x(x + 2)$
o $3a(2a - 1)$ **p** $4y(2 - 3y)$
q $-3a(a - 7)$

p. 19

21 Expand and simplify:

a $2(a + 3) + 4(a + 1)$
b $5(y + 2) + 2(3y + 1)$
c $3(4m - 1) + 2(6m - 1)$
d $7(5a - 7) + 3(2a - 6)$
e $4(2y - 1) - 5(y + 1)$
f $6(4 - 2y) - 3(1 - y)$
g $5(2a - 1) - (a + 7)$
h $6(1 - y) - (3y - 1)$
i $4(2p - 1) - p(p + 1)$
j $x(2x - 1) - x(3 - x)$

p. 19

22 Expand and simplify:

a $3 + 2(a - 1)$ **b** $4 + 6(2y - 7)$
c $2 - 4(m - 1)$ **d** $6 - (4r - 7)$
e $4b - (3 - 5b)$

p. 19

23 Find the sum of:

a $2a + 1$ and $3a + 4$
b $5x - 7$ and $3x + 11$

p. 19

24 Find the difference between:

a $2m + 1$ and $m + 6$
b $5y - 7$ and $3y - 4$
c $7q$ and $4 - 2q$
d $5p - 1$ and $3 - 4p$

p. 19

25 Expand and simplify:

a $(x + 4)(x + 2)$ **b** $(y + 5)(y - 1)$
c $(c + 3)(c + 1)$ **d** $(2y - 1)(y + 3)$
e $(3n + 6)(n - 1)$ **f** $(q - 2)(3q + 6)$
g $(2t - 1)(3t + 5)$ **h** $(4x - 3)(2x - 7)$
i $(2w - 5)(3w + 2)$ **j** $(4y - 7)(2y - 3)$

p. 20

26 Expand and simplify:

a $(x + 3)^2$ **b** $(b - 1)^2$
c $(2a - 3)^2$ **d** $(4y + 5)^2$
e $(4 - n)^2$

p. 21

27 Expand and simplify:

a $(x + 2)(x - 2)$ **b** $(y + 3)(y - 3)$
c $(2a - 1)(2a + 1)$ **d** $(5 - 2y)(5 + 2y)$
e $(3a + 5b)(3a - 5b)$

p. 21

28 Factorise:

a $4y - 8$ **b** $3a - 12b$
c $6m - 10$ **d** $12p - 8$
e $4ab - 6a$ **f** $5xy - 15x$
g $6pq - 4q$ **h** $a^2 - 3a - ab$
i $y^2 - 6y$ **j** $2a^2 - 12ab$

pp. 21–22

29 Factorise:

a $m(x + y) + n(x + y)$
b $p(m + n) + q(m + n)$
c $ab + ac + 2b + 2c$
d $3p + 3q + ap + aq$
e $ca - cb - da + db$
f $xy - y^2 - 7x + 7y$
g $6a - ct + ac - 6t$
h $g^3 - 3g^2 + 2g - 6$

p. 22

30 Factorise:

a $p^2 - 16$ **b** $y^2 - 36$
c $m^2 - 121$ **d** $4a^2 - b^2$
e $9y^2 - 16$ **f** $25x^2 - 64y^2$
g $81 - 121m^2$ **h** $2p^2 - 2$
i $3z^2 - 27$ **j** $10h^2 - 490$

p. 22

31 Factorise:

a $x^2 + 4x + 3$ **b** $y^2 + 3y + 2$
c $a^2 - a - 20$ **d** $y^2 - 3y - 18$
e $m^2 + 2m - 15$ **f** $k^2 + 5k - 6$
g $x^2 - 3x + 2$ **h** $y^2 - 7y + 12$
i $t^2 - 6t + 9$ **j** $y^2 - 8y + 16$

pp. 23–24

Go to p. 217 for **Quick Answers** or to pp. 227–229 for **Worked Solutions**

For a complete understanding of this topic, you must be able to:

✓	Complete arithmetic questions algebraically		pp. 15–16
✓	Substitute into algebraic expressions		p. 16
✓	Simplify algebraic expressions, including algebraic fractions		pp. 17–19
✓	Remove grouping symbols and simplify resulting expressions		p. 19
✓	Expand binomial products, including difference of two squares and perfect squares		pp. 20–21
✓	Factorise algebraic expressions.		pp. 21–24

Now you are ready to do the tests!

Intermediate Test

Algebraic Techniques

(35 marks)

1 Write down the:

a sum of $3x$ and $5x$ (1 mark)
b product of $6x$ and y (1 mark)
c cents in $\$4x$ (1 mark)

2 If $m = -6$ and $n = 4$, evaluate:

a $2mn$ (1 mark)
b $5 + m + n$ (1 mark)
c $m^2 - n^2$ (1 mark)
d $\dfrac{m+n}{m-n}$ (1 mark)

3 Simplify the expressions:

a $7m - m + 4m$ (1 mark)
b $9y \times 4y$ (1 mark)
c $5ab - 4a \times 3b$ (1 mark)
d $(-2y) \times (-8y)$ (1 mark)
e $21xy \div 7x$ (1 mark)
f $9ab \div 18a^2$ (1 mark)

4 Simplify the fractions:

a $\dfrac{42xy}{7xy^2}$ (1 mark)
b $\dfrac{2m}{3} - \dfrac{3m}{2}$ (1 mark)
c $\dfrac{11}{2x} - \dfrac{7}{3x}$ (1 mark)
d $\dfrac{2x}{5} \times \dfrac{10x}{4}$ (1 mark)
e $\dfrac{2x}{5} \div \dfrac{10x}{4}$ (1 mark)

5 Expand and simplify:

a $5(3x - 2y)$ (1 mark)
b $5 - 2(3x - 5)$ (2 marks)
c $3(4x - 3) - 4(3x - 4)$ (2 marks)

6 Expand and simplify:

a $(x + 5)(x - 3)$ (1 mark)
b $(2x - 5)(2x + 5)$ (1 mark)
c $(2x + 5)^2$ (1 mark)

7 Factorise the following completely:

a $7y - 21$ (1 mark)
b $5m^2 + 15mn$ (1 mark)
c $m^2 - 9$ (1 mark)
d $a^2 + 3a - 40$ (1 mark)
e $y^2 - 11y + 18$ (1 mark)
f $y^2 + 2xy + 9y + 18x$ (2 marks)
g $a^3 - 3a^2 - 4a$ (2 marks)

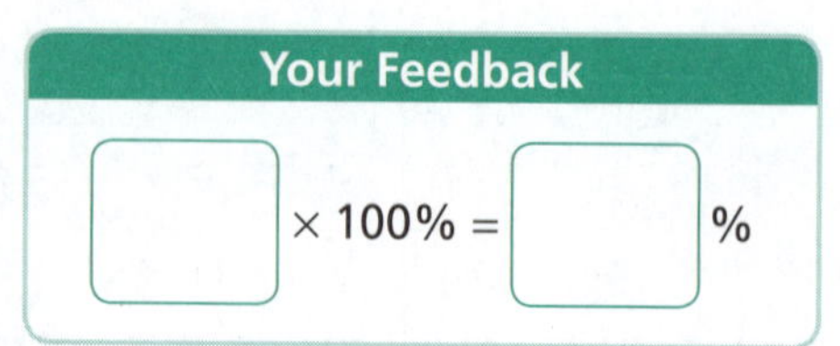

QA PAGE 221

WS PAGE 260

Advanced Test

Algebraic Techniques

(35 marks)

1 Write down the:

a difference between $8y$ and $5y$ (1 mark)
b product of $8a$ and $3a$ (1 mark)
c dollars in $3x$ cents (1 mark)

2 Simplify the expressions:

a $7t - 4t - 5t$ (1 mark)
b $7m \times (-6n)$ (1 mark)
c $18m^2 - 7m \times 5m$ (1 mark)
d $42ab \div 14a$ (1 mark)
e $(-3m) \times (-9mn)$ (1 mark)

3 Simplify the fractions:

a $\dfrac{35ab^3}{7ab}$ (1 mark)
b $\dfrac{5a}{2} - \dfrac{3a}{5}$ (1 mark)
c $\dfrac{7}{m} - \dfrac{5}{3m}$ (1 mark)
d $\dfrac{15}{xy} \times \dfrac{3x}{5}$ (1 mark)
e $\dfrac{25y^2}{14x} \div \dfrac{5xy}{7x^2}$ (1 mark)

4 Expand and simplify:

a $7m(5 - 3m)$ (1 mark)
b $15t - 8(2t - 5)$ (2 marks)
c $6(2a + 5) - 5(3a - 2)$ (2 marks)

5 Expand and simplify:

a $(3x - 2)(x + 5)$ (1 mark)
b $(5x + 3)(5x - 3)$ (1 mark)
c $(5x - 3)^2$ (1 mark)
d $(2x + 3)(x - 2) - 2x^2 + 6$ (2 marks)
e $(2x - 1)^2 - (2x + 1)^2$ (2 marks)

6 Factorise the following completely:

a $54 - 9a$ (1 mark)
b $m^2 - 64$ (1 mark)
c $27y^2 - 9y$ (1 mark)
d $p^2 + p - 42$ (1 mark)
e $p^2 - p - 72$ (1 mark)
f $10x^2 - 10x - 200$ (1 mark)
g $m^2 - 8m - 7tm + 56t$ (2 marks)
h $a^4 - b^4$ (2 marks)

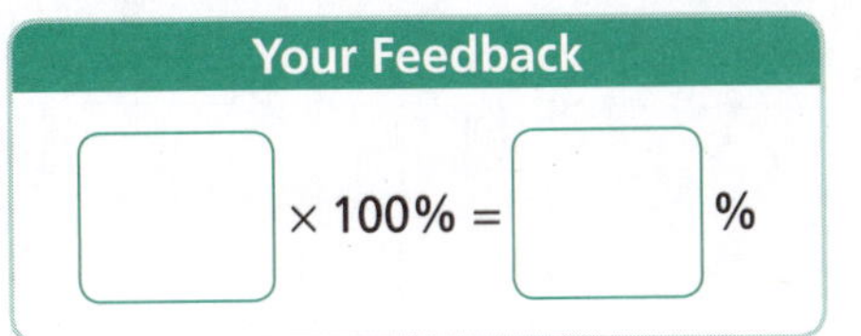

QA PAGE 221

WS PAGE 260

Chapter 3
Indices

Introduction

For x^n, x is the base, and n is the index.

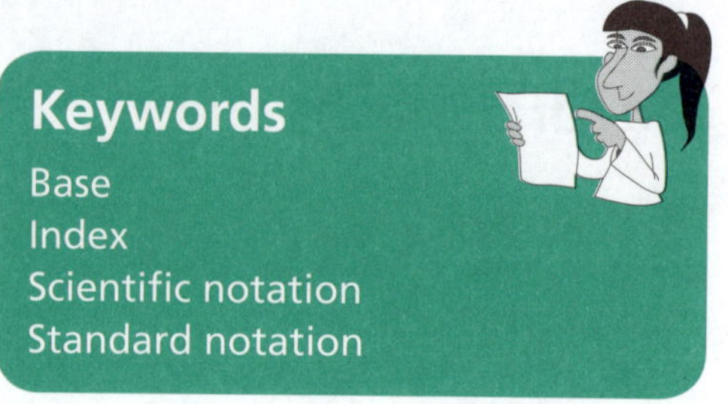

Keywords
Base
Index
Scientific notation
Standard notation

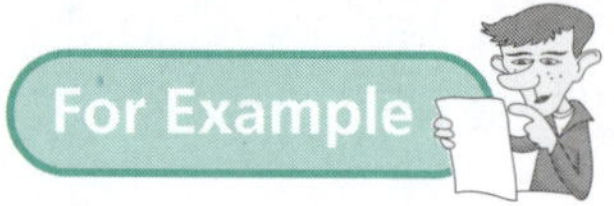

1 Use your calculator to evaluate 3^7.

S We use our x^y button. To find $3^7 = 2187$, key 3 [x^y] 7 [=]

Factorising integers

A factor tree can be used to express an integer as a product of its prime factors.

1 Express as a product of their prime factors in index form:

a 200

S

200
25 8
5 5 4 2
5 5 2 2 2

$\therefore 200 = 5^2 \times 2^3$

b 432

S

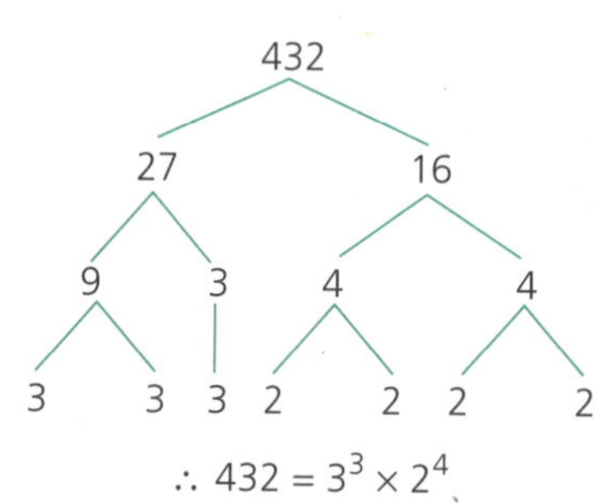

$\therefore 432 = 3^3 \times 2^4$

Index rules

We can use the following four rules:

$x^a \times x^b = x^{a+b}$

$x^a \div x^b = x^{a-b}$

$(x^a)^b = x^{ab}$

$x^0 = 1$ (**Note:** $x \neq 0$)

For Example

1 Simplify:

a $p^4 \times p^2$

S $p^4 \times p^2 = p^{(4+2)}$
$= p^6$

b $p^6 \div p$

S $p^6 \div p = p^6 \div p^1$
$= p^{6-1}$
$= p^5$

c $4^5 \div 4^3$

S $4^5 \div 4^3 = 4^{5-3}$
$= 4^2$
$= 16$

d $12a^4b \div 3a^3$

S $12a^4b \div 3a^3 = 4ab$

e $(x^2)^4$

S $(x^2)^4 = x^{2\times 4}$
$= x^8$

f $(a^{\frac{1}{2}})^3$

S $(a^{\frac{1}{2}})^3 = a^{\frac{3}{2}}$

g $(4y^2)^3$

S $(4y^2)^3 = 4^3 \times (y^2)^3$
$= 64y^6$

h $4x^0$

S $4x^0 = 4 \times x^0$
$= 4 \times 1$
$= 4$

i $(3a)^0 + 3a^0$

S $(3a)^0 + 3a^0 = 1 + 3 \times 1$
$= 1 + 3$
$= 4$

Negative powers

We can use the following rule:

$$x^{-a} = \frac{1}{x^a}$$

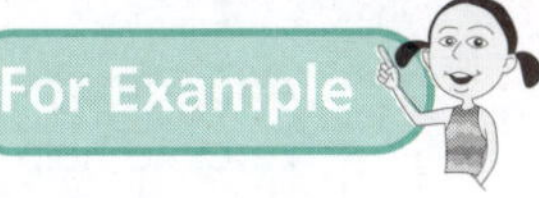

For Example

1 Simplify:

a 4^{-2}

Note: the negative power is the reciprocal.

S $4^{-2} = \frac{1}{4^2} = \frac{1}{16}$

b $\left(\frac{1}{2}\right)^{-3}$

S $\left(\frac{1}{2}\right)^{-3} = \frac{1}{\left(\frac{1}{2}\right)^3} = \frac{1}{\frac{1}{8}}$
$= 1 \div \frac{1}{8}$
$= 1 \times \frac{8}{1}$
$= 8$

c $3x^{-2}$

S $3x^{-2} = 3 \times x^{-2}$
$= 3 \times \frac{1}{x^2}$
$= \frac{3}{x^2}$

d $(5x^3)^{-2}$

S $(5x^3)^{-2} = \frac{1}{(5x^3)^2}$
$= \frac{1}{25x^6}$

e $\frac{x^2}{x^5}$

S $\frac{x^2}{x^5} = x^2 \div x^5$
$= x^{-3}$
$= \frac{1}{x^3}$

Scientific notation

Very large or very small numbers are written as a product of a number between 1 and 10, and a power of ten. This is sometimes referred to as **standard notation**.

1 Express in scientific notation:

a 181.4

S $181.4 = 1.814 \times 100$
$= 1.814 \times 10^2$ ← ∴ 2 places

The number of decimal places moved by the decimal point equals the power of ten.

b 2 000 000

S $2\,000\,000 = 2 \times 1\,000\,000$
$= 2 \times 10^6$ ← ∴ 6 places

c 0.0215

S $0.0215 = 2.15 \div 100$
$= 2.15 \times \frac{1}{100}$
$= 2.15 \times 10^{-2}$ ← ∴ 2 places

'Negative power' means 'reciprocal'.

d 0.0001

S $0.0001 = 1 \div 10\,000$
$= 1 \times \frac{1}{10\,000}$
$= 1 \times 10^{-4}$ ← ∴ 4 places

2 Express in ordinary, or decimal, notation:

a 4.271×10^2

S $4.271 \times 10^2 = 4.271 \times 100$
$= 427.1$

The power of ten equals the number of decimal places moved by the decimal point.

b 3.08×10^4

S $3.08 \times 10^4 = 3.08 \times 10\,000$ ← ∴ 4 places
$= 30\,800$

Add some zeros to help.

c 1.56×10^{-2}

S $1.56 \times 10^{-2} = 1.56 \times \frac{1}{100}$
$= 1.56 \div 100$ ← ∴ 2 places
$= 0.0156$

d 5×10^{-3}

S $5 \times 10^{-3} = 5 \times \frac{1}{1000}$
$= 5 \div 1000$ ← ∴ 3 places
$= 0.005$

Scientific notation and the calculator

The calculator expresses very large and very small numbers in standard notation.

1 Calculate the following expressions, leaving your answer correct to three significant figures.

a $37\,649 \times 28\,547 \times 486$

S $37\,649 \times 28\,547 \times 486$

$= $ 5.223362775 11 ← Calculator display.

$= 5.223\,362\,775 \times 10^{11}$

$= 5.22 \times 10^{11}$

(correct to three significant figures)

b $(0.0023)^4$

S $(0.0023)^4 = $ 2.79841 −11

$= 2.798\,41 \times 10^{-11}$

$= 0.000\,000\,000\,027\,9841$

$= 0.000\,000\,000\,0280$

(correct to three significant figures)

Or

$= 2.80 \times 10^{-11}$

(correct to three significant figures)

c 5^{15}

S $5^{15} = $ 3.051757813 10

$= 3.051\,757\,813 \times 10^{10}$

$= 30\,517\,578\,130$

$= 30\,500\,000\,000$

(correct to three significant figures)

Or

$= 3.05 \times 10^{10}$

(correct to three significant figures)

Practise Practise

1 Express as a product of their prime factors in index form:
a 144 **b** 400
c 676
p. 30

2 Simplify:
a $p^5 \times p^3$ **b** $y^6 \times y$
c $a^4 \times a^3 \times a^2$ **d** $c^4 \times c^3$
e $x^6 \times x \times x^3$ **f** $y^9 \times y^5$
pp. 30–31

3 Simplify:
a $4y^3 \times 2y^4$ **b** $7a^4 \times 2a$
c $-2y^4 \times -3y^2$ **d** $-5m^3 \times 10m^4$
e $6y^3 \times (-2y^4)$ **f** $10a^2b^3 \times 2a^4b$
pp. 30–31

4 Simplify:
a $w^7 \div w^3$ **b** $x^7 \div x^5$
c $p^9 \div p$ **d** $\frac{y^6}{y^3}$
e $\frac{a^8}{a^5}$ **f** $\frac{m^4n^3}{m^2n}$
pp. 30–31

5 Simplify:
a $4y^3 \div 2y^2$ **b** $15c^4 \div 5c^2$
c $42p^9 \div 6p^5$ **d** $10a^4b \div 5a^2b$
e $-5xy^3 \div 5y$ **f** $-10c^2d^3 \div -10cd$
g $\frac{10y^4z}{5y^2}$ **h** $\frac{48a^3b^2}{8ab}$
i $\frac{24p^7q^5}{18p^4q^3}$
pp. 30–31

6 Simplify:
a $p^4q^3 \times p^2q \div pq$
b $10x^7y^5 \div 2x^3y \times x^4y$
pp. 30–31

7 Simplify:
a $\frac{4a \times 3a^2}{2a}$ **b** $\frac{12p^3 \times 2p^4}{8p^7}$
c $\frac{10m^7n^3 \times 4m^2}{20m^3n^2}$
pp. 30–31

8 Simplify:
a $(y^3)^2$ **b** $(a^4)^3$
c $(m^6)^5$ **d** $(3a^4)^2$
e $(2y^5)^3$ **f** $(7a^2b^3)^2$
g $(-3m^4)^2$ **h** $(-5p^6)^2$
i $(-a^4b^3)^3$
pp. 30–31

9 Simplify:
a $y^3 \div y^5$ **b** $a^6 \div a^{11}$
c $y^5 \div y^6$ **d** $x^{-3} \div x^{-3}$
e $b^5 \div b^{-3}$ **f** $g^{-4} \div g^{-2}$
g $k^2 \div k^{-3} \div k^3$ **h** $(p^3)^{-2}$
i $(m^{-2})^3$ **j** $(2p^{-5})^2$
k $10a^4b \div 5a^3b^4$ **l** $15p^4 \div -3p^6$
pp. 31–32

10 Evaluate:
a 3^0 **b** $(3x)^0$ **c** $(10a^3 \times 2a^3)^0$
pp. 30–32

11 If $p = 2$, evaluate:
a p^0 **b** $2p^0$ **c** $(2p)^0$
pp. 30–31

12 Express in scientific notation:
a 4280 **b** 360 000 **c** 500
p. 32

13 Express in scientific notation:
a 0.068 **b** 0.000 35 **c** 0.000 008 6
p. 32

14 Express in decimal form:
a 2.6×10^3 **b** 1.51×10^4 **c** 4.63×10^7
p. 32

15 Express in decimal form:
a 5.8×10^{-2} **b** 7.06×10^{-3} **c** 8.4×10^{-5}
p. 32

16 Calculate, leaving your answer in scientific notation, correct to three significant figures:
a $\sqrt{0.025}$ **b** 6.32^5
c 6985×321 **d** $\frac{0.256}{965}$
e $\frac{\sqrt{521}}{0.075}$ **f** $\frac{\sqrt{3.016 \times 99.3}}{56.98^2}$
p. 33

17 The distance of a light year is approximately 9.4605×10^{15} metres. If the Bubble Nebula is 11 000 light years from Earth, express the distance in metres, in scientific notation, correct to three significant figures.
p. 33

18 65 billion neutrinos per second pass through every cm^2 of the Earth. How many neutrinos pass through each cm^2 every day? Leave your answer in scientific notation.
p. 33

Go to p. 217 for **Quick Answers** or to p. 230 for **Worked Solutions**

For a complete understanding of this topic, you must be able to:

✓	Use a factor tree to express an integer as a product of its prime factors		p. 30
✓	Apply the index rules to simplify algebraic expressions		pp. 30–31
✓	Use negative indices to simplify algebraic expressions		pp. 31–32
✓	Express numbers in scientific notation.		pp. 32–33

Now you are ready to do the tests!

Intermediate Test

Indices

(50 marks)

1 Express each of the following as products of their prime factors, expressed in index form:
a 500 (2 marks)
b 1296 (2 marks)

2 Simplify:
a $4x^2 \times 3x^4$ (1 mark)
b $10a^3 \times 4a^2$ (1 mark)
c $5m^3n^2 \times 3m^4n^5$ (1 mark)
d $3p^3q^4 \times 6p^2q^6$ (1 mark)

3 Simplify:
a $8b^5 \div 2b^3$ (1 mark)
b $14a^3 \div 7a^2$ (1 mark)
c $25k^5 \div 5k^2$ (1 mark)
d $30g^4 \div 6g$ (1 mark)

4 Simplify:
a $(2b^3)^2$ (1 mark)
b $(3w^5)^2$ (1 mark)
c $(8a^6)^2$ (1 mark)
d $(-3y^4)^2$ (1 mark)
e $(-2m^3)^3$ (1 mark)

5 Simplify:
a $\frac{8a^2}{4a}$ (1 mark)
b $\frac{20k^5}{10k}$ (1 mark)
c $\frac{14m^3}{7m^2}$ (1 mark)
d $\frac{b^7c^2}{b^5c}$ (1 mark)

6 Simplify:
a 4^0 (1 mark)
b a^0 (1 mark)
c $5p^0$ (1 mark)
d $(3 + 8)^0$ (1 mark)

7 Simplify:
a $(a^6)^0$ (1 mark)
b $(m^0)^6$ (1 mark)

8 Simplify:
a 5^{-2} (1 mark)
b 6^{-2} (1 mark)
c 2^{-3} (1 mark)
d $\left(\frac{1}{5}\right)^{-1}$ (1 mark)
e $\left(\frac{1}{3}\right)^{-2}$ (1 mark)

9 Simplify:
a $x^{-2} \times x^{-2}$ (1 mark)
b $a^{-4} \times a^2$ (1 mark)
c $g^2 \times g^{-2}$ (1 mark)
d $3a^5 \times 2a^{-2}$ (1 mark)
e $7q^3 \times 5q^{-5}$ (1 mark)
f $p^3 \div p^{-2}$ (1 mark)
g $d^{-4} \div d^3$ (1 mark)
h $y^2 \div y^{-2}$ (1 mark)
i $8a^2 \div 4a^{-3}$ (1 mark)
j $15x^3 \div 5x^{-1}$ (1 mark)

10 Express in scientific notation:
a 35 000 (1 mark)
b 262 400 (1 mark)
c 0.003 11 (1 mark)
d 0.805 (1 mark)

11 Express as an ordinary number:
a 2.57×10^5 (1 mark)
b 1.91×10^4 (1 mark)
c 2.531×10^{-2} (1 mark)
d 8.4×10^{-3} (1 mark)

Your Feedback

☐ × 100% = ☐ %

QA PAGE 221

WS PAGE 260

Advanced Test

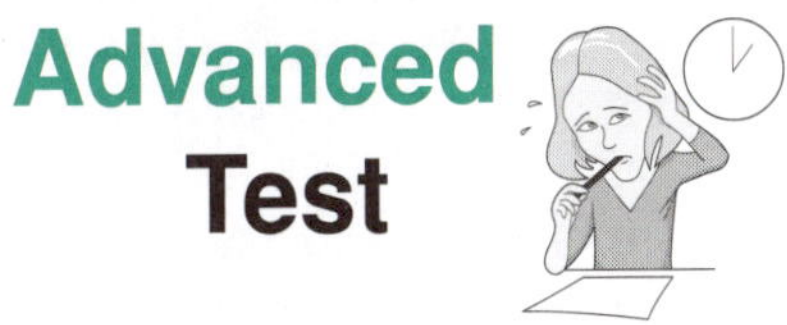

45 min

Indices

(50 marks)

1 Simplify:

a $3x^2y^5 \times 2x^5y^4$ (1 mark)
b $12a^2b^8 \times 5a^3b^5$ (1 mark)
c $70p^3q^2r \times 7pq^4r^3$ (1 mark)
d $7x^4yz^2 \times 9xy^3z^9$ (1 mark)

2 Simplify:

a $18a^7 \div 2a^5$ (1 mark)
b $25p^3q^2 \div 5p^2q$ (1 mark)
c $32a^8b^3 \div 4a^7b^3$ (1 mark)
d $24m^3n^4 \div 8mn^4$ (1 mark)
e $9x^3y^2 \div 6xy$ (1 mark)
f $15pq^4 \div 25pq^6$ (1 mark)

3 Simplify:

a $\frac{12b^2}{3b}$ (1 mark)
b $\frac{50g^9}{10g^2}$ (1 mark)
c $\frac{a^7b^6}{a^5b^2}$ (1 mark)
d $\frac{12m^3n^2}{3m^2n^2}$ (1 mark)
e $\frac{-5a^2bc}{5a^2b}$ (1 mark)

4 Simplify:

a $(2y^3)^2$ (1 mark)
b $(5c^4)^2$ (1 mark)
c $(-3m^5)^2$ (1 mark)

5 Simplify:

a 3^0 (1 mark)
b $(a + b)^0$ (1 mark)
c $(4a)^0 - 4a^0$ (1 mark)
d $10p^3q^2 \div 2p^3q^2$ (1 mark)
e $(a^0)^6$ (1 mark)
f $(3h^0)^4$ (1 mark)
g $(2s^7)^0$ (1 mark)
h $(5m^3)^2 \div (5m^3)^0$ (1 mark)

6 Write the following with positive integers:

a x^{-2} (1 mark)
b $6a^{-2}$ (1 mark)
c $(2t)^{-3}$ (1 mark)
d $\left(\frac{1}{p}\right)^{-1}$ (1 mark)
e $\left(\frac{3}{q}\right)^{-2}$ (1 mark)
f $\frac{4}{a^{-2}}$ (1 mark)
g $3b^{-2}c^4$ (1 mark)
h $\frac{2}{3}a^3b^{-4}$ (1 mark)

7 Simplify:

a $x^{-4} \times x^3$ (1 mark)
b $a^7 \times a^{-7}$ (1 mark)
c $5p^{-2} \times 3p^{-2}$ (1 mark)
d $12m^3 \div 3m^{-5}$ (1 mark)
e $10z^{-3} \div 2z^{-6}$ (1 mark)
f $(3m^{-5})^{-2}$ (1 mark)

8 Simplify:

a $\frac{10a^2b^4}{3c^4d^3} \times \frac{6c^3d^2}{5a^2b^2}$ (2 marks)
b $\frac{8x^2y^3}{9p^5q} \div \frac{4x^5y^4}{15p^7q^2}$ (2 marks)

9 Calculate, leaving your answer in scientific notation, correct to three significant figures:

a $\frac{\sqrt{11.7 - 3.84}}{12.91 - 8.004}$ (2 marks)
b $\sqrt{\frac{19.3105 - 11.02}{3.1 \div 42.87}}$ (2 marks)
c $\frac{(3.24 \times 10^5)^3}{6.2 \times 10^2}$ (2 marks)

Your Feedback

☐ × 100% = ☐ %

QA PAGE 221

WS PAGE 261

Chapter 4
Equations

Simple equations and inequations

Keywords

Coefficient	Literal
Elimination	Quadratic
Equation	Simultaneous
Factorise	Solution
Formula	Subject
Formulae	Substitution
Inequality	Variable

Solutions of equations

The main rules to remember are:

- Pronumeral on one side and number on the other side.
- To *change sides* means we have to *change signs*.

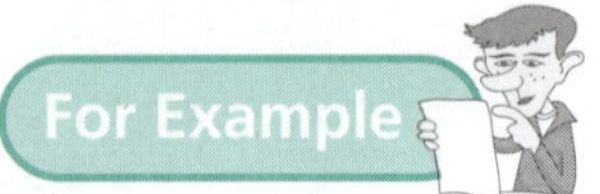

1 Solve:

a $3x - 1 = x + 2$

(this is $+3x$; this is -1; this is $+x$; this is $+2$)

S

$$3x - x = 2 + 1$$
$$2x = 3$$
$$\frac{{}^{1}\not{2}x}{\not{2}_{1}} = \frac{3}{2}$$ ← We divide by the coefficient (number in front of *x*).
$$x = 1\frac{1}{2}$$

b $5y - 7 = 3(2y + 4)$ ← First we expand the bracket.

S

$$5y - 7 = 6y + 12$$
$$5y - 6y = 12 + 7$$
$$-y = 19$$
$$\frac{-y}{-1} = \frac{19}{-1}$$ ← The coefficient of *y* is -1
$$y = -19$$

c $\frac{3a - 4}{2} = 8$

We have to get rid of denominators, so we multiply both sides by 2.

S

$${}^{1}\not{2}\left[\frac{3a - 4}{\not{2}_{1}}\right] = 8 \times 2$$
$$3a - 4 = 16$$
$$3a = 16 + 4$$
$$3a = 20$$
$$\frac{{}^{1}\not{3}a}{\not{3}_{1}} = \frac{20}{3}$$
$$a = 6\frac{2}{3}$$

We can leave this as a mixed numeral or $6.\dot{6}$.

d $\frac{5a - 1}{2} = \frac{4a + 8}{3}$

Best to multiply by lowest common denominator.

S

$${}^{3}\not{6}\left[\frac{5a - 1}{\not{2}_{1}}\right] = {}^{2}\not{6}\left[\frac{4a + 8}{\not{3}_{1}}\right]$$
$$3(5a - 1) = 2(4a + 8)$$
$$15a - 3 = 8a + 16$$
$$15a - 8a = 16 + 3$$
$$7a = 19$$
$$\frac{{}^{1}\not{7}a}{\not{7}_{1}} = \frac{19}{7}$$
$$a = 2\frac{5}{7}$$

Note that we could have jumped to this step by cross-multiplying: ⤫

Solution of inequalities

Algebraic expressions with an inequality symbol have a range of solutions. An important rule is:

- When multiplying or dividing an inequality by a negative, we must reverse the inequality symbol.

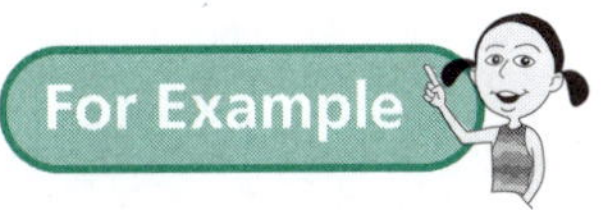

1 Solve:

a $3x - 4 > x + 7$

S
$$3x - x > 7 + 4$$
$$2x > 11$$
$$\frac{2x}{2} > \frac{11}{2}$$
$$x > 5\frac{1}{2}$$

b $x + 7 \geq 3x - 4$

S
$$x - 3x \geq -4 - 7$$
$$-2x \geq -11$$
$$\frac{-2x}{-2} \leq \frac{-11}{-2}$$
$$x \leq 5\frac{1}{2}$$

When dividing by a negative we reverse the inequality.

c $7 - 5x > 12$

S
$$-5x > 12 - 7$$
$$-5x > 5$$
$$\frac{-5x}{-5} < \frac{5}{-5}$$
$$x < -1$$

Graphing on the number line

When graphing a solution on a number line, a closed circle is used to denote inclusion (e.g. $x \geq 3$), while an open circle is used to denote an exclusion (e.g. $x > 3$).

For Example

1 Graph on separate number lines:

a $x \geq 2$

S

We use a closed circle to denote the inclusion of 2.

Hint: the arrow-head looks like the inequality symbol.

b $x < -1$

S

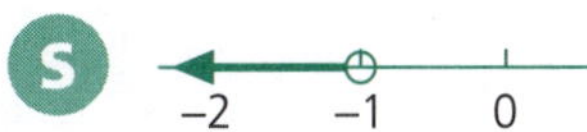

We use an open circle to denote the *exclusion* of −1.

c $-2 < x \leq 1$

S

We want all values of x greater than −2 but less than or equal to 1.

2 Write down the inequality that corresponds to the graph:

a 

S $x < 2$

b

S $-2 < x \leq 4$

c

S $x < 0, x > 3$

3 Solve, and graph your solution on a number line:

a $2x - 4 > x - 1$

S
$$2x - x > -1 + 4$$
$$x > 3$$

b $4 \leq 3(4 - 5x)$

S
$$4 \leq 12 - 15x$$
$$15x \leq 12 - 4$$
$$15x \leq 8$$
$$\frac{15x}{15} \leq \frac{8}{15}$$
$$x \leq \frac{8}{15}$$

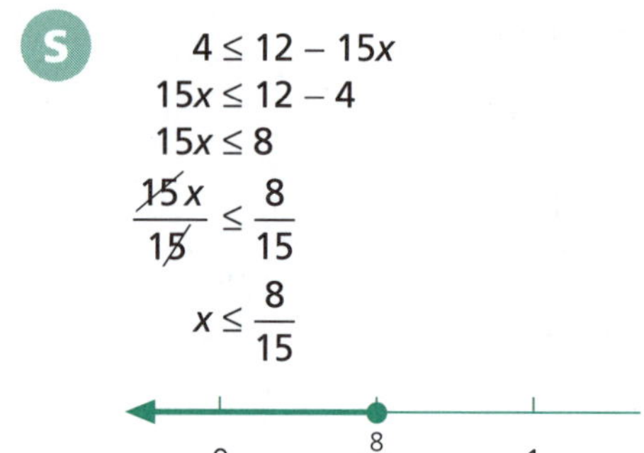

Formulae

Formulae are widely used and usually represent physical quantities. Unlike equations, formulae involve more than one pronumeral. The subject of a formula is the pronumeral on the left-hand side of the equals sign (the pronumeral by itself). A formula can also be called a literal equation.

Substitution into a formula

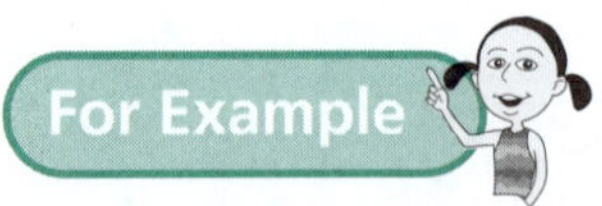

1 If $S = n(n + 1)$, find S when $n = 16$.

S
$$\begin{aligned} S &= n(n + 1) \\ &= 16(16 + 1) \\ &= 16(17) \\ &= 272 \end{aligned}$$

2 If $A = P\left(1 + \frac{r}{100}\right)^n$, find A when:
$P = 10\,000$, $r = 20$, $n = 2$.

S
$$\begin{aligned} A &= P\left(1 + \frac{r}{100}\right)^n \\ &= 10\,000\left(1 + \frac{20}{100}\right)^2 \\ &= 10\,000\,(1.2)^2 \\ &= 14\,400 \end{aligned}$$

3 If $T = \frac{n}{2}[2a + (n - 1)d]$, find T if:
$a = 6$, $d = 3$, $n = 10$.

S
$$\begin{aligned} T &= \frac{n}{2}[2a + (n - 1)d] \\ &= \frac{10}{2}[2(6) + (10 - 1) \times 3] \\ &= 5(12 + 9 \times 3) \\ &= 5(12 + 27) \\ &= 5(39) \\ &= 195 \end{aligned}$$

Changing the subject of a formula (solving literal equations)

A formula such as $v^2 = u^2 - 2as$ has v^2 as the subject and we can easily find v, if we know u, a and s, by simple substitution (and taking the square root). However, to find u, a or s it is best to change the subject to the unknown pronumeral. This is known as solving a literal equation.

For Example

1 Make y the subject:

a $2x + y = 7$

S $y = 7 - 2x$

b $3x - y = 5$

S $3x - 5 = y$

$y = 3x - 5$

Taking y to the RHS changes its sign. Then we just swap sides.

Note: we could have left the y on the LHS as $-y$ and then divide through by -1.

c $4x - 3y = 12$

S
$$4x - 12 = 3y$$
$$3y = 4x - 12$$
$$\frac{\cancel{3}y}{\cancel{3}} = \frac{4x - 12}{3}$$
$$y = \frac{4x - 12}{3}$$

d $xy - 7 = 3a + 2b$

S
$$xy = 3a + 2b + 7$$
$$\frac{\cancel{x}y}{\cancel{x}} = \frac{3a + 2b + 7}{x}$$
$$y = \frac{3a + 2b + 7}{x}$$

Divide both sides by x.

e $3(4x - 2y) = 13x + 1$

S
$$12x - 6y = 13x + 1$$
$$12x - 13x - 1 = 6y$$
$$6y = -x - 1$$
$$\frac{\cancel{6}y}{\cancel{6}} = \frac{-x - 1}{6}$$
$$y = \frac{-x - 1}{6}$$

Expand the bracket first.

f $A = \frac{3xy}{7}$

S
$$7[A] = \left[\frac{3xy}{\cancel{7}_1}\right]\cancel{7}_1$$
$$7A = 3xy$$
$$3xy = 7A$$
$$\frac{\cancel{3}\cancel{x}y}{\cancel{3}\cancel{x}} = \frac{7A}{3x}$$
$$y = \frac{7A}{3x}$$

Multiply both sides by 7.

g $P = \frac{3x - 4}{y}$

S
$$y[P] = \left[\frac{3x - 4}{\cancel{y}_1}\right]\cancel{y}_1$$
$$Py = 3x - 4$$
$$\frac{\cancel{P}y}{\cancel{P}} = \frac{3x - 4}{P}$$
$$y = \frac{3x - 4}{P}$$

h $x = \sqrt{\frac{A}{y}}$

S
$$x^2 = \frac{A}{y}$$
Squaring both sides.
$$x^2y = A$$
Multiplying both sides by y.
$$y = \frac{A}{x^2}$$
Dividing both sides by x^2.

Solving problems involving formulae

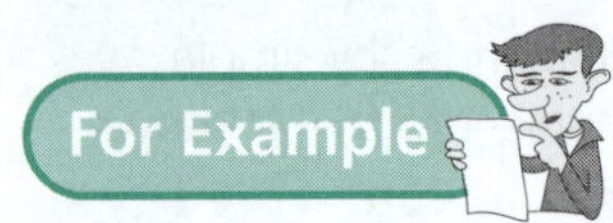

1 Write down the formula for the curved surface area A of a cylinder:

$A = 2\pi rh$

a Find the area in terms of π if the radius is 12 cm and the height is 10 cm.

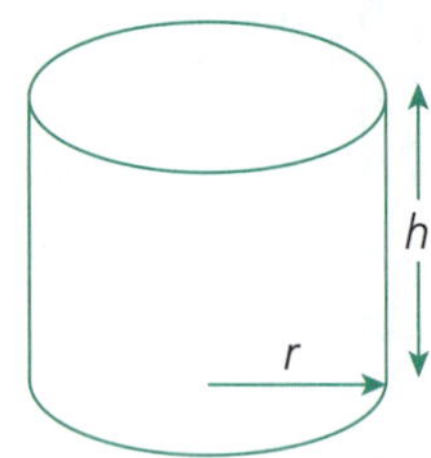

S $A = 2\pi rh$

$A = 2 \times \pi \times 12 \times 10$
$= 240\pi$

The area is 240π cm^2.

b Find the height h if the area A of the curved surface is 132π cm^2 and the radius is 6 cm.

S
$$\begin{aligned} A &= 2\pi rh \\ 132\pi &= 2 \times \pi \times 6 \times h \\ 132\pi &= 12\pi h \\ 12\pi h &= 132\pi \\ \frac{12\pi h}{12\pi} &= \frac{132\pi}{12\pi} \\ h &= 11 \end{aligned}$$

The height is 11 cm.

2 Write down the formula for the area A of the annulus (below) if the radius of the larger circle is R while the radius of the smaller circle is r.

That is,
$A = \pi R^2 - \pi r^2$
$= \pi(R^2 - r^2)$.

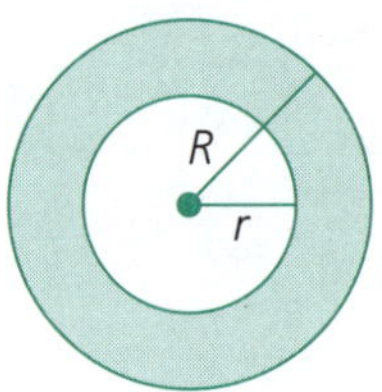

a Find the area of an annulus where $R = 6.2$ cm and $r = 4.7$ cm, correct to three significant figures.

S
$$\begin{aligned} A &= \pi(R^2 - r^2) \\ &= \pi(6.2^2 - 4.7^2) \\ &= \pi(38.44 - 22.09) \\ &= \pi(16.35) \\ &= 51.365\,039\,89 \\ &= 51.4\text{, to three significant figures.} \end{aligned}$$

The area is 51.4 cm^2.

b Find the radius of the smaller circle if the radius of the larger circle is 10 cm and the area of the annulus is 36π cm^2.

S
$$\begin{aligned} A &= \pi(R^2 - r^2) \\ 36\pi &= \pi(10^2 - r^2) \\ 36\pi &= \pi(100 - r^2) \end{aligned}$$

Divide by π:
$$\begin{aligned} 36 &= 100 - r^2 \\ r^2 &= 100 - 36 \\ r^2 &= 64 \\ r &= 8\ (r > 0) \end{aligned}$$

The radius of the smaller circle is 8 cm.

Simultaneous equations

If an equation has two unknowns, for example $x + y = 4$, there is an infinite number of solutions. If there is a pair of equations with two unknowns we may be able to find a solution to satisfy both equations at the same time, that is, simultaneously.

There are three methods of solution: graphically, algebraically through *substitution*, and algebraically through *elimination*. The latter two are treated below.

The substitution method

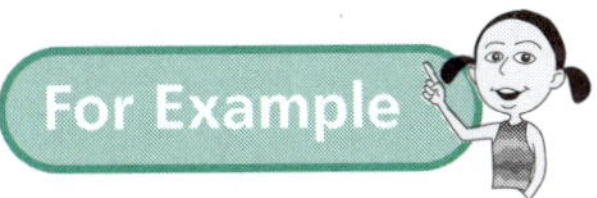

1 Solve the simultaneous equations:

a $2x + y = 5$ (1)
$5x - 3y = 7$ (2)

We number our equations to help.

From (1), make y the subject:
$y = 5 - 2x$ (3)

It is better here to choose y as it is a monic term (coefficient of 1).

Now substitute into (2):

$$5x - 3(5 - 2x) = 7$$
$$5x - 15 + 6x = 7$$
$$11x - 15 = 7$$
$$11x = 7 + 15$$
$$11x = 22$$
$$x = 2$$

Now we substitute $x = 2$ in (3).

Equation (3) is 'simpler'.

$$y = 5 - 2\,(2)$$
$$y = 5 - 4$$
$$\therefore y = 1$$

Therefore $x = 2$, $y = 1$.
[*Or* (2, 1) as an ordered pair.]

It is good to write the complete solution here.

Note: we could check this answer by substituting into both equations, (1) and (2).

b $3x - 4y = 11$
$2x + 3y = -4$

S $3x - 4y = 11$ (1)
$2x + 3y = -4$ (2)

From (2),

$$2x = -4 - 3y$$
$$x = \frac{-4 - 3y}{2}$$

Not as easy as Example 1a; we choose the term with smallest coefficient.

Substituting in (1):

$$3\left(\frac{-4 - 3y}{2}\right) - 4y = 11$$

By multiplying through by 2

$$3(-4 - 3y) - 8y = 22$$
$$-12 - 9y - 8y = 22$$
$$-12 - 17y = 22$$
$$-17y = 22 + 12$$
$$-17y = 34$$
$$y = -2$$

By dividing through by −17

Substituting $y = -2$ in (2):

$$2x + 3(-2) = -4$$
$$2x - 6 = -4$$
$$2x = -4 + 6$$
$$2x = 2$$
$$x = 1$$

Therefore, $x = 1$, $y = -2$.

The elimination method

The second example above could have been solved in a simpler way by using the elimination method. One of the pronumerals is eliminated.

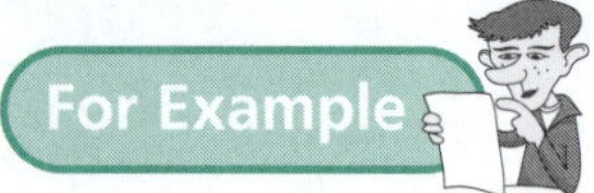

1 Solve:

a $2x - 3y = 3$ (1)
$4x + 3y = 15$ (2)

S Add equations (1) and (2):
$6x = 18$
$x = 3$

The pronumeral y has been eliminated by adding.

From here the method is the same as the substitution method.

Substituting in (2):
$4(3) + 3y = 15$
$12 + 3y = 15$
$3y = 15 - 12$
$3y = 3$
$y = 1$

Remember that we can substitute into either equation—choose the simpler.

Therefore, $x = 3$ and $y = 1$.

b $3x - 4y = 11$ (1)
$2x + 3y = -4$ (2)

S $2 \times (1)$ $6x - 8y = 22$ (3)
$3 \times (2)$ $6x + 9y = -12$ (4)

We chose to eliminate the pronumeral x—the lowest common multiple is 6.

Subtracting (4) from (3):
$-17y = 34$
$y = -2$

Substituting in (2):
$2x + 3(-2) = -4$
$2x - 6 = -4$
$2x = -4 + 6$
$2x = 2$
$x = 1$

Therefore $x = 1$, $y = -2$.

Using equations to solve problems

Problems can be solved by introducing pronumerals to form equations.

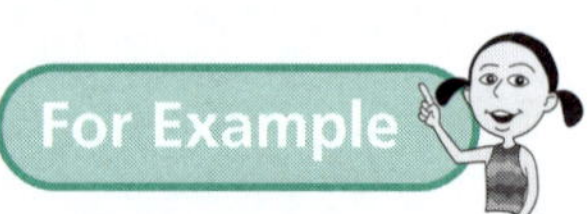

1 The sum of two consecutive even numbers is 22. Find the numbers.

S Let the numbers be x and $x + 2$.
Then $x + x + 2 = 22$
$2x + 2 = 22$
$2x = 20$
$x = 10$

Therefore, $x = 10$, $x + 2 = 12$, and so the numbers are 10 and 12.

2 Find two numbers such that their sum is 20, while half their difference is 1.

S Let the numbers be x and y.
Then $x + y = 20$ (1)
$\frac{1}{2}(x - y) = 1$ (2)
$2 \times (2)$ $x - y = 2$ (3)
$[(1) - (3)]$ $2y = 18$
$y = 9$

Substituting $y = 9$ in (1):
$x + 9 = 20$
$x = 20 - 9$
$x = 11$

Therefore, $x = 11$, $y = 9$, and so the numbers are 11 and 9.

3 If the sides of an isosceles triangle are $(x + 20)$ cm, $(3x - 16)$ cm and $(x + 2)$ cm, what are the possible values of x?

S Two of the three sides are equal. We form equations and solve them to find value(s) of x, but there is no point in setting $x + 20 = x + 2$ as this is nonsense.

$$\begin{aligned} 3x - 16 &= x + 20 \\ 3x - x &= 20 + 16 \\ 2x &= 36 \\ x &= 18 \end{aligned} \quad \text{or} \quad \begin{aligned} 3x - 16 &= x + 2 \\ 3x - x &= 2 + 16 \\ 2x &= 18 \\ x &= 9 \end{aligned}$$

But $x \neq 9$ as the sides are then 11 cm, 11 cm and 29 cm, $\therefore$ not a triangle.

Therefore, $x = 18$ is the only solution.

4 John is 10 years older than Peter, but 25 years ago, John was twice Peter's age. Find Peter's present age.

S Let Peter's age $= x$
John's age $= x + 10$

Twenty-five years ago:
Peter $= x - 25$
John $= x + 10 - 25 = x - 15$

Therefore 25 years ago:

$$\begin{aligned} 2(x - 25) &= x - 15 \\ 2x - 50 &= x - 15 \\ 2x - x &= 50 - 15 \\ x &= 35 \end{aligned}$$

Therefore Peter's present age is 35 years.

5 Daniel has 22 coins, each being either a one-dollar coin or a two-dollar coin. He has a total of $35. How many of each coin does he have?

S Let x be the number of $1 coins, and y be the number of $2 coins.

The equations are:
$x + y = 22$ [he has 22 coins] (1)
$x + 2y = 35$ [as $1 and $2] (2)
[(2) – (1)] $y = 13$

Substituting $y = 13$ in (1):
$x + 13 = 22$
$x = 9$

Therefore he has nine $1 coins and thirteen $2 coins.

6 Find the value of x and y for the rectangle:

S

$$\begin{aligned} 3x + 4y &= 14 && (1) \\ 2x - y &= 2 && (2) \\ 4 \times (2) \therefore\ 8x - 4y &= 8 && (3) \\ [(1) + (3)] \quad 11x &= 22 \\ x &= 2 \end{aligned}$$

Substituting $x = 2$ in (2)

$$\begin{aligned} 4 - y &= 2 \\ -y &= 2 \\ -y &= -2 \\ y &= 2 \end{aligned}$$

Therefore $x = 2$ and $y = 2$.

7 Three peaches and four plums cost $2.75 while five peaches and two plums cost $2.95. Find the price of each.

S Let a = price of a peach, b = price of a plum.

$$\begin{aligned} \therefore \quad 3a + 4b &= 275 && (1) \\ 5a + 2b &= 295 && (2) \\ 2 \times (2) \quad 10a + 4b &= 590 && (3) \\ \therefore\ [(3) - (1)] \quad 7a &= 315 \\ \therefore \quad a &= 45 \end{aligned}$$

Substituting $a = 45$ in (1)

$$\begin{aligned} \therefore \quad 135 + 4b &= 275 \\ \therefore \quad 4b &= 275 - 135 \\ 4b &= 140 \\ b &= 35 \end{aligned}$$

$\therefore$ Peach costs 45c, plum costs 35c.

Quadratic equations

A quadratic equation is in the form $ax^2 + bx + c = 0$, where $a \neq 0$.

Using the result that if $ab = 0$, then either $a = 0$ or $b = 0$ (or $a = b = 0$), we attempt to factorise our quadratic expression and then solve the equation.

In this book we will focus on solving quadratic equations by first factorising them.

Note: students should be confident with Chapter 2 work on factorising quadratic trinomials.

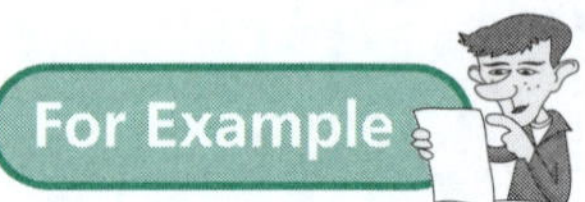

1 Solve:

a $x^2 - 5x - 6 = 0$

S $(x - 6)(x + 1) = 0$

$x = 6, -1$

We could check by substitution.

This means that $x = 6$ or -1.

b $x^2 - 4x = -4$

S $x^2 - 4x + 4 = 0$

$(x - 2)(x - 2) = 0$

$x = 2$

We must have the quadratic equalling zero.

c $x^2 = 5x$

S $x^2 - 5x = 0$

$x(x - 5) = 0$

$x = 0, 5$

d $6 + x - x^2 = 0$

S Multiply through by -1.

$x^2 - x - 6 = 0$

$(x - 3)(x + 2) = 0$

$x = 3, -2$

e $x^2 - 5x + 6 = 2$

S $x^2 - 5x + 6 - 2 = 0$

$x^2 - 5x + 4 = 0$

$(x - 4)(x - 1) = 0$

$x = 4, 1$

We must always make one side equal to zero.

Practise
Practise

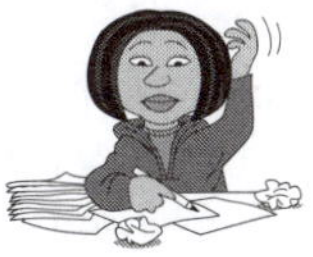

1 Solve:

a $4x - 2 = 3x + 6$
b $5a - 4 = 3a - 12$
c $4y = 2y + 11$
d $5(a + 1) = 3(a + 6)$
e $4 + 2(x + 1) = 12$
f $7(2y - 3) = 4(2y + 12)$
g $4x - 2 = 7x + 10$
h $3a - 4 = 16a$
i $\frac{5y - 2}{3} = 6$
j $\frac{4y - 2}{3} = \frac{y + 6}{2}$
k $\frac{5x}{2} - 3x = 4$
l $\frac{2a}{3} - 4 = \frac{6a}{5}$
m $\frac{4 - 3y}{-2} = 10$
n $5 - \frac{4x}{3} = 3x$
o $\frac{4}{x} = 12$
p $\frac{3}{y} - 2 = 4$ p. 38

2 Solve:

a $4x - 2 < 12$
b $3a - 4 \geq 2a + 1$
c $\frac{5y - 4}{2} > 4$
d $5 - 4y < 3$
e $\frac{7y - 2}{4} < -3$
f $5y - 2 \geq 7y + 12$
g $6y - 4 \leq 7y + 2$
h $\frac{4y - 2}{-3} > 4$ p. 39

3 Graph on separate number lines:

a $x \leq -2$
b $x > 3$
c $x < \frac{2}{3}$
d $-4 < x \leq 2$
e $x < 2, x \geq 3$ pp. 39–40

4 Write down the inequality that corresponds to the graph:

a
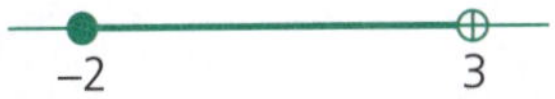

b
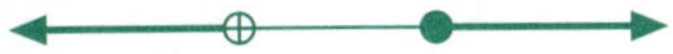

pp. 39–40

5 Solve, and graph your solution on a number line:

a $3x - 2 > x + 4$
b $4 - 2x \geq 12$ pp. 39–40

6 **a** If $D = ST$, find D when $S = 240$ and $T = 3$.

b If $A = \frac{h}{2}(a + b)$, find A when $h = 12$, $a = 6$ and $b = 9$.

c If $S = V(1 - r)^n$, find S when $V = 1200$, $r = 0.03$ and $n = 4$ (correct to 2 dec. pl.).

d If $A = \pi(R^2 - r^2)$, find A when $R = 11.6$ and $r = 6.7$ (correct to 2 dec. pl.).

e If $B = \frac{10N - 7.5H}{6.8M}$, find B if $N = 3$, $H = 2$, $M = 82$ (correct to 2 dec. pl.) p. 40

7 The volume, V, of a cone is found using the formula $V = \frac{4\pi r^2 h}{3}$. What is the volume of a cone with radius 12 cm and height 15 cm? Give your answer correct to three decimal places. p. 40

8 The formula $C = \frac{5}{9}(F - 32)$ is used to change temperatures expressed in Fahrenheit (F) to Celsius (C). What is the temperature in degrees Celsius (°C) when it is 113 °F? p. 40

9 Make y the subject of these equations:

a $2x + y = 13$
b $3x - y = 11$
c $5x + 4y = 12$
d $7x - 3y = 18$
e $3xy = 19$
f $\frac{x}{y} = 5$ p. 41

10 **a** If $v = u + at$, find t, to two decimal places, if $v = 12$, $a = 9.8$, $u = 6$.

b If $D = \frac{S}{T}$, find S, if $D = 120$, $T = 4.5$.

c If $V = \frac{1}{3}lbh$, find l, to two decimal places, if $V = 320$, $b = 60$, $h = 13\frac{1}{2}$.

d If $K = \frac{4V}{T}$, find T, to two decimal places, if $K = 7.26$, $V = 12.43$.

e If $V = \pi r^2 h$, find h, to two decimal places, if $V = 126$, $\pi = 3.14$, $r = 4$.

f If $w = \frac{3}{4}gk$, find k, to two decimal places, if $w = 7.4$, $g = 9.31$. p. 41

11 **a** The sum of three consecutive whole numbers is 93. Using an equation, find the numbers.

b

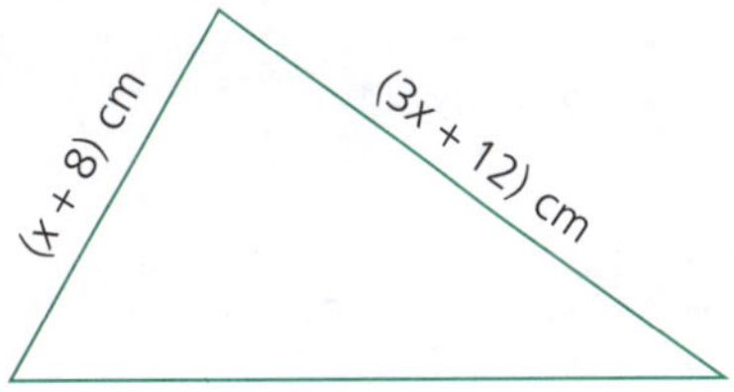

The triangle has a perimeter of 38 cm. Write an expression for the perimeter in terms of x, and use it to find the value of x.

c

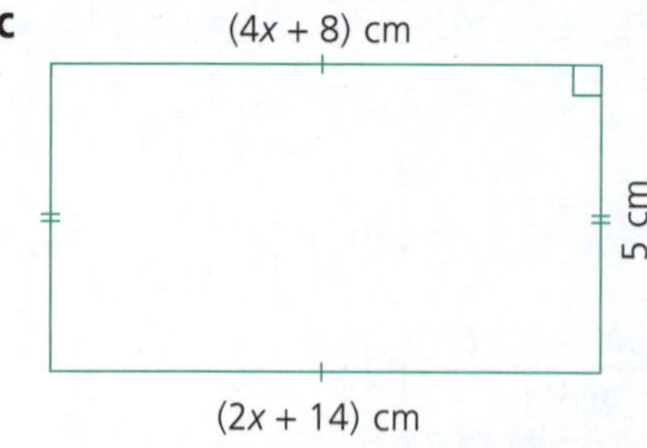

Solve an equation to find the value of x for this rectangle.

d When a number is doubled, the result is the same as the sum of the number and eight. Find the number.

e The product of three and the sum of four and a number is equal to the difference between the number and two. Find the number.

f Fiona is now twice as old as she was 15 years ago. Letting x be her age 15 years ago, find her age now.

g Daniel is twice as old as Laura now, but 4 years ago he was three times as old as Laura. Find Laura's present age. p. 42

12 Solve, using the substitution method:

a $2a + b = 1$
$a + 2b = 8$

b $3m - 2n = 7$
$2m + n = 0$ pp. 43–44

13 Solve the following, using the elimination method:

a $p - 3q = 2$
$2p + 3q = 13$

b $x + 2y = 3$
$3x + 4y = 7$ p. 44

14 **a** The sum of two numbers is twelve and their difference is two. By solving a pair of simultaneous equations, find the numbers.

b Find the values of x and y:

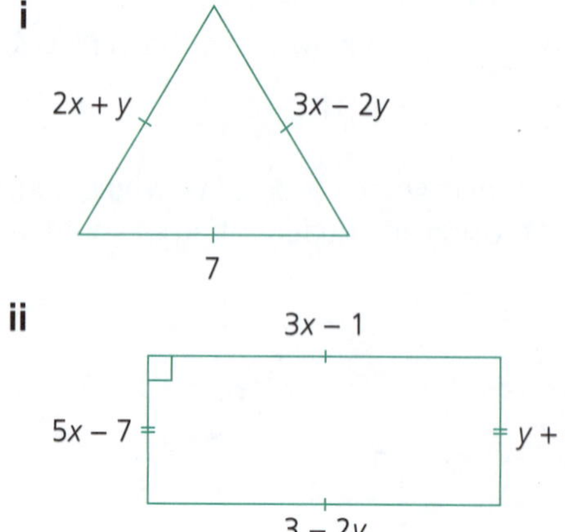

c Six pens and five pencils cost \$2.50 while three pens and two pencils cost \$1.15. What will be the cost of a pen and four pencils?

d Two hamburgers and a can of drink cost \$10 while three hamburgers and two cans of drinks cost \$15.90. Find the cost of a hamburger. p. 44–45

15 Solve:

a $(x - 1)(x + 3) = 0$ **b** $(p + 5)(p - 2) = 0$
c $a(a - 2) = 0$ p. 46

16 Solve for x:

a $x^2 + 3x + 2 = 0$ **b** $m^2 - 4m + 3 = 0$
c $k^2 - 7k + 12 = 0$ **d** $b^2 + 3b - 4 = 0$
e $t^2 - 4t - 12 = 0$ **f** $w^2 - 2w - 15 = 0$
g $c^2 + 2c = 0$ **h** $y^2 - 16 = 0$
i $q^2 - 7q = 0$ **j** $h^2 = 10h$
k $5 - 4q - q^2 = 0$ **l** $x^2 - 9x + 25 = 5$ p. 46

Go to p. 218 for **Quick Answers** or to pp. 231–234 for **Worked Solutions**

For a complete understanding of this topic, you must be able to:

✓	Solve simple equations		p. 38
✓	Solve inequations and graph the solution on the number line		pp. 39–40
✓	Substitute into formulae		p. 40
✓	Change the subject of the formula		p. 41
✓	Solve problems by using formulae		p. 42
✓	Solve simultaneous equations		pp. 42–44
✓	Use equations to solve problems		pp.44–45
✓	Solve quadratic equations by factorising.		p. 46

Now you are ready to do the tests!

Intermediate Test

Equations

(50 marks)

1 Solve:

a $3x + 1 = 7$ (1 mark)

b $2a - 3 = a + 12$ (2 marks)

c $\frac{2a - 1}{5} = 3$ (2 marks)

d $\frac{5x}{2} + 3x = 7$ (2 marks)

e $2(a - 3) = 12$ (2 marks)

f $4(5p - 2) + 3p = 19$ (2 marks)

g $2(2y - 1) - 3(y + 3) = 16$ (3 marks)

2 Solve:

a $3a + 2 > a + 10$ (2 marks)

b $2p - 3 \leq p - 5$ (2 marks)

c $\frac{2a - 1}{5} \geq 3$ (2 marks)

d $q - 5 \leq 3q - 9$ (2 marks)

3 Solve and graph your solution on a number line:

a $2x + 6 \leq x - 2$ (2 marks

b $x - 3 \leq 4x - 6$ (2 marks

4 If $P = 2(l + b)$, find:

a P when $l = 6$ and $b = 4$ (2 marks

b l when $P = 20$ and $b = 3$ (2 marks

5 Make y the subject in the following equations:

a $5x + 2y = 11$ (2 marks)

b $3x - 7y = 10$ (2 marks)

c $\frac{2x - y}{3} = 5$ (2 marks)

6 The sum of two consectutive whole numbers is 25. Let the first number be x.

a Write down the second number. (1 mark)

b Use the equation $x + (x + 1) = 25$ to find the two numbers. (2 marks)

7

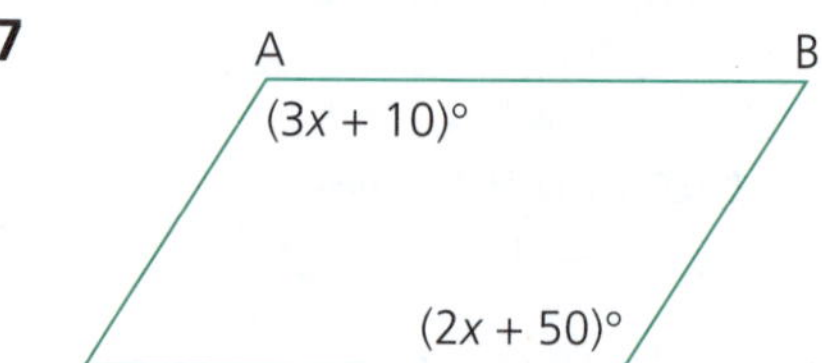

ABCD is a parallelogram.

a Find the value of x. (2 marks)

b Find the size of $\angle DAB$. (1 mark)

8 Solve the pair of simultaneous equations:

$2x + y = 5$

$2x + 3y = 7$ (2 marks)

9 Solve:

a $x^2 - 6x + 5 = 0$ (2 marks)

b $x^2 - 2x - 8 = 0$ (2 marks)

c $x^2 - 5x = 0$ (2 marks)

Your Feedback

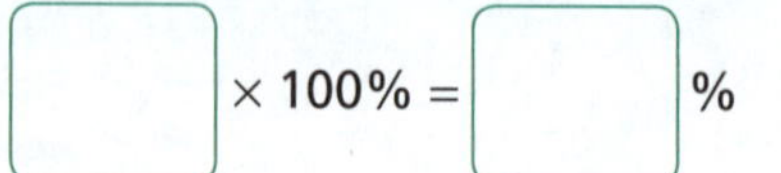

QA PAGE 222

WS PAGE 262

Advanced Test

Equations

(50 marks)

1 Solve:

a $3x - 4 = 4x - 2$ (2 marks)

b $1 - x = 3x + 5$ (2 marks)

c $3(4x - 3) + 2x = 11$ (2 marks)

d $5(3a - 1) + 4(2a + 1) = 10$ (2 marks)

e $2(3y - 2) - (4y + 3) = 18$ (2 marks)

2 Solve:

a $\dfrac{2p-3}{2} = 3p$ (2 marks)

b $\dfrac{2a-1}{5} = \dfrac{a-3}{2}$ (2 marks)

c $\dfrac{5x}{2} + \dfrac{x}{3} = 2$ (2 marks)

3 Solve and graph your solution on a number line:

a $2x + 2 \leq 4x - 6$ (3 marks)

b $\dfrac{1-3x}{5} > 2$ (3 marks)

4 Graph on a number line the inequality $-2 \leq x < 6$. (1 mark)

5

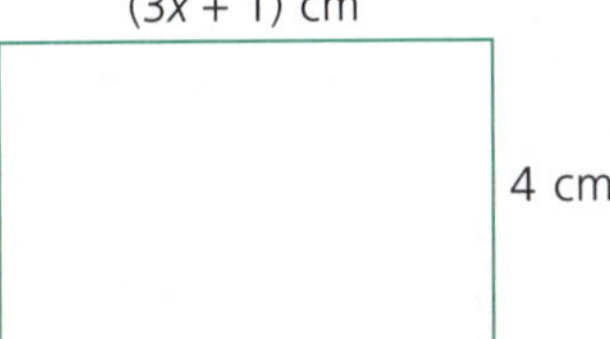

The area of the rectangle is 64 cm^2.

a Write an equation for the area of the rectangle in terms of x. (1 mark)

b Find the value of x. (2 marks)

c Write down the numerical value of the perimeter of the rectangle. (2 marks)

6 Make y the subject of each of the following:

a $3ax - 5by = z$ (2 marks)

b $T = \sqrt{\dfrac{ay}{m}}$ (2 marks)

c $P = \dfrac{4a+b}{y}$ (2 marks)

d $K = \dfrac{3xy-2}{a}$ (2 marks)

7 The area of an annulus is given by the formula $A = \pi(R^2 - r^2)$, where A is the area, R is the radius of the larger circle and r the radius of the smaller circle. If the area of the annulus is 148.3 cm^2, and the radius of the smaller circle is 4 cm, what is the radius of the larger circle, correct to two decimal places? (3 marks)

8 Solve:

$4x = y + 3$

$2x + y = 3$ (2 marks)

9 A wrap and two bottles of water cost \$6.50, and two wraps and three bottles of water cost \$11.00. By solving a pair of simultaneous equations, find the cost of one wrap. (3 marks)

10 Solve:

a $x^2 - 6x + 5 = 0$ (1 mark)

b $x^2 = 2x + 15$ (2 marks)

11 The dimensions of a rectangle are $(x + 5)$ cm and $(x - 2)$ cm. If the area of the rectangle is 18 cm^2, what is the perimeter? (3 marks)

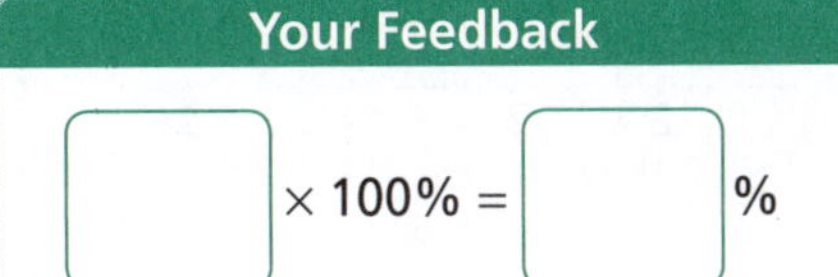

QA PAGE 222

WS PAGE 262

Chapter 5
Linear Relationships

Points and lines

Coordinates of a point

Each point is located by assigning an ordered pair to it. This consists of an ***x* value** and a ***y* value**.

Remember: (x, y) means 'x comes before y'.

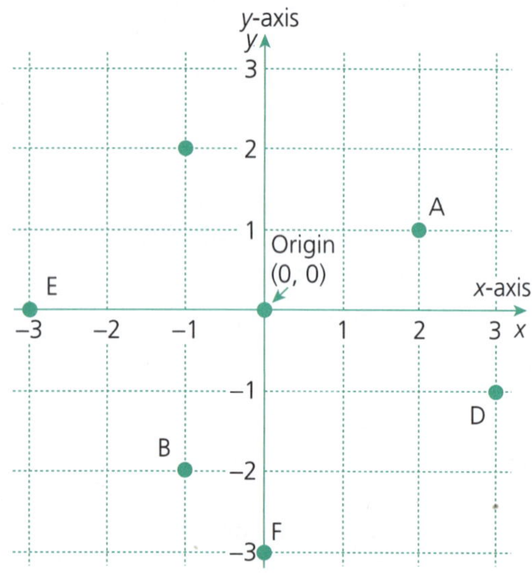

Note:
- on the x-axis, all y values are zero.
- on the y-axis, all x values are zero.

Keywords

Axes, Axis, Collinear, Coordinates, Dependent variable, Distance, Gradient, Graph, Horizontal, Independent variable, Intercept, Linear, Midpoint, Ordered pair, Origin, Parallel, Perpendicular, Substitute, Vertices

From the diagram:

A has coordinates (2, 1) ← Ordered pair or coordinates for A.

↑ x value ↑ y value

B is (–1, –2), C is (–1, 2), D is (3, –1).

E is (–3, 0) — y value is zero.

F is (0, –3) — x value is zero.

Distance between two points

A has coordinates (1, 2) while B is (5, 4).

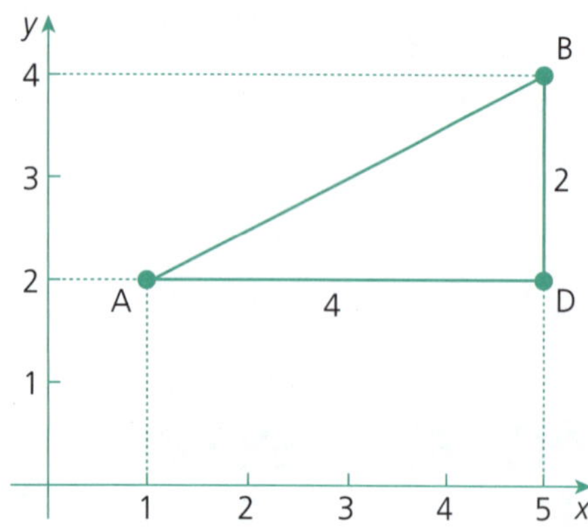

Form a right-angled triangle ABD.
By observation, AD = 4, BD = 2.

Then by using Pythagoras' Theorem (Chapter 8):

$$AB^2 = 2^2 + 4^2$$
$$= 4 + 16$$
$$= 20$$
$$AB = \sqrt{20}$$

1. Plot points.
2. Form right-angled ⊿.
3. Read off lengths.
4. Pythagoras' Theorem.

The distance between A and B is $\sqrt{20}$ units.

Midpoint of an interval

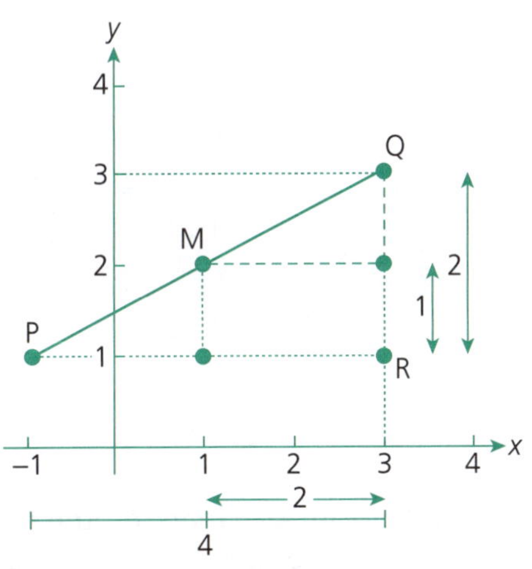

P has the coordinates (–1, 1), Q is (3, 3).

Form the right-angled triangle PQR.

Horizontal distance PR = 4, half-way is 2.

Vertical distance QR = 2, half-way is 1.

M is the midpoint. It is across 2 from P and up 1 from R.

M has the coordinates (1, 2).

Gradient of a line

The gradient or slope of a line is given by:

$$\frac{\text{Vertical rise}}{\text{Horizontal run}} \quad \text{or} \quad \frac{\text{Rise}}{\text{Run}}.$$

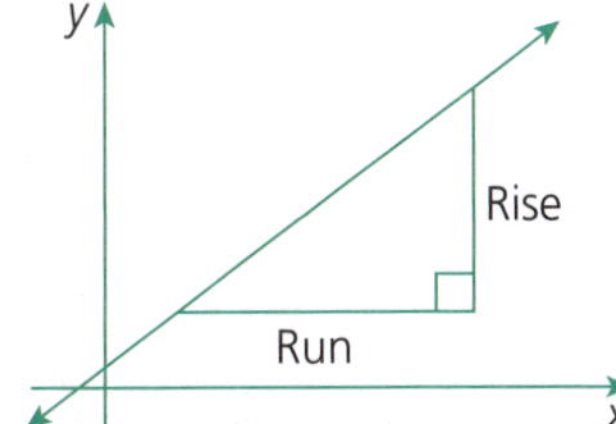

A line can have either a positive gradient (sloping upwards to the right), or a negative gradient (sloping downwards to the right).

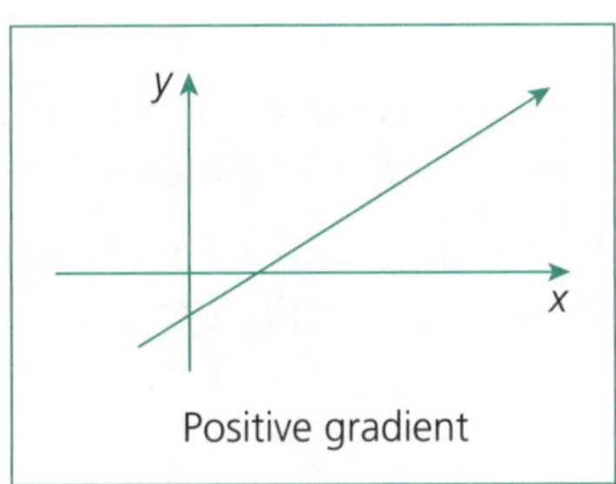

Positive gradient

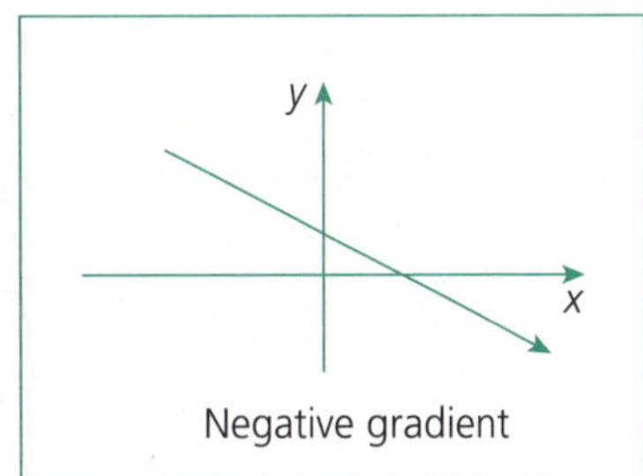

Negative gradient

Finding the gradient from a diagram

To find the gradient of a line from a diagram, a right-angled triangle is needed. This usually is already there. It is the triangle formed by the line and the coordinate axes.

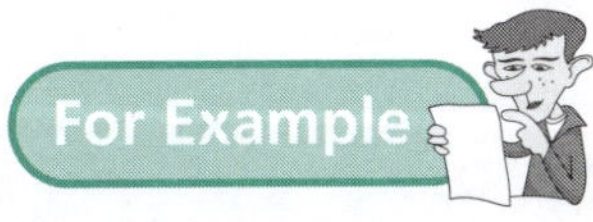

1

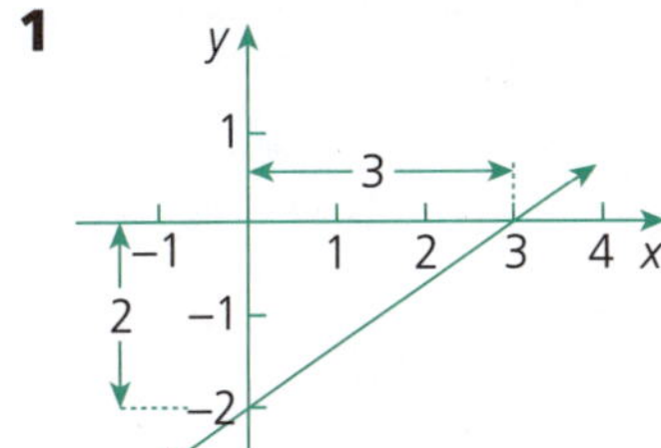

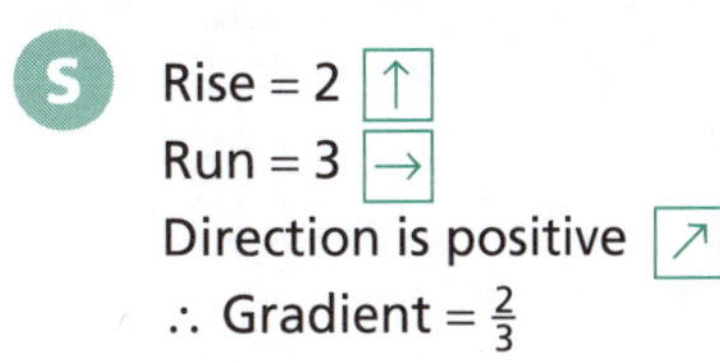

Rise = 2 [↑]

Run = 3 [→]

Direction is positive [↗]

∴ Gradient = $\frac{2}{3}$

2

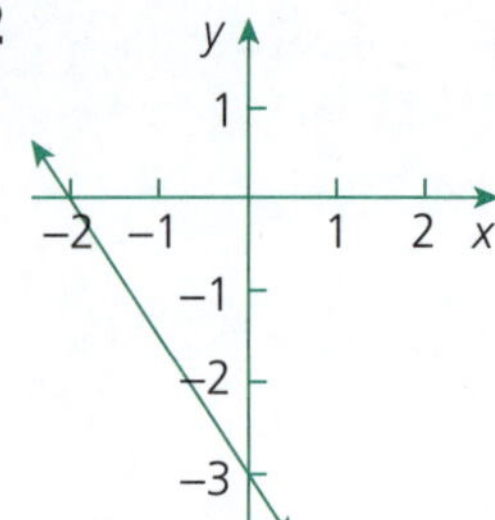

S Rise = 3 ↓
Run = 2 →
Direction is negative ↘
$\therefore$ Gradient $= -\frac{3}{2}$

3 Find the gradient of the interval joining (–2, –3) and (2, 1).

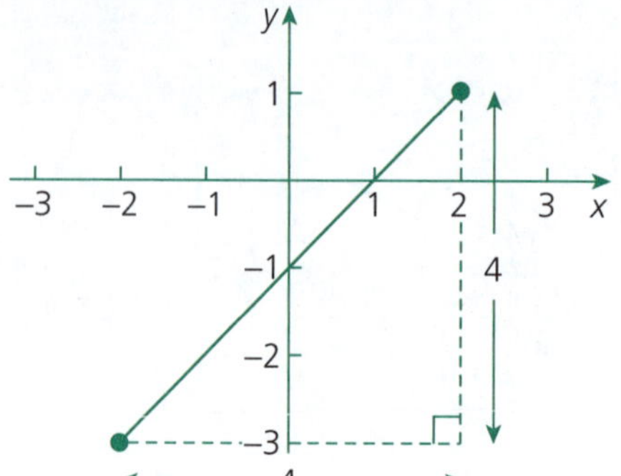

S **Draw a diagram and mark on the points. Join them and complete a right-angled triangle. The question can now be answered from the triangle.**

Rise = 4 ↑
Run = 4 →
Direction is positive ↗
$\therefore$ Gradient $= \frac{4}{4} = 1$

Straight lines

Graphing a straight line

x	–2	–1	0	1	2	3
y	–3	–1	1	3	5	7

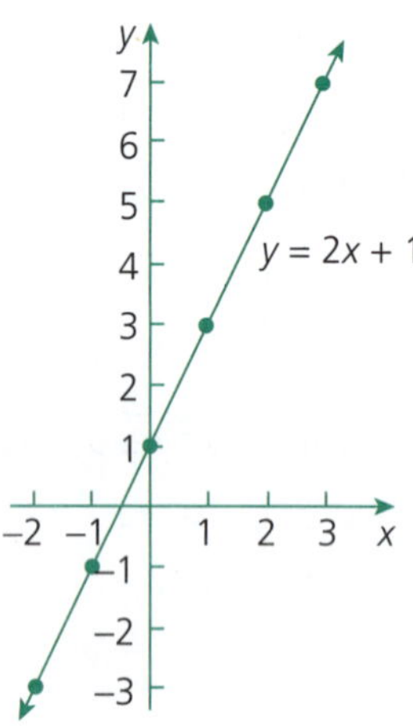

By completing the table (at left) of values for $y = 2x + 1$, a set of ordered pairs is generated. These can be plotted and joined to show the graph of $y = 2x + 1$. It is a straight line.

All equations of the form $y = mx + b$ represent straight lines. Calculation of three points is sufficient to draw the line.

The table

x	0	1	2
y			

is the easiest to work with.

The independent variable is x.

The dependent variable is y.

1 By first completing a table of values, sketch the line $y = 2x - 1$.

x	0	1	2
y	−1	1	3

- Draw the line completely through the number plane.
- Label the line.

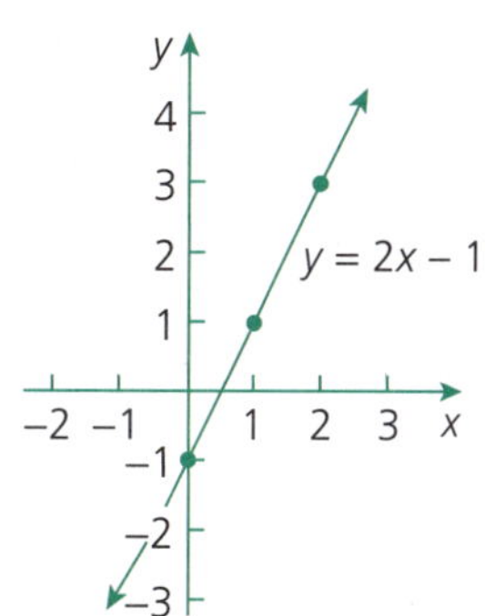

Lines parallel to the coordinate axes

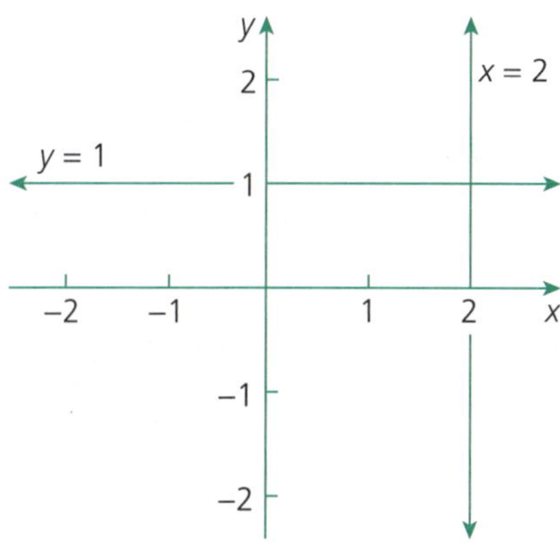

Equations that look like $x = a$ (a is a constant) are lines parallel to the y-axis through $x = a$. Equations that look like $y = b$ (b is a constant) are lines parallel to the x-axis through $y = b$.

For example, consider $x = 2$ and also $y = 1$.

All points on $x = 2$ have an x coordinate of 2, that is (2, 1), (2, 2), (2, −1), etc., while all points on $y = 1$ have a y coordinate of 1, that is (0, 1), (2, 1), (−2, 1), etc.

Line parallel to x-axis: $y = b$
Line parallel to y-axis: $x = a$

Equation of x-axis: $y = 0$
Equation of y-axis: $x = 0$

This means that:

All y values on the x-axis are zero: (2, 0), (−2, 0), (3, 0), etc.
All x values on the y-axis are zero: (0, −1), (0, 2), (0, 4), etc.

This allows you to calculate the points at which any straight line cuts the x- and y-axes, given its equation. These values are known as the ***x*- and *y*-intercepts**, usually designated as a and b.

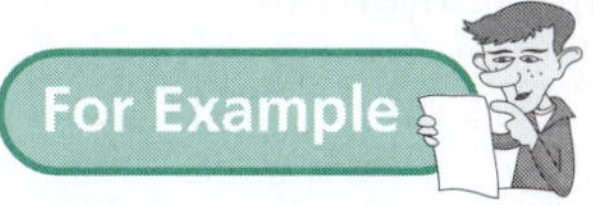

1 Find the x and y intercepts for the line with equation $2x + y = 6$. Hence, or otherwise graph the line.

$2x + y = 6$
When $y = 0$, $2x = 6$
$x = 3$
When $x = 0$, $y = 6$
x-intercept: $x = 3$
y-intercept: $y = 6$

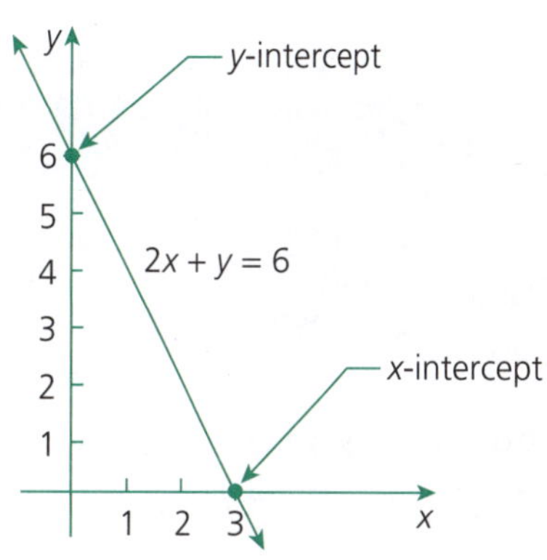

The general form $Ax + By + C = 0$

$Ax + By + C = 0$, where $A \geq 0$, and A, B and C are integers.

This is a preferred form for expressing the equation of a line.

The equation $y = mx + b$ (the gradient–intercept form)

When the equation of a straight line is written in the form $y = mx + b$, then m is the gradient of the line and b is the y-intercept (the y value where the line cuts the y-axis).

> $y = mx + b$
> ↑ gradient ↑ y-intercept

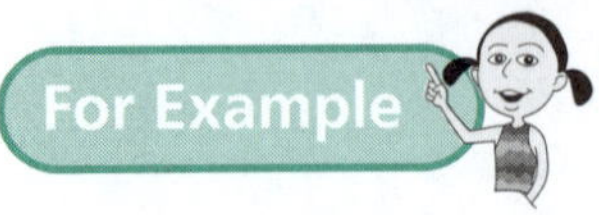

1 Find the equation of the line with gradient $\frac{2}{3}$ passing through the y-axis at (0, 4).

$m = \frac{2}{3}$, $b = 4$

The equation is $y = mx + b$

that is, $y = \frac{2}{3}x + 4$.

2 From the given diagrams, write down the equations of the lines.

a

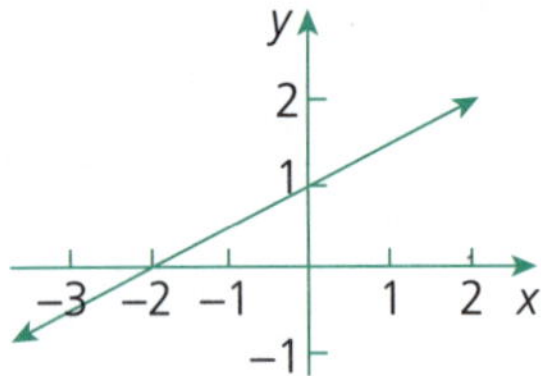

$m = \frac{\text{Rise}}{\text{Run}} = \frac{1}{2}$ positive ↗

$b = 1$

The equation is $y = mx + b$

that is, $y = \frac{1}{2}x + 1$.

b

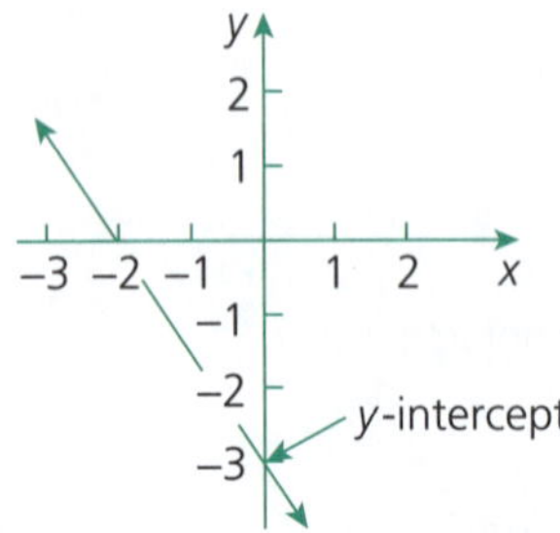

$m = \frac{\text{Rise}}{\text{Run}} = -\frac{3}{2}$ negative ↘

$b = -3$

The equation is $y = mx + b$

that is, $y = -\frac{3}{2}x - 3$.

3 For the equation $y = 3x - 1$, write down the gradient and the y-intercept.

From $y = 3x - 1$

$m = 3$, $b = -1$.

> Compare with $y = mx + b$.

The gradient is 3, the y-intercept is -1.

4 By writing the equations in the form $y = mx + b$, find the gradient and y-intercept of the following straight lines:

> The equations must be transformed to look like $y = mx + b$.

a $4y = 6x + 8$

$4y = 6x + 8$

$\therefore y = \frac{6}{4}x + 2$

that is, $y = \frac{3}{2}x + 2$

> Compare with $y = mx + b$.

$\therefore m = \frac{3}{2}$, $b = 2$.

The gradient is $\frac{3}{2}$, the y-intercept is 2.

b $2x + 4y = 3$

S

$$2x + 4y = 3$$
$$\therefore \quad 4y = -2x + 3$$
$$\therefore y = -\frac{2}{4}x + \frac{3}{4}$$
$$\therefore y = -\frac{1}{2}x + \frac{3}{4}$$

Compare with $y = mx + b$.

$\therefore m = -\frac{1}{2}, b = \frac{3}{4}$.

The gradient is $-\frac{1}{2}$, the y-intercept is $\frac{3}{4}$.

c $x - 2y = 8$

S

$$x - 2y = 8$$
$$x - 8 = 2y$$
$$2y = x - 8$$
$$y = \frac{1}{2}x - 4$$

Compare with $y = mx + b$.

$\therefore m = \frac{1}{2}, b = -4$.

That is, the gradient is $\frac{1}{2}$, the y-intercept is -4.

5 Use the gradient and y-intercept to sketch the lines from the following information.

a Gradient 3, y-intercept -1.

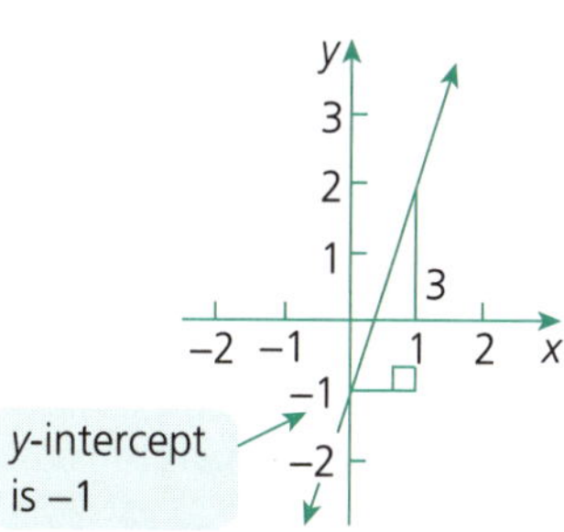

$m = 3$ positive slope ↗

$\therefore \frac{\text{Rise}}{\text{Run}} = \frac{3}{1}$,

that is, up 3, across 1, ↗

y-intercept $= -1$.

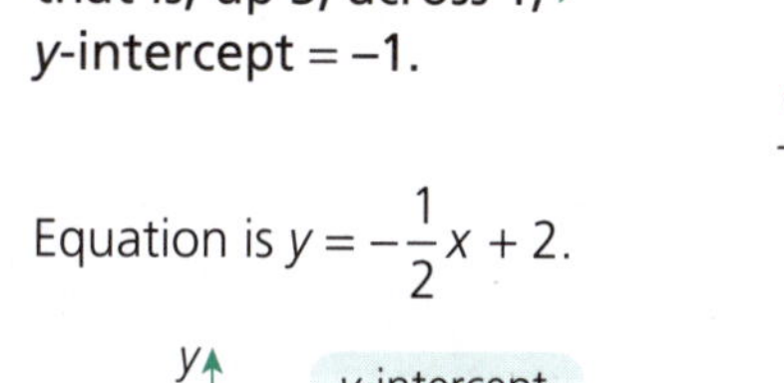

b Equation is $y = -\frac{1}{2}x + 2$.

S

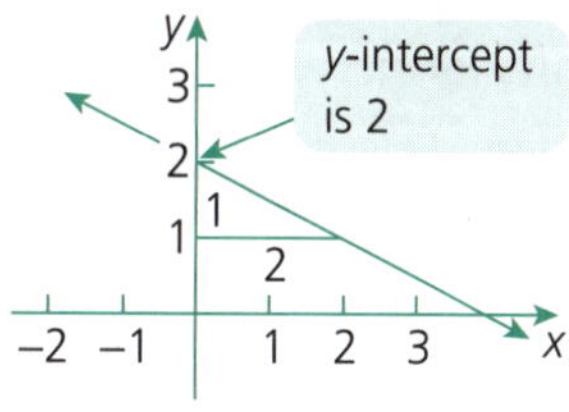

$y = -\frac{1}{2}x + 2$

$m = -\frac{1}{2}, b = 2$ From $y = mx + b$.

Then $\frac{\text{Rise}}{\text{Run}} = -\frac{1}{2}$, negative slope ↘

that is, down 1, across 2, ↘

y-intercept $= 2$.

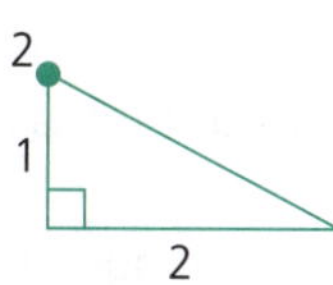

Using formulae for distance, midpoint and gradient

Given any two points on the number plane, (x_1, y_1) and (x_2, y_2):

Distance between the points will be given by	$d = \sqrt{(x_2 - x_1)^2 + (y_2 - y_1)^2}$
The **midpoint** of the interval will be given by	$\text{MP} = \left(\frac{x_1 + x_2}{2}, \frac{y_1 + y_2}{2}\right)$
The **gradient** of the line through the two points is	$m = \frac{y_2 - y_1}{x_2 - x_1}$

For Example

1 Consider the points A(–2, –1) and B(3, 3).

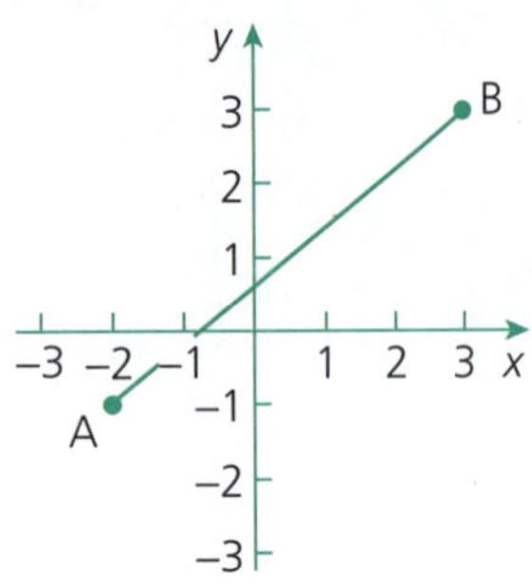

a Calculate the distance AB.

S $d = \sqrt{(x_2 - x_1)^2 + (y_2 - y_1)^2}$

$= \sqrt{(3 + 2)^2 + (3 + 1)^2}$

$= \sqrt{5^2 + 4^2}$

$= \sqrt{25 + 16}$

$= \sqrt{41}$ This is the exact distance AB.

AB is $\sqrt{41}$ units long.

Note: this is simply an application of Pythagoras' Theorem.

b Calculate the midpoint of AB.

S $MP = \left(\frac{x_1 + x_2}{2}, \frac{y_1 + y_2}{2}\right)$

$= \left(\frac{-2 + 3}{2}, \frac{-1 + 3}{2}\right)$

$= \left(\frac{1}{2}, 1\right)$

Midpoint is $\left(\frac{1}{2}, 1\right)$.

Average of *x*-coordinates, average of *y*-coordinates.

c Calculate the gradient of AB.

S $m = \frac{y_2 - y_1}{x_2 - x_1}$

$= \frac{3 - (-1)}{3 - (-2)}$

$= \frac{4}{5}$

Gradient is $\frac{4}{5}$.

The gradient is always left as a simple fraction. Do not convert either to a decimal or to a mixed number.

2 Consider the following cases (see the diagram below):

a The points P(1, 3) and Q(–1, 3).

b The points P(1, 3) and S(1, –3).

S **Distances, midpoints and gradients can be read from the diagram when they are on lines parallel to the axes.**

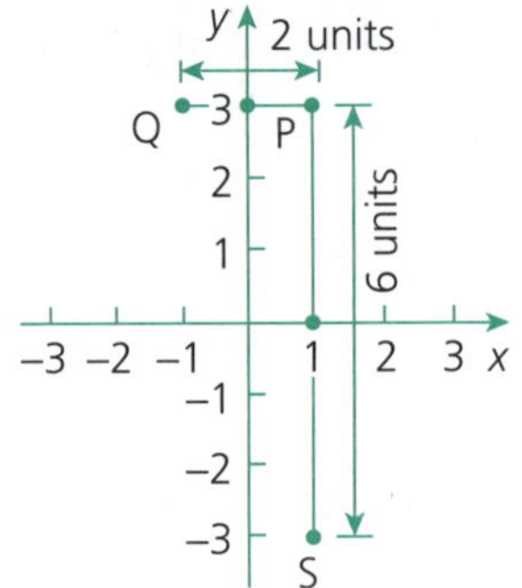

Distance PQ = 2 units
Midpoint PQ = (0, 3)
Gradient of PQ = 0
Distance PS = 6 units
Midpoint PS = (1, 0).
Gradient of PS = undefined

PQ has slope zero.

PS has indeterminate or infinite slope.

3 Plot the points A(–2, –2), B(2, –2) and C(0, 3) on a number plane.

a Prove that length AC is $\sqrt{29}$ units.

S

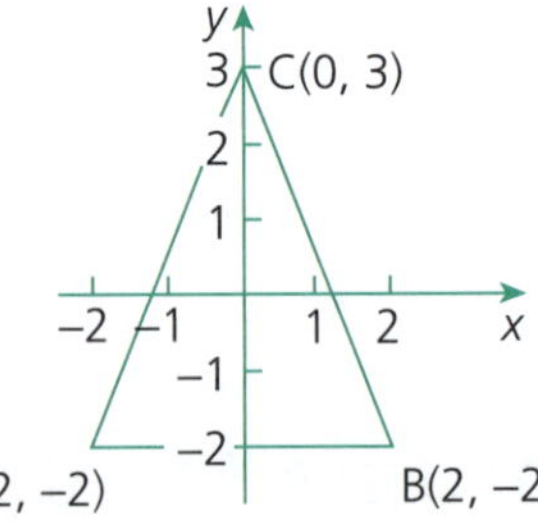

Draw a clear sketch.

$d = \sqrt{(x_2 - x_1)^2 + (y_2 - y_1)^2}$

$= \sqrt{(0 + 2)^2 + (3 + 2)^2}$

$= \sqrt{4 + 25}$

$= \sqrt{29}$ **units**

AC is $\sqrt{29}$ units.

b Prove that △ABC is isosceles.

S For △ABC to be isosceles, two sides must have the same length. Check the length of BC.

$$\begin{aligned} BC &= \sqrt{(x_2 - x_1)^2 + (y_2 - y_1)^2} \\ &= \sqrt{(0-2)^2 + (3+2)^2} \\ &= \sqrt{4 + 25} \\ &= \sqrt{29} \text{ units} \end{aligned}$$

$\therefore$ AC = BC (both $\sqrt{29}$)

That is, △ABC is isosceles.

c Prove that the midpoint of AB lies on the *y*-axis.

S
$$\begin{aligned} \text{Midpoint} &= \left(\frac{x_1 + x_2}{2}, \frac{y_1 + y_2}{2}\right) \\ &= \left(\frac{-2+2}{2}, \frac{-2+-2}{2}\right) \\ &= (0, -2) \end{aligned}$$

The point (0, –2) lies on the *y*-axis as the *x*-coordinate is 0.

The condition for a point to lie on a line

A point lies on a line (or a curve) if the coordinates of the point satisfy the equation of the line (or curve). **Collinear points** occur when two or more points lie on the same line.

1 Show that (2, –1) lies on the line $y = 2x - 5$.

S Substitute $x = 2$ and $y = -1$ into the equation.
LHS = –1, RHS = $2 \times 2 - 5$
$= -1$.

$-1 = -1$

Therefore (2, –1) satisfies the equation.
The point (2, –1) lies on the line $y = 2x - 5$.

2 Determine whether the line $4x + y = -1$ passes through (–5, 17).

S Substitute $x = -5$ and $y = 17$ into the equation.
LHS = $4 \times (-5) + 17$, RHS = –1
$= -20 + 17$
$= -3$

$-3 \neq -1$

(–5, 17) does not satisfy the equation.
Therefore the line $4x + y = -1$ does not pass through (–5, 17).

3 The point (–3, *k*) lies on the line $x + 2y = 5$. Find the value of *k*.

S If a point lies on a line, its coordinates must satisfy the equation of the line.
Substitute $x = -3$, $y = k$ into:
$x + 2y = 5$
Thus $-3 + 2k = 5$
$2k = 8$
$k = 4$

Solve the simple equation for *k*.

4 The points (2, 1) and (–1, –1) lie on the line $2ax - by = 5$. Find *a* and *b* and hence the equation of the line.

S The point (2, 1) lies on $2ax - by = 5$, (Substitute $x = 2$, $y = 1$)

$\therefore 4a - b = 5$ (1)

The point (–1, –1) lies on $2ax - by = 5$, (Substitute $x = -1$, $y = -1$)

$\therefore -2a + b = 5$ (2)

Add (1) + (2): $2a = 10$
$\therefore a = 5$
Substitute $a = 5$ in (2)
$-10 + b = 5$
$\therefore b = 15$
Then $a = 5$, $b = 15$.
Therefore the equation is $10x - 15y = 5$, that is, $2x - 3y = 1$.

Parallel and perpendicular lines

If two lines have gradients m_1 and m_2 respectively, the condition for the lines to be:

> **1** parallel ∥ is: $m_1 = m_2$
>
> **2** perpendicular ⊥ is: $m_1 = -\frac{1}{m_2}$ or $m_1 m_2 = -1$ (Negative reciprocal)

1 Show that l_1, $2x - y = 7$ is parallel to l_2, $y = 2x + 1$, but perpendicular to l_3, $x + 2y - 3 = 0$.

S **Let m_1, m_2, m_3 be the gradients of lines l_1, l_2, l_3 respectively.**

Line 1 (l_1)

$2x - y = 7$

$\Rightarrow \quad y = 2x - 7$

$\therefore m_1 = 2$

Line 2 (l_2)

$y = 2x + 1$

$\therefore m_2 = 2$

Therefore $2x - y = 7$ is parallel to $y = 2x + 1$. (Both have a gradient of 2.)

> Each equation must be expressed in the form $y = mx + b$ in order to find gradients. The gradients are then compared using the above criteria.

Line 3 (l_3)

$x + 2y - 3 = 0$

$2y = -x + 3$

$y = -\frac{1}{2}x + \frac{3}{2}$

$\therefore m_3 = -\frac{1}{2}$

> y is made the subject of the equation in each case.

Compare m_1 and m_3.

$2 \times -\frac{1}{2} = -1 \quad (m_1 \times m_3 = -1)$

Therefore $2x - y = 7$ is perpendicular to $x + 2y - 3 = 0$.

2 Show that △ABC is right-angled, given that the coordinates of A, B and C are (–3, 4), (2, 1) and (–1, –4) respectively. Which angle is the right angle?

S **This question could be answered using Pythagoras' Theorem but it is easier to consider the gradients.**

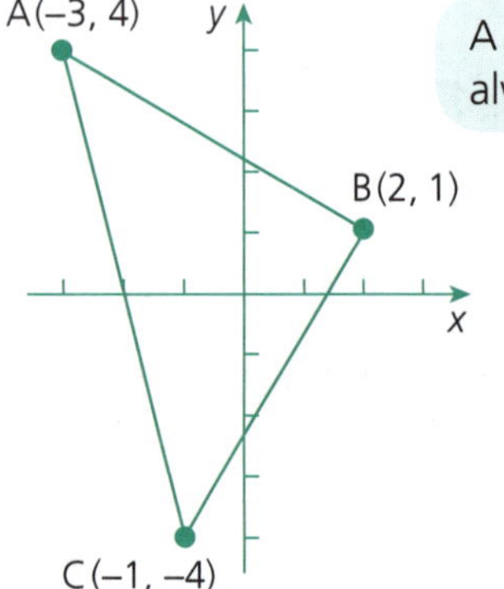

> A sketch is always valuable!

Using $m = \frac{y_2 - y_1}{x_2 - x_1}$ in each case:

Gradient AB $= \frac{4 - 1}{-3 - 2} = \frac{3}{-5} = -\frac{3}{5}$

Gradient BC $= \frac{1 - (-4)}{2 - (-1)} = \frac{1 + 4}{2 + 1} = \frac{5}{3}$

Gradient AC $= \frac{4 - (-4)}{-3 - (-1)} = \frac{4 + 4}{-3 + 1} = \frac{8}{-2}$
$= -4$

Now compare the gradients.

$m_{AB} \times m_{BC} = -\frac{3}{5} \times \frac{5}{3} = -1$

$\therefore$ AB ⊥ BC

That is, the triangle is right-angled at B.

Equation of a line parallel or perpendicular to another line

Remember, the equation of the line can be written in the form $y = mx + b$, where m is the gradient and b is the y-intercept. If the gradient is known and the line passes through a point, then using substitution, the y-intercept can be found.

1 Find the equation of the line with gradient 3, and passing through the point (2, 5).

S Using $y = mx + b$,

$y = 3x + b$

Subs. (2, 5): $5 = 3(2) + b$

$5 = 6 + b$

$b = -1$

$\therefore y = 3x - 1$

2 Find the equation of the line parallel to $y = 4 - 3x$ and passing through the point (2, 1).

S Gradient of $y = 4 - 3x$ is -3.

$\therefore m = -3$

Using $y = mx + b$,

$y = -3x + b$

Subs. (2, 1): $1 = -3(2) + b$

$1 = -6 + b$

$b = 7$

$\therefore y = -3x + 7$

3 Find the equation of the line parallel to $2x - y - 3 = 0$ and passing through the point (–4, 1).

S Rewrite $2x - y - 3 = 0$ as $y = 2x - 3$

$\therefore m = 2$

Using $y = mx + b$,

$y = 2x + b$

Subs. (–4, 1): $1 = 2(-4) + b$

$1 = -8 + b$

$b = 9$

$\therefore y = 2x + 9$

4 Find the equation of the line perpendicular to $3x + 5y - 1 = 0$ and passing through the point (3, 4).

S First, find the gradient of $3x + 5y - 1 = 0$:

$5y = -3x + 1$

$y = \dfrac{-3x + 1}{5}$

$\therefore$ Gradient $= \dfrac{-3}{5}$

$\therefore$ Gradient of perpendicular is $\dfrac{5}{3}$

$\therefore m = \dfrac{5}{3}$

Using $y = mx + b$,

$y = \dfrac{5}{3}x + b$

Subs. (3, 4): $4 = \dfrac{5}{3}(3) + b$

$4 = 5 + b$

$b = -1$

$\therefore y = \dfrac{5}{3}x - 1$

5 Find the equation of the line perpendicular to $2x - 3y - 7 = 0$ and passing through the x-axis at $x = 2$.

S First, find the gradient of $2x - 3y - 7 = 0$:

$3y = 2x - 7$

$y = \dfrac{2x - 7}{3}$

$\therefore$ Gradient $= \dfrac{2}{3}$

$\therefore$ Gradient of perpendicular is $-\dfrac{3}{2}$

$\therefore m = -\dfrac{3}{2}$

Using $y = mx + b$,

$y = -\dfrac{3}{2}x + b$

Subs. (2, 0): $0 = -\dfrac{3}{2}(2) + b$

$0 = -3 + b$

$b = 3$

$\therefore y = -\dfrac{3}{2}x + 3$

Applications of linear graphs to word problems

In the general form of linear graphs, $y = mx + b$, the pronumerals x and y are known to be **variables**. The independent variable is x and the dependent variable is y. On a graph, the independent variable is graphed on the horizontal axis, and the dependent variable graphed on the vertical axis.

In word problems, the letters used relate to the quantities measured in the question. For example, a hire car may charge a price (P) based on the distance of the fare (d). The rule (or equation, or formula) could be expressed as $P = 2d + 5$, where d is the independent variable and P is the dependent variable.

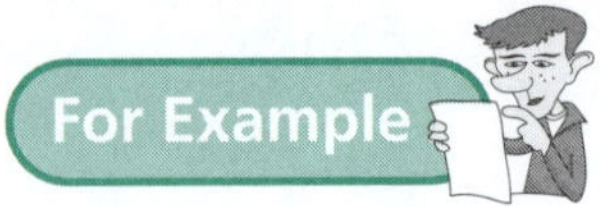

1 An electrician charges a call-out fee of \$40 plus an hourly rate of \$60. Let C = total cost and n = number of hours.

a Write a rule that relates the total cost with the number of hours, identifying the independent and dependent variables.

S **$C = 60n + 40$, where n is independent variable and C is dependent variable.**

b Complete a table for the rule.

S

n	0	2	4	6
C	40	160	280	400

c Draw a graph for the rule.

S

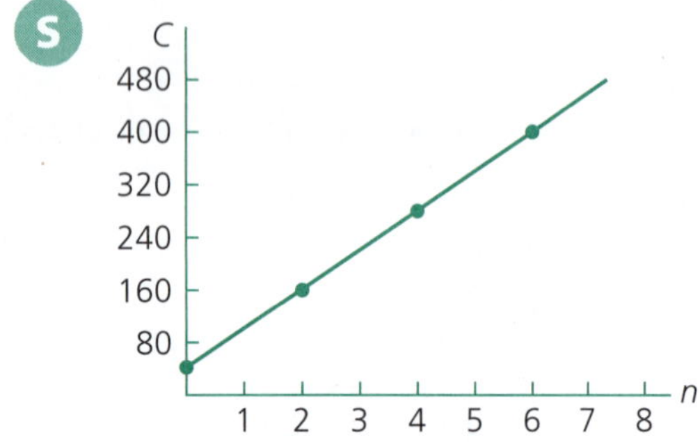

d What is the gradient and vertical intercept for the line graph?

S **The gradient is 60 and the intercept is 40.**

e If the electrician is employed for 3.5 hours, what is the total cost?

S **Let $n = 3.5$, then** $C = 60 \times 3.5 + 40$
$= 250$

The electrician charges \$250.

f If the cost is \$340, what is the number of hours the electrician is employed?

S **Let $C = 340$, then** $340 = 60n + 40$
$60n = 300$
$n = 5$

The electrician is employed for 5 hours.

Practise Practise

1 From the given diagram, write down the ordered pair for the points A, B, C, D, E, F, G, H, I, J, K and L.

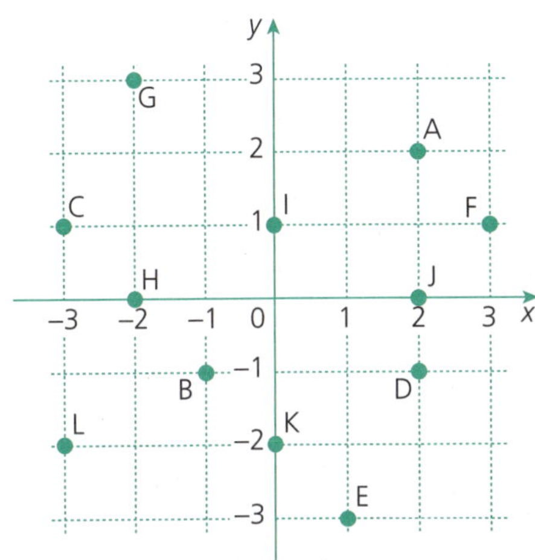

p. 52

2 Find the length of the interval AB in the following diagrams. (Leave it in exact form.)

a

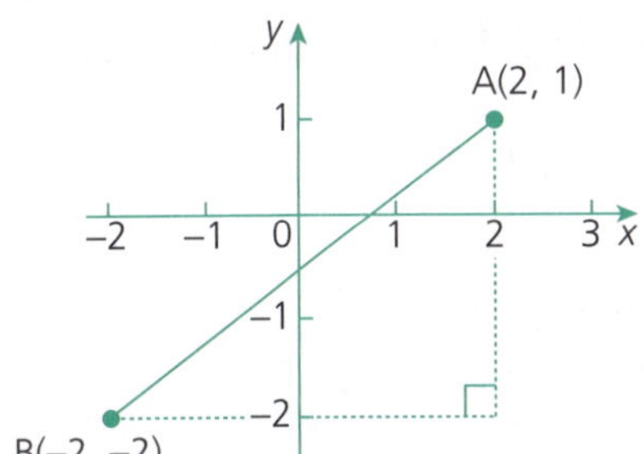

b

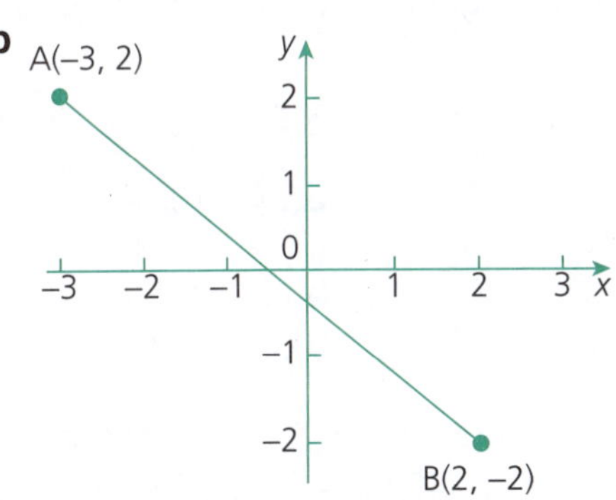

p. 52

3 By plotting the points on a number plane, find the distance between the following pairs of points:

a (–3, 2) (2, 1) **b** (2, –3) (1, 4)
c (4, 2) (–1, –3) **d** (5, 1) (0, 0)
e (3, –2) (–2, –2) **f** (2, 4) (2, –1)

(Leave the answer as a surd if necessary.)

p. 52

4 From the diagrams below, find the coordinates of the midpoint of AB.

a

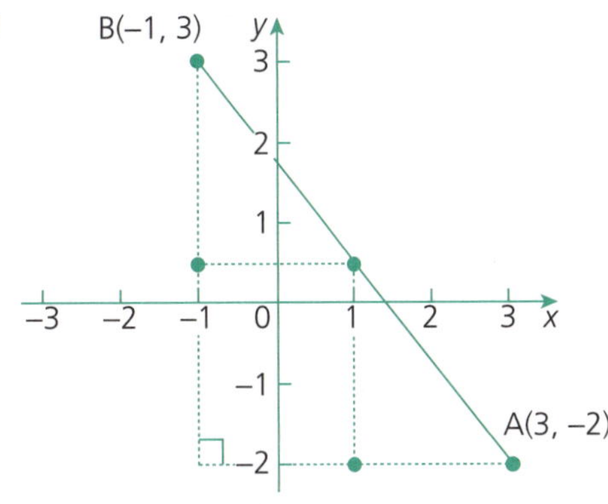

b

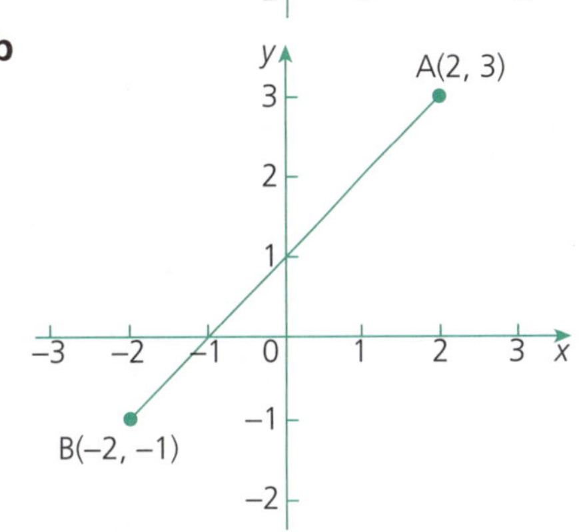

p. 53

5 By plotting the points on the number plane, find the midpoint of the interval joining the following points:

a (–3, 2) (2, 4) **b** (2, 3) (–2, 1)
c (4, 2) (–4, –2) **d** (4, –2) (0, 0)
e (3, –2) (–1, –2) **f** (2, 4) (2, –1)

p. 53

6 From the diagrams, find the gradients of the following lines:

a

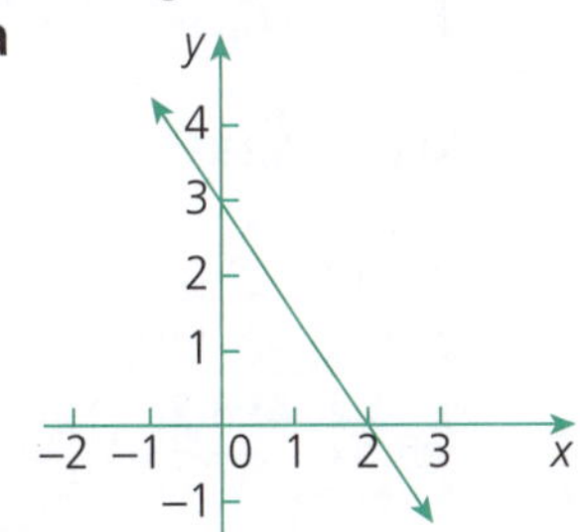

b

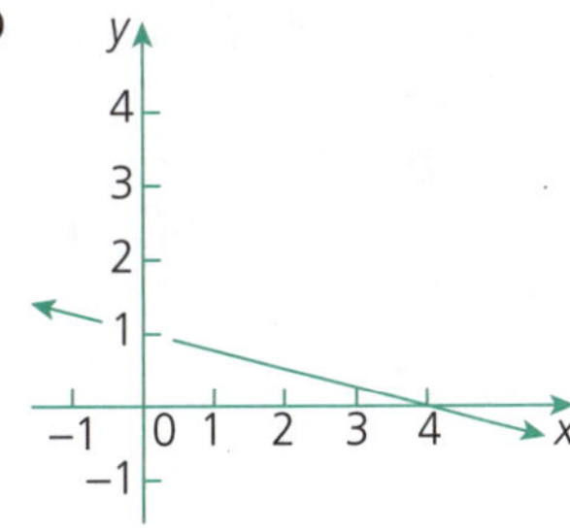

c

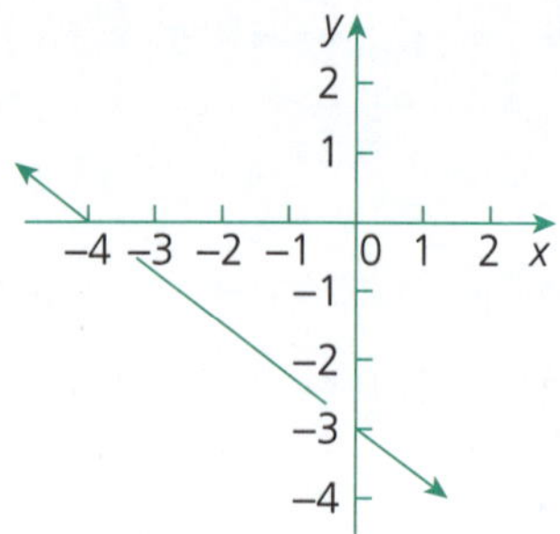

d

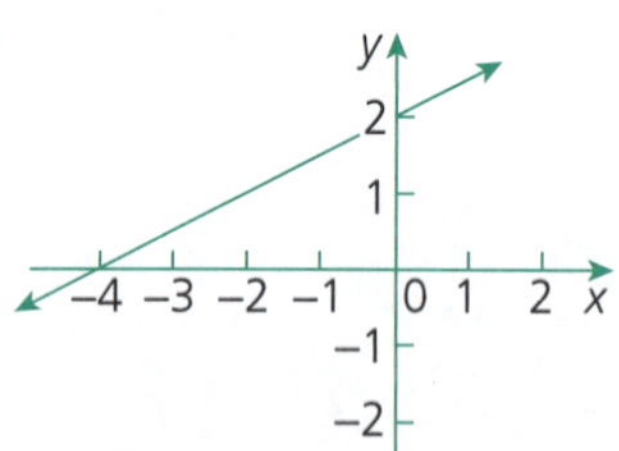

e

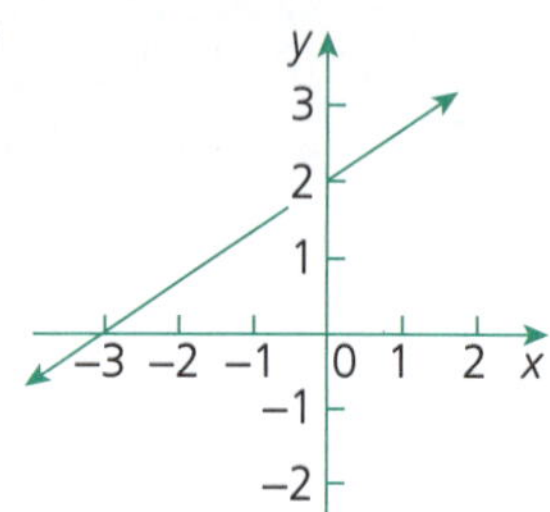

f

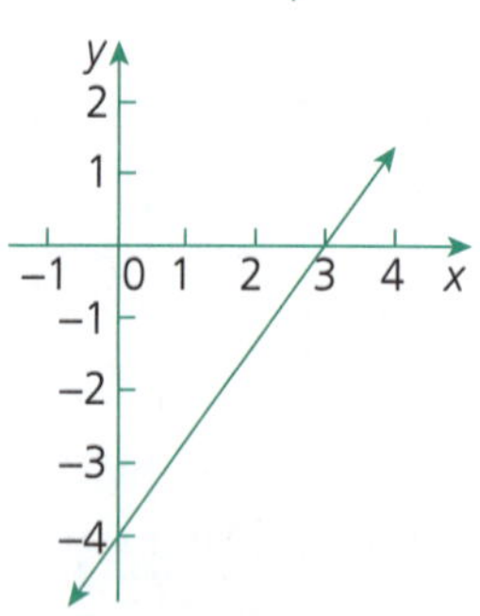

p. 53

7 By plotting the points on a number plane, find the gradient of the line passing through the following points:

a (−3, 2) (2, 4) **b** (2, −3) (−2, 1)
c (4, 2) (−4, −2) **d** (4, −2) (0, 0)
e (3, −2) (−1, −2) **f** (2, 4) (2, −1) p. 53

8 Use the distance formula to calculate the length AB, given:

a A(−3, 2) B(2, 4) **b** A(2, −3) B(−2, 1)
c A(4, 2) B(−4, −2) **d** A(4, −2) B(0, 0)
e A(3, −2) B(−1, −2) **f** A(2, 4) B(2, −1)

pp. 57–59

9 Use the midpoint formula to find the midpoint of the following intervals AB:

a A(−3, 2) B(1, 4) **b** A(2, −3) B(−2, 1)
c A(4, 2) B(−4, −2) **d** A(4, −2) B(0, 0)
e A(3, −2) B(−1, −2) **f** A(2, 4) B(2, −1)

pp. 57–59

10 Use the gradient formula to find the gradient of the line through the following points A and B:

a A(−3, 2) B(2, 4) **b** A(2, −3) B(−2, 1)
c A(4, 2) B(−4, −2) **d** A(4, −2) B(0, 0)
e A(3, −2) B(−1, −2) **f** A(2, 4) B(2, 1)

pp. 57–59

11 Given the points A(2, −3), B(−3, 1) and C(1, 4) find the:

a distances AB and AC
b midpoints of AC and BC
c gradients of BA and CB. pp. 57–59

12 By completing a table of values, sketch the following lines:

a $y = 2x$ **b** $y = -2x$
c $y = -2x + 1$ **d** $y = -2x - 1$
e $y = \frac{1}{2}x + 2$ **f** $y = \frac{1}{2}x - 2$
g $y = \frac{1}{2}x$ **h** $y = -\frac{1}{2}x$ pp. 54–55

13 Sketch on the same number plane:

a $x = 4$ **b** $y = 4$
c $x = -3$ **d** $y = -2$ p. 55

14 Write down the equation of the line with:

a gradient 4, y-intercept −1
b gradient −2, y-intercept 3
c $m = \frac{1}{2}, b = 2$
d $m = \frac{3}{4}, b = -4$ p. 55

15 Write down the gradient and y-intercept for the following equations:

a $y = 3x + 7$ **b** $y = -4x + \frac{1}{2}$
c $y = 4 - 3x$ **d** $5y = 10x + 2$
e $3y = 2x - 6$ **f** $2x + y = 3$
g $2x - y = 3$ **h** $3x + 2y = 6$
i $2x - 3y = 6$ **j** $4x = 2y + 7$ pp. 56–57

16 From the following diagrams, find the gradient of the line and hence write down its equation:

a

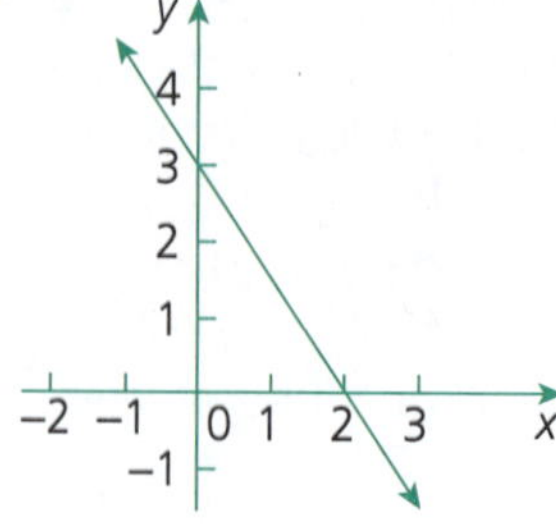

b

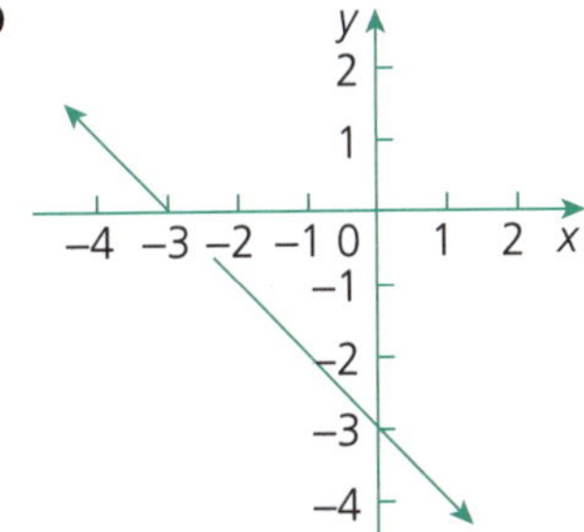

c

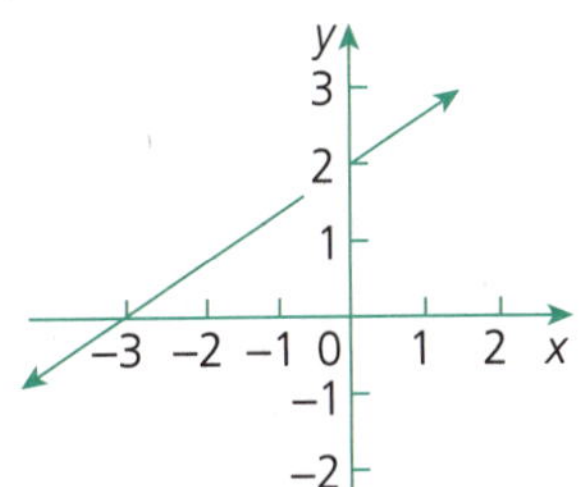

d

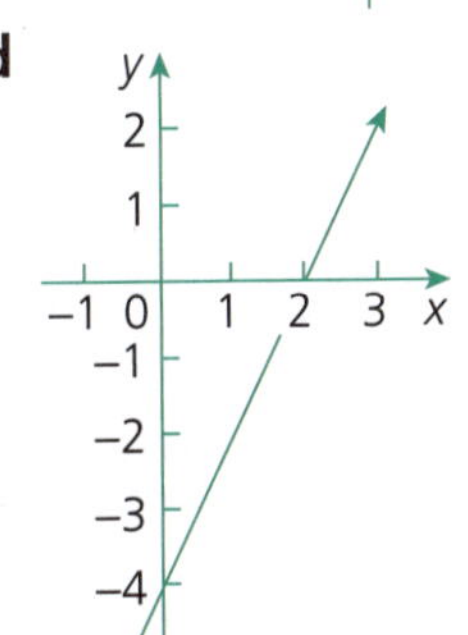

e

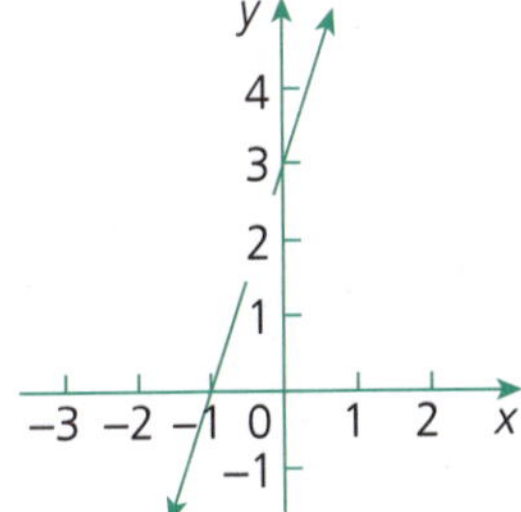

f

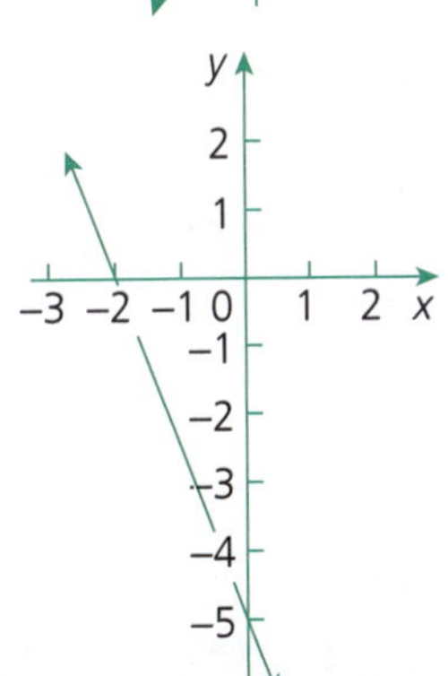

pp. 56–57

17 Sketch the following lines, given the information:

a gradient −2, y-intercept 4

b $m = \frac{1}{3}, b = 2$

c $m = -\frac{2}{3}, b = -3$

d gradient 4, $b = -1$ pp. 56–57

18 Which of the following lines are parallel and why?

a $y = \frac{1}{2}x - 3$ **b** $2y = x + 5$

c $4y = 2x$ **d** $y = 2x + \frac{1}{2}$

e $x + 2y = 1$ **f** $x - 2y = 0$

Also, which lines pass through the origin? pp. 59–60

19 Which of the following points lies on the line $y = 3x - 1$?

a $(-1, -4)$ **b** $(4, -1)$

c $(1, -4)$ **d** $(2, 5)$

e $(\frac{1}{3}, 0)$ **f** $(0, 1)$ p. 59

20 On which of the following lines does $(-2, 1)$ lie?

a $y = 2x + 3$ **b** $y = 2x + 5$

c $x + 2y = 0$ **d** $x - 2y + 4 = 0$ p. 59

21 Plot the points A(−2, 4), B(4, 2) and C(0, −2) on the number plane. Show that the:

a △ABC is isosceles.

b gradient of AB is $-\frac{1}{3}$

c midpoint of AC is (−1, 1). pp. 57–59

22 **a** Find the equation of the line passing through (0, 4) with gradient −3.

b Find the equation of the line which is parallel to $y = 4x - 7$ and passes through (0, −4). p. 61

23 State whether the following pairs of lines are perpendicular:

a $2x - 3y = 7$ and $6x - 12y - 5 = 0$

b $x + 2y - 3 = 0$ and $y = 2x + 1$

c $2x + 3y = 6$ and $4y = 6x - 5$ p. 60

24 If $2x - ky = 5$ is parallel to $3x + 4y - 7 = 0$, find the value of k. p. 60

25 Which of the points A(−3, 7), B(3, −7) or C(−11, 23) lie on the line $2x + y = 1$? p. 59

26 Show that the points (−1, 3), (1, 1) and (4, −2) are collinear. p. 59

27 The point $(2d, 5)$ lies on the line $2x - y - 8 = 0$. Find d. p. 59

28 Find the equation of the line:

a with gradient 3 and passing through (2, 5)

b with gradient −2 and passing through (0, 2)

c with gradient $\frac{1}{2}$ and passing through (2, 1)

d with gradient $\frac{-2}{3}$ and passing through (3, −2)

e parallel to $y = 3x - 2$ and passing through (4, 6)

f parallel to $2x - y + 9 = 0$ and passing through (3, 4)

g parallel to $3x + 2y - 5 = 0$ and passing through (–4, 2)

h perpendicular to $x + y - 2 = 0$ and passing through (–3, –2)

i perpendicular to $4x - 2y = 1$ and passing through (6, 1)

j perpendicular to $3x + 5y + 1 = 0$ and passing through (3, 4).

p. 61

29 A plumber charges a call-out fee of \$50 plus \$70 per hour of work.

a Using C = total cost in dollars and n = number of hours, write a rule linking C and n.

b Complete the table.

n	0	2	4	6
C				

c Use the rule to draw a graph that relates the number of hours worked and the total cost.

d Find the cost if the plumber was hired for:

- **i** 3 hours
- **ii** 8 hours.

e How long is the plumber employed if the charge is:

- **i** \$400?
- **ii** \$750?

p. 62

30 The profit made by a charity selling pens is given by the formula $P = 3n - 2800$, where P = profit in dollars and n = number of pens sold.

a Which pronumeral is:

- **i** independent?
- **ii** dependent?

b Complete the table and draw a graph to represent the relationship between n and P.

n	1000	1500	2000	2500	3000
P					

c How much profit is made if 3250 pens are sold?

d How many pens are sold if the profit is \$10 160?

p. 62

Go to p. 218 for **Quick Answers** or to pp. 234–238 for **Worked Solutions**

For a complete understanding of this topic, you must be able to:

✓	Find the distance between two points using a diagram		p. 52
✓	Find the midpoint of an interval joining two points using a diagram		p. 53
✓	Find the gradient of an interval using a diagram		pp. 53–54
✓	Graph a straight line including those parallel to axes		pp. 54–55
✓	Find intercepts on the axes made by a straight line		p. 56
✓	Use $y = mx + b$ to find the gradient of a line, or use the gradient to determine the equation		p. 56
✓	Use formulae to determine the distance, midpoint and gradient between two points		pp. 57–59
✓	Determine whether a point lies on a line		p. 59
✓	Determine whether two lines are parallel, perpendicular or neither		p. 60
✓	Find the equation of a line with a known gradient passing through a point		p. 61
✓	Apply linear graphs to word problems.		p. 62

Now you are ready to do the tests!

Intermediate Test

Linear Relationships

(40 marks)

1 Plot the points A(–3, –2) and B(1, 1) on a number plane.
a Use Pythagoras' Theorem to calculate the distance AB.
b Find the midpoint of AB. (3 marks)

2 For the equation $y = 2x - 3$ complete this table:

x	0	1	2
y			

Then sketch the line $y = 2x - 3$ on a number plane. (4 marks)

3 On a number plane, sketch the graph of $x = 4$. (1 mark)

4 Find the points at which the line $2x + y = 6$ cuts the x and y axes. (2 marks)

5 For the line $y = \frac{2}{3}x - 4$, write down the gradient of the line and the y-intercept. (2 marks)

6 The gradient of a line is –4 and it cuts the y-axis at (0, 1). Write down the equation of this line. (2 marks)

7 From this diagram write down the:

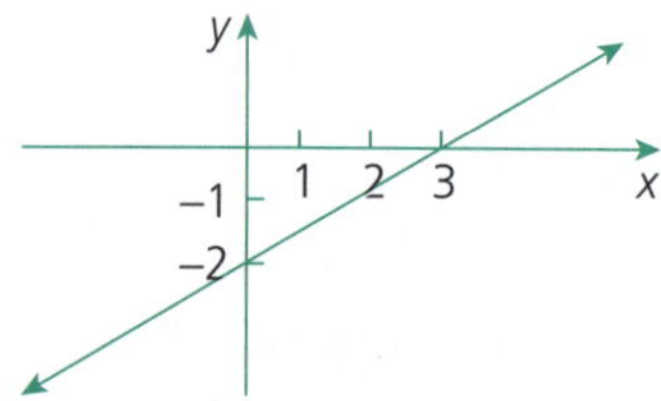

a gradient of the line
b y-intercept
c equation of the line. (3 marks)

8 The gradient of a line is $\frac{1}{2}$ and its y-intercept is 1. Sketch this line on the number plane. (2 marks)

9 For the points M(2, 5) and N(–1, 1), calculate the:
a distance MN
b midpoint of MN
c gradient of MN. (6 marks)

10 Show that the point (3, 1) lies on the line $y = 2x - 5$. (2 marks)

11 Determine whether the lines $y = \frac{1}{2}x - 5$ and $x = 2y - 3$ are parallel. (3 marks)

12 Find the equation of the line with gradient –2 that passes through the point (–3, –1). (2 marks)

13 What is the equation of the line that passes through the point (1, 6) and which is parallel to the line $y = 4x - 2$? (2 marks)

14 The cost of a hire car is \$15 plus a charge of \$3 per kilometre. Using C to represent the total cost in dollars and n to represent the distance in kilometres:
a Write a formula linking C and n. (1 mark)
b Complete the table:

n	0	4	8	12
C				

(1 mark)

c Use the rule to draw a graph that relates the number of kilometres and the cost. (2 marks)

d Find the cost of hiring the car for a distance of 25 km. (2 marks)

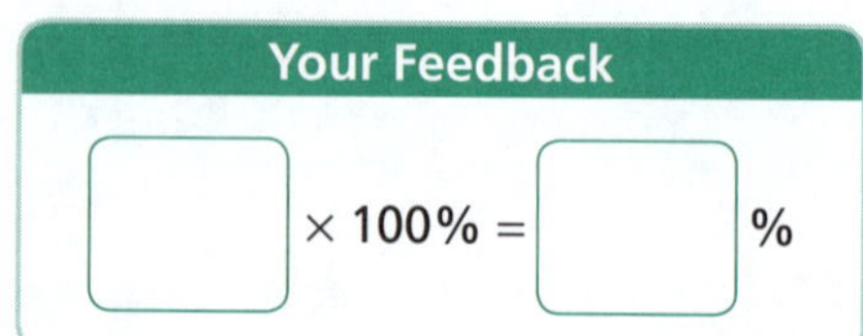

QA PAGE 222
WS PAGE 263

Advanced Test

Linear Relationships

(45 marks)

1 On a number plane, plot the points P(–2, 5), Q(1, –3) and R(3, 1), then:

a calculate the distance PQ

b find the gradient of QR. (4 marks)

2 Complete this table of values for the equation $2x + y = 3$:

x	0	1	2
y			

Sketch the line $2x + y = 3$ on the number plane. (3 marks)

3 On the number plane plot the line $y = -2$. (1 mark)

4 Find the x and y-intercepts for the $3x - 2y = 4$. (2 marks)

5 For the line $2x + 3y = 6$, find the gradient of the line and the y-intercept. (3 marks)

6 From the diagram write down the:

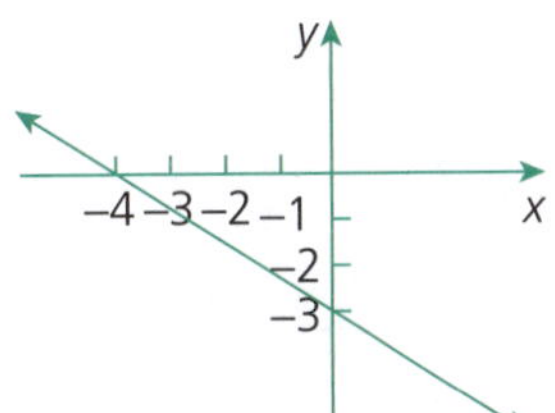

a gradient of the line (1 mark)

b y-intercept (1 mark)

c equation of the line. (1 mark)

7 A particular line cuts the x-axis at (1, 0), the y-axis at (0, –2) and has a gradient of 2. Write down the equation of this line. (2 marks)

8 For the points A(4, 1), B(–1, 3) and C(–3, –4), calculate the:

a distance AC

b midpoint of BC

c gradient of AB. (6 marks)

9 If the point $(-2, 3k)$ lies on the line $3x + 2y = 9$, find the value of k. (3 marks)

10 Find the equation of the line parallel to $3y = 12x - 4$ and passing through (0, –1). (3 marks)

11 **a** On the same number plane graph the lines $y = 2x + 1$ and $y = 4 - x$ and find the point of intersection of the two lines. (3 marks)

b Use your diagram in part **a** to solve the simultaneous equations:

$2x - y = -1$

$x + y = 4$ (1 mark)

12 Given A(2, 8) and B(–4, 2), find the equation of the line that passes through the midpoint of AB that is perpendicular to AB. (4 marks)

13 If the lines $2x - y - 3 = 0$ and $4x + ky + 11 = 0$ are parallel, find the value of k. (3 marks)

14 A party hire firm charges \$120 plus \$12 per guest to provide seating and table decorations for functions.

a Using C = total cost in dollars and n = number of guests, write a formula to relate the cost and the number of guests. (1 mark)

b Draw a graph for the formula. (2 marks)

c Find the cost if 33 guests were invited to the party. (1 mark)

Your Feedback

☐ × 100% = ☐ %

QA PAGE 222

WS PAGE 264

Chapter 6
Non-linear Relationships

The parabola

The parabola is the most important non-linear graph. The trajectory of a ball thrown into the air, or the trail when water is sprayed from a hose, are examples of parabolas in the real world. The parabola is the name given to the curve with the basic rule of $y = x^2$. When a parabola is drawn, the following features should be noted: maximum or minimum, axis of symmetry, x-intercepts and y-intercept.

Keywords

Asymptote	Maximum
Centre	Minimum
Circle	Non-linear
Concave	Parabola
Continuous	Radius
Constant	Reflected
Dilated	Symmetry
Discontinuity	Translated
Exponential	Undefined
Hyperbola	

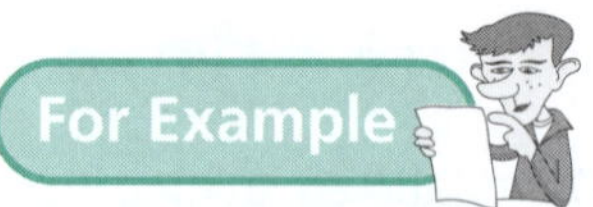

1 Complete the table for $y = x^2$ and draw a graph of the parabola.

x	−3	−2	−1	0	1	2	3
y							

S

x	−3	−2	−1	0	1	2	3
y	9	4	1	0	1	4	9

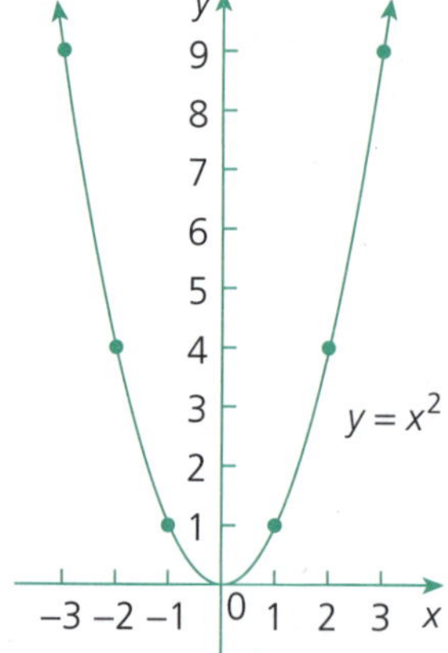

Notice the following features:

- Minimum at (0, 0)
- Axis of symmetry is y-axis (or $x = 0$)
- x-intercept is 0
- y-intercept is 0

Other forms of the parabola

When the basic equation $y = x^2$ is changed, the parabola may be **dilated** (fatter or thinner), **reflected** about the x-axis or **translated** to the left/right.

$y = ax^2$, where $a > 0$

The coefficient of the x^2 (the number in front of x^2) determines the steepness, or the direction, of the parabola.

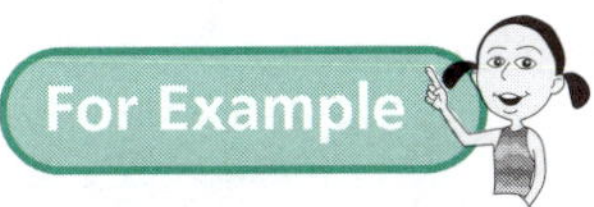

1 On the same set of axes, graph $y = x^2$, $y = 2x^2$, $y = \frac{1}{2}x^2$.

S Notice the following features:

- The higher the value of a, the steeper the parabola.
- Minimum at (0, 0), axis of symmetry is y-axis (or $x = 0$), x-intercept and y-intercept are both 0.
- When $a > 0$, the parabola is concave up.

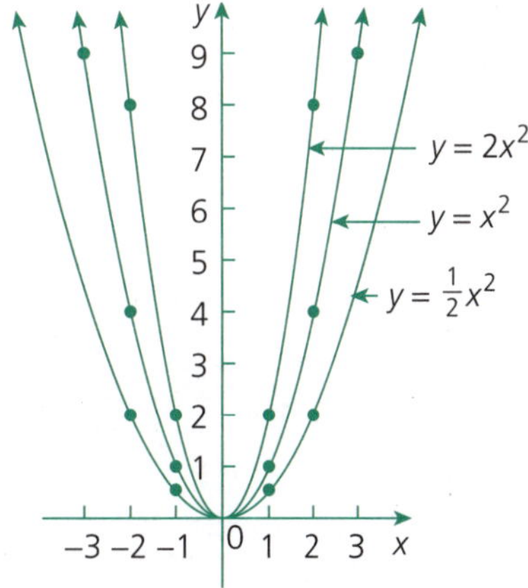

$y = ax^2$, where $a < 0$

The coefficient of the x^2 (the number in front of x^2) determines the steepness, or the direction, of the parabola.

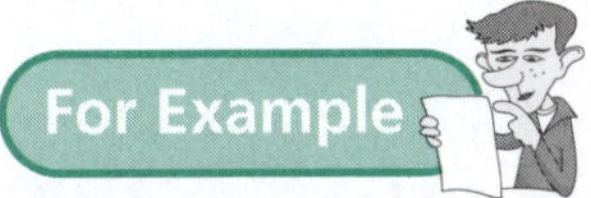

1 On the same set of axes, graph $y = x^2$, $y = -x^2$, $y = -2x^2$.

S Notice when $a < 0$, the parabola is reflected down (concave down), and the parabola has a maximum.

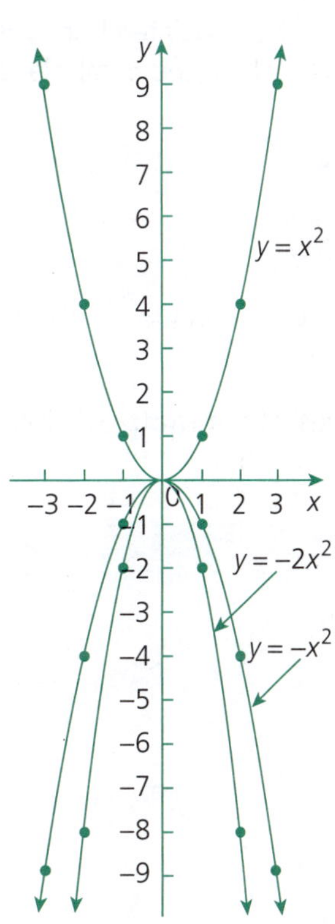

$y = x^2 + c$

The basic parabola ($y = x^2$) is translated vertically. The value of the constant c determines the y-intercept of the parabola. This is the point where the parabola crosses the y-axis.

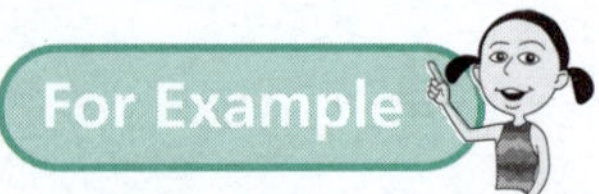

1 On the same set of axes, graph $y = x^2$, $y = x^2 + 3$, $y = x^2 - 6$.

S **Notice the value of the constant term is the y-intercept.**

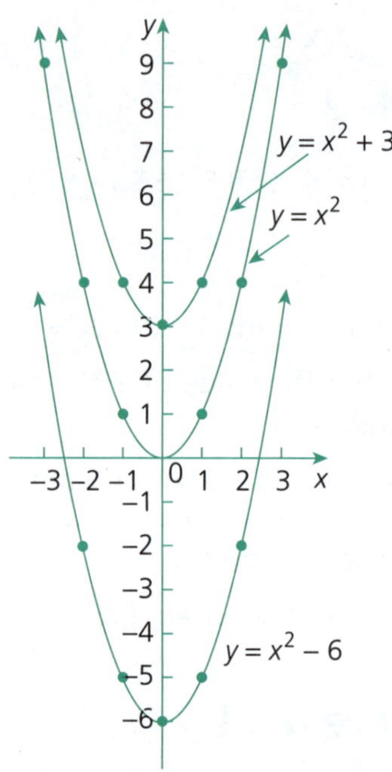

$y = (x - b)^2$

The basic parabola ($y = x^2$) is translated horizontally. The value of the constant b determines the x-intercept of the parabola. This is the point where the parabola touches the x-axis.

For Example

1 On the same set of axes, graph $y = x^2$, $y = (x - 3)^2$, $y = (x + 2)^2$.

S **Notice the axis of symmetry of the parabola is at $x = b$.**

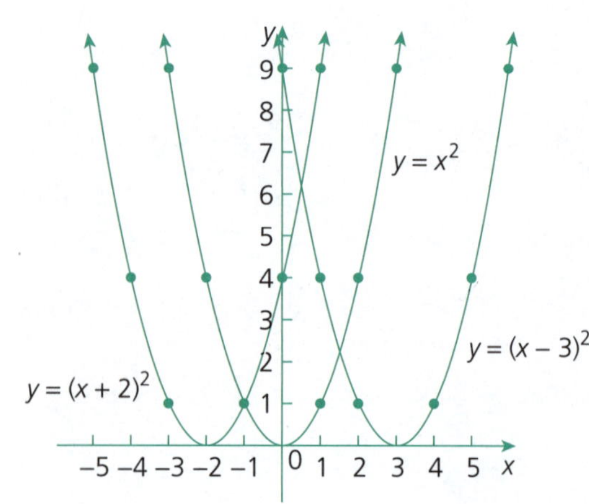

$y = (x - b)^2 + k$

The basic parabola ($y = x^2$) is translated horizontally and vertically. The values of the constants b and k are used to locate the point where the parabola turns. This turning point is called the maximum, or minimum.

1 On the same number plane, sketch $y = x^2$, $y = (x - 3)^2$ and $y = (x - 3)^2 + 2$.

S **To sketch $y = (x - 3)^2 + 2$, shift $y = x^2$ to the right 3 units and then move up 2 units.**

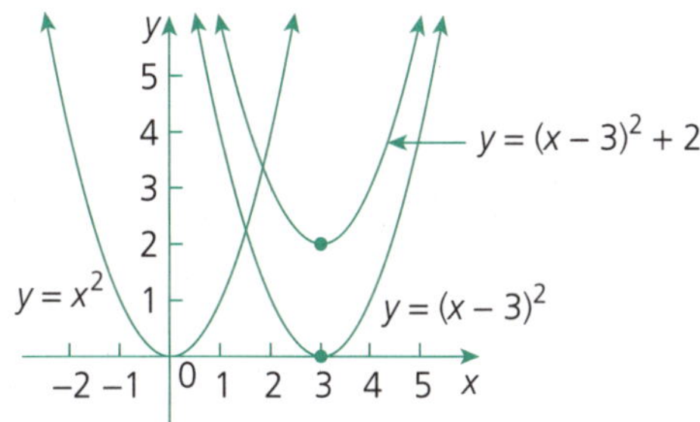

$y = (x - r)(x - s)$

In this case the parabola cuts the x-axis at $x = r$ and $x = s$.

1 Graph $y = (x - 2)(x + 4)$.

S **This means the parabola cuts the x-axis at $x = 2$ and $x = -4$, and as positive x^2, the parabola is concave up.**

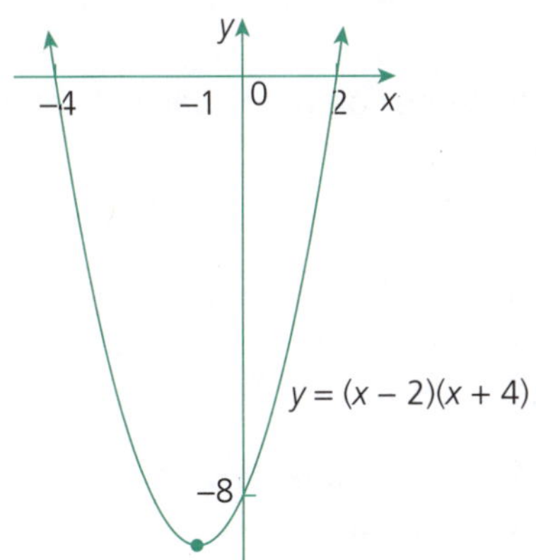

Notice the following features:

- By substituting $x = 0$ into the equation, then $y = -8$. This means the y-intercept is -8.
- The axis of symmetry is the middle of the two x-intercepts. As the middle of 2 and -4 is -1, the axis of symmetry is $x = -1$.

2 Graph $y = x^2 - 2x - 15$.

S **First, factorise the quadratic $x^2 - 2x - 15$:**
$y = (x - 5)(x + 3)$

This means the parabola cuts the x-axis at $x = 5$ and $x = -3$, and as positive x^2, the parabola is concave up.

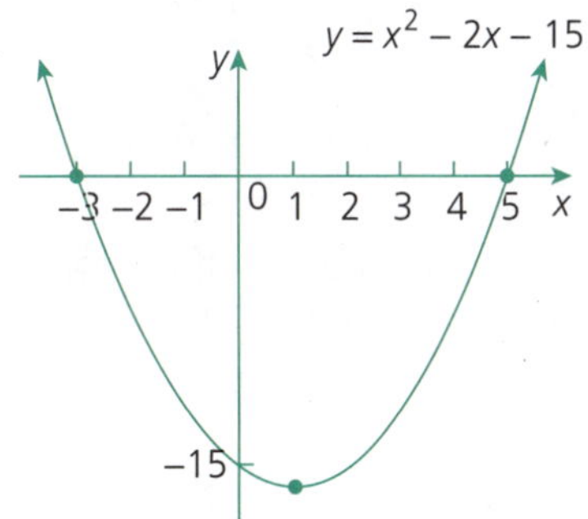

Notice the following features:

- By substituting $x = 0$ into the equation, then $y = -15$. This means the y-intercept is -15.
- The axis of symmetry is the middle of the two x-intercepts. As the middle of -3 and 5 is 1, the axis of symmetry is $x = 1$.

Determining other points on the parabola

Substitution is used to find points on the parabola.

For Example

1 Determine whether (2, 5) lies on the parabola $y = x^2 + 1$.

S Substitute (2, 5) into $y = x^2 + 1$:

$$5 = (2)^2 + 1$$
$$= 5$$

$\therefore$ (2, 5) lies on the parabola.

2 The parabola $y = 3x^2 + 4$ passes through the point $(1, a)$. Find the value of a.

S Substitute $(1, a)$ into $y = 3x^2 + 4$:

$$a = 3(1)^2 + 4$$
$$= 7$$

$\therefore a = 7$

Finding the parabola equation from given information

The graph of the parabola can be determined using its features.

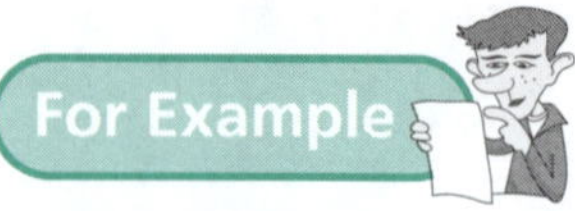

1 Find the equation of the parabola of the form $y = ax^2$.

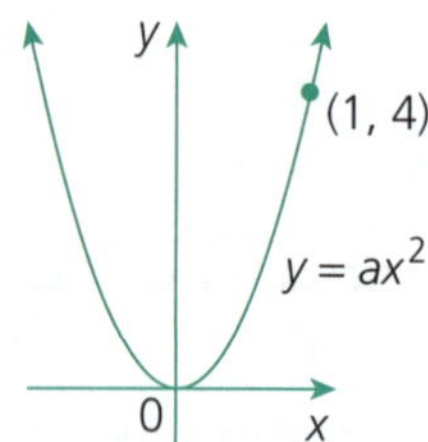

S Substitute (1, 4) into $y = ax^2$

$$4 = a(1)^2$$
$$a = 4$$

$\therefore y = 4x^2$

2 Find the equation of the parabola of the form $y = ax^2 + c$.

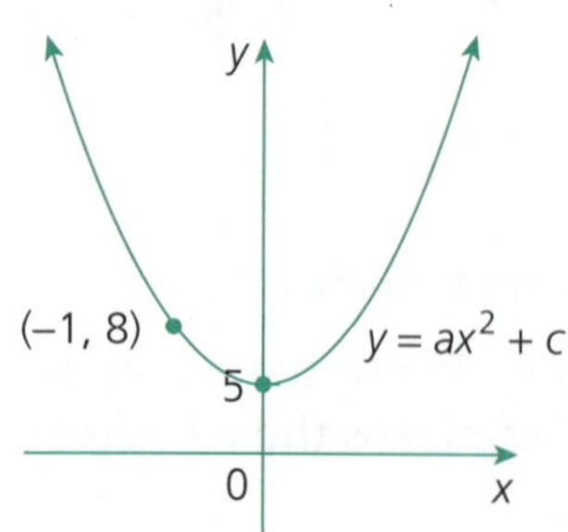

S The parabola has a y-intercept of 5. This means the equation is of the form $y = ax^2 + 5$.

Substitute (–1, 8) into $y = ax^2 + 5$

$$8 = a(-1)^2 + 5$$
$$8 = a + 5$$
$$a = 3$$

$\therefore y = 3x^2 + 5$

3 Find the equation of the parabola of the form $y = a(x - b)^2$.

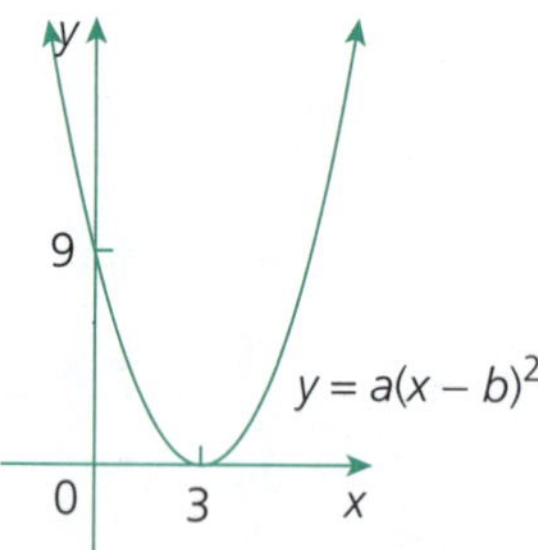

S The parabola has an x-intercept of 3. This means the equation is of the form $y = a(x - 3)^2$.

Substitute (0, 9) into $y = a(x - 3)^2$

$$9 = a(0 - 3)^2$$
$$9 = 9a$$
$$a = 1$$

$\therefore y = (x - 3)^2$

4 Find the equation of the parabola of the form $y = a(x - r)(x - s)$.

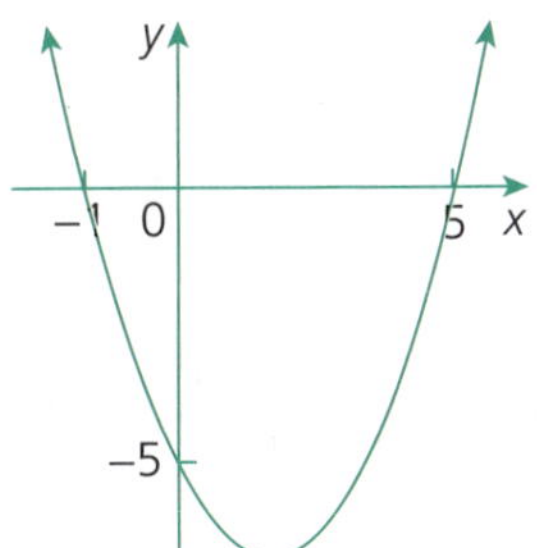

S The parabola has x-intercepts of −1 and 5. This means the equation is of the form $y = a(x + 1)(x - 5)$.

The parabola has a y-intercept of −5.
Substitute (0, −5) into $y = a(x + 1)(x - 5)$

$$-5 = a(0 + 1)(0 - 5)$$
$$-5 = -5a$$
$$a = 1$$

$\therefore\ y = (x + 1)(x - 5)$

5 Find the equation of the parabola of the form $y = a(x - r)(x - s)$.

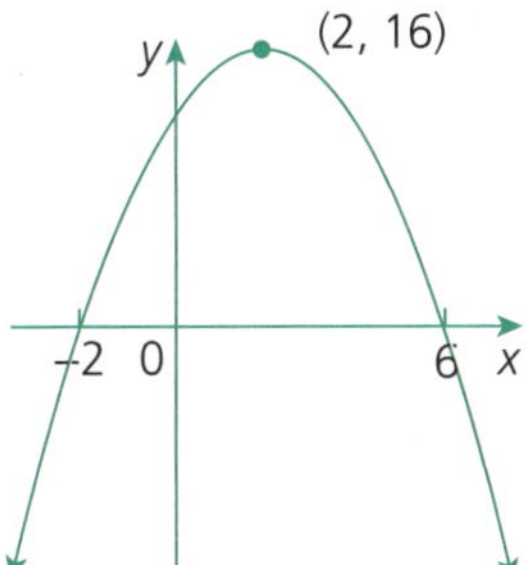

S The parabola has x-intercepts of −2 and 6. This means the equation is of the form $y = a(x + 2)(x - 6)$.

The parabola passes through the point (2, 16).
Substitute (2, 16) into $y = a(x + 2)(x - 6)$

$$16 = a(2 + 2)(2 - 6)$$
$$16 = -16a$$
$$a = -1$$

$\therefore\ y = -(x + 2)(x - 6)$

This can be rewritten as $y = -(x^2 - 4x - 12)$, or $y = 12 + 4x - x^2$.

The exponential

The equation of a simple exponential is of the form $\boldsymbol{y = a^x}$ and always passes through the point (0, 1). As x is getting smaller (or a large negative), the value of y is approaching 0. This means that $y = 0$ (or the x-axis) is an **asymptote** for the curve.

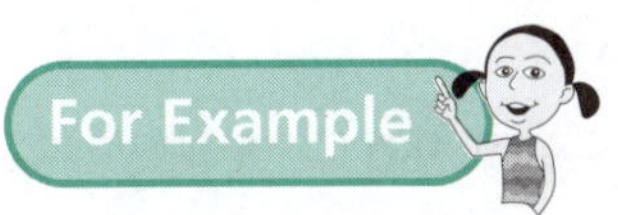

1 Complete the table for $y = 2^x$ and draw a graph of the exponential.

x	−3	−2	−1	0	1	2	3
y							

S

x	−3	−2	−1	0	1	2	3
y	$\frac{1}{8}$	$\frac{1}{4}$	$\frac{1}{2}$	1	2	4	8

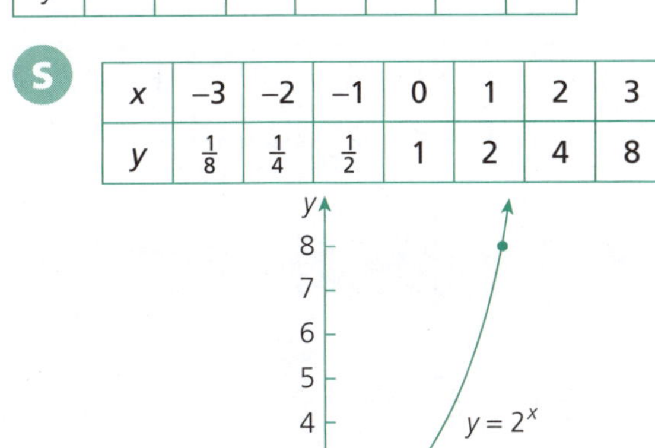

2 On the same number plane, sketch $y = 3^x$ and $y = 3^{-x}$.

S

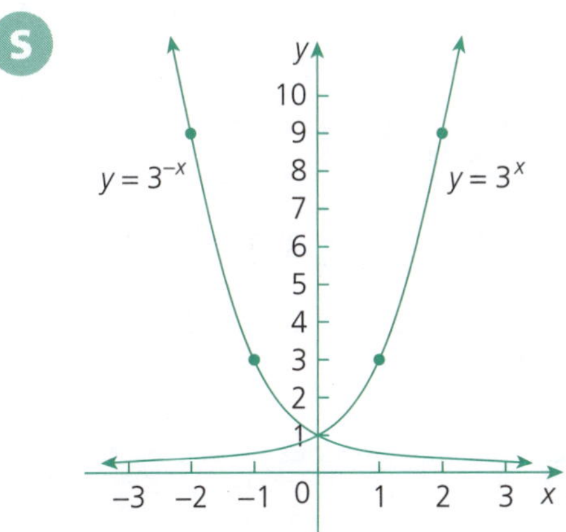

The circle

The equation of a circle with **centre (0, 0)** and **radius *r*** is given by the formula $\mathbf{x^2 + y^2 = r^2}$.

1 Find the radius of the circle with equation $x^2 + y^2 = 25$.

S $r^2 = 25$

$r = \sqrt{25}$

$= 5$

$\therefore$ Radius is 5 units.

2 Graph the circle with equation $x^2 + y^2 = 16$.

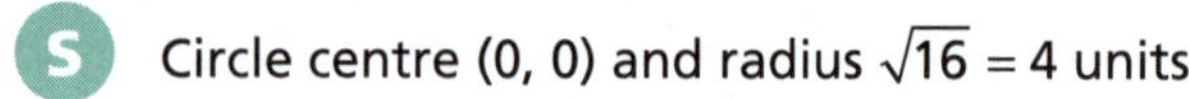

S Circle centre (0, 0) and radius $\sqrt{16} = 4$ units.

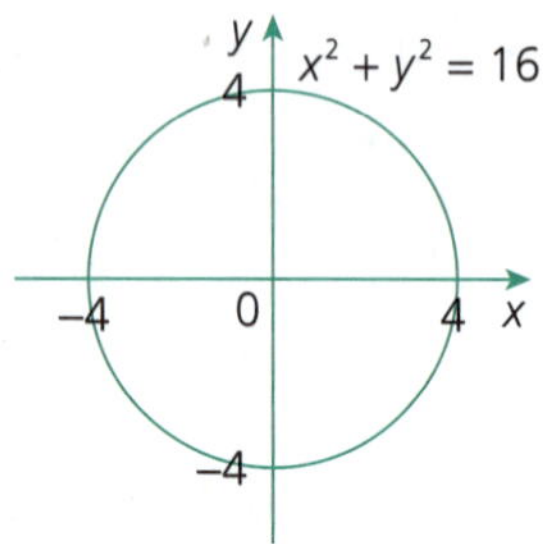

3 Write the equation of the circle graphed below.

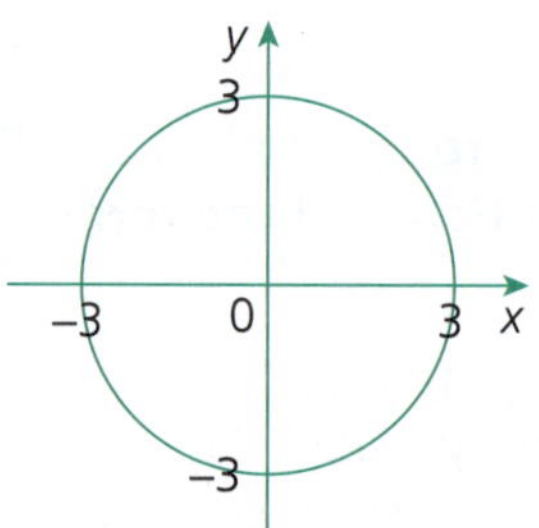

S Circle centre (0, 0) and radius 3

$\therefore x^2 + y^2 = 9$.

The hyperbola

The equation of a simple hyperbola is of the form $\mathbf{y = \frac{c}{x}}$, or $\mathbf{xy = c}$. The hyperbola is not a continuous curve and has a point of discontinuity. This occurs at x = 0, when $\frac{1}{x}$ is **undefined**.

1 Complete the table for $y = \frac{1}{x}$ and draw a graph of the hyperbola.

x	−2	−1	$-\frac{1}{2}$	0	$\frac{1}{2}$	1	2
y							

S

x	−2	−1	$-\frac{1}{2}$	0	$\frac{1}{2}$	1	2
y	$-\frac{1}{2}$	−1	−2	undefined	2	1	$\frac{1}{2}$

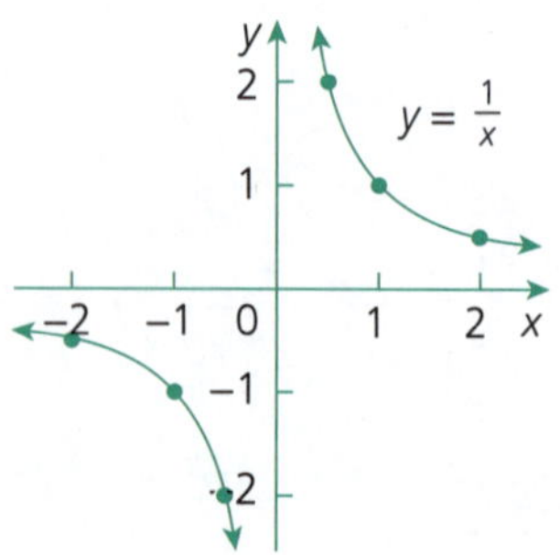

2 On a number plane, sketch $xy = 4$.

If $xy = 4$, then $y = \dfrac{4}{x}$.

x	–4	–2	–1	0	1	2	4
y	–1	–2	–4	undefined	4	2	1

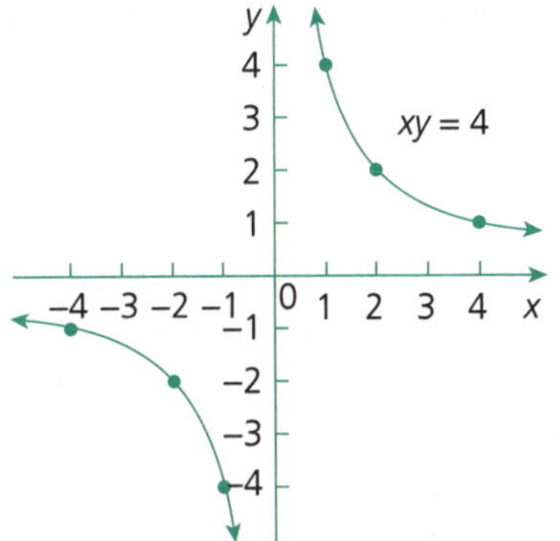

Intersection of lines and curves

To find the point of intersection of lines and curves, form an equation and solve.

1 Find the point of intersection of the parabola $y = x^2 - 3x - 4$ and the line $y = x + 1$.

As both equations equal y, then write

$$x^2 - 3x - 4 = x + 1$$
$$x^2 - 4x - 5 = 0$$
$$(x - 5)(x + 1) = 0$$
$$x = 5 \text{ and } -1$$

Subs. $x = 5$ in $y = x + 1$:

$$y = 6$$

Subs. $x = -1$ in $y = x + 1$:

$$y = 0$$

$\therefore$ Points of intersection are (5, 6) and (–1, 0).

Note:

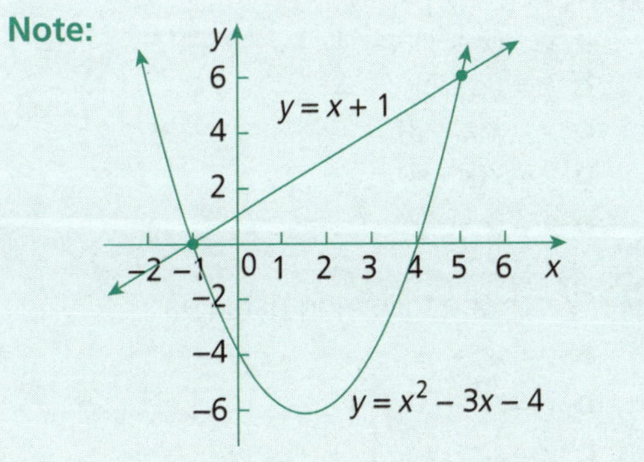

Practise Practise

1 For each of the parabolas, complete a table of values and then graph each on separate number planes:
a $y = x^2 + 1$
b $y = 2x^2$
c $y = 2 - x^2$ p. 70

2 Determine which of the following parabolas are concave up or concave down?
a $y = x^2 + 8$
b $y = 3 - x^2$
c $y = -5x^2$
d $y = 2x^2 + x - 9$ p. 71

3 Decide whether the following parabolas have a maximum or a minimum:
a $y = 3x^2$
b $y = 5 - x^2$
c $y = x^2 - 4x + 5$
d $y = 3 + x - x^2$ p. 71

4 On the same set of axes, graph the following pairs of parabolas:
a $y = x^2$ and $y = 3x^2$
b $y = x^2$ and $y = -3x^2$
c $y = x^2 + 3$ and $y = x^2 - 2$
d $y = 5 - x^2$ and $y = -x^2 - 4$
e $y = x^2$ and $y = (x - 4)^2$
f $y = (x + 1)^2$ and $y = (x - 1)^2$ pp. 71–72

5 Sketch the following parabolas on the same number plane:
$y = x^2$, $y = (x - 2)^2$ and $y = (x - 2)^2 + 3$ pp. 71–73

6 Sketch the parabolas:
a $y = (x + 1)(x - 4)$
b $y = (x - 2)(x - 6)$
c $y = x(x - 2)$
d $y = x(x + 4)$
e $y = (x + 4)^2$ p. 73

7 Sketch the following parabolas:
a $y = x^2 - 9$
b $y = x^2 - 3x - 4$
c $y = x^2 - x - 6$
d $y = x^2 + 2x + 1$
e $y = -x^2 + 4x - 4$ p. 73

8 What is the axis of symmetry of:
a $y = (x + 2)(x - 2)$
b $y = (x + 4)(x + 2)$
c $y = (x - 3)(x + 1)$
d $y = x(x + 4)$
e $y = -x(x - 8)$ p. 73

9 Determine whether the following points lie on the parabola $y = x^2 - 4$:
a (2, 0)
b (–1, 3)
c (3, 5) p. 74

10 Which of the following parabolas passes through the point (2, 3)?
a $y = x^2 - 3x + 1$
b $y = x^2 + 4x - 9$
c $y = x^2 - x - 2$ p. 74

11 The parabola $y = 2x^2 - 3$ passes through the point $(1, b)$. What is the value of b? p. 74

12 The point (1, 3) lies on the parabola $y = x^2 - cx - 2$. What is the value of c? p. 74

13 Find the equations of the following parabolas:

a

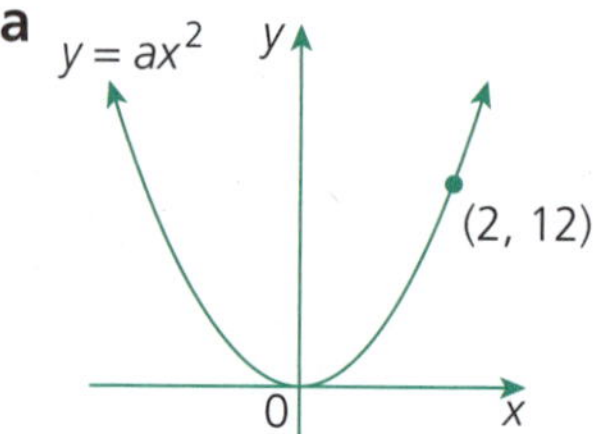

b

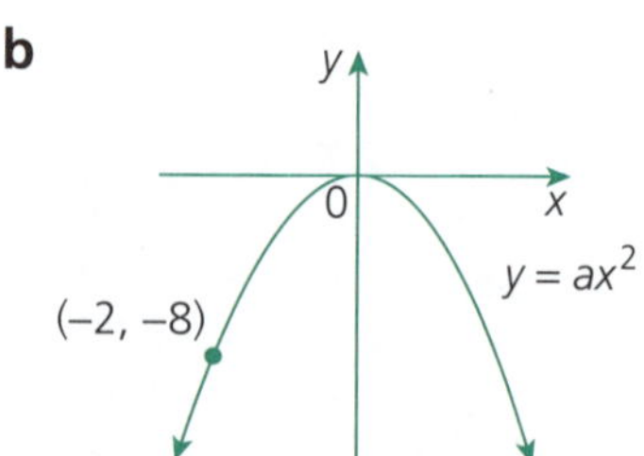

c

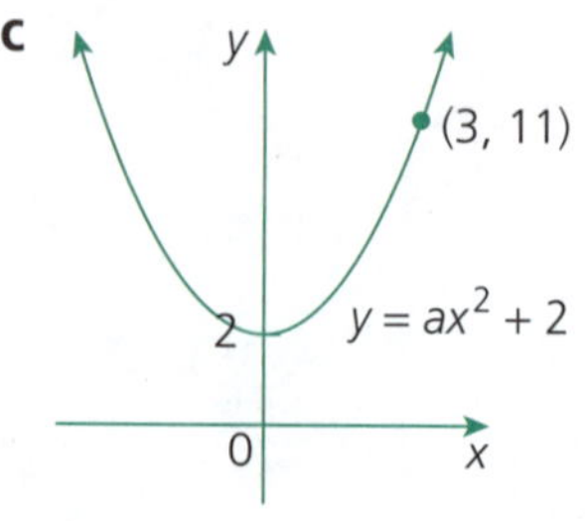

d

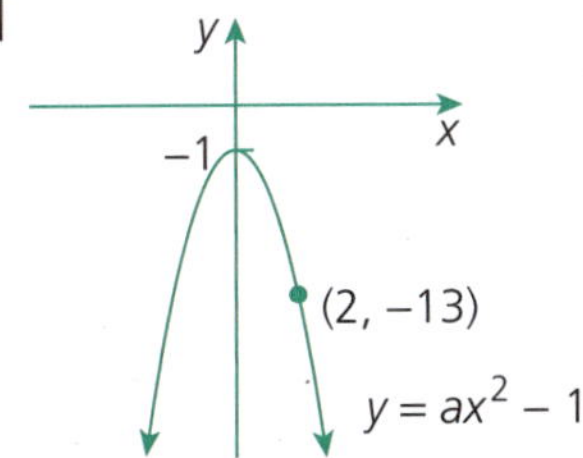

e

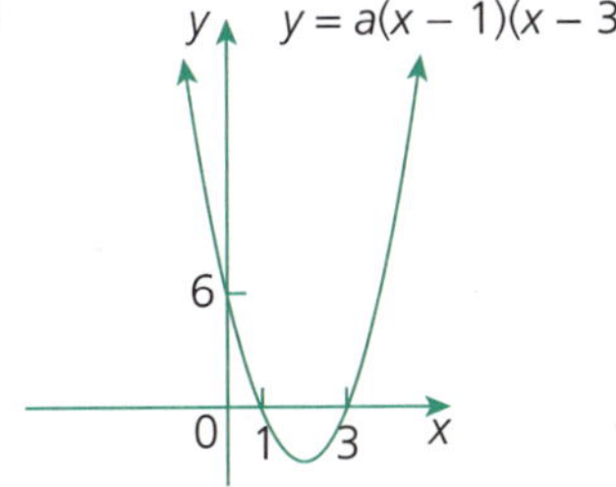

f

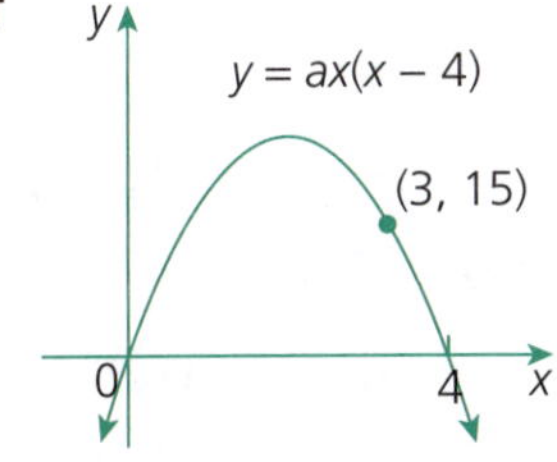

pp. 74–75

14 Complete the table for $y = 4^x$ and draw a graph.

x	−2	−1	0	1	2
y					

p. 75

15 On the same number plane, draw graphs of the exponentials $y = 2^x$ and $y = 2^{-x}$. p. 75

16 Find the radius of the circles with equations:

a $x^2 + y^2 = 49$
b $x^2 + y^2 = 81$
c $x^2 + y^2 - 16 = 0$
d $x^2 + y^2 = 7$
e $x^2 + y^2 = 29$ p. 76

17 Graph the following circles:

a $x^2 + y^2 = 9$
b $x^2 + y^2 = 5$ p. 76

18 Write the equations of these circles:

a

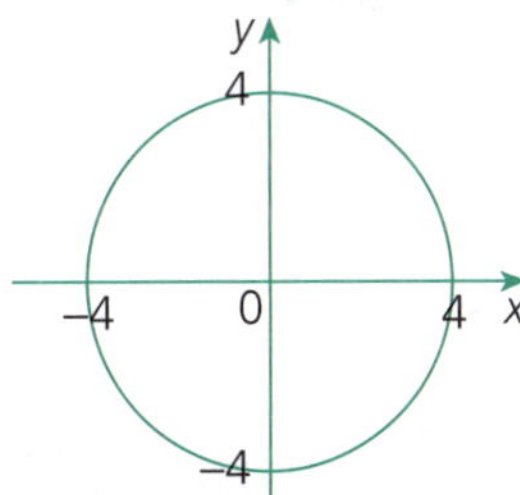

b

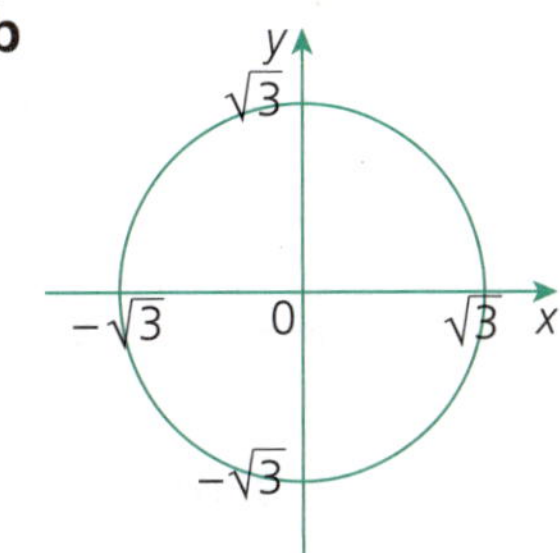

p. 76

19 Complete the table to graph $y = \frac{2}{x}$, and graph on a number plane.

x	−2	−1	$-\frac{1}{2}$	0	$\frac{1}{2}$	1	2
y							

pp. 76–77

20 On separate number planes, graph $xy = 3$ and $xy = -3$. pp. 76–77

21 Find the points of intersection of the following:

a $y = x^2$ and $y = x + 2$
b $y = x^2 + 2$ and $y = 5x - 2$ p. 77

Go to p. 218 for **Quick Answers** or to pp. 238–241 for **Worked Solutions**

For a complete understanding of this topic, you must be able to:

✓	Complete a table of values to draw the graph of a parabola		p. 70
✓	Sketch parabolas of the form $y = ax^2$, $y = ax^2 + c$, $y = (x - b)^2$, $y = (x - b)^2 + k$, $y = (x - r)(x - s)$		pp. 71–73
✓	Determine whether a particular point lies on a parabola		p. 74
✓	Find the equation of a parabola from given information		pp. 74–75
✓	Complete a table of values to draw the graph of an exponential of the form $y = a^x$		p. 75
✓	Find the radius of a circle of the form $x^2 + y^2 = r^2$		p. 76
✓	Graph a circle with an equation of the form $x^2 + y^2 = r^2$		p. 76
✓	Use the graph of a circle to determine its equation		p. 76
✓	Complete a table of values to draw the graph of an hyperbola of the form $y = \frac{c}{x}$ or $xy = c$		pp. 76–77
✓	Use algebra to find the point of intersection of a parabola and a line.		p. 77

Now you are ready to do the tests!

Intermediate Test

Non-linear Relationships

(20 marks)

1 Complete the table for $y = 3x^2 - 4$ and draw a graph of the parabola. (3 marks)

x	-2	-1	0	1	2
y					

2 Match up the equation of the curve with its graph: (8 marks)

a $y = x^2 - 4$ **b** $y = \frac{4}{x}$

c $y = 4^x$ **d** $x^2 + y^2 = 4$

e $y = 4x^2$ **f** $y = -\frac{4}{x}$

g $x^2 + y^2 = 16$ **h** $y = (x - 4)^2$

A

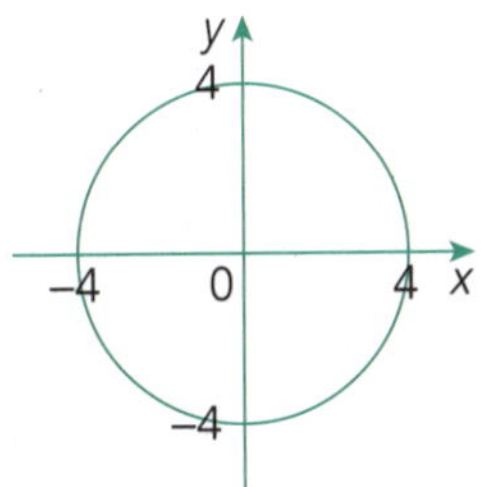

B

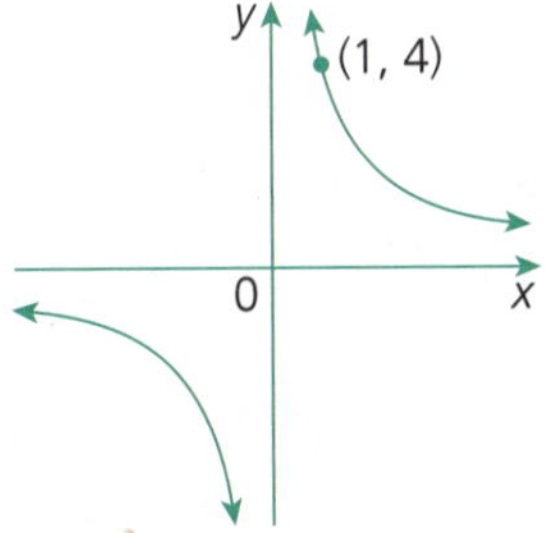

C

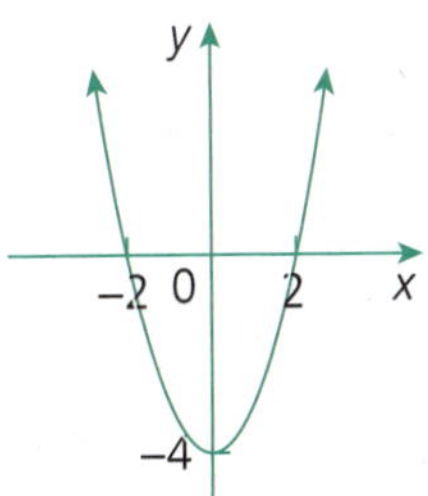

D

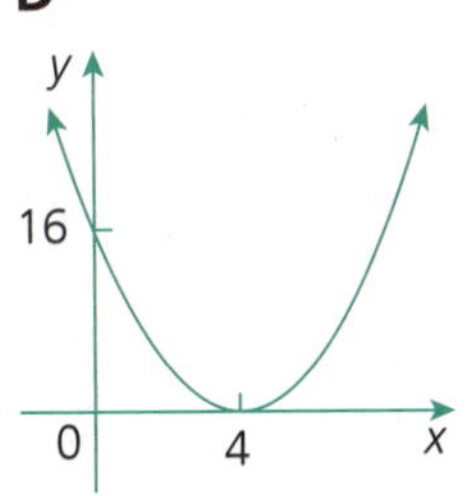

E

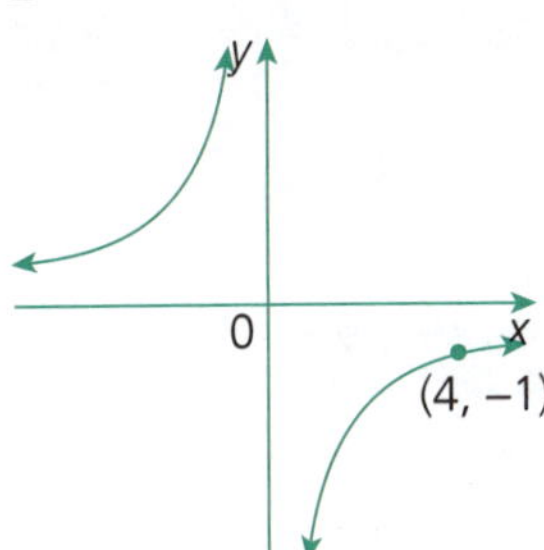

F

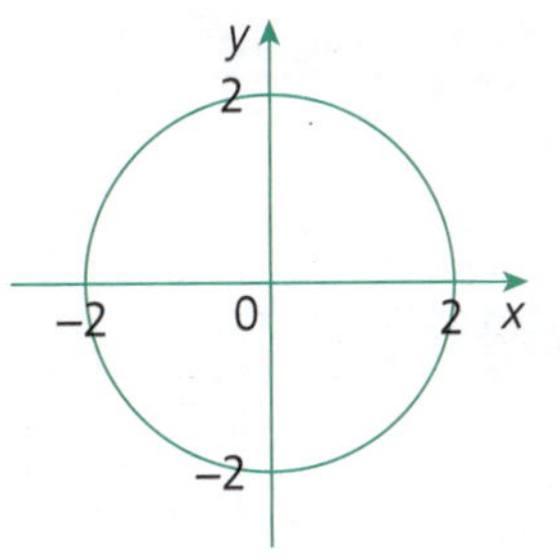

G

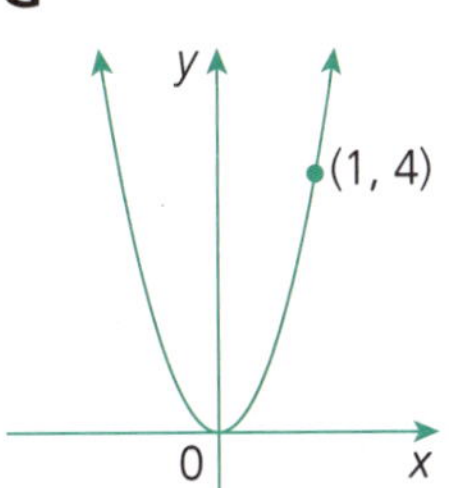

H

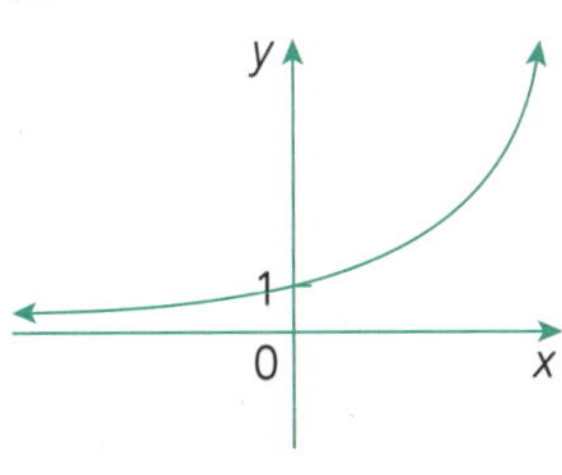

3 On the same number plane draw the graphs of $y = x^2$, $y = -x^2$ and $y = 3 - x^2$. (3 marks)

4 Draw a graph of a parabola, concave up, with a minimum of (3, 0).
Give a possible equation of the curve. (2 marks)

5 The hyperbola $xy = 4$ passes through the point $(8, a)$. What is the value of a? (2 marks)

6 Which of these curves passes through the point (2, 1)? (1 mark)

A $y = x^2 + 1$ **B** $y = 2 - x^2$

C $y = \frac{2}{x}$ **D** $y = 2^x$

7 Which of the equations represents a straight line? (1 mark)

A $y = 4 - x^2$ **B** $y = 4 - x$

C $y = 4^x$ **D** $y = \frac{4}{x}$

Your Feedback

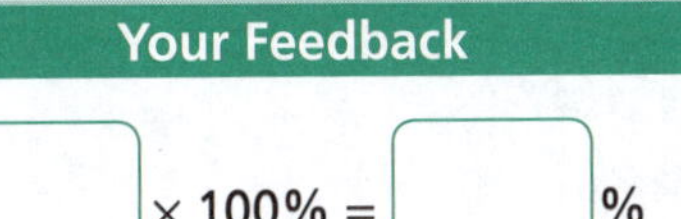

QA PAGE 222

WS PAGE 265

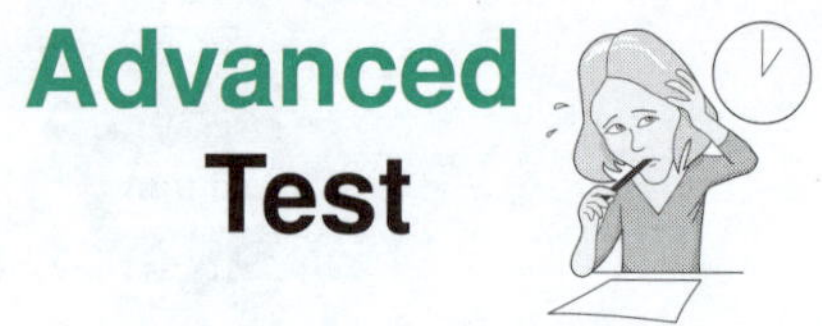

Non-linear Relationships

(20 marks)

1 Draw sketches of the following parabolas, showing x- and y-intercepts: (8 marks)

a $y = (x - 1)(x + 2)$
b $y = x(x - 3)$
c $y = (2 - x)(x + 3)$
d $y = x^2 - 3x - 4$

2 Find the equation of these curves: (8 marks)

a

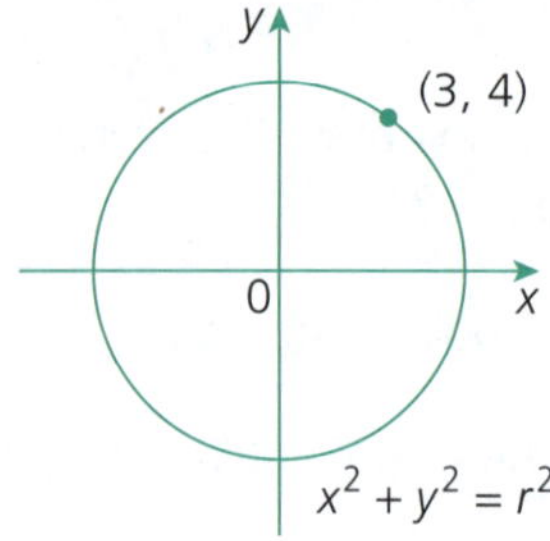

b

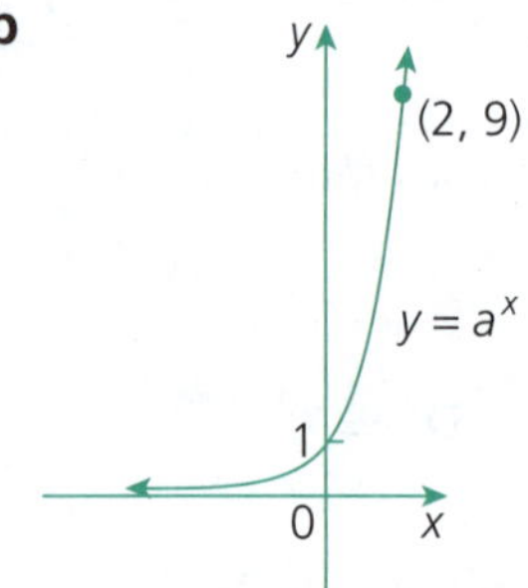

c

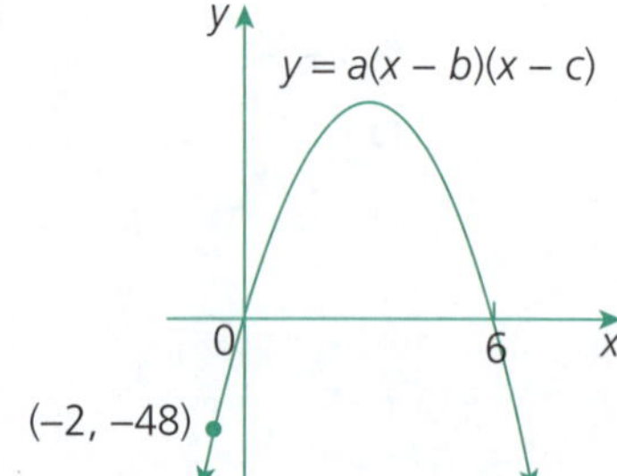

d

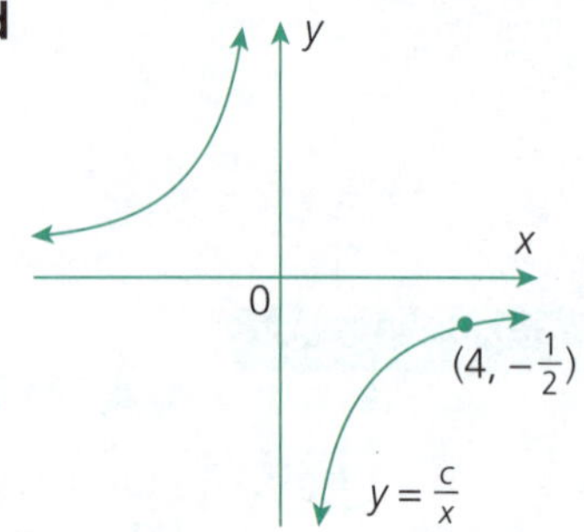

3 Find the points of intersection of: (4 marks)

a parabola $y = x^2 - 5x + 1$ and line $y = 2x - 5$
b parabolas $y = 2x^2 + 3x + 4$ and $y = x^2 - x + 9$.

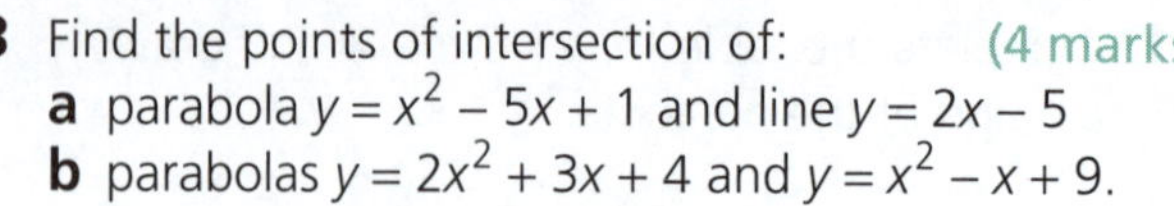

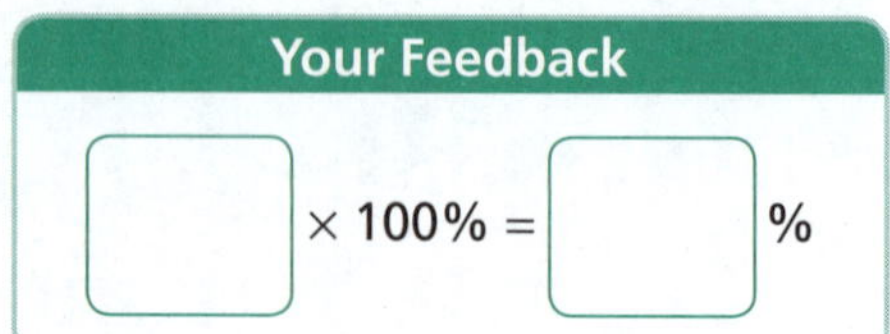

QA PAGE 222

WS PAGE 266

Chapter 7
Rates and Proportion

Converting between rates

A rate is the comparison of two quantities of different units.

Keywords

Constant	Increases
Conversion	Inverse proportion
Decreases	Negative
Direct proportion	Positive
Gradient	Rate

1 Convert 60 km/h to m/s.

S Change kilometres to metres and hours to seconds:

60 km = 60 000 m
1 h = 60 min = 3600 seconds

60 km/h = 60 000 m/3600 s

Divide by 3600:

$\therefore$ 60 km/h = $16\frac{2}{3}$ m/s

2 Convert 10 m/s to km/h.

S Change metres to kilometres and seconds to hours:

Multiply by 60, and then multiply by 60 again.

10 m/s = 36 000 m/h
Now, 36 000 m = 36 km
$\therefore$ 10 m/s = 36 km/h

3 Convert 8 km/L to L/100 km.

S Rewrite 8 km/L as 1 L/8 km.

Divide by 8 then multiply by 100:
$1 \div 8 \times 100 = 12.5$
$\therefore$ 8 km/L = 12.5 L/100 km

Direct proportion

A **proportion** is a statement that two ratios are equal. For example, $a{:}b = c{:}d$ can be written as the proportion $\frac{a}{b} = \frac{c}{d}$.

Consider a pair of similar triangles:

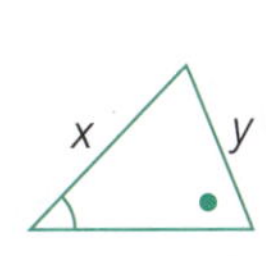

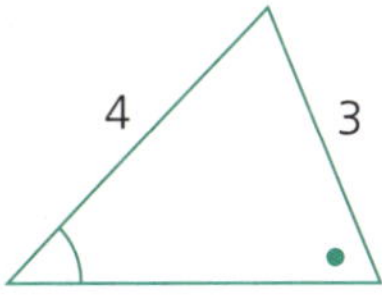

Using matching sides in proportion, $\frac{y}{3} = \frac{x}{4}$.

This can be rewritten as $y = \frac{3x}{4}$, which means that **y is directly proportional to x**.

This can be expressed as $y \propto x$, or $y = kx$, where k is the **constant of proportionality (or the rate of change)**.

(In the case above, $k = \frac{3}{4}$).

As the **rate of change is constant**, the graph is a **straight line** passing through the origin with gradient k.

When a graph is drawn linking two variables that are directly proportional, the gradient will be **positive**.

Everyday examples of events that are in direct proportion are:

- The more hours worked, the more income is earned.
- As the side length of a square paddock decreases, the perimeter of the paddock decreases.

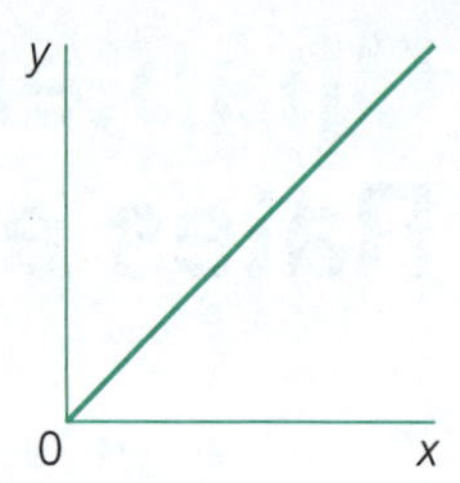

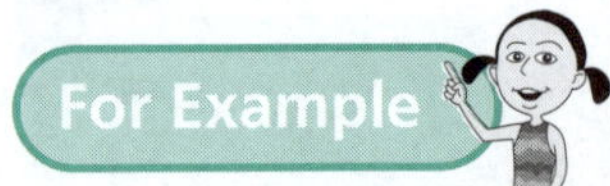

1 The table shows the height (h in cm) of a plant over time (t in weeks).

t	0	10	20	30
h	0	30	60	90

a Write a rule linking h and t.

S $\therefore h = 3t$

b Complete: As t increases, h __________.

S **As t increases, h increases.**

c What is the height of the plant after 50 weeks?

S **Subs. $t = 50$ in $h = 3t$:**
$h = 3 \times 50$
$= 150$
$\therefore$ **Plant is 150 cm high.**

e Draw a graph of h against t for values of $t \leq 100$.

S

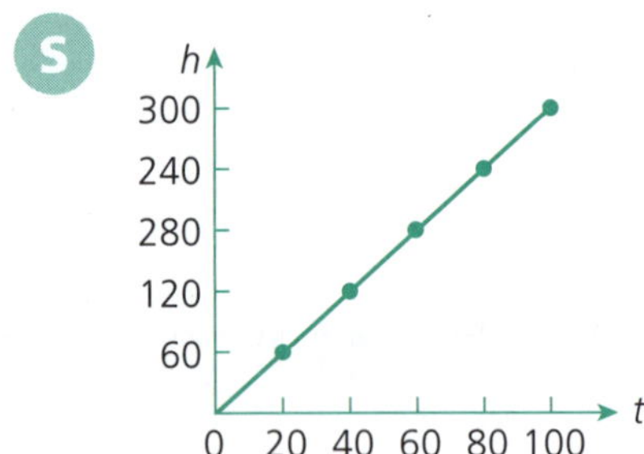

f What is the gradient of the graph?

S $\frac{\text{Rise}}{\text{Run}} = \frac{300}{100} = 3$

2 A hose is used to fill an empty container with water. It takes 4 minutes to fill the container with 120 L of water.

a At what rate is the water filling the container, in L/min?

S **120 L in 4 minutes means 30 L/min.**

b Write a rule linking V (volume in litres) and t (time in minutes).

S $V = 30t$

c Complete: As t increases, V __________.

S **As t increases, V increases.**

d Draw a graph of volume against time for $t \leq 20$.

S

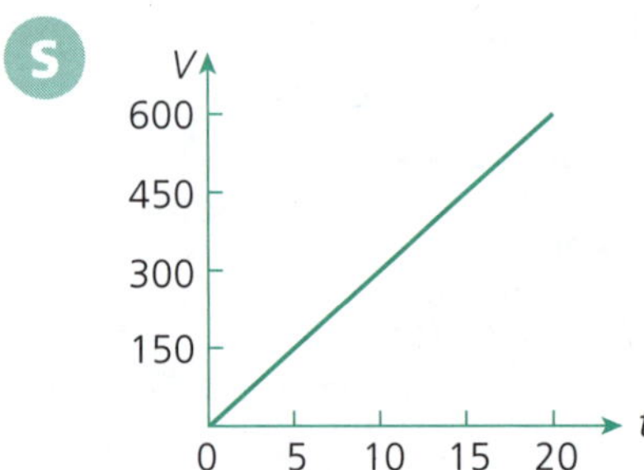

e Write down the gradient.

S **Gradient = 30**

f How much water is in the container after an hour?

S **Subs. $t = 60$ in $V = 30t$:**
$V = 30 \times 60$
$= 1800$
$\therefore$ **1800 litres**

g How long would it take to fill a tank with a capacity of 2400 litres?

S **Subs. $V = 2400$ in $V = 30t$:**
$2400 = 30t$
$t = \frac{2400}{30}$
$= 80$
$\therefore$ **80 minutes**

3 The graph shows the distance in kilometres (d) travelled by a motorist over a number of hours (h).

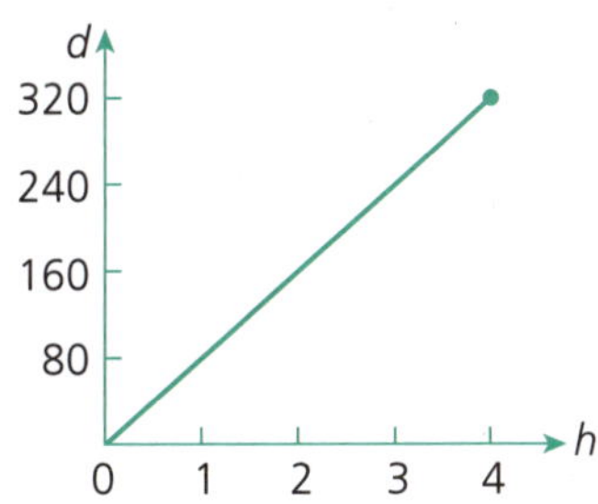

a What is the gradient of the graph?

$$\text{Gradient} = \frac{\text{Rise}}{\text{Run}} = \frac{320}{4} = 80$$

b What is represented by the gradient?

The gradient is the rate of change of the distance over time.

This is the speed of the motorist (80 km/h).

c Write a rule linking d with h.

$d = 80h$

d How far will the motorist travel in $2\frac{1}{2}$ hours?

Subs $h = 2\frac{1}{2}$ in $d = 80h$:

$$d = 80 \times 2\frac{1}{2} = 200$$

$\therefore$ Motorist travels 200 km.

Conversion graphs

A conversion graph is used to change one quantity to another when there is a change in the units.

1 Jannah converted speeds recorded in m/s into km/h and recorded them in the table below.

m/s	10	20	30	40	50
km/h	36	72	108	144	180

Draw a graph to record the data.

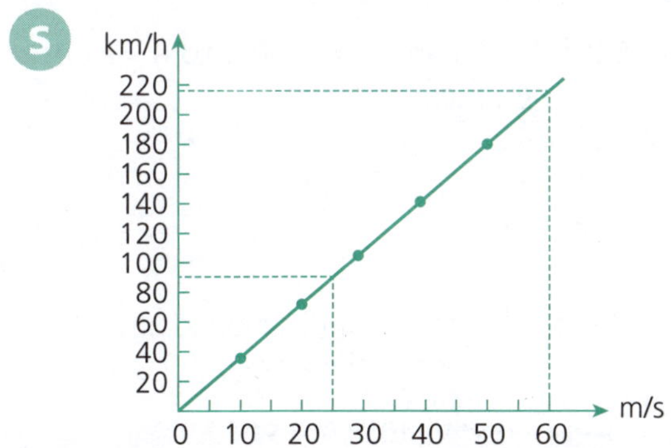

2 Adam exchanged $100 Australian dollars for $80 US. He knows that $80 US is equal to 70 euros.

a Draw currency conversion graphs for $US against $A, and euros € against $US.

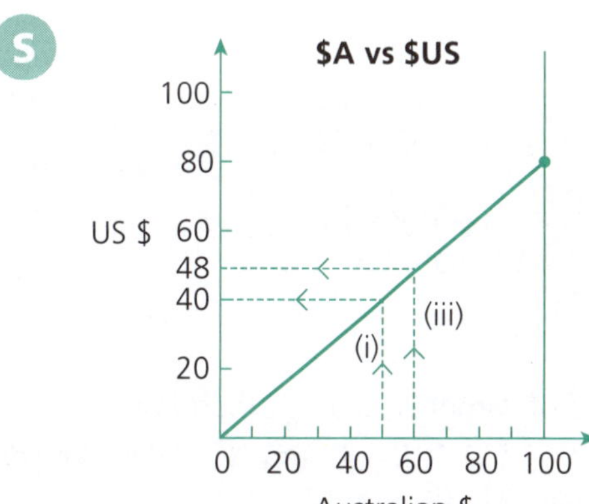

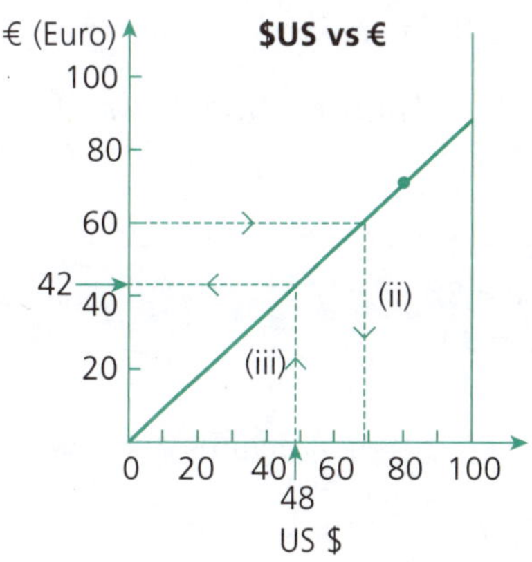

b Use the graphs to convert:

i \$A50 to US dollars.

S **\$A50 = \$US40**

ii 60 euros to US dollars.

S **€60 = \$US69 (nearest whole)**

iii \$A60 to euros.

S **\$A60 = \$US48 = €42**

Problem solving using direct proportion

If one quantity (y) is directly proportional to another quantity (x), then $y \propto x$, or $y = kx$, where k is the constant of proportionality.

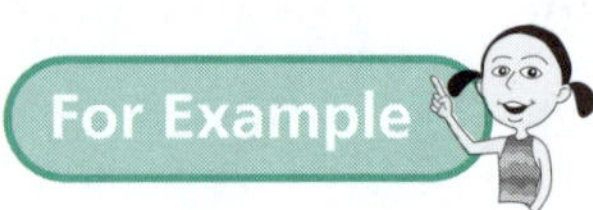

1 It is known that $y \propto x$ (y is proportional to x) and if $y = 24$ then $x = 8$.

a Write an equation that links y and x using the constant of proportionality k.

S $y = kx$

b Find the value of k.

S Subs. $y = 24$ and $x = 8$ in $y = kx$:
$24 = k \times 8$
$8k = 24$
$k = 3$

c Find the value of y when $x = 12$.

S Subs. $x = 12$ in $y = 3x$:
$y = 3 \times 12$
$= 36$

2 The cost \$$C$ of carpet is directly proportional to the length n of the carpet in metres. Twelve metres of carpet costs \$1176.

a Write a formula for C in terms of n.

S C is directly proportional to n.
Subs. $C = 1176$ and $n = 12$ into $C = kn$:
$1176 = k(12)$
$12k = 1176$
$k = 98$
$\therefore C = 98n$

b Find the cost of 17 metres of carpet.

S Subs. $n = 17$ into $C = 98n$:
$C = 98 \times 17$
$= 1666$
∴ Carpet costs \$1666.

c What length of carpet can be purchased for \$1391.60?

S Subs. $C = 1391.6$ into $C = 98n$:
$1391.6 = 98n$
$n = \frac{1391.6}{98}$
$= 14.2$
∴ Carpet is 14.2 metres long.

3 The height of a certain termite mound is directly proportional to the square root of the number of termites. The height of this mound is 60 cm when the number of termites is 3600.

a By letting h = height of the mound in metres and n = number of termites, find an equation for h in terms of n.

S h is directly proportional to $\sqrt{n}$.

Subs. $h = 60$ and $n = 3600$ into $h = k\sqrt{n}$:
$60 = k\sqrt{3600}$
$60k = 60$
$k = 1$
$\therefore h = \sqrt{n}$

b What is the height of this mound, in centimetres, when there are 6400 termites?

S Subs. $n = 6400$ into $h = \sqrt{n}$:
$h = \sqrt{6400}$
$= 80$
∴ Mound is 80 cm high.

Inverse (indirect) proportion

If y is inversely (indirectly) proportional to x, we write $\boldsymbol{y} \propto \frac{\mathbf{1}}{\boldsymbol{x}}$, or $\boldsymbol{y} = \frac{\boldsymbol{k}}{\boldsymbol{x}}$, where k is the constant of proportionality.

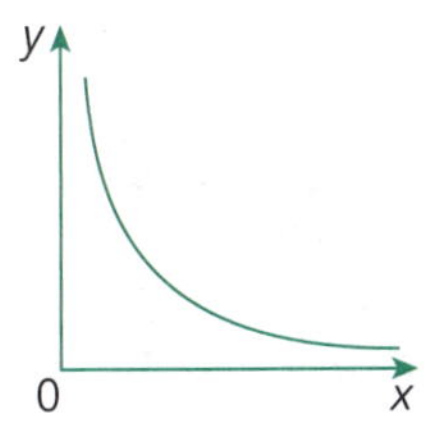

When a graph is drawn linking two variables that are inversely proportional, the gradient will be **negative**.

Everyday examples of events that are in inverse proportion are:

- The faster a car travels, the less time for the trip.
- The fewer people mow a lawn, the longer the job will take.

1 The time (T) taken by workers (n) to paint a wall is detailed in the table below:

n	1	2	3	6	9	18
T	18	9	6	3	2	1

Use the table to draw a graph and explain why T is inversely proportional to n.

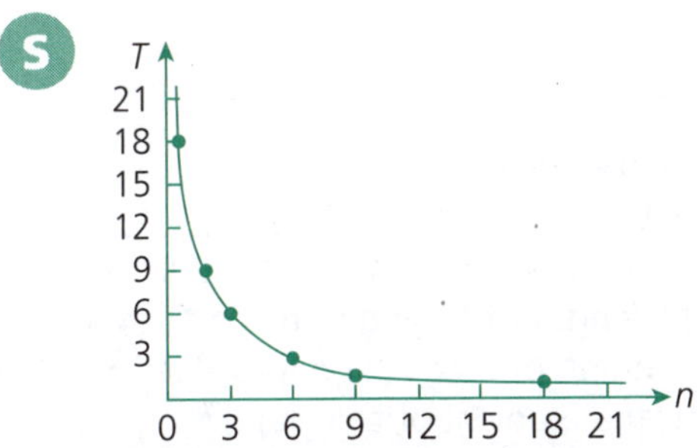

T is inversely proportional to n because as n increases, T decreases.

Practise Practise

1 Convert the following rates:

a 100 km/h to m/s **b** 40 m/s to km/h
c 20 km/L to L/100 km **d** 10 L/100 km to km/L
e 28 cm/s to km/h **f** 80c/100 g to kg/$
g 160c/L to L/$100 **h** 5 mL/min to h/L
i 2 L/h to s/mL

p. 83

2 The table shows the cost of garden edging (C dollars) over different lengths (d metres).

d	0	5	10	15	20
C	0	60	120	180	240

a Write a rule linking C and d.
b Complete: As d increases, C __________.
c What is the cost of 35 metres of edging?
d Find the length of edging if the cost is $504.
e Draw a graph of C against d for values of $d \leq 30$.
f What is the gradient of the graph?

pp. 83–85

3 The table below shows the wage (W) for hours worked (n) by Miriam.

n	0	8	16	24	32	40
W	0	132	264	396	528	660

a Write a rule linking W and n.
b Complete: As n increases, W __________.
c How much does Miriam earn when she works 30 hours?
d If Miriam earns $594, how many hours does she work?
e Draw a graph of W against n for values of $n \leq 40$.
f What is the gradient of the graph?

pp. 83–85

4 The table shows the distance a motorist travelled (D) for each litre of petrol (n).

n	0	5	10	15	20
D	0	45	90	135	180

a Write a rule linking D and n.
b How far does the motorist travel on 35 litres?
c The petrol tank has a capacity of 60 litres. What fraction of petrol remains after the motorist fills the tank and then travels 405 kilometres?

pp. 83–85

5 The table shows the conversion of temperatures between Celsius (°C) and Fahrenheit (°F).

°C	0	10	20	30	40
°F	32	50	68	86	104

a Draw the conversion graph.
b Use the graph to convert:
  i 38 °C to °F
  ii 59 °F to °C.

pp. 85–86

6 Daniel used a fitness book published in the United States to plan his training schedule. Using the conversion of 1 mile = 1.6 kilometres, he completed the table below.

Miles	5	10	15	30
Kilometres	8	16	24	48

a Draw the conversion graph.
b Use the graph to convert:
  i 20 miles to kilometres
  ii 40 kilometres to miles.

pp. 85–86

7 Mitch used the conversion of 1 kg = 2.2 pounds to make conversions between metric and imperial measures of mass.
a Complete the table below.

Kilograms	0	5	10	15	20
Pounds					

b Draw the conversion graph.
c Use the graph to convert:
  i 8 kilograms to pounds
  ii 150 kilograms to pounds.

pp. 85–86

8 Currency conversion graphs are drawn below to change Australian dollars ($A) to euros (€), and euros to British pounds (£).

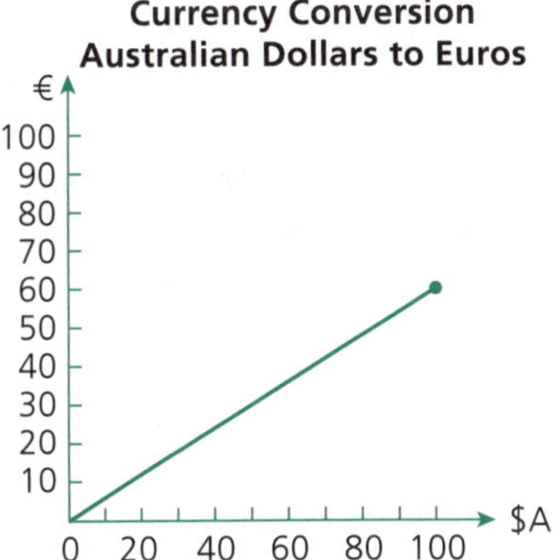

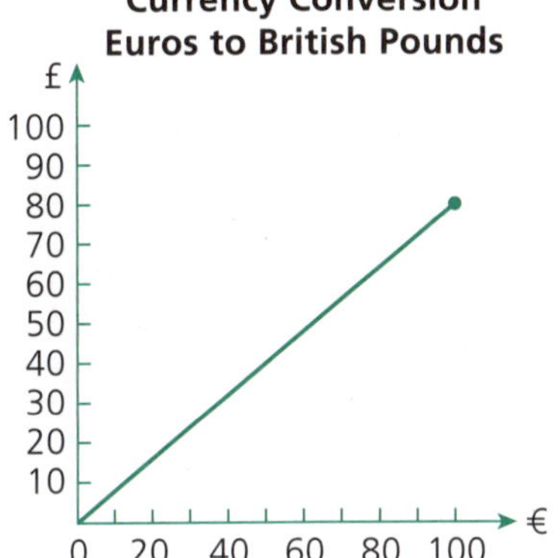

a Complete:
  i $A100 = € ________
  ii €100 = £ ________

b Use the graphs to convert:
  i $A60 to euros
  ii €42 to Australian dollars
  iii €60 to British pounds
  iv £20 to euros.

c Make the following conversions:
  i $A5000 into euros
  ii €2400 into Australian dollars
  iii €750 into British pounds
  iv £100 into euros.

d Use the conversion graphs to change:
  i $A100 into British pounds
  ii £60 into Australian dollars. pp. 85–86

9 It is known that y is directly proportional to x and if $y = 12$ then $x = 4$.

a Write an equation that links y and x using the constant of proportionality k.
b Find the value of k.
c What is the value of y when $x = 20$?
d Find the value of x when $y = 18$. p. 86

10 It is known that $P \propto t$ and when $P = 10$ then $t = 2$.

a Write an equation that links P and t using the constant of proportionality k.
b Find the value of k.
c If $t = 16$, find the value of P.
d When $P = 30$, what is the value of t. p. 86

11 Tessa's wage is directly proportional to the number of hours she works. When she works 26 hours she is paid $468. Let n = number of hours worked and W = wages in dollars.

a By finding the constant of proportionality, write an equation that links W and n.
b If Tessa worked for 27 hours, find her wage.
c If the wage is $648, how many hours did she work? p. 86

12 The price of a Christmas pudding is directly proportional to its mass. The price of a pudding with a mass of 1.5 kg is $7.80. Let m = mass of pudding in kg and P = price of the pudding in $.

a By finding the constant of proportionality, write an equation that links P and m.
b What is the price of a pudding with a mass of 3.6 kg?
c Find the mass of a pudding if it costs $14.56. p. 86

13 In the table below, it is known that y is directly proportional to x.

x	1.3		4.5
y		10.44	16.2

a By finding the constant of proportionality, write an equation that links y and x.
b Complete the table. p. 86

14 The horizontal force, F Newtons, needed to push a wooden block across a horizontal surface is directly proportional to its mass, m kg. It is known that it takes a force of 36 Newtons to move a block with mass 3 kg.

a By finding the constant of proportionality, write an equation that links F and m.
b How much force is required if the block has a mass of 6.2 kg?
c What is the mass of the block if the force applied was 43.2 Newtons? p. 86

15 Given that y is directly proportional to $\sqrt{x}$ and if $y = 24$ then $x = 9$:

a Write an equation that links y and $\sqrt{x}$ using the constant of proportionality k.
b Find the value of k.
c What is the value of y when $x = 16$?
d Find the value of x when $y = 240$. p. 86

16 Given that y is directly proportional to x^2 and if $y = 48$ then $x = 4$:

a Write an equation that links y and x^2 using the constant of proportionality k.

b Find the value of k.

c What is the value of y when $x = 6$?

d Find the value of x when $y = 588$ (if $x > 0$). p. 86

17 The table below shows the time taken (T hours) for a train to travel 600 kilometres at various average speeds (s km/h):

s	30	60	120	200	240	300
T	20	10	5	3	2.5	2

Use the table to draw a graph and explain why T is inversely proportional to s. p. 87

Go to p. 219 for **Quick Answers** or to pp. 241–242 for **Worked Solutions**

For a complete understanding of this topic, you must be able to:

✓	Convert between rates		p. 83
✓	Use a table of values to establish a rule linking two variables		pp. 83–85
✓	Draw and use conversion graphs		pp. 85–86
✓	Solve problems involving direct proportion		p. 86
✓	Recognise the difference between direct and indirect proportion.		p. 87

Now you are ready to do the tests!

Intermediate Test

Rates and Proportion

(20 marks)

1 Human hair grows at an average of 0.4 mm/day. Express this rate in cm/year. (1 mark)

2 Convert:
a 20 grams/m^2 to kg/hectare
b 0.08 cents/cm^3 to \$/m^3. (2 marks)

3 Usain Bolt has run 100 metres in 9.58 seconds. A cheetah has run the same distance in 5.95 seconds. Find the difference between the two speeds, to the nearest km/h. (3 marks)

4 The graph below shows the relationship between distance travelled (D km) over time (h hours) when a car is driven.

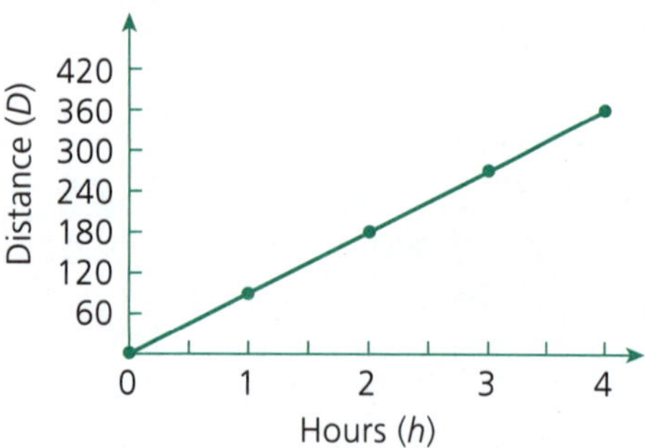

a Use the graph to complete the table.

h	0	1	2	3	4
D					

b Write the rule that links D and h.
c What is the speed of the car?
d How far will the car have travelled after 75 minutes?
e On the same graph, draw a second line to represent the distance travelled for a second car that averages 60 km/h.
f Find the distance between the two cars after travelling for 3.5 hours.
g How much longer will it take the second car to travel 180 km? (7 marks)

5 The cost (\$$C$) of building a single-storey house on a concrete slab is directly proportional to the area (A) of the floor space of the house in square metres. It costs \$235 000 to build a house with a floor space of 200 m^2.
a Find an equation for C in terms of A.
b Find the cost of a house with a floor space of 310 m^2. (3 marks)

6 The distance (d), in kilometres, to the horizon is directly proportional to the square root of the height (h), in metres, of the observer's eye above sea level.
A person whose eyes are 1.69 m above the ground can see 4.16 km out to sea.
a Find an equation for d in terms of h.
b How high above sea level, in metres, would a person's eyes need to be to see a boat that is 8 km out to sea?
c If the person stood on a headland so that their eyes were 18 metres above sea level, can they see an island 12 kilometres out to sea? Justify your answer with calculations. (4 marks)

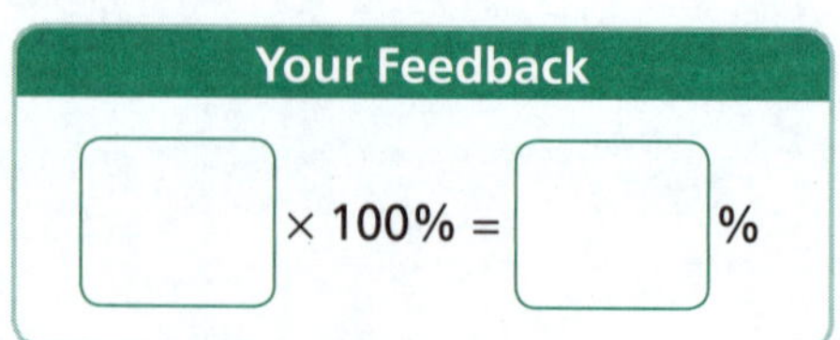

QA PAGE 223
WS PAGE 266

Advanced Test

Rates and Proportion

(20 marks)

1 125 mL of water is poured into a container. Twenty grams of salt is added to the water. What is the concentration of salt in the water in grams/L? (1 mark)

2 Given that y is directly proportional to x, what is the effect on y when x is doubled? (1 mark)

3 Given that y is directly proportional to x^2, what is the effect on y when x is doubled? (1 mark)

4 In the table, y is directly proportional to $\sqrt{x}$.

x	4	16	49	
y		8		24

a Find the constant of proportionality k.
b Complete the table. (5 marks)

5 Two trucks, A and B, leave the same location at the same time and travel the same road. The distance (D) in km and the time of travel (t) in hours of the trucks are recorded in the graph below.

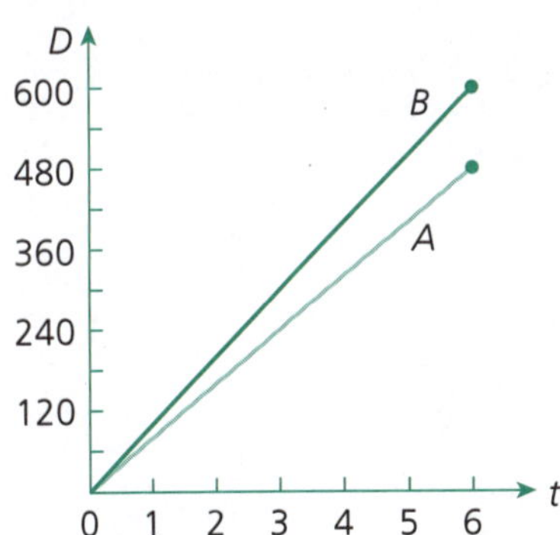

a As D and t are in direct proportion, write an equation for each truck, linking D and t.
b How much further has B travelled than A after 4 hours?
c When are the two trucks 60 km apart? (6 marks)

6 Gina plans to buy some silver pendants. The cost (\$$C$) varies directly with the square of the length (n cm) of the pendant. A pendant of length 3 cm costs \$64.80.
a Write an equation linking C and n.
b What is the length of the pendant that costs \$45?
c What is the price of a pendant that is 4.3 cm in length? (4 marks)

7 The cost (\$$C$) of a circular mirror is directly proportional to the square of the mirror's diameter (d cm). A circular mirror with a diameter of 25 cm is \$62.50. What is the cost of a mirror with a diameter of 40 cm? (2 marks)

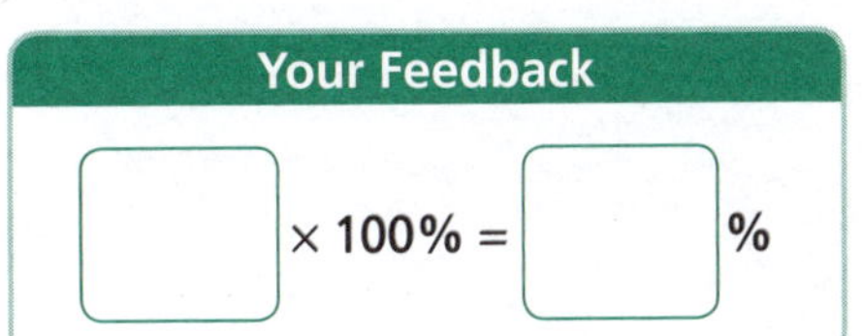

QA PAGE 223
WS PAGE 267

Chapter 8
Pythagoras and Trigonometry

Introduction to Pythagoras' Theorem

In any right-angled triangle the square on the hypotenuse is equal to the sum of the squares on the other two sides.

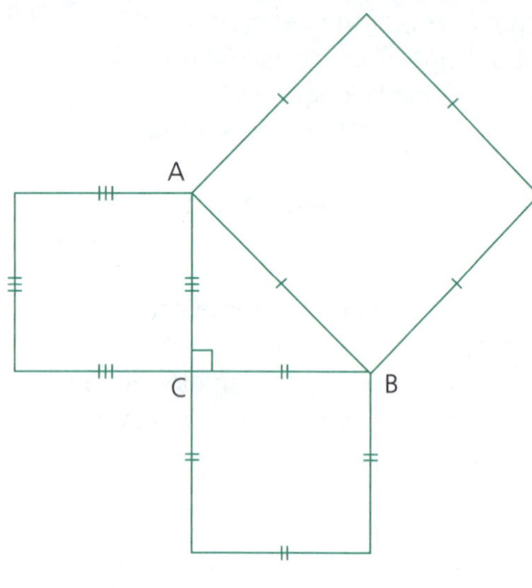

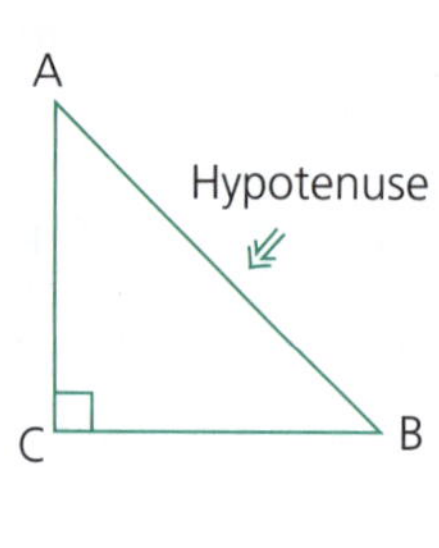

Keywords

Adjacent	Ratio
Bearing	Reciprocal
Cos θ	Right-angled
Cosine	Sin θ
Degree	Sine
Depression	Square
Elevation	Square root
Exact	Surd
Horizontal	Tan θ
Hypotenuse	Tangent
Inclination	Theorem
Minute	Triad
Opposite	Triangle
Pythagoras	Trigonometric
Pythagorean	

The hypotenuse is the longest side of the right-angled triangle—it is always the side opposite the right-angle.

In simpler terms:

$c^2 = a^2 + b^2$

where c is the length of the hypotenuse.

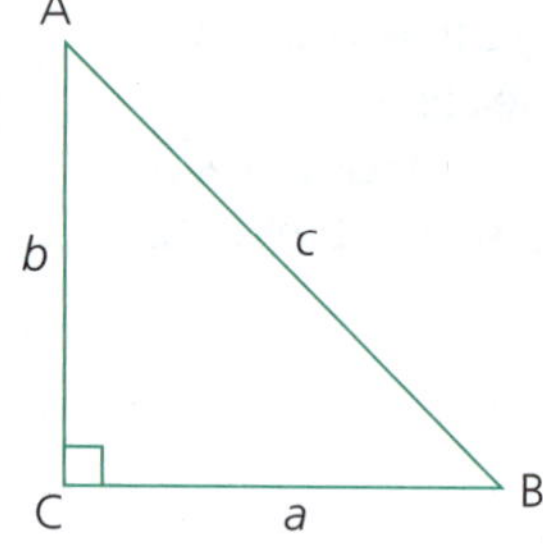

In the following diagrams, state Pythagoras' Theorem using the markings on the diagrams.

1

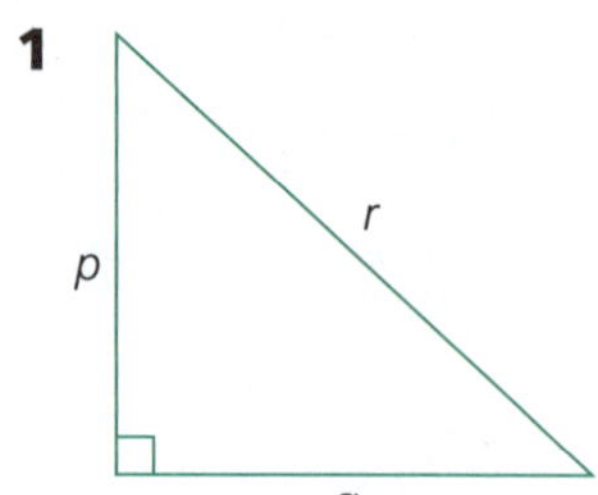

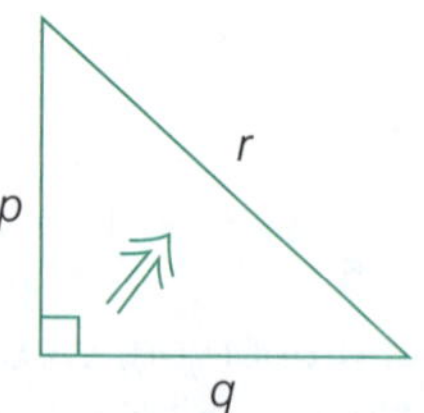

Locate the hypotenuse—draw an arrow directly across the triangle from the right angle to the opposite side.

Hypotenuse = r

Then:

$r^2 = p^2 + q^2$

The hypotenuse squared = sum of the squares of the other two sides.

2

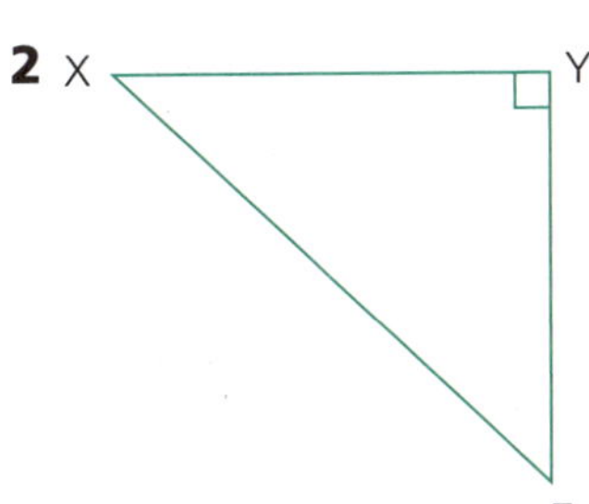

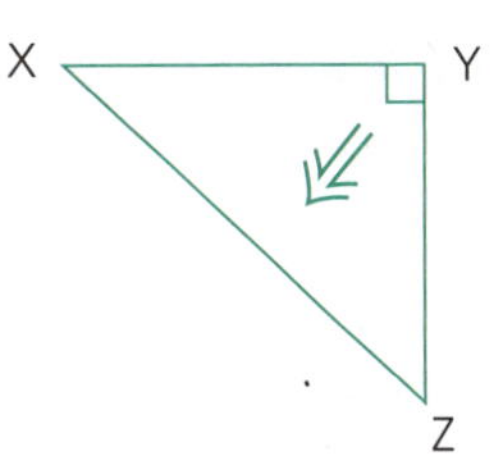

Sides are XZ, ZY, XY.
Hypotenuse is XZ (see arrow).
Then $XZ^2 = XY^2 + YZ^2$.

Finding the length of the hypotenuse

Given the lengths of any two sides of a right-angled triangle, it is possible to calculate the length of the third side.

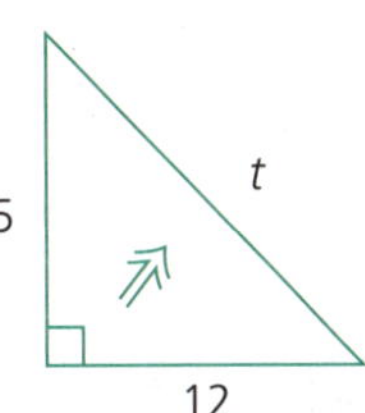

Consider this triangle. Find the value of t.

State Pythagoras' Theorem for the triangle. (Hypotenuse = t.)

$$\begin{aligned} t^2 &= 5^2 + 12^2 \\ &= 25 + 144 \\ &= 169 \end{aligned}$$

Evaluate $5^2 + 12^2$

$$\begin{aligned} \therefore t &= \sqrt{169} \\ &= 13 \end{aligned}$$

Calculate $\sqrt{169}$

1 Calculate the value of y in this triangle:

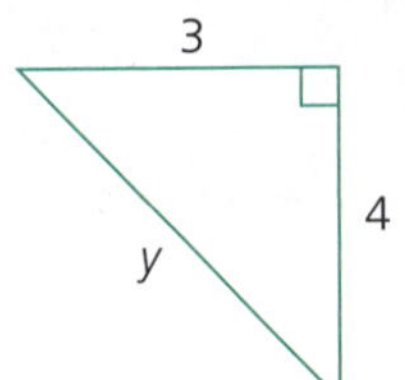

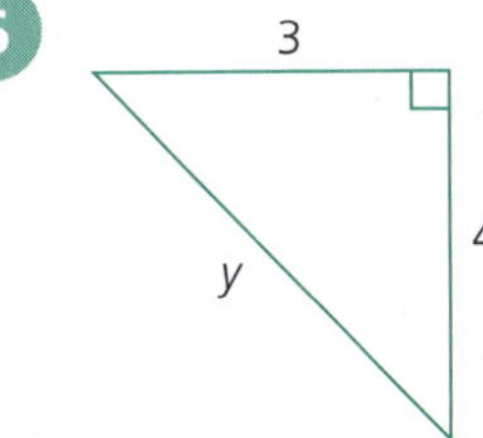

$$\begin{aligned} y^2 &= 3^2 + 4^2 \\ &= 9 + 16 \\ &= 25 \end{aligned}$$

$$\begin{aligned} \therefore y &= \sqrt{25} \\ &= 5 \end{aligned}$$

2 Find the length of the hypotenuse:

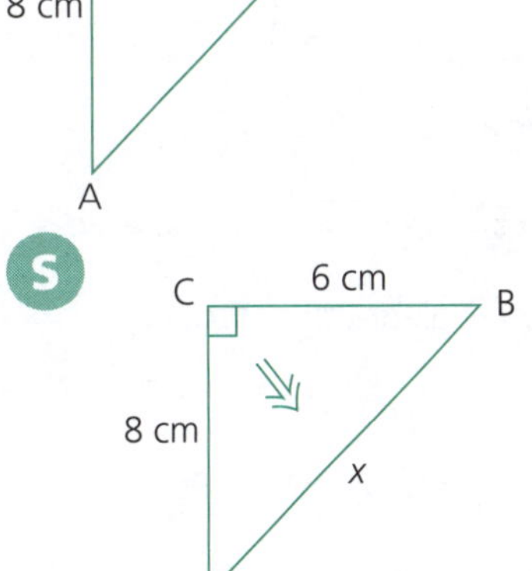

S

C 6 cm B, 8 cm, x, A

Hypotenuse = AB

Let length of AB be x cm.

$$\begin{aligned} \therefore x^2 &= 6^2 + 8^2 \\ &= 36 + 64 \\ &= 100 \end{aligned}$$

$$\begin{aligned} \therefore x &= \sqrt{100} \\ &= 10 \end{aligned}$$

$\therefore$ AB is 10 cm long.

3 Calculate the length PQ in this triangle:

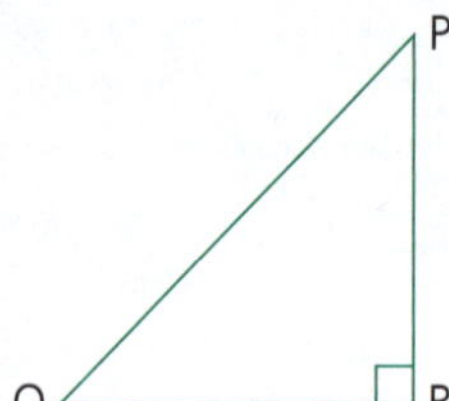

PR = 10 cm
QR = 24 cm

S

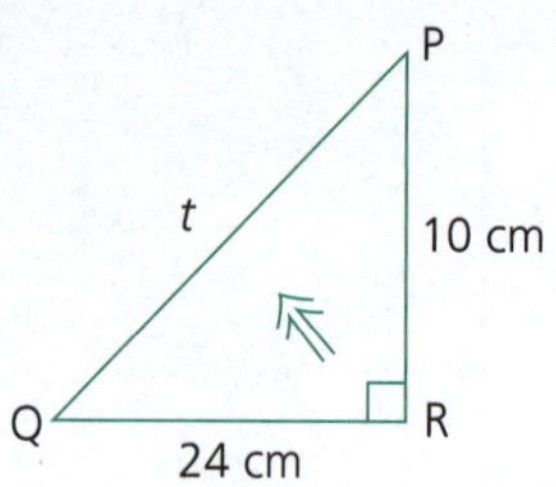

PQ is the hypotenuse.

Let length of PQ be *t* cm.

$$t^2 = 10^2 + 24^2$$
$$= 100 + 576$$
$$= 676$$
$$\therefore t = \sqrt{676}$$
$$= 26$$

∴ Length of PQ is 26 cm.

Finding the length of a side not the hypotenuse

Two sides, one of which is the hypotenuse are known and we need to find the length of the third side.

Consider this example:

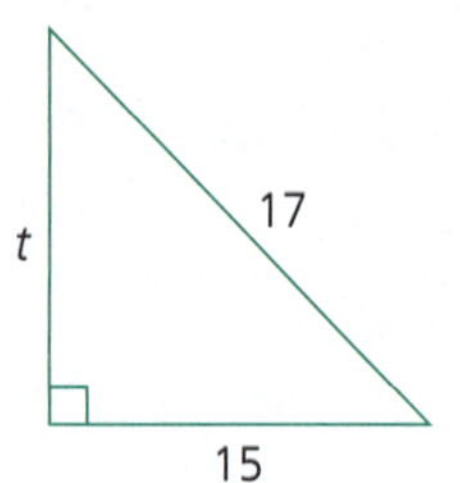

Note the hypotenuse is 17 units.

Using Pythagoras' Theorem:

$t^2 + 15^2 = 17^2$

Keep the letter on the left side of the equation.

$$\therefore t^2 = 17^2 - 15^2$$
$$= 289 - 225$$
$$= 64$$

Move 15^2 to right side. It then is subtracted.

$$\therefore t = \sqrt{64}$$
$$= 8$$

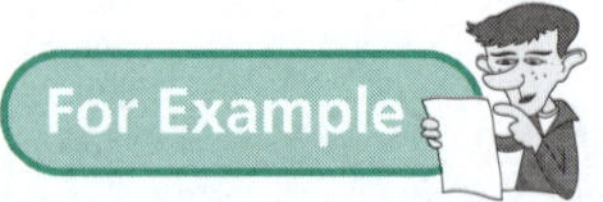

1 Find the value of *t* in this diagram:

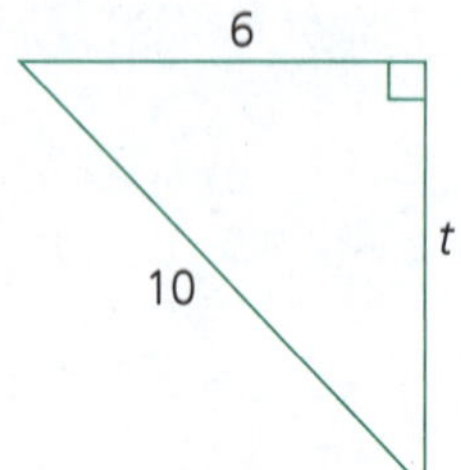

S

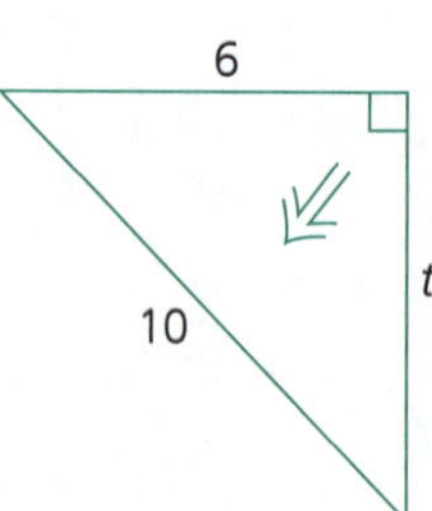

Hypotenuse = 10 units

$$\therefore t^2 + 6^2 = 10^2$$
$$t^2 = 10^2 - 6^2$$
$$= 100 - 36$$
$$= 64$$
$$\therefore t = \sqrt{64}$$
$$= 8$$

2 Find the length of the side YZ given that XY = 15 cm and XZ = 12 cm:

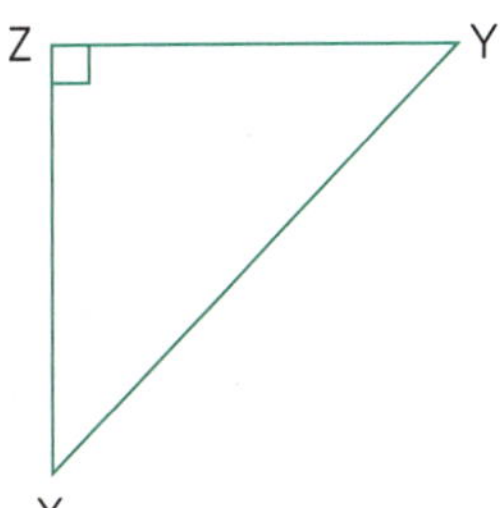

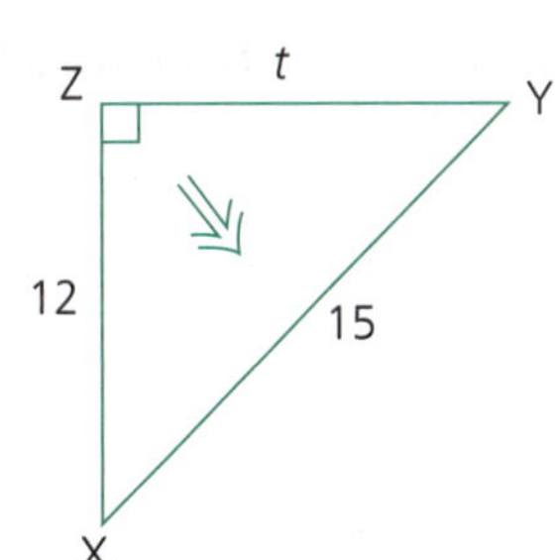

Let YZ = t

Hypotenuse = 15

$$\therefore\ t^2 + 12^2 = 15^2$$
$$t^2 = 15^2 - 12^2$$
$$= 225 - 144$$
$$= 81$$
$$\therefore\ t = \sqrt{81}$$
$$= 9$$

YZ is 9 cm long.

Using Pythagoras' Theorem to prove a triangle is right-angled

For △ABC to be right-angled, we must be able to prove that $AB^2 = BC^2 + AC^2$.

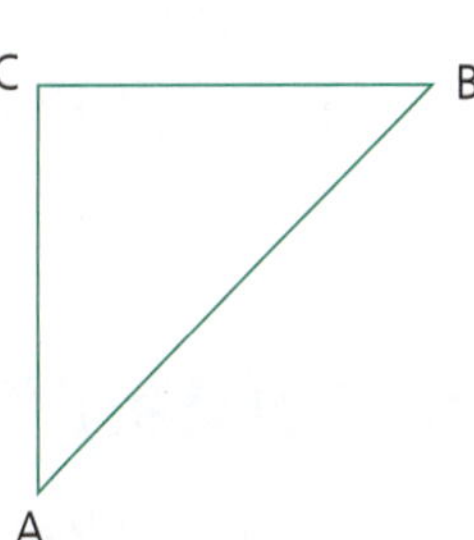

For Example

1 Show that the following triangle is right-angled:

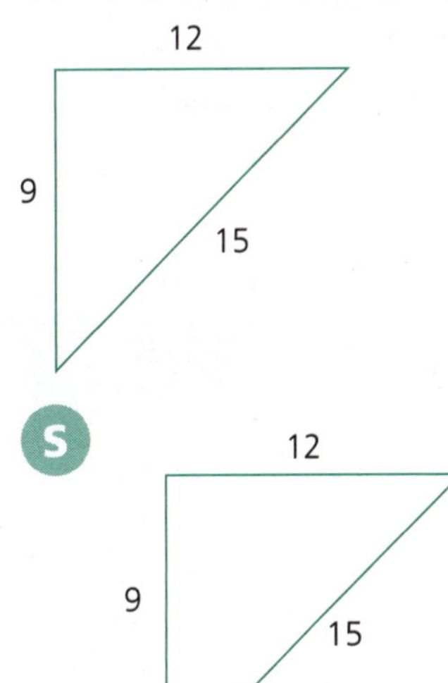

Hypotenuse is the longest side, i.e. 15 units.

We have to prove that:

$15^2 = 9^2 + 12^2$

Now LHS $= 15^2$
$= 225$

LHS means 'left-hand side' of expression.

RHS $= 9^2 + 12^2$
$= 81 + 144$
$= 225$
$=$ LHS

RHS means 'right-hand side' of expression.

$\therefore\ 15^2 = 9^2 + 12^2$

Then △ is right-angled.

2 Prove $\triangle PQR$ is right-angled given that PR = 5 cm, RQ = 12 cm and PQ = 13 cm.

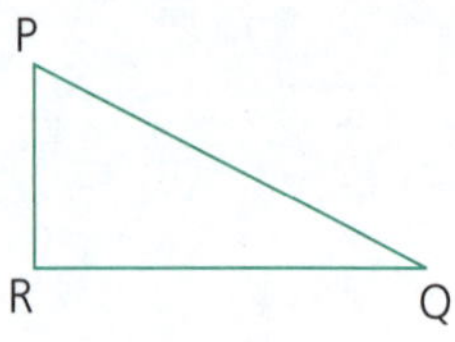

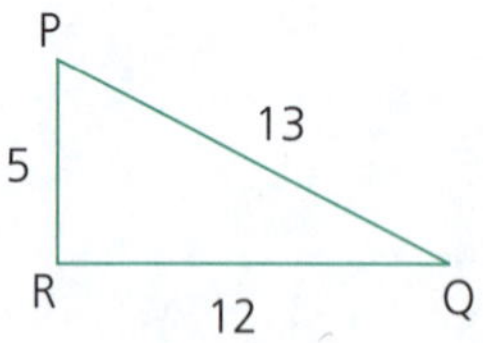

Hypotenuse = 13

For $\triangle PQR$ to be right-angled, we must prove that:

$$13^2 = 5^2 + 12^2$$

$$\begin{aligned} \text{LHS} &= 13^2 \\ &= 169 \end{aligned}$$

$$\begin{aligned} \text{RHS} &= 5^2 + 12^2 \\ &= 25 + 144 \\ &= 169 \\ &= \text{LHS} \end{aligned}$$

$\therefore\ 13^2 = 5^2 + 12^2$

i.e. $\triangle PQR$ is right-angled.

3 Prove $\triangle XYZ$ is right-angled, given that XY is 25 cm, ZY is 20 cm and XZ is 15 cm long.

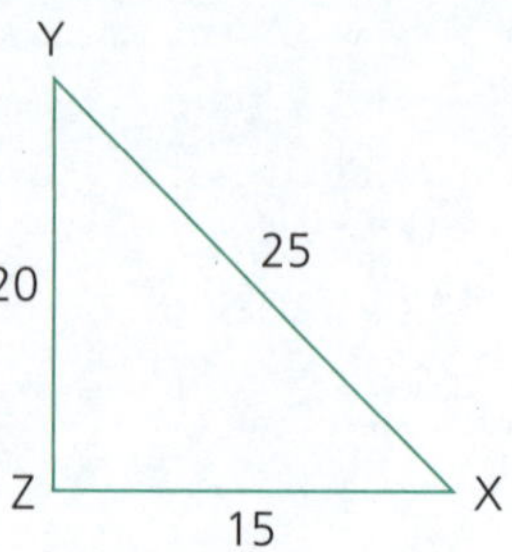

Hypotenuse is XY, i.e. 25 cm (the longest side).

For $\triangle XYZ$ to be right-angled, we must prove that:

$$25^2 = 20^2 + 15^2$$

$$\begin{aligned} \text{LHS} &= 25^2 \\ &= 625 \end{aligned}$$

$$\begin{aligned} \text{RHS} &= 20^2 + 15^2 \\ &= 400 + 225 \\ &= 625 \\ &= \text{LHS} \end{aligned}$$

$\therefore\ 25^2 = 20^2 + 15^2$

i.e. $\triangle XYZ$ is right-angled.

Pythagorean triads (or triples)

1 The numbers $\{a, b, c\}$ form a Pythagorean triad if:

a $c^2 = a^2 + b^2$

b a, b, and c are positive whole numbers, that is, {3, 4, 5} is a Pythagorean triad because $3^2 + 4^2 = 5^2$.

$$\begin{aligned} \text{LHS} &= 3^2 + 4^2 \\ &= 9 + 16 \\ &= 25 \end{aligned} \qquad \begin{aligned} \text{RHS} &= 5^2 \\ &= 25 \\ &= \text{LHS} \end{aligned}$$

2 But {1.2, 1.6, 2} is *not* a Pythagorean triad, even though $2^2 = 1.2^2 + 1.6^2$, because 1.2 and 1.6 are not whole numbers.

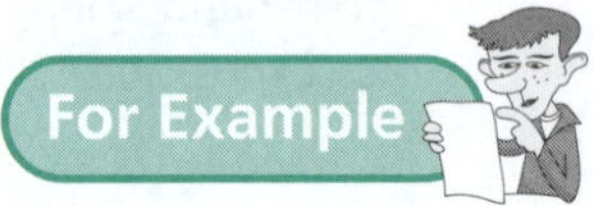

1 Find the value of t such that:

a $\{18, 24, t\}$ is a Pythagorean triad.

S

$$\begin{aligned} t^2 &= 18^2 + 24^2 \\ &= 900 \end{aligned}$$

$$\begin{aligned} \therefore\ t &= \sqrt{900} \\ &= 30 \end{aligned}$$

b $\{t, 12, 13\}$ is a Pythagorean triad.

S $\therefore t^2 + 12^2 = 13^2$

$t^2 = 13^2 - 12^2$

$= 169 - 144$

$= 25$

$\therefore t = \sqrt{25}$

$= 5$

Using the calculator for Pythagoras' Theorem

Not all numbers have square roots such as $\sqrt{25} = 5$ or $\sqrt{49} = 7$.

Sometimes a Pythagoras question results in an answer such as $\sqrt{10}$, $\sqrt{13}$, etc. These numbers do not have whole number square roots. However, these answers are the **exact** solution to the Pythagoras question.

The answer to these questions can either be left in **exact** form $\sqrt{\ }$ or evaluated correct to a given number of decimal places (approximate answer) using the square root function on your calculator.

Exact form is often stated as **surd** form. That is, an answer left as $\sqrt{17}$ (for example) has been left in exact form or surd form.

For Example

1 Calculate the value of t from the following diagrams correct to one decimal place.

a

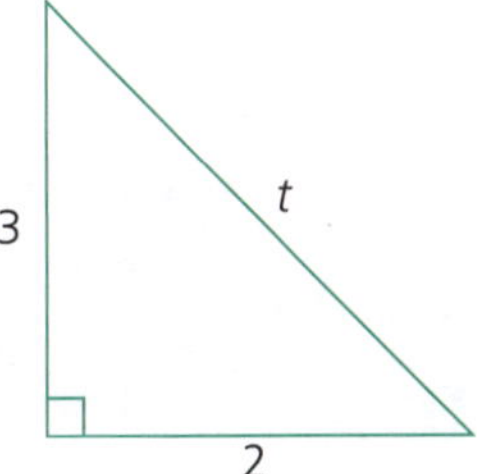

S

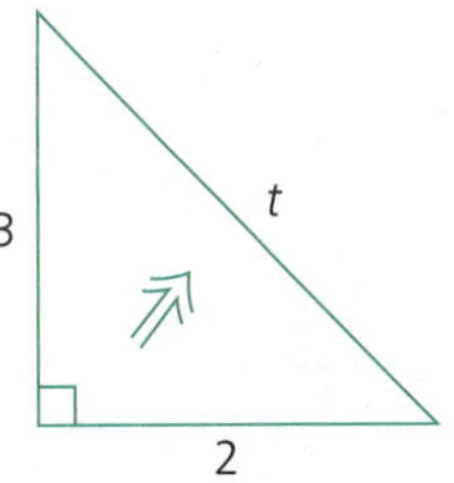

Using Pythagoras' Theorem:

$t^2 = 3^2 + 2^2$

$= 9 + 4$

$= 13$

$\therefore t = \sqrt{13}$ (Exact solution)

$= 3.605\,551\,3$

$= 3.6$ (to 1 decimal place)

b

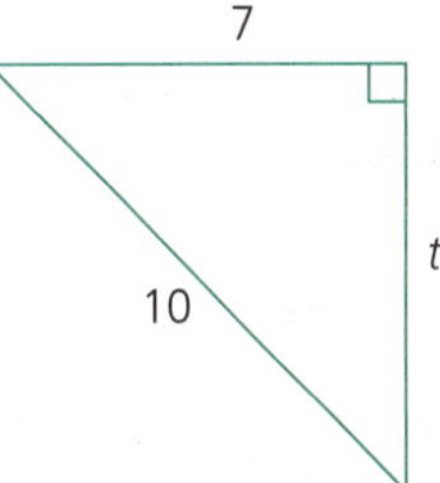

S

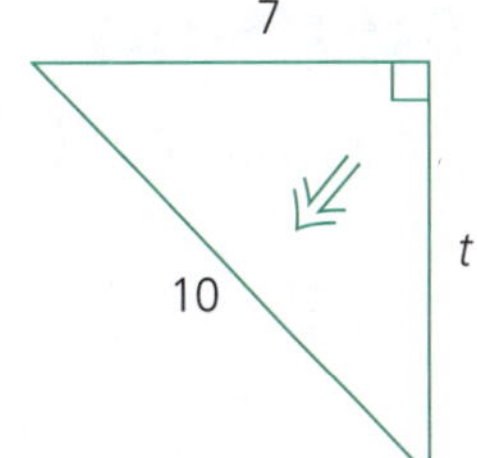

$t^2 + 7^2 = 10^2$

$\therefore t^2 = 10^2 - 7^2$

$= 100 - 49$

$= 51$

$\therefore t = \sqrt{51}$ (Exact solution)

$= 7.141\,428\,4$

$= 7.1$ (to 1 decimal place)

2 Given that AB = 7 cm, AC = 4 cm, calculate the length of BC correct to one decimal place:

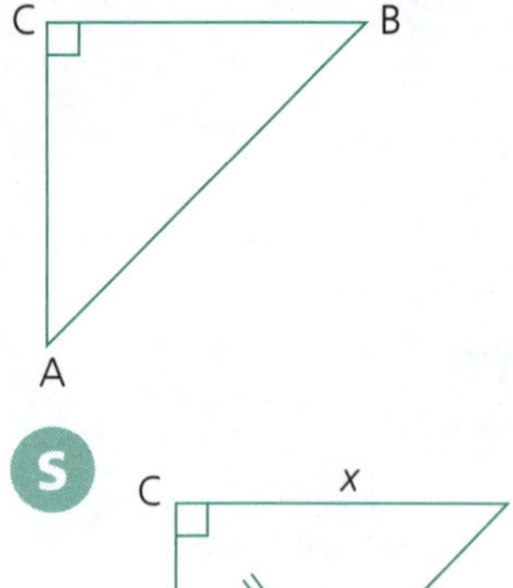

S

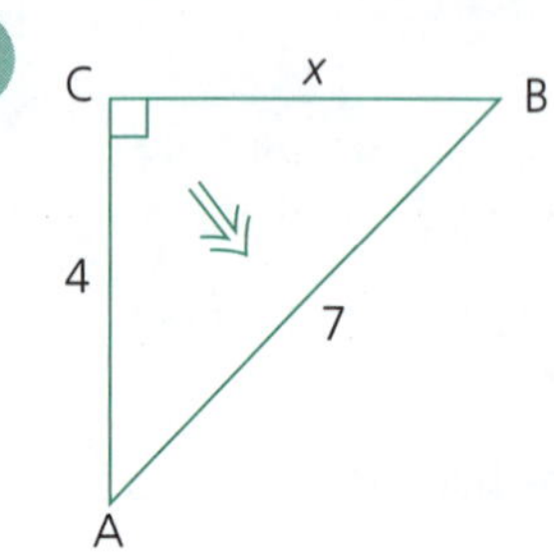

Let BC = x cm (AB is hypotenuse)

$\therefore x^2 + 4^2 = 7^2$

$x^2 = 7^2 - 4^2$

$= 49 - 16$

$= 33$

$\therefore x = \sqrt{33}$ Exact solution

$= 5.744\,562\,6$

$= 5.7$ (to 1 decimal place)

That is, BC is 5.7 cm long.

3 In the ΔPQR, PR = 5 cm, RQ = 4 cm and $\angle PRQ = 90°$. Calculate the length of PQ correct to three significant figures.

S Let PQ be y cm (PQ is hypotenuse):

$\therefore y^2 = 5^2 + 4^2$

$= 25 + 16$

$= 41$

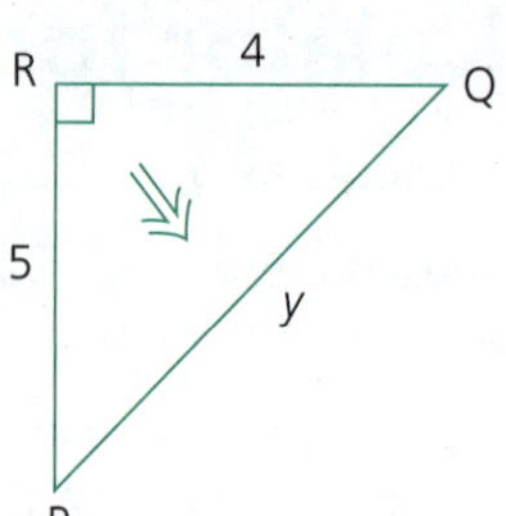

$\therefore y = \sqrt{41}$ Exact solution

$= 6.403\,124\,2$

$= 6.40$ (to three significant figures)

PQ is 6.40 cm long.

Problem solving using Pythagoras' Theorem

To be able to use Pythagoras' Theorem to solve a problem a right angle must be present.

For Example

1

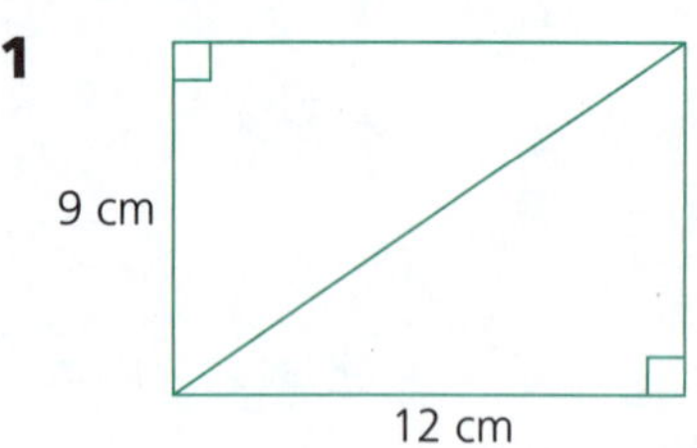

A rectangle has sides of 9 cm and 12 cm. Calculate the length of the diagonal.

S

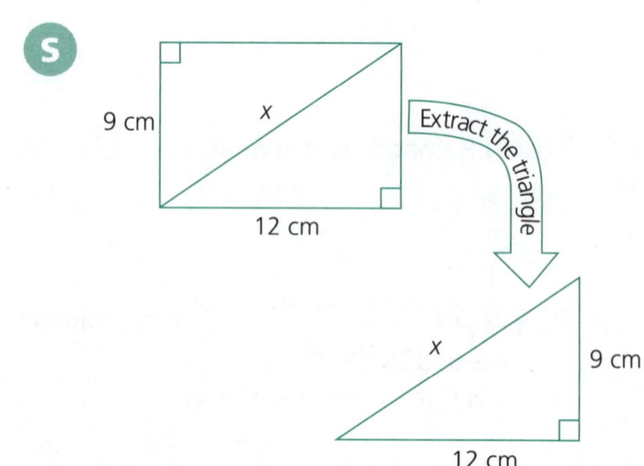

Let diagonal have length x cm.

From the triangle, x is the hypotenuse:

$$\therefore x^2 = 9^2 + 12^2$$
$$= 81 + 144$$
$$= 225$$

$$\therefore x = \sqrt{225}$$
$$= 15$$

The diagonal has length 15 cm.

2

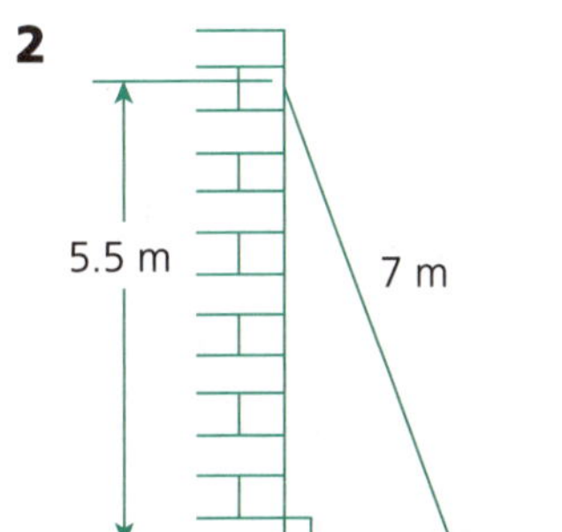

A 7-metre ladder must reach 5.5 metres up a brick wall. How far from the wall must the foot of the ladder be placed? (Answer correct to one decimal place.)

Consider only the triangle.
Let the unknown side be d m.

The hypotenuse = 7 m.

By Pythagoras' Theorem:

$$d^2 + 5.5^2 = 7^2$$

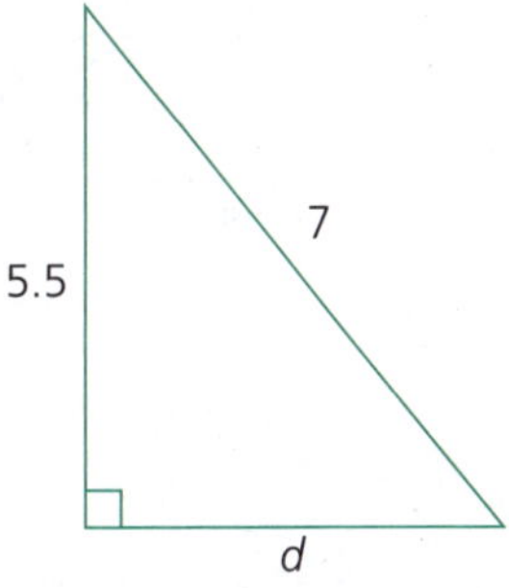

$$\therefore d^2 = 7^2 - 5.5^2$$
$$= 49 - 30.25$$
$$= 18.75$$

$$\therefore d = \sqrt{18.75}$$
$$= 4.330\,127$$
$$= 4.3 \text{ (1 decimal place)}$$

The foot of the ladder must be placed 4.3 metres from the wall.

3

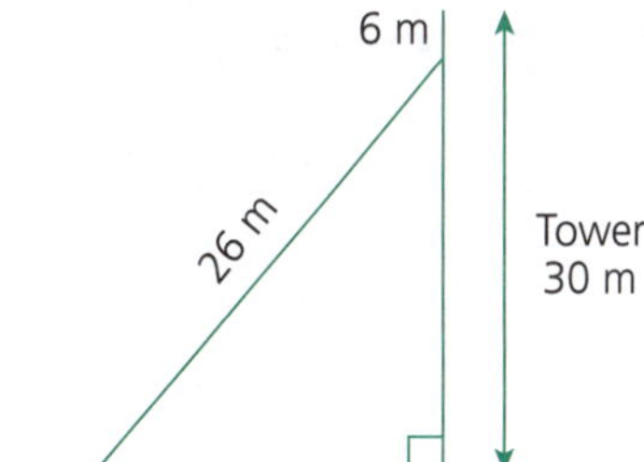

A support wire is to be attached 6 metres from the top of a 30-metre tower. If the length of wire used is 26 metres long, how far from the base of the tower must the wire be attached?

$30 - 6 = 24$

$\therefore$ The wire is attached 24 m above the base.

Let the distance from the base be y metres.

Hypotenuse = 26 metres.

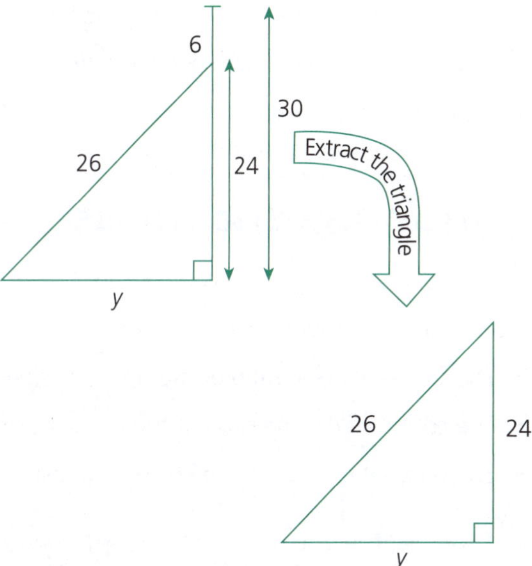

By Pythagoras' Theorem:

$$y^2 + 24^2 = 26$$

$$\therefore y^2 = 26^2 - 24^2$$
$$= 676 - 576$$
$$= 100$$

$$\therefore y = \sqrt{100}$$
$$= 10$$

The wire is attached 10 metres from the base of the tower.

Introduction to trigonometry

In any right-angled triangle the sides are given special names to identify their positions relative to a particular angle.

If one of the acute angles of a right-angled triangle is $\theta°$, the three sides of the triangle are named with reference to this angle as **opposite**, **adjacent** and **hypotenuse**, as shown below:

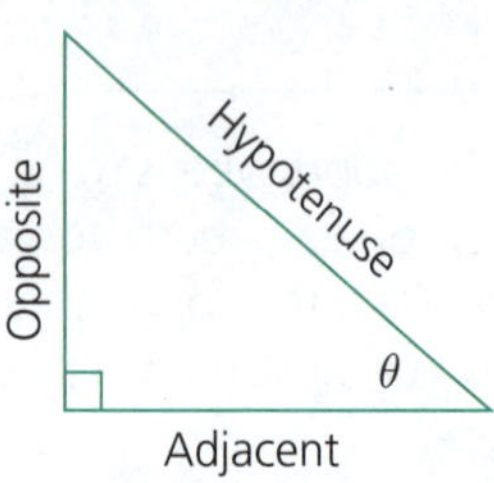

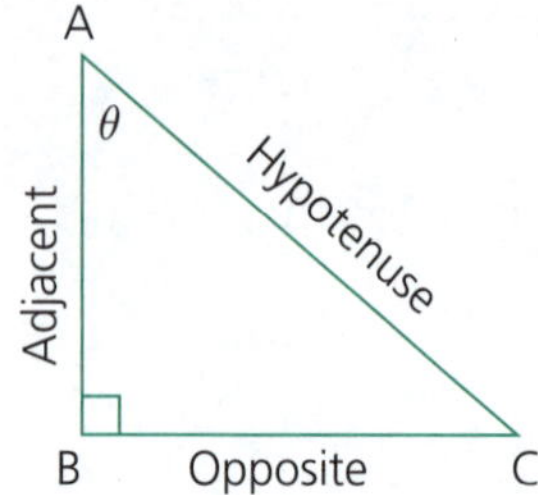

Notes
- The hypotenuse is always opposite the right-angle.
- The side opposite the given angle is the opposite side.
- The other side is the adjacent side.

For example, in the $\triangle ABC$, $\hat{B} = 90°$ and $\hat{A} = \theta°$.

The hypotenuse is side AC (which is opposite the right-angle).

The opposite is side BC (which is opposite θ).

The adjacent is side AB (which is the other side).

Trigonometric ratios

In any right-angled triangle ABC, where $\hat{C} = \theta°$, $\hat{B} = 90°$,

we define three trigonometric ratios:

- The sine of θ is the ratio of the opposite side to the hypotenuse.

- The cosine of θ is the ratio of the adjacent side to the hypotenuse.
- The tangent of θ is the ratio of the opposite side to the adjacent side.

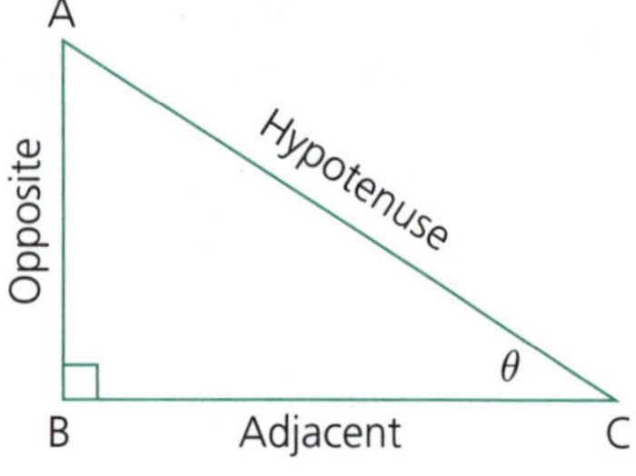

Notes
- The sine of θ is abbreviated to $\sin\theta$.
- The cosine of θ is abbreviated to $\cos\theta$.
- The tangent of θ is abbreviated to $\tan\theta$.

Summary of ratios

In any right-angled triangle:

$$\sin\theta = \frac{\text{Opposite}}{\text{Hypotenuse}} \quad \text{(SOH)}$$

$$\cos\theta = \frac{\text{Adjacent}}{\text{Hypotenuse}} \quad \text{(CAH)}$$

$$\tan\theta = \frac{\text{Opposite}}{\text{Adjacent}} \quad \text{(TOA)}$$

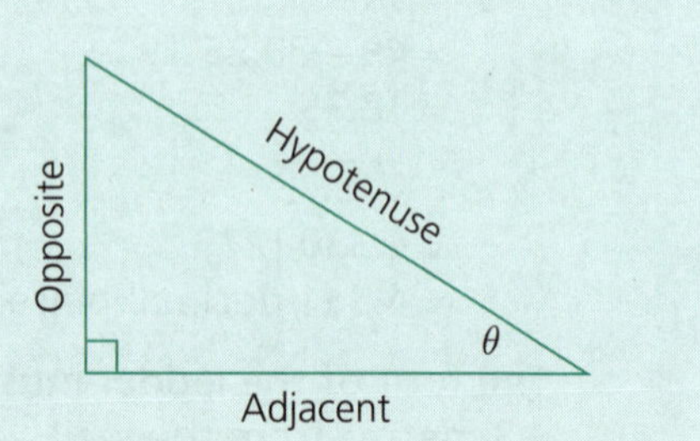

A method of remembering the three ratios is to learn the mnemonic:

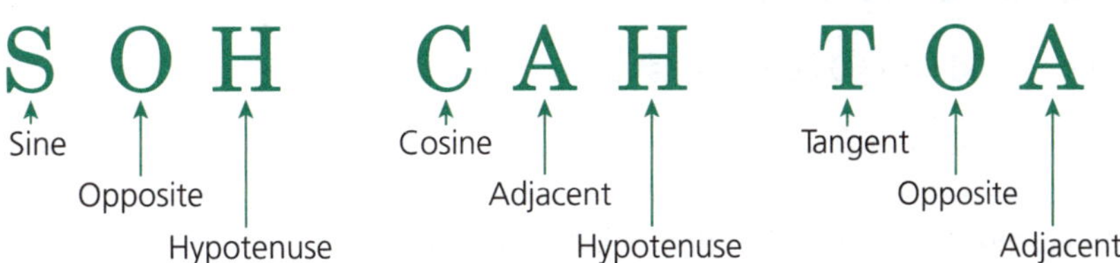

1 What is the value of $\tan\theta$?

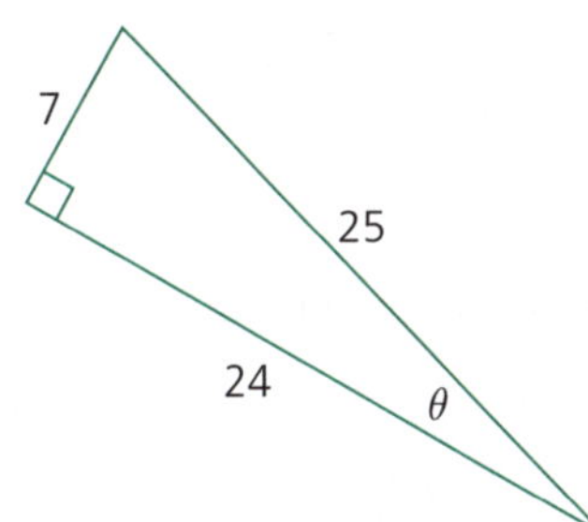

S

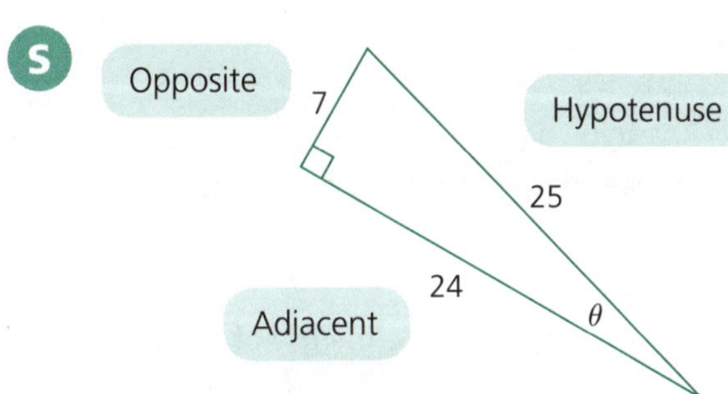

$$\tan\theta = \frac{\text{Opposite}}{\text{Adjacent}}$$ TOA

$$\therefore \tan\theta = \frac{7}{24}$$

Note: name the sides with reference to the given angle.

2 For the given triangle, find the value of:

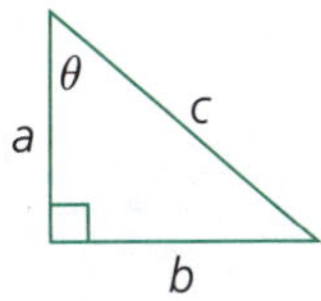

a $\sin\theta$

S

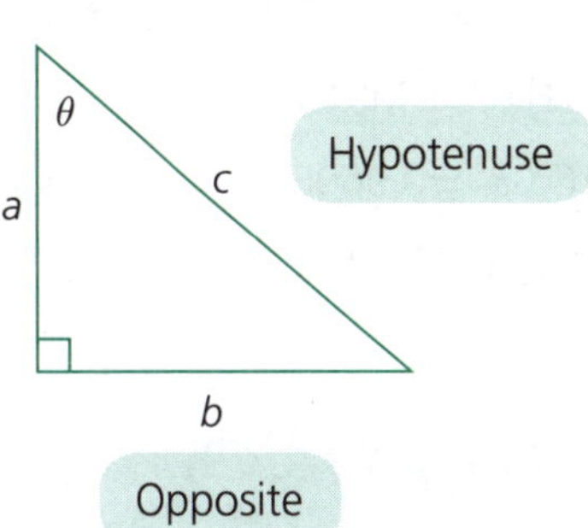

Opposite

You must name the sides with reference to the given angle θ.

$$\sin\theta = \frac{\text{Opposite}}{\text{Hypotenuse}}$$ SOH

$$\therefore \sin\theta = \frac{b}{c}$$

b $\cos\theta$

S

$$\cos\theta = \frac{\text{Adjacent}}{\text{Hypotenuse}}$$ CAH

$$\therefore \cos\theta = \frac{a}{c}$$

c $\tan\theta$

S

$$\tan\theta = \frac{\text{Opposite}}{\text{Adjacent}}$$ TOA

$$\therefore \tan\theta = \frac{b}{a}$$

Degrees and minutes

A degree may be subdivided into 60 equal parts called minutes:

$1° = 60'$ (minutes)

$1' = 60''$ (seconds)

That is, an angle could measure 60°48′ (60 degrees 48 minutes).

We can find the value of trigonometric ratios for angles measured in degrees and minutes.

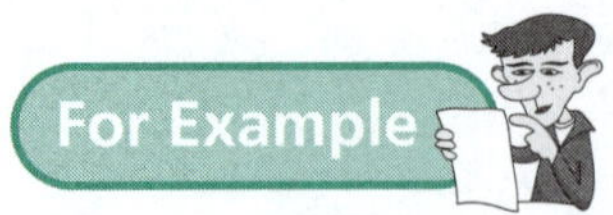

$\sin 60°48' = 0.8729$ (to 4 decimal places)

[SIN] 60 [DMS] 48 =

$\cos 16°08' = 0.9606$ (to 4 decimal places)

[COS] 16 [DMS] 8 =

$\tan 78°39' = 4.9819$, [TAN] 78 [DMS] 39 =

or, given the ratio, the angle can be found in degrees and minutes.

Make sure your calculator is in degree mode.

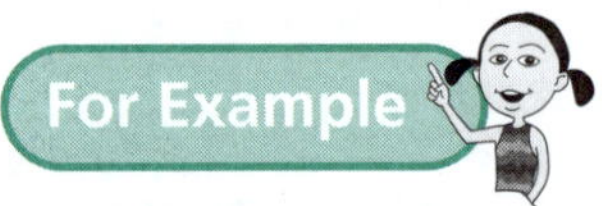

If θ is acute, find θ to the nearest minute, given:

a $\tan\theta = 0.72$

$\therefore \theta = 35°45'$

[2ndF] [TAN⁻¹] 0.72 [2ndF] [DMS] =

b $\cos\theta = 0.9$

$\therefore \theta = 25°51'$

[2ndF] [COS⁻¹] 0.9 [2ndF] [DMS] =

25° 50′ 30″ ← When this number is 30 or more, the number of minutes is raised by one.

c $\tan\theta = \frac{3}{4}$

$\therefore \theta = 35°52'$

Note the use of brackets.

[2ndF] [TAN⁻¹] [(] 3 [÷] 4 [)] [2ndF] [DMS]

Notes on the use of calculators

1. Before using your calculator for trigonometry make sure it is in degree mode. On screen it should say DEG.
2. The trigonometric functions on the calculator are:

   [SIN] [COS] [TAN]
3. [INV] or [SHIFT] or [2nd F] all mean the same thing.
4. [D°M′S] or [° ′ ″] mean the same thing.
5. The calculator steps for trigonometry vary for different models and different brands of calculators.

   Consult your calculator handbook. For this reason *most* calculator sequences have *not* been included in this study guide.
6. Care must be taken when finding an angle in some situations. For example, when you find that you have a statement such as $\cos\theta = \frac{4.8}{7.9}$, use grouping symbols, i.e. [2ndF] [COS⁻¹] [(] 4.8 [÷] 7.9 [)] [=] [2ndF] [DMS] (gives angle in degrees and minutes).

Using trigonometric ratios to find sides and angles

Finding the lengths of sides

In any right-angled triangle, given one side and an acute angle, we can find the length of another side by using one of the trigonometric ratios.

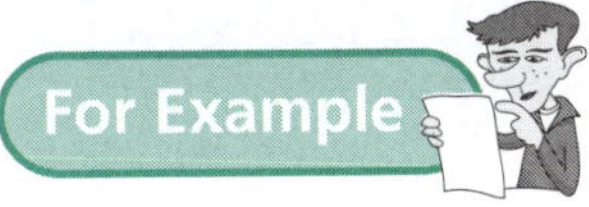

1 Find the value of x, correct to two decimal places.

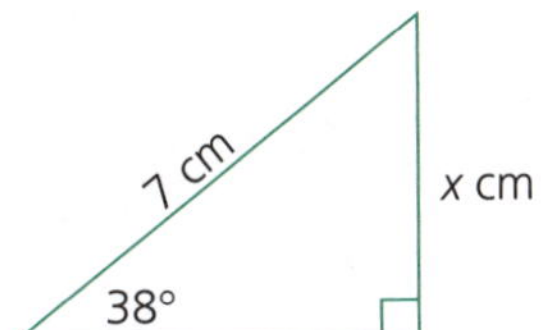

S First you must name the sides with reference to the given angle.

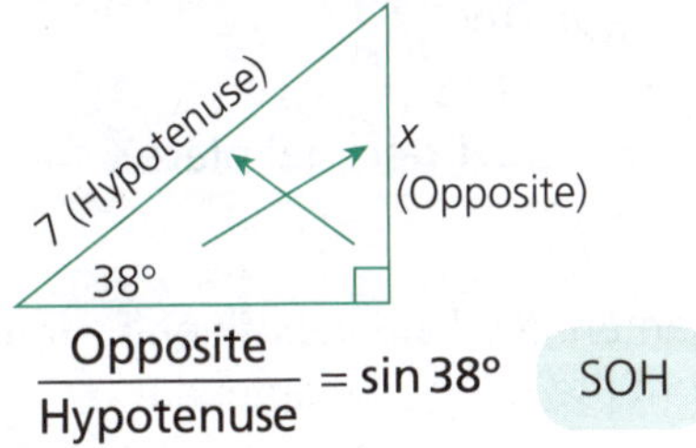

$$\frac{\text{Opposite}}{\text{Hypotenuse}} = \sin 38° \quad \text{SOH}$$

$$\therefore \frac{x}{7} = \sin 38°$$
$$x = 7 \times \sin 38°$$
$$= 4.31 \text{ cm (to 2 decimal places)}$$

2 Find the value of x, correct to one decimal place.

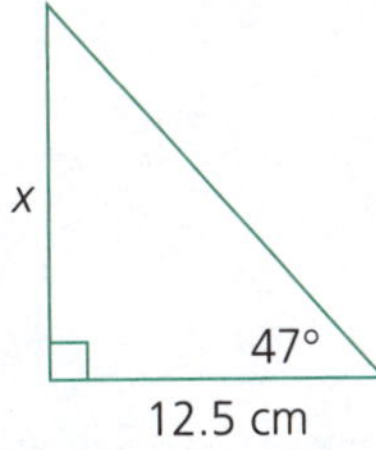

S Name the sides with reference to the given angle.

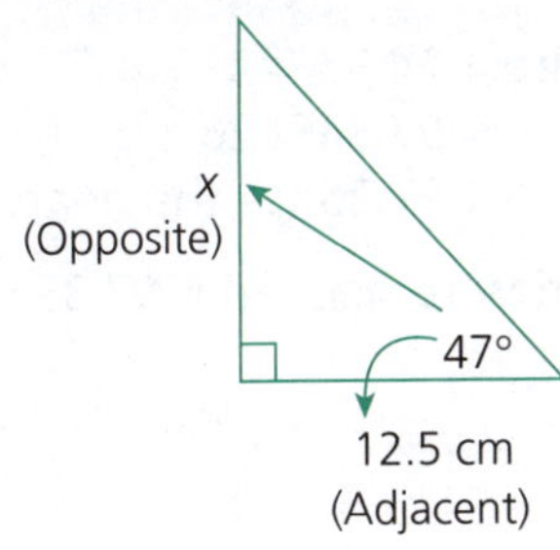

$$\frac{\text{Opposite}}{\text{Adjacent}} = \tan 47° \quad \text{TOA}$$

$$\therefore \frac{x}{12.5} = \tan 47°$$
$$x = 12.5 \times \tan 47°$$
$$= 13.4 \text{ cm (to 1 decimal place)}$$

3 Find a, correct to two decimal places.

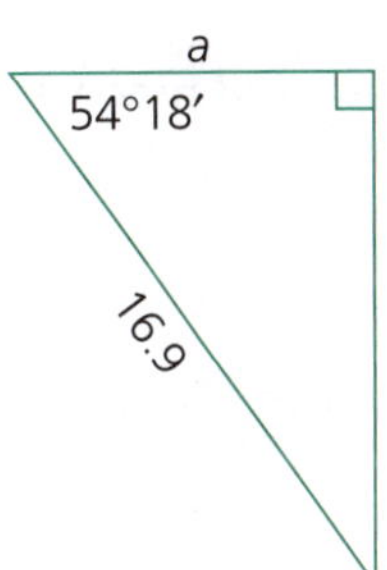

S Name the sides with reference to the given angle.

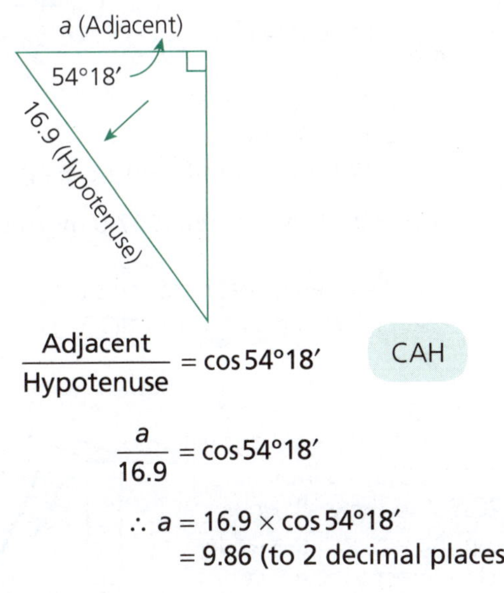

$$\frac{\text{Adjacent}}{\text{Hypotenuse}} = \cos 54°18' \quad \text{CAH}$$

$$\frac{a}{16.9} = \cos 54°18'$$
$$\therefore a = 16.9 \times \cos 54°18'$$
$$= 9.86 \text{ (to 2 decimal places)}$$

4 Find the value of x, correct to two decimal places.

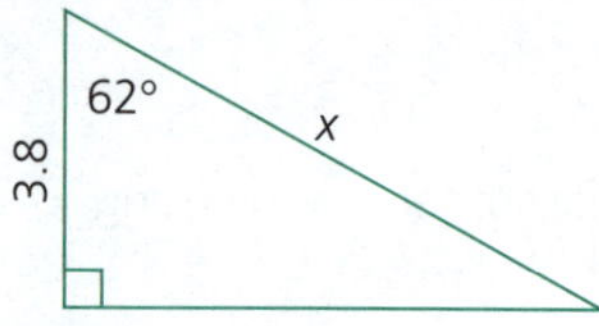

S First you must name the sides with reference to the given angle.

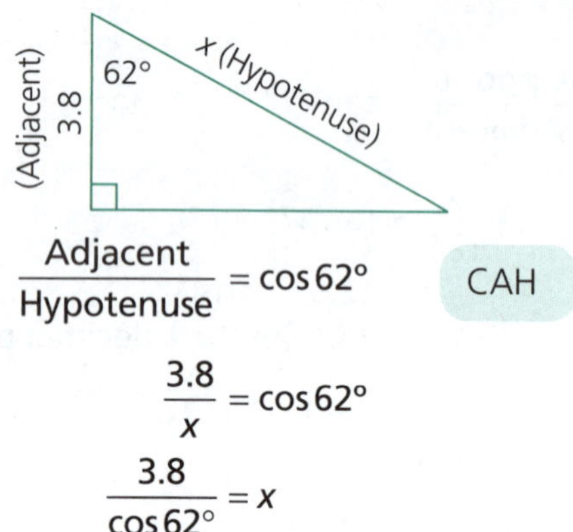

$$\frac{\text{Adjacent}}{\text{Hypotenuse}} = \cos 62°$$

CAH

$$\frac{3.8}{x} = \cos 62°$$

$$\frac{3.8}{\cos 62°} = x$$

$$\therefore x = 8.09 \text{ (to 2 decimal places)}$$

5 When the angle of elevation of the sun is 63°14′ a tree casts a shadow 8.6 m long. Calculate the height of the tree correct to three significant figures.

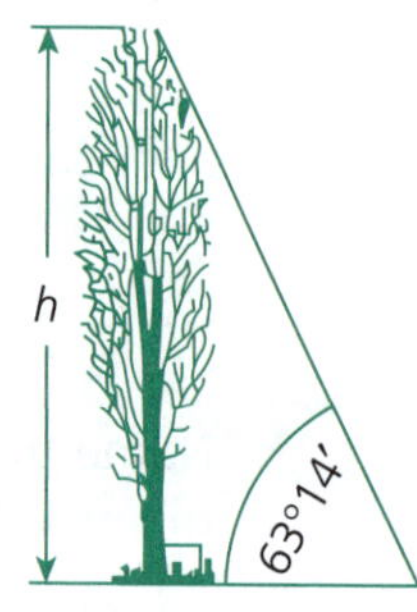

S $\dfrac{h}{8.6} = \tan 63°14'$

$$\therefore h = 8.6 \times \tan 63°14'$$
$$= 17.049746$$
$$= 17.0 \text{ (to 3 significant figures)}$$

The height of the tree is 17.0 metres.

Angle given is that between the sun's rays and the horizontal.

6 A 3-metre ladder leans against a brick wall and is inclined at 28°35′ to the wall. How far up the wall will the ladder reach (correct to one decimal place)?

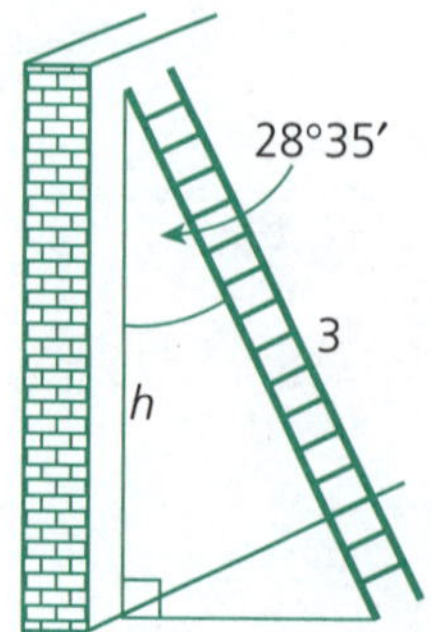

S $\dfrac{h}{3} = \cos 28°35'$

$$\therefore h = 3 \times \cos 28°35'$$
$$= 2.634366$$
$$= 2.6 \text{ (to 1 decimal place)}$$

The ladder reaches 2.6 metres up the wall.

7 Find the hypotenuse y.

S

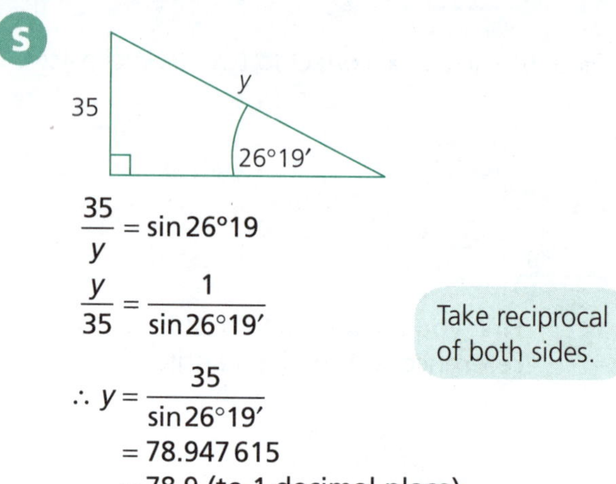

$$\frac{35}{y} = \sin 26°19$$

$$\frac{y}{35} = \frac{1}{\sin 26°19'}$$

Take reciprocal of both sides.

$$\therefore y = \frac{35}{\sin 26°19'}$$
$$= 78.947615$$
$$= 78.9 \text{ (to 1 decimal place)}$$

8 A 25-metre antenna is held in position by four wires fixed 7 m from the top of the antenna.

If the wires are attached so as to make an angle of 42°18′ with the antenna, calculate the length of wire required to the nearest cm.

S

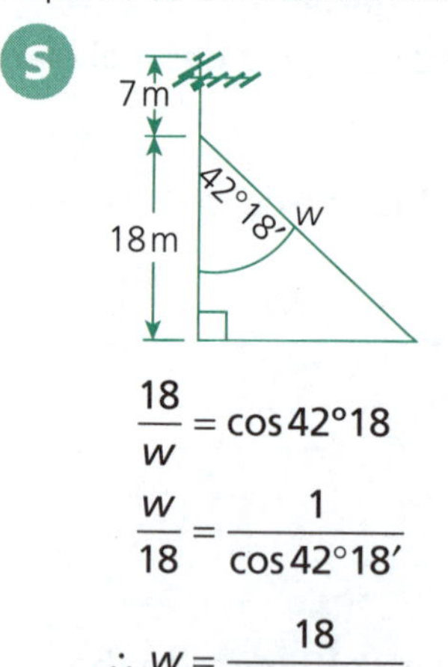

$$\frac{18}{w} = \cos 42°18$$

$$\frac{w}{18} = \frac{1}{\cos 42°18'}$$

$$\therefore w = \frac{18}{\cos 42°18'}$$
$$= 24.336457$$

Do not round until you have finished.

$$\text{Length of wire} = 24.336457 \times 4$$
$$= 97.345826$$
$$= 97.35 \text{ (to the nearest cm)}$$

The length of wire needed is 97.35 metres.

Finding sizes of angles given two sides

In any right-angled triangle, given any two sides, we can use one of the three trigonometric ratios to find an acute angle of the triangle.

Remember, before using the trigonometric functions on the calculator, make sure your calculator is set on degree mode.

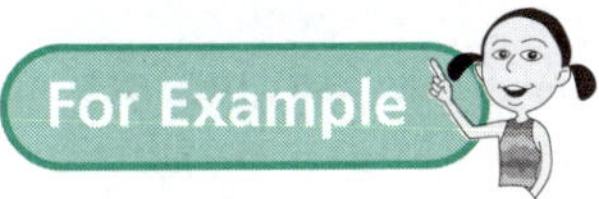

1 Use the calculator to find the value of x to the nearest degree:

a $\cos x = 0.723$

S $x \approx 44°$

b $\sin x = 0.9413$

S $x \approx 70°$

c $\tan x = \frac{3}{4}$

S $x \approx 37°$

2ndF TAN (3 ÷ 4) =

2 Find θ, to the nearest degree.

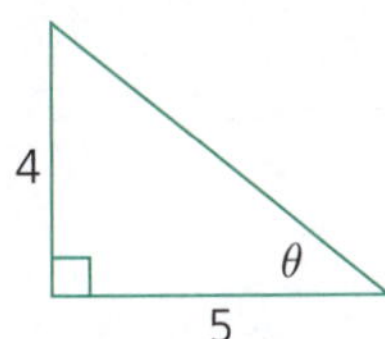

S **First name the sides with reference to the angle θ.**

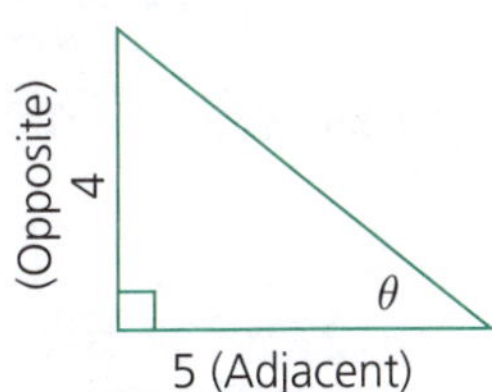

$\tan\theta = \frac{\text{Opposite}}{\text{Adjacent}}$ TOA

$= \frac{4}{5} \rightarrow \theta = 39°$

2ndF TAN (4 ÷ 5) =

(or use fraction key)

3

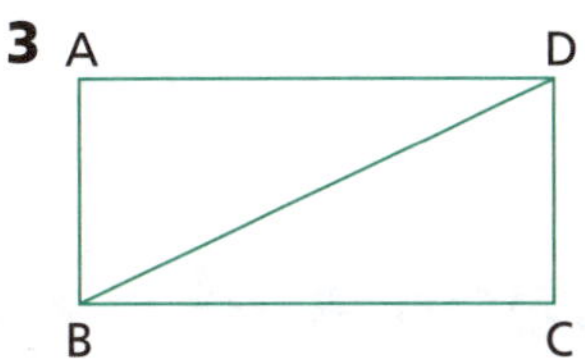

In the diagram, ABCD is a rectangle. If AB = 3 cm and BD = 7 cm, find the size of ∠DBC to the nearest minute.

S **Let ∠DBC = $x°$, then name the sides with reference to the angle $x°$.**

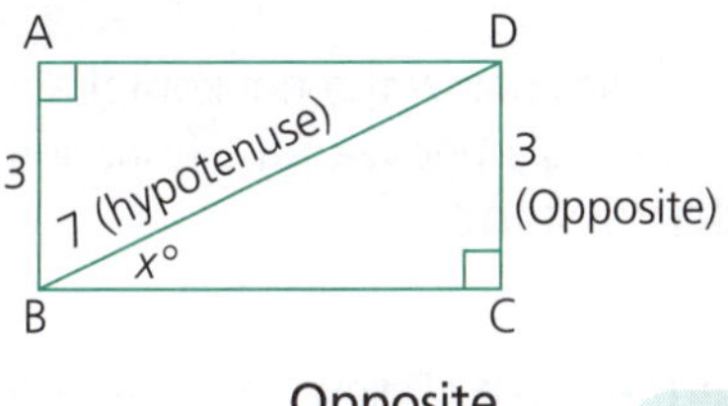

$\sin x = \frac{\text{Opposite}}{\text{Hypotenuse}}$ SOH

$= \frac{3}{7}$

$\therefore x = 25°23'$

$\therefore \angle DBC = 25°23'$

4

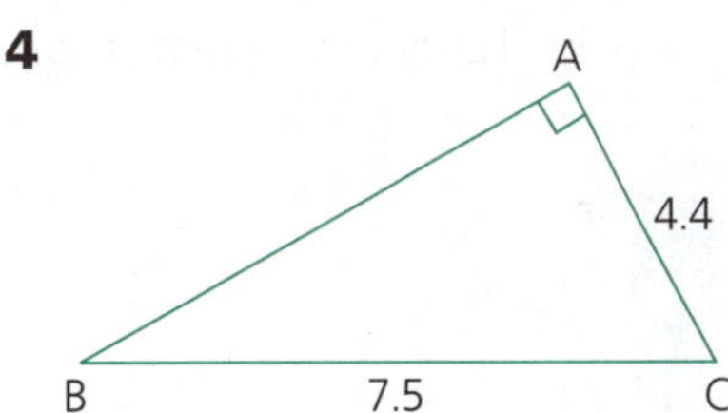

Find the size of ∠ABC and ∠ACB (to the nearest degree).

S Let $\angle ABC = \theta°$

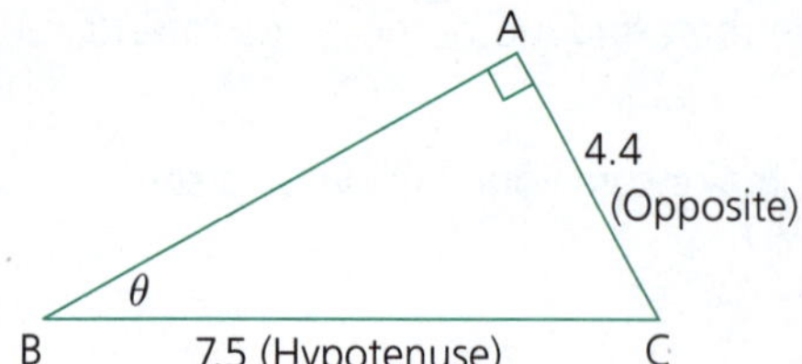

Using trigonometry:

$$\sin\theta = \frac{\text{Opposite}}{\text{Hypotenuse}} \quad \text{SOH}$$
$$= \frac{4.4}{7.5}$$

$\therefore \theta = 36°$ (nearest degree)

$\therefore \angle ABC = 36°$

To find the size of $\angle ACB$ we use the fact that the angle sum of a triangle is 180°.

$\angle ACB + 36° + 90° = 180°$

$\therefore \angle ACB = 54°$

Angles of elevation and depression

When an object P is viewed upwards from a point O on the horizontal, the angle between the horizontal and the line of sight is called the **angle of elevation**.

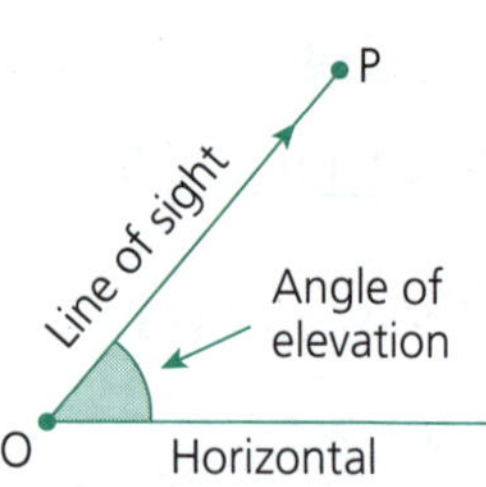

When the object P lies below the horizontal and is viewed from a point O on the horizontal, the angle between the horizontal and the line of sight is called the **angle of depression**.

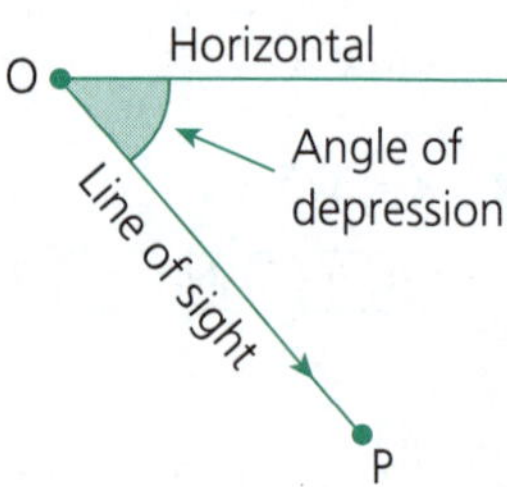

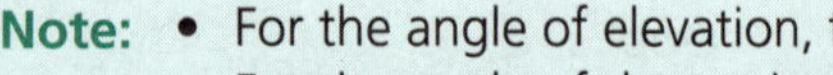

Note:
- For the angle of elevation, the observer is looking up.
- For the angle of depression, the observer is looking down.

1 Find the height of a tree that has an angle of elevation of 40°15′ at a point 12 metres from its base.

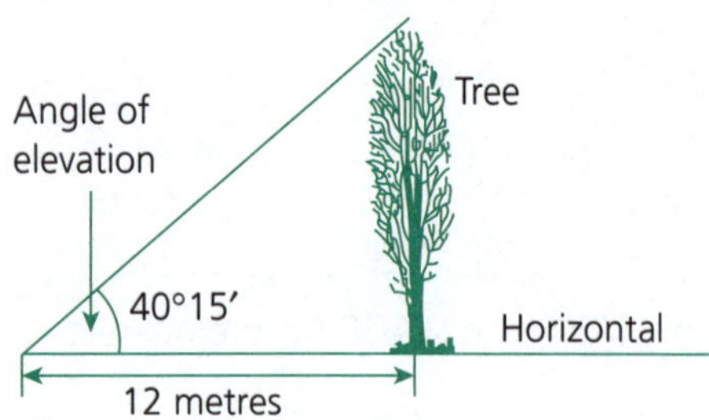

S Remember, when dealing with problems, draw a simple diagram to represent the given situation.

Let the height of the tree be h metres.

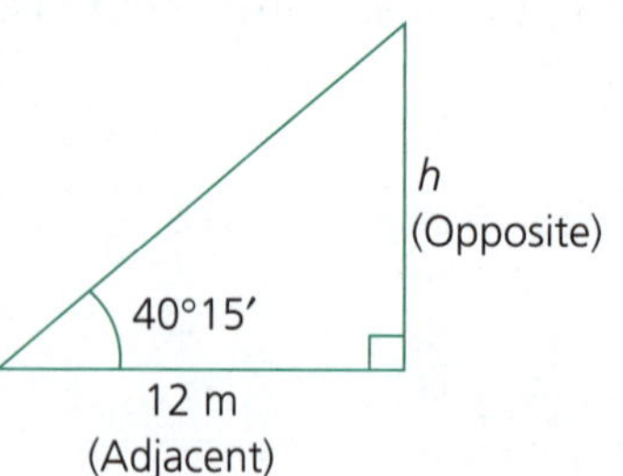

$$\frac{h}{12} = \tan 40°15' \quad \text{TOA}$$

$\therefore h = 12 \times \tan 40°15'$

$= 10.16\ldots$ (using the calculator)

Hence, the height of the tree is approximately 10 metres.

2 From the top of a cliff 50 metres above sea level, the angle of depression of a fishing boat out at sea is 32°. How far is the fishing boat from the base of the cliff? (Answer to the nearest metre.)

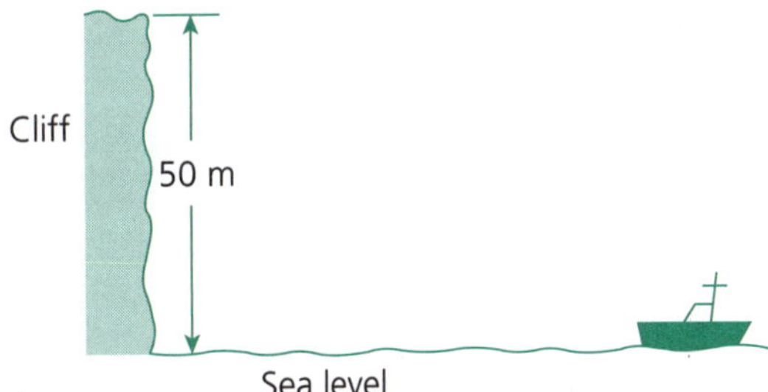

S **Draw a neat diagram.**

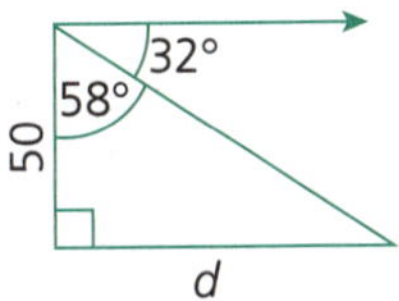

Let the distance of the fishing boat from the base of the cliff be *d* metres.

Then

$$\frac{d}{50} = \tan 58^\circ$$

TOA

$$\begin{aligned} d &= 50 \times \tan 58^\circ \\ &= 80.016726 \\ &= 80 \text{ (nearest whole)} \end{aligned}$$

The fishing boat is approximately 80 metres away from the base of the cliff.

3 A ladder of length 2.5 metres is leaning against a wall. The foot of the ladder is 1.4 metres from the base of the wall. Find the angle between the ladder and the wall. (Answer to the nearest minute.)

S **Draw a neat sketch to represent the situation. Let the angle between the ladder and the wall be θ°.**

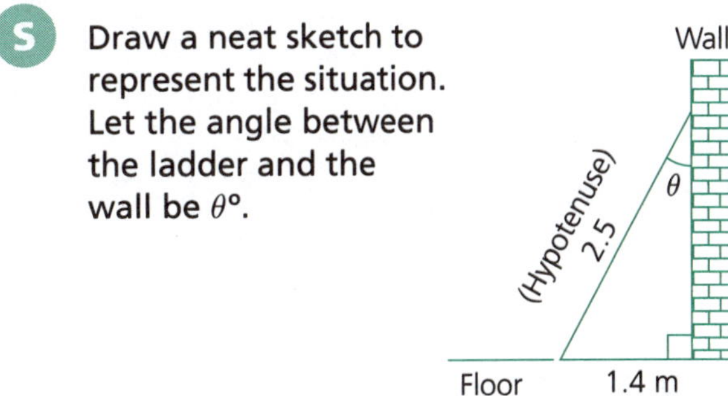

$$\frac{\text{Opposite}}{\text{Hypotenuse}} = \sin\theta$$

TOA

$$\sin\theta = \frac{1.4}{2.5}$$

$$\therefore \theta = 34°3'$$

Therefore the angle between the wall and the ladder is approximately 34°3′.

Bearings

Bearings are measured **clockwise** from due **north** and are written as three-digit numbers.

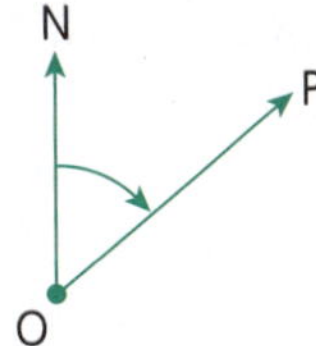

In the figure, the bearing of P from O is 047°. This means that ∠NOP = 47°.

In the following examples find the bearing of P from O.

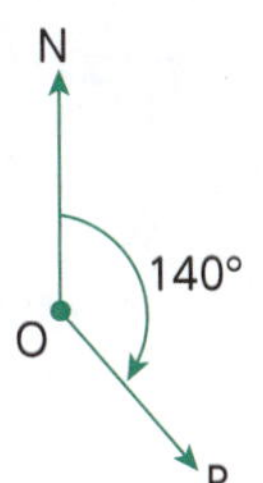

The bearing of P from O is 140°.

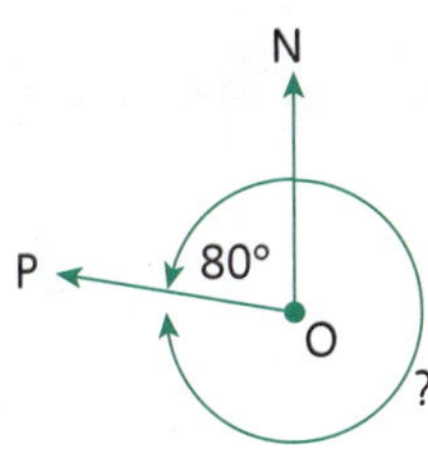

360° – 80° = 280°

The bearing of P from O is 280°.

Also:

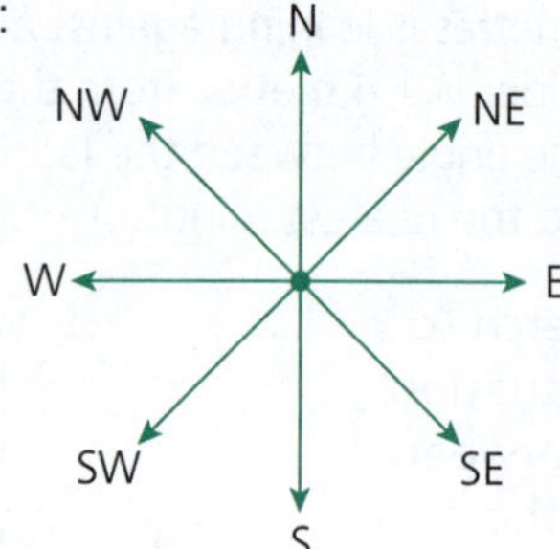

The bearing of due E is 090°.
The bearing of due S is 180°.
The bearing of due W is 270°.
The bearing of NE is 045°.
The bearing of SE is 135°.
The bearing of SW is 225°.
The bearing of NW is 315°.

The angle between W and N is 90°, between W and NW is 45°, SW and NW is 90°, and so on.

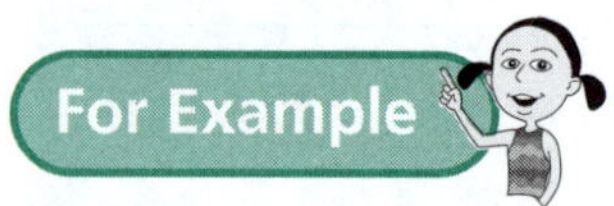

1 Draw a diagram to illustrate a bearing of:

a 150°

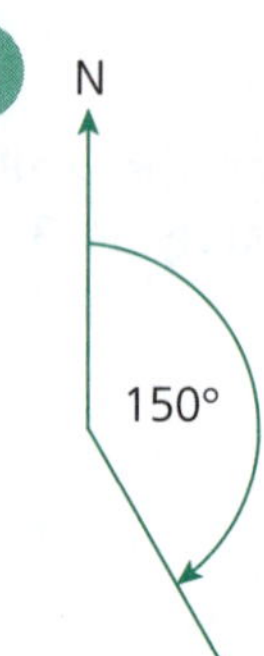

b 260°

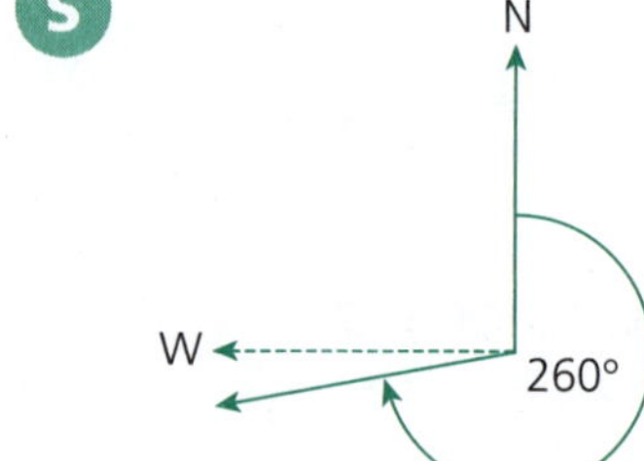

2 Write down the bearing illustrated in the following diagrams:

a

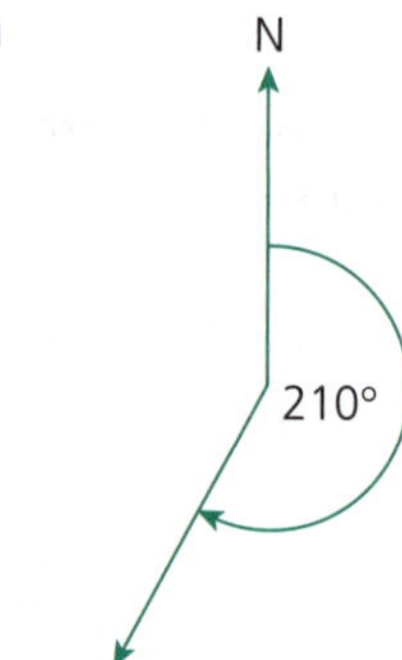

S 210°

b

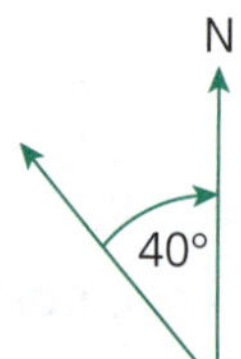

S Bearing = 360° – 40° = 320°

c

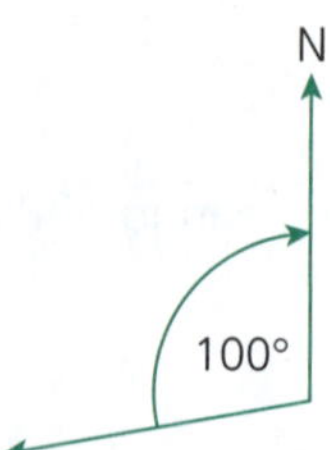

S Bearing = 360° – 100° = 260°

d

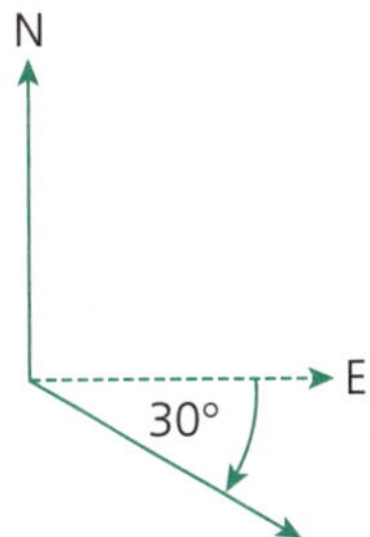

S Bearing = 90° + 30° = 120°

3 A ferry leaves Minmi Breakwater and steams 12 km on a bearing of 114° to a point S. Find the distance that the ferry is south of Minmi.

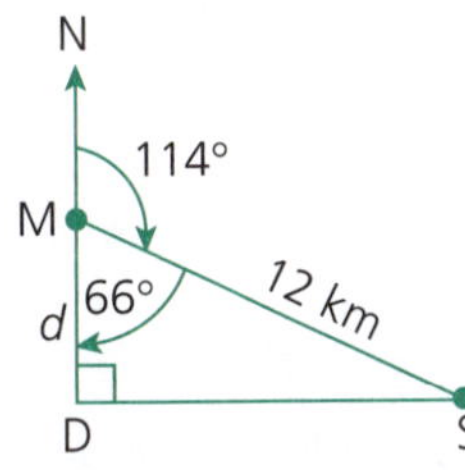

S The distance required is MD; call it d.

$\angle NMS = 114°$ (bearing)

$\therefore \angle DMS = 180° - 114°$ ($\angle$ sum of line)
$= 66°$

Then $\dfrac{d}{12} = \cos 66°$

$\therefore d = 12 \times \cos 66°$
$= 4.880\,8397$
$= 4.9$ (to 1 decimal place)

The ferry is 4.9 km south of Minmi.

4 Little Red Riding Hood walked 2.4 km due south from her home (H) to an old tree (T) and then due west for 0.9 km. Calculate Little Red Riding Hood's bearing from home (to the nearest degree).

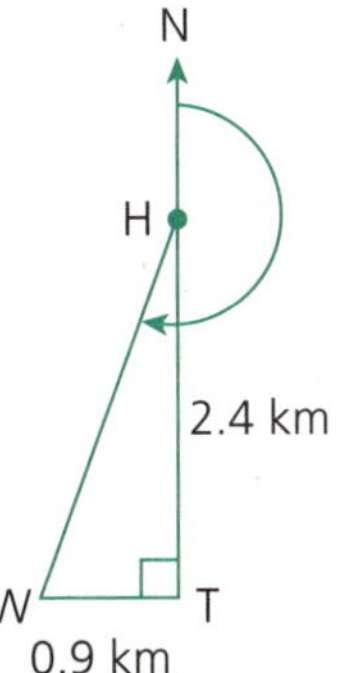

S The angle required is $\angle NHW$. This angle represents the bearing of W from H. But we must first find $\angle WHT$. Call it $\theta°$.

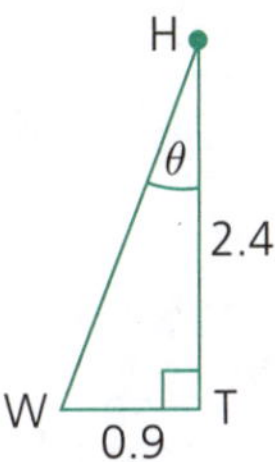

From $\triangle HTW$,

$$\tan\theta = \frac{0.9}{2.4} = 0.375$$

$\therefore \theta = 21°$ (to nearest degree)

$\therefore \angle NHW = 21° + 180°$
$= 201°$

The bearing from home is 201° (to the nearest degree).

Practise Practise

Pythagoras' Theorem

1 Calculate the size of the marked side:

a

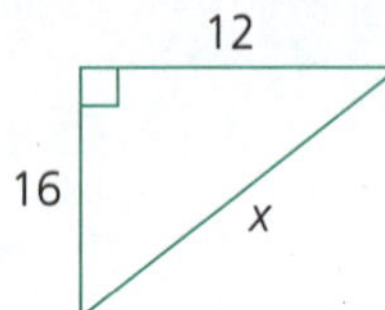

b

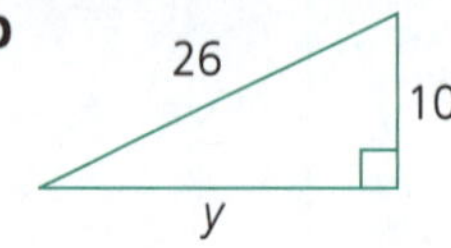

c

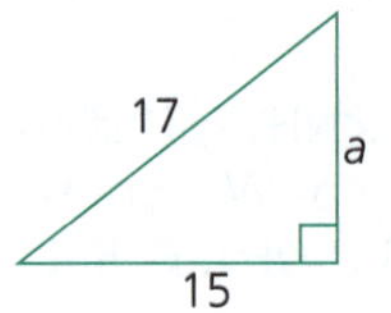

d

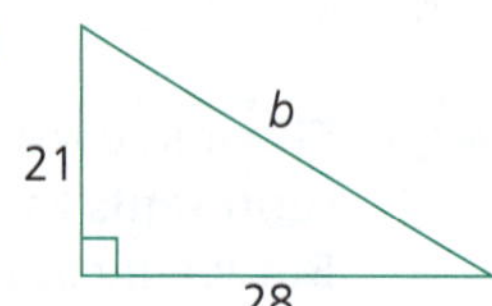

e

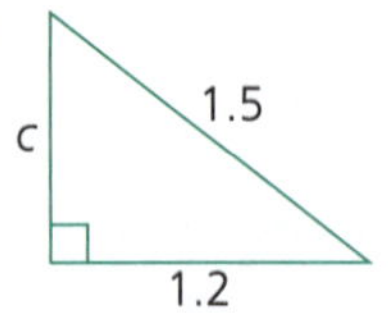

f

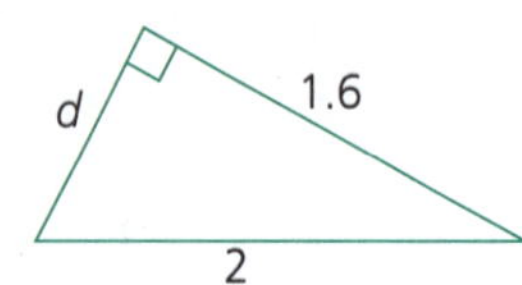

g

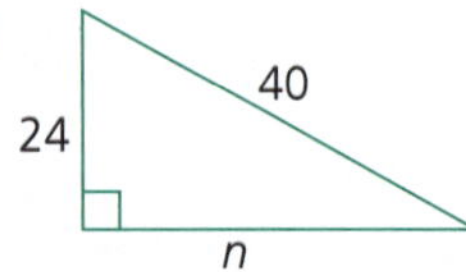

h

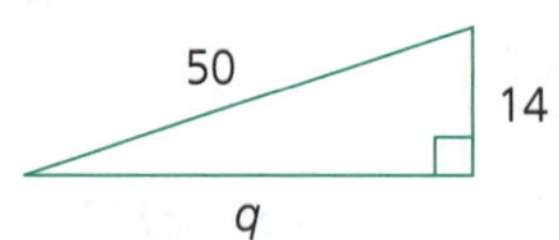

i

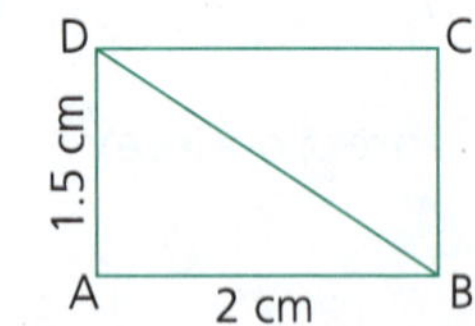

ABCD is a rectangle. Find length of BD.

j

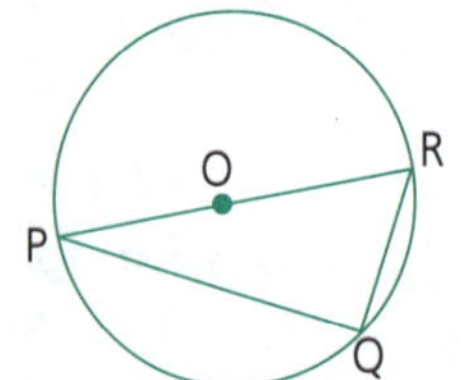

PR is a diameter of the circle centre O. $\angle PQR = 90°$. If OP is 25 cm and QR is 30 cm, calculate the length of PQ.

pp. 94–97

2 Decide which of the following triangles are right-angled. Give reasons:

a

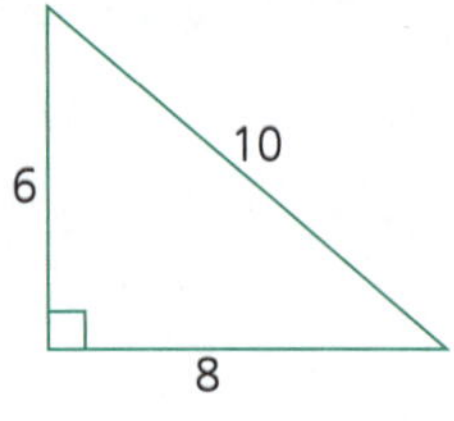

b

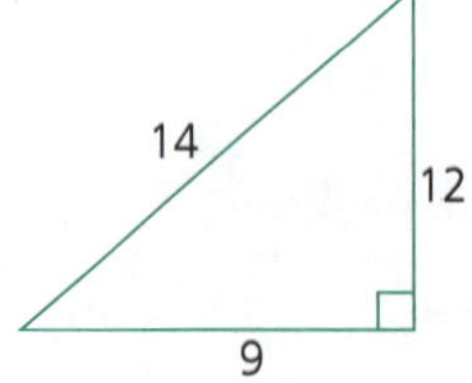

c

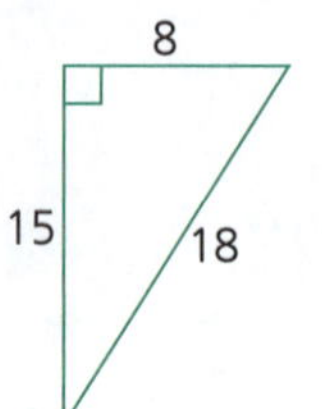

d

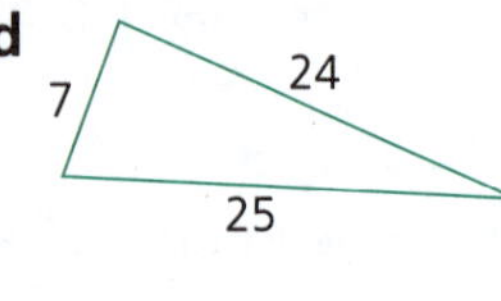

e

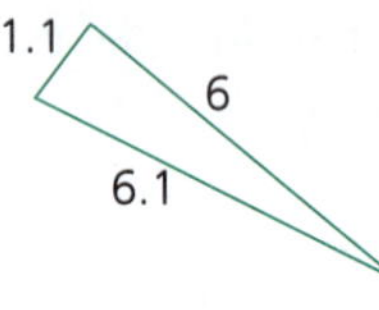

f

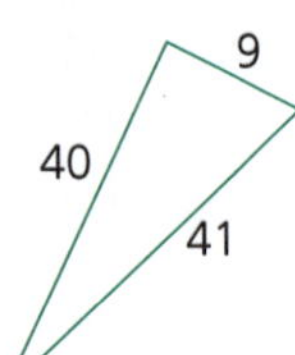

g

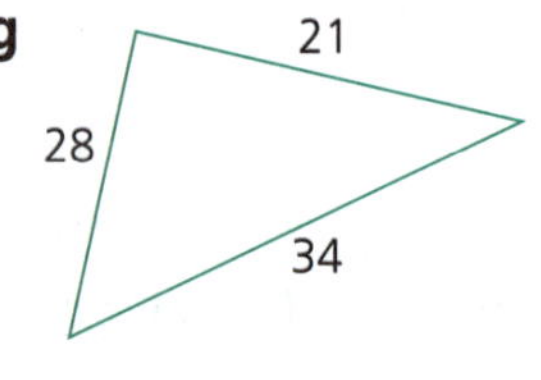

h

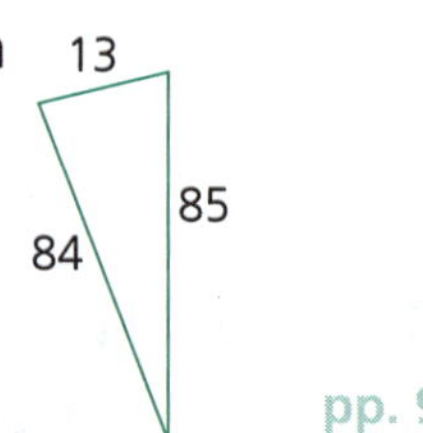

pp. 97–98

3 Complete the Pythagorean triads:

a $\{7, 24, a\}$
b $\{9, 12, y\}$
c $\{18, n, 30\}$
d $\{9, t, 41\}$
e $\{13, n, 85\}$
f $\{21, 72, p\}$

pp. 98–99

4 Find the value of the pronumeral (correct to one decimal place):

a

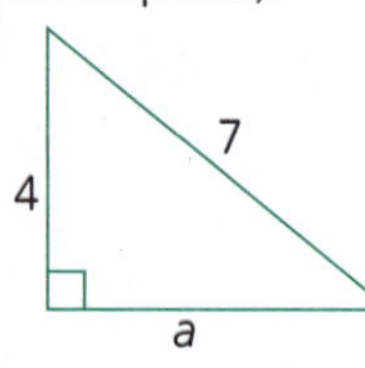

b

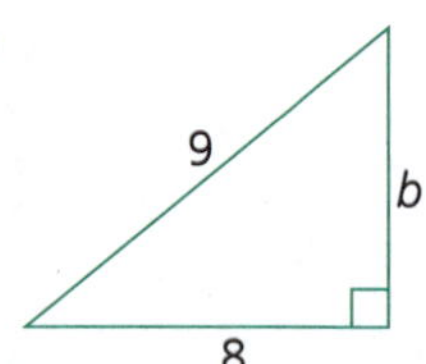

c

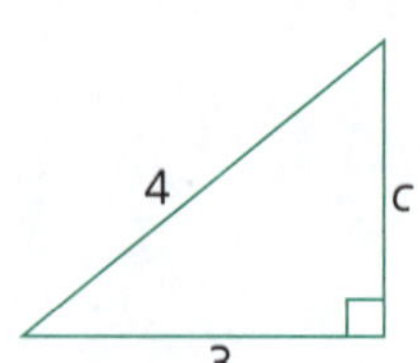

d

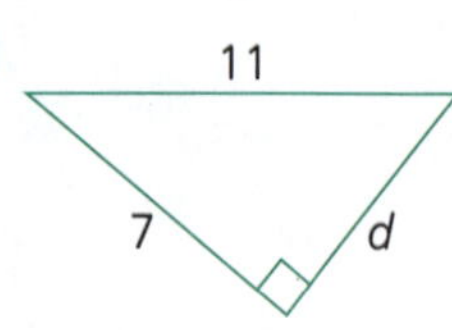

e

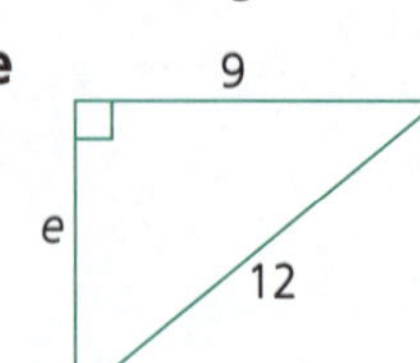

f

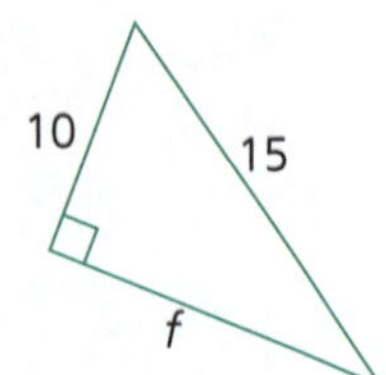

g

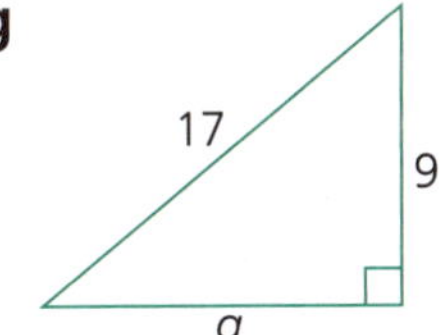

h

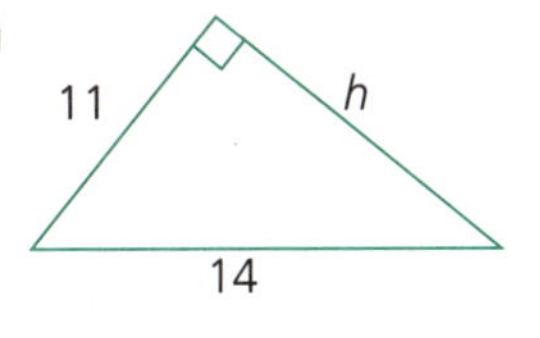

pp. 99–100

5 Calculate the exact length of the marked side (correct to one decimal place):

a

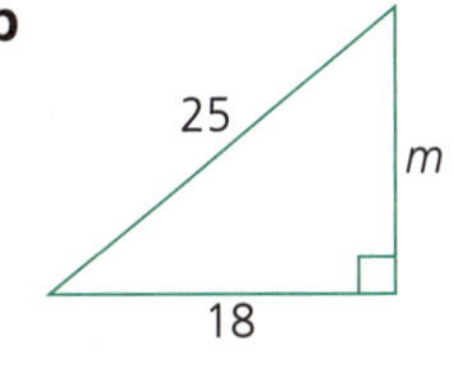

b

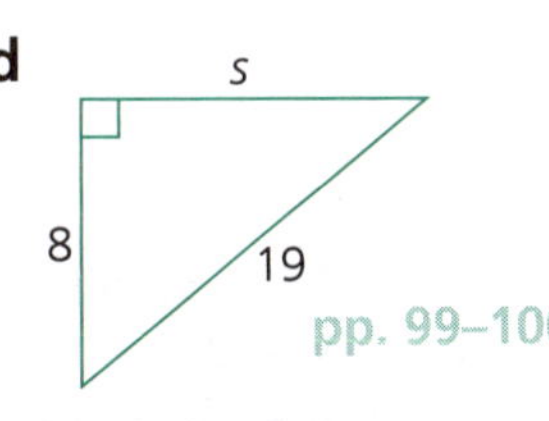

c

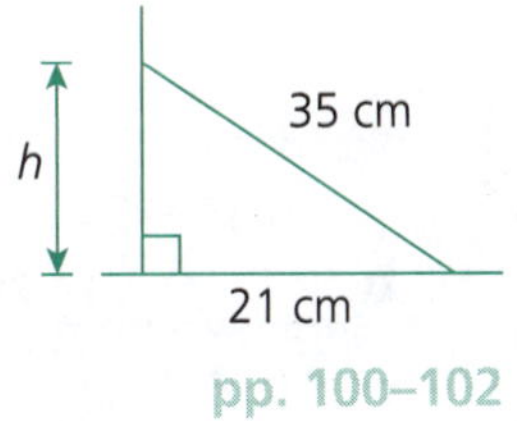

d

pp. 99–100

6 A rectangle is 16 cm long and 12 cm wide. Calculate the length of a diagonal of the rectangle. pp. 100–102

7 Calculate the length of a diagonal of a square with sides of 3 cm. Answer correct to one decimal place. pp. 100–102

8 A metal brace 35 cm long is screwed into place so that one end is 21 cm from a wall. How far up the wall does the brace reach.

pp. 100–102

9 An 8-metre ladder reaches 7.5 metres up a vertical brick wall. How far is the foot of the ladder from the wall? Answer correct to one decimal place.

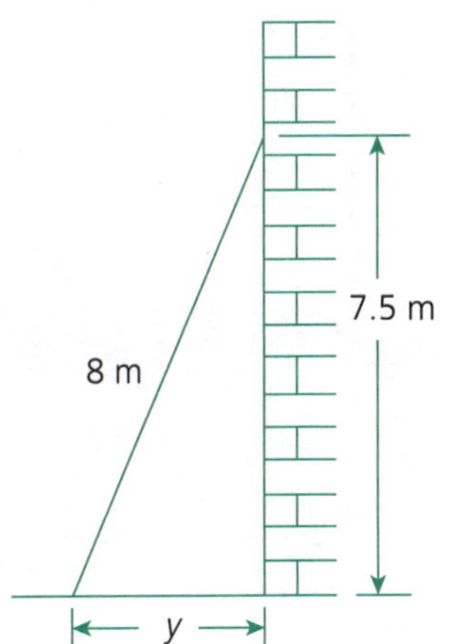

pp. 100–102

10 A pollution tower 45 metres high is supported by wire 40 metres long. Each wire is attached at ground level 21 metres from the base of the tower. How far from the top of the tower are the wires attached? Answer correct to four significant figures.

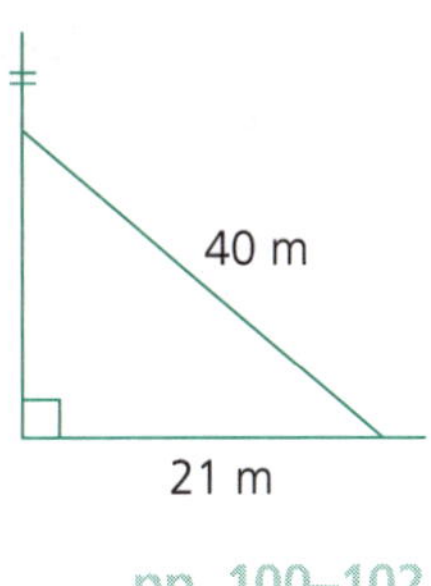

pp. 100–102

11 The square MNOP has diagonal of length 18 cm. Calculate the length of a side of the square. Answer correct to one decimal place.

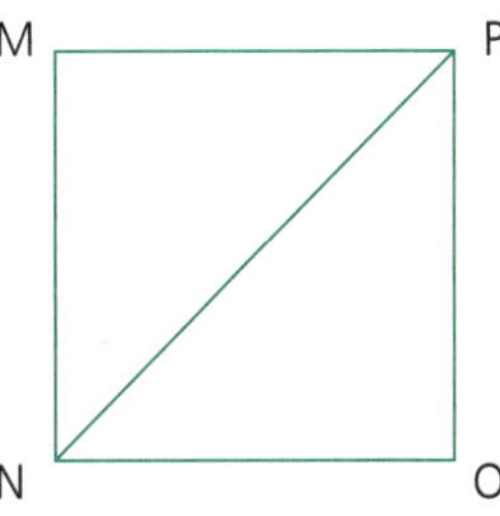

pp. 100–102

12 A rectangle CDEF has length 20 cm and diagonal 25 cm long. Calculate the width of the rectangle.

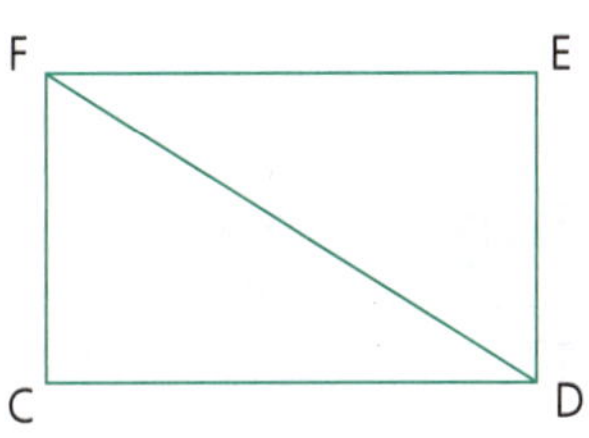

pp. 100–102

13 LMNO is a rhombus with diagonals 24 cm and 10 cm. Calculate the length of each side of the rhombus.

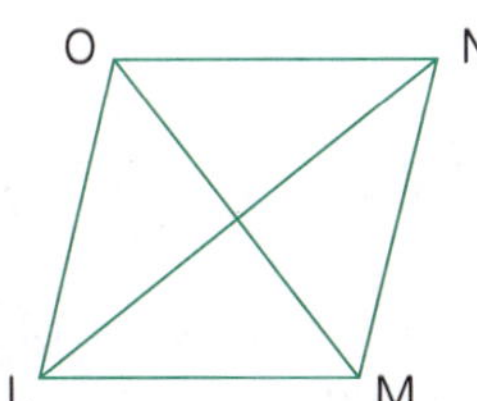

pp. 100–102

Note: the diagonals of a rhombus bisect each other at right angles. All sides of a rhombus are of equal length.

14 Calculate the value of x and hence calculate the length y. Give your answer for y correct to three significant figures.

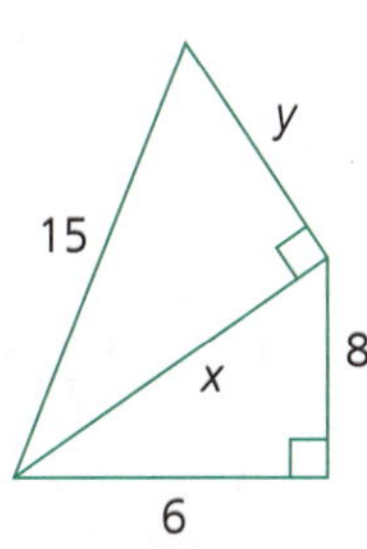

pp. 100–102

15 A sail is to be cut in the shape of a right-angled triangle. If the longest side is 25 metres and the shortest is 7 metres, calculate the length of the third side.

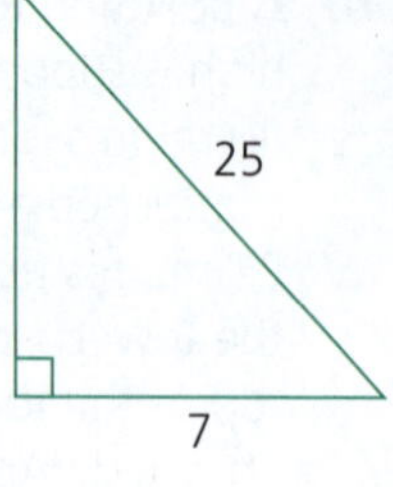

pp. 100–102

16 Calculate the perimeter of the triangle PQR.

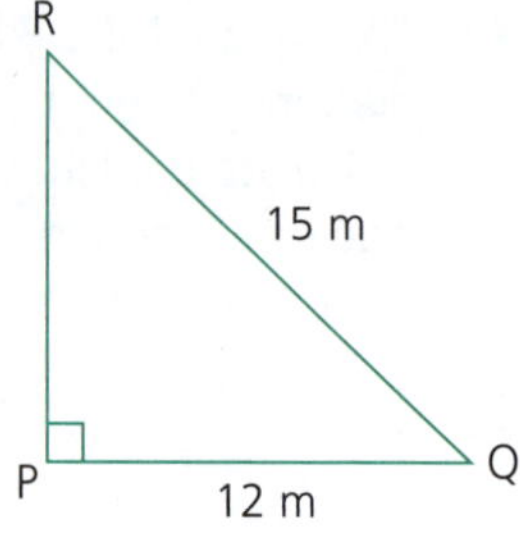

pp. 100–102

Trigonometry

17

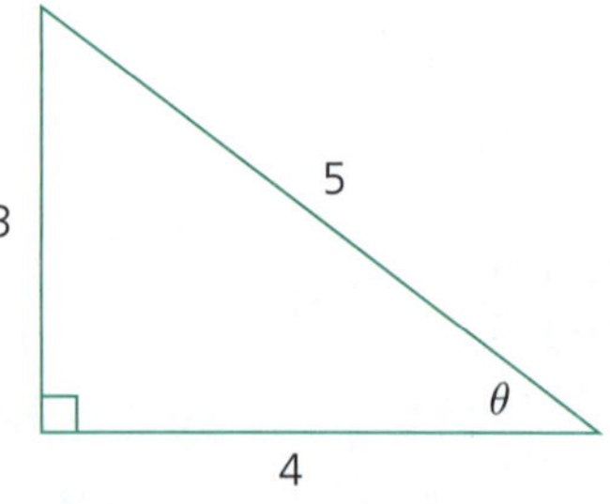

What is the value of:

a $\sin\theta$?
b $\cos\theta$?
c $\tan\theta$?

pp. 102–103

18

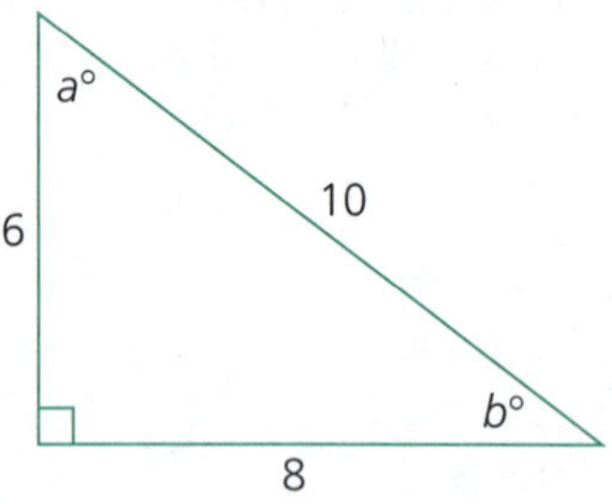

Find the value of:

a $\sin a°$?
b $\sin b°$?
c $\cos a°$?

pp. 102–103

19

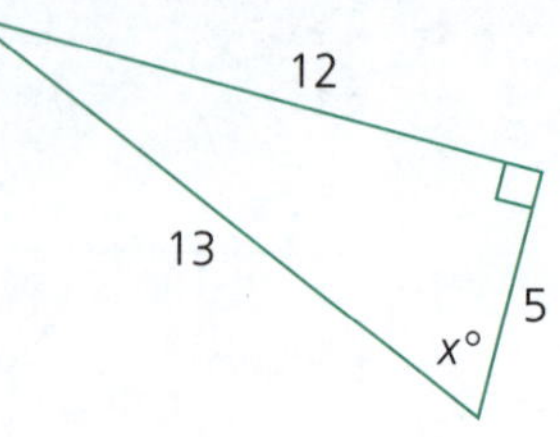

Find the value of $\tan x$.

pp. 102–103

20 Use your calculator to write, correct to four decimal places:

a cos 16°48′
b sin 29°33′
c tan 78°15′

p. 104

21 Simplify:

a 180° – 79°16′
b 90° – 38°14′
c 60°14′ + 32°58′
d 180° – (36°29′ + 39°46′)
e 180° + 67°24′

p. 104

22 Determine the length of the side marked x in each of the following triangles (answers correct to two decimal places):

a

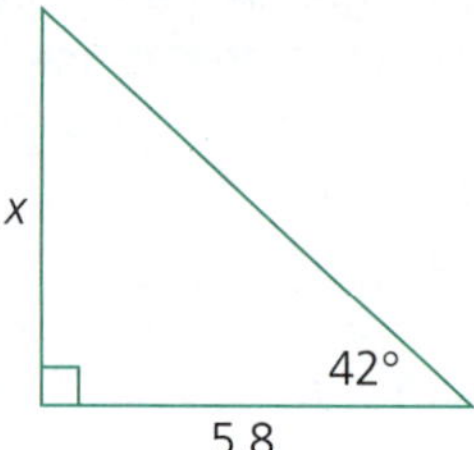

b

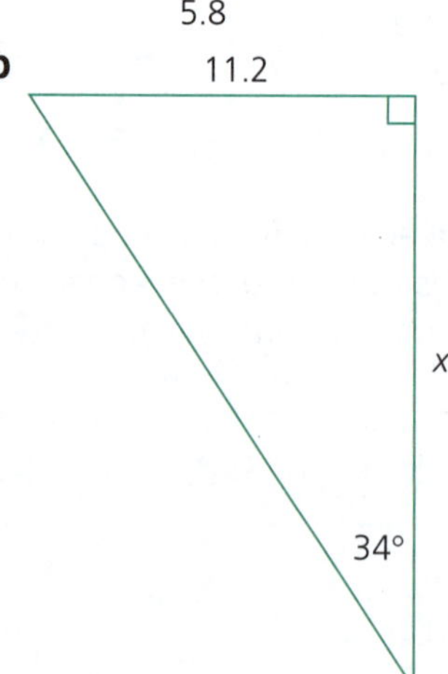

c

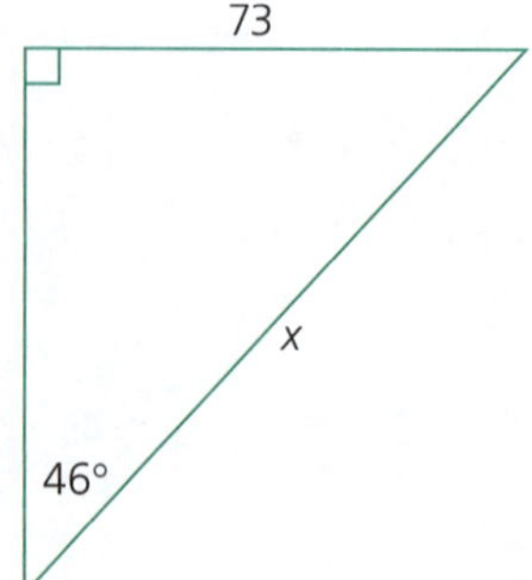

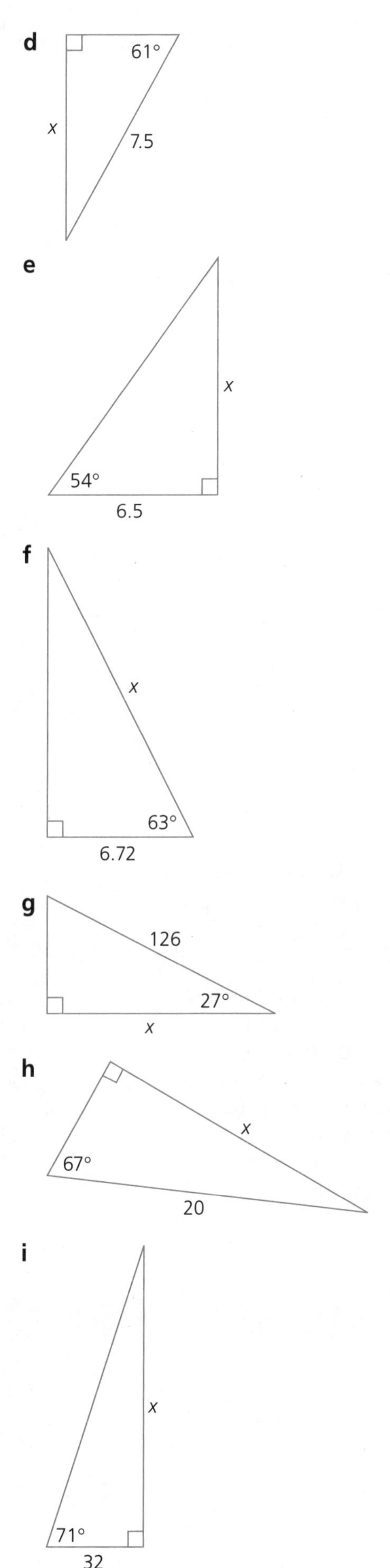

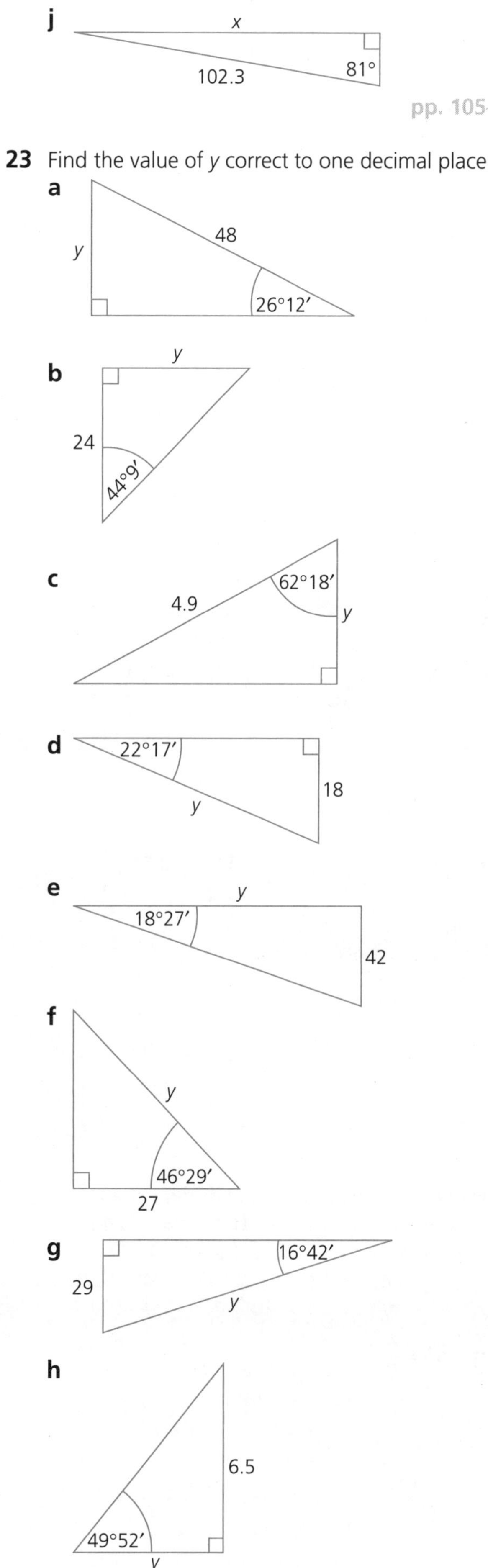

pp. 105–106

23 Find the value of *y* correct to one decimal place:

i

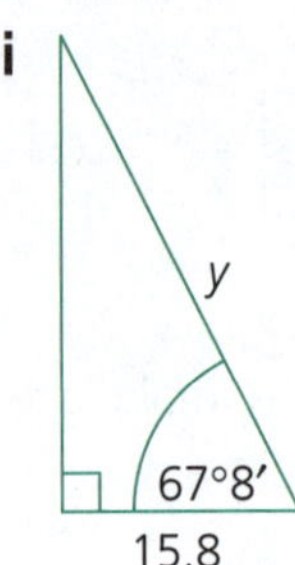

pp. 105–106

24 Find the size of the angle marked x in each of the following triangles (answers to the nearest degree):

a

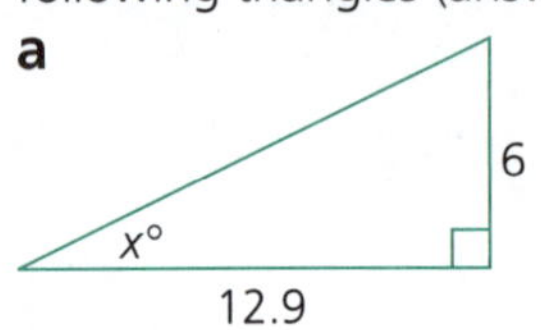

b

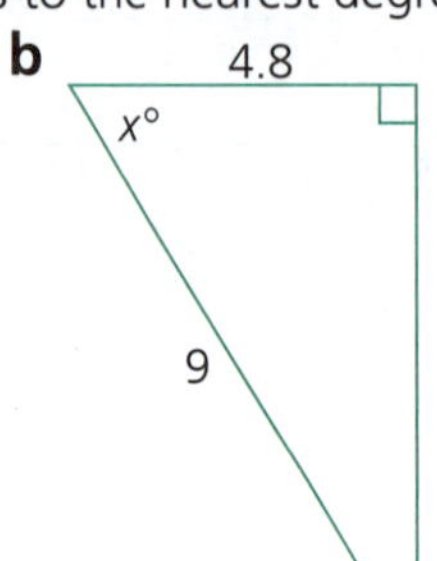

c

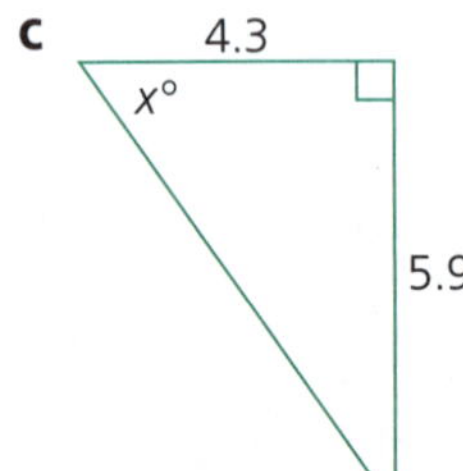

d

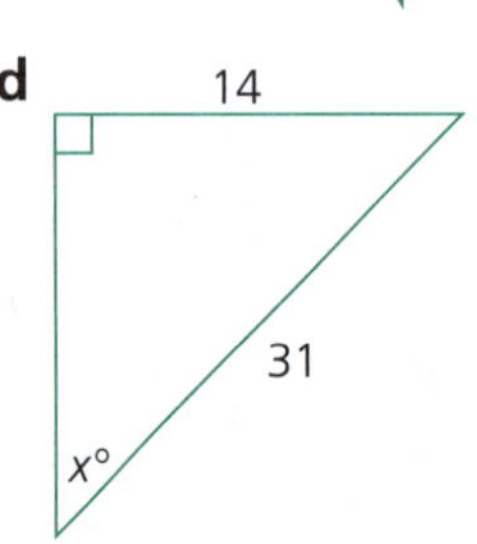

e

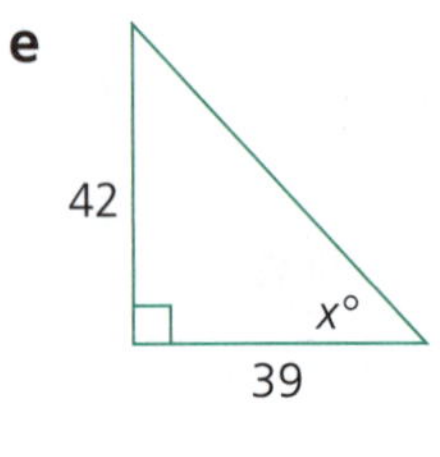

f

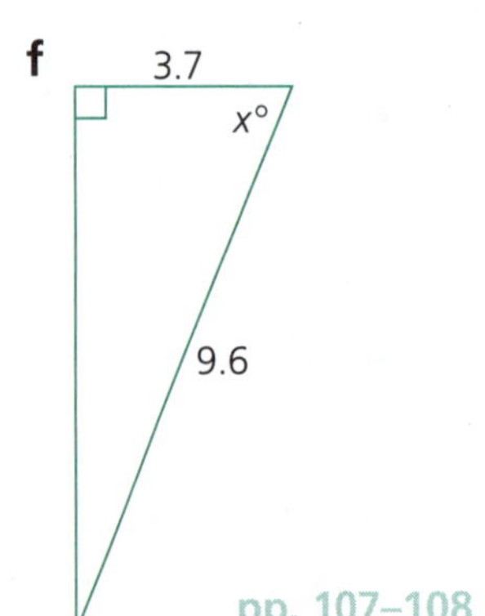

pp. 107–108

25 Find θ to the nearest minute if $0° \leq \theta \leq 90°$:

a $\sin\theta = 0.8127$ **b** $\cos\theta = 0.2243$

c $\tan\theta = 1.8$ **d** $\sin\theta = 0.002$

e $\tan\theta = \dfrac{4}{7}$ **f** $\cos\theta = \dfrac{4}{7}$

g $\sin\theta = \dfrac{4}{7}$

pp. 107–108

26 Find the size of the marked angle, correct to the nearest minute:

a

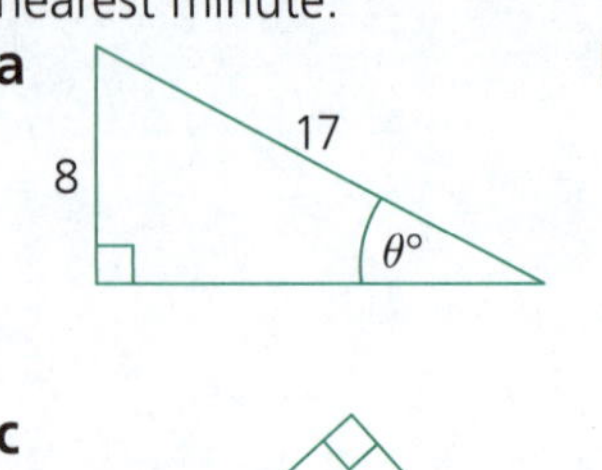

b

c **d** **e**

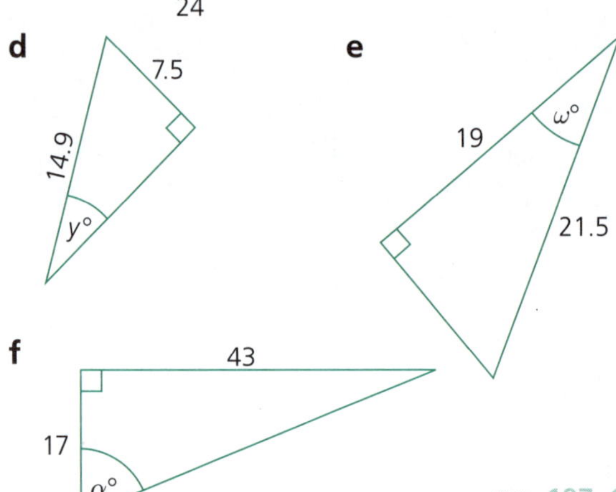

f

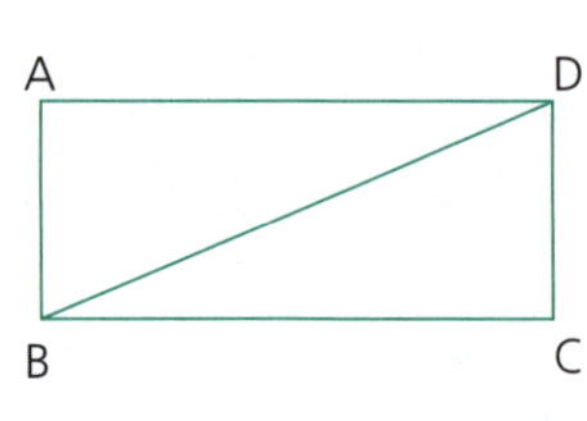

pp. 107–108

27 In the diagram, ABCD is a rectangle. If AB = 5 cm and AD = 13 cm, find the size of ∠ABD to the nearest degree.

A D B C

pp. 107–108

28 A ladder of length 2.3 metres is leaning against a wall. The foot of the ladder is 1.2 metres from the base of the wall. Find the angle between the ladder and the wall. (Answer to the nearest degree.)

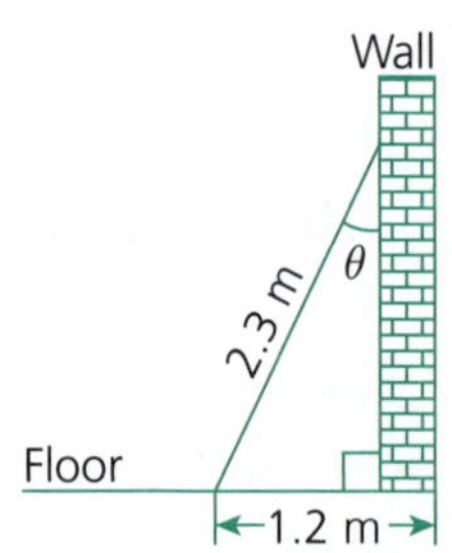

pp. 107–108

29 O is the centre of the circle. AB is a diameter. The radius of the circle is 6 cm and ∠ABC = 28°. Find the length of side BC, correct to one decimal place.

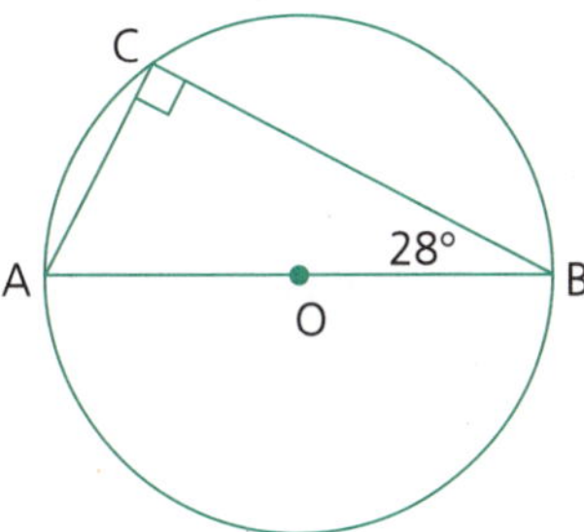

pp. 107–108

30 From the top of a tower 72 metres high, the angle of depression of a woman is 68°. How far is the woman from the base of the tower? (Answer to the nearest metre.)

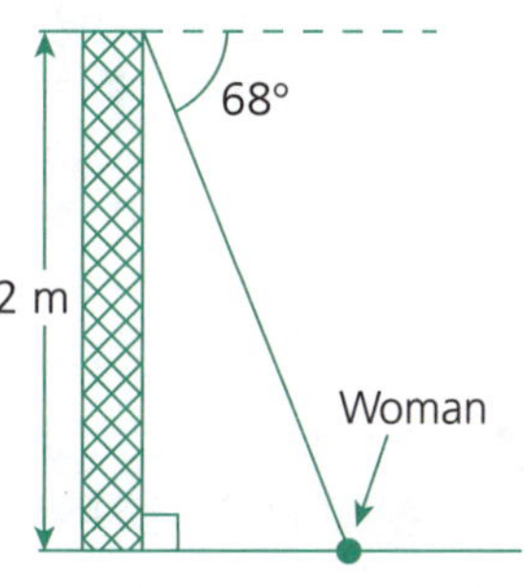

pp. 108–109

31 Find the height of a building that has an angle of elevation of 32° at a point 40 metres from its base. (Answer correct to the nearest metre.)

pp. 108–109

32 Find the height *h* of a flagpole that has an angle of elevation of 47° at a point 8 metres from the base of the pole. (Answer correct to two decimal places.)

pp. 108–109

33 A metal ladder rests against a wall and the ladder makes an angle of 62° with the ground. If the foot of the ladder is 2.5 metres from the wall, find the height that the ladder reaches up the wall. (Answer correct to one decimal place.)

pp. 108–109

34 Captain Hook, from the bow of his ship, notes that the angle of depression of a crocodile is 18°20′. If the bow of the ship is 3.2 m above water level, calculate the distance of the crocodile from the ship (correct to one decimal place).

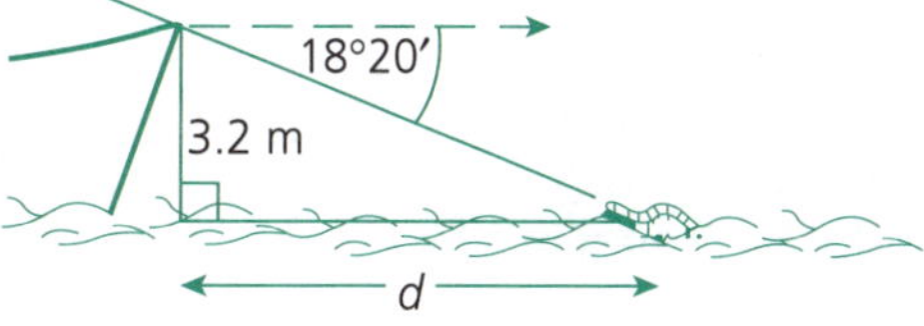

pp. 108–109

35 Marissa measures the angle of elevation of a 35 m high building from a position 100 m from the base of the building. Calculate the angle of elevation, correct to the nearest minute.

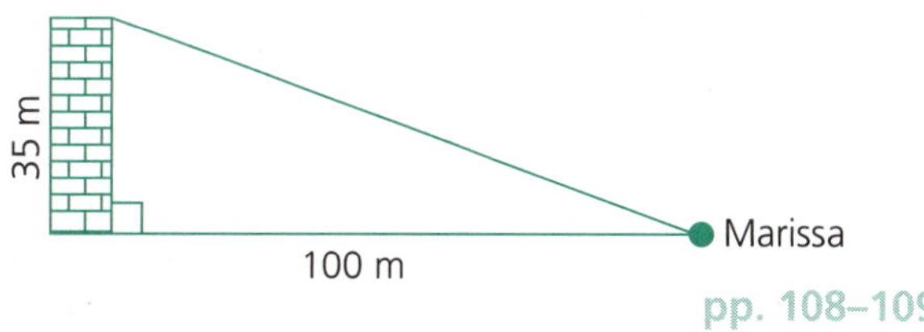

pp. 108–109

36 Two sails on a yacht are cut so that AD = 2.8 m, ∠DAB = 48°9′, BE = 0.9 m, and DC = 3.7 m.

Calculate the length BD from △ADB and hence calculate ∠DCE to the nearest minute.

pp. 108–109

37 The rugby league goalposts at a stadium are 3.5 m high. Peter observed that the shadow cast by the posts was 5 m long. Calculate the angle of elevation of the sun (to the nearest degree).

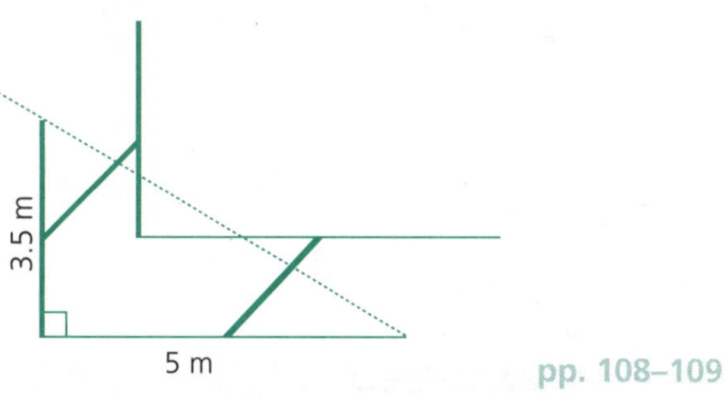

pp. 108–109

38 Marko and Greg set up garbage cans 4 m apart for soccer posts. If Marko is directly in line with one post, for what range of angles will the soccer ball pass between the posts if he is:
a 10 m **b** 15 m
from the nearest post?
(Answer to the nearest minute.)

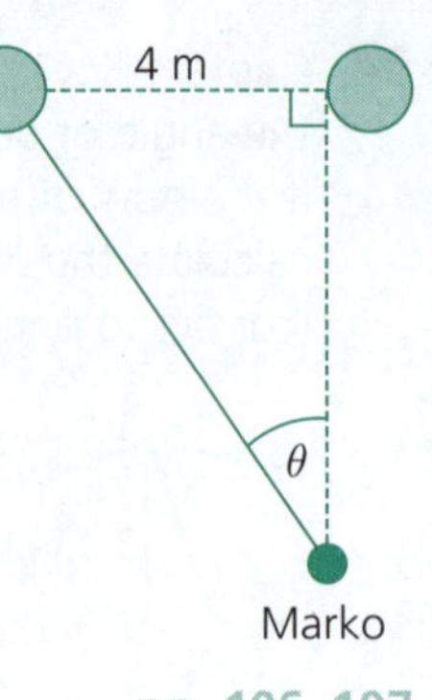

pp. 106–107

39 In the following diagrams, write down the bearing of A from B:

a

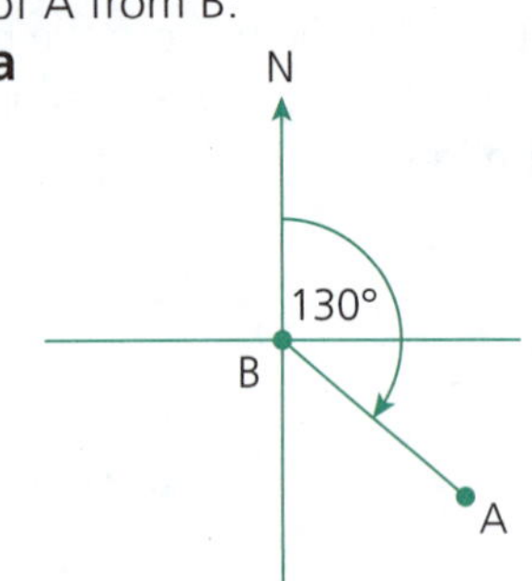

b

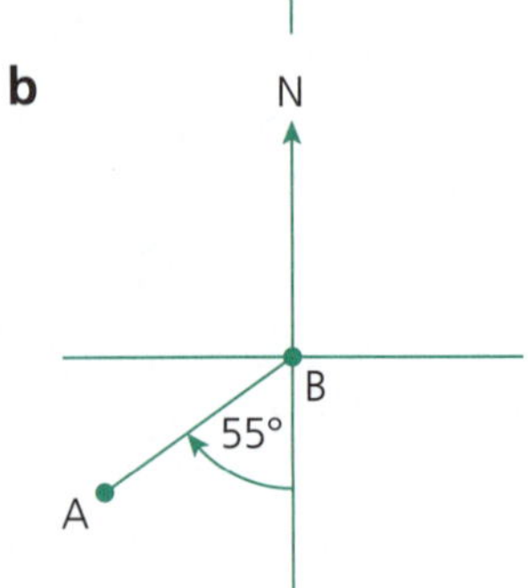

c

d

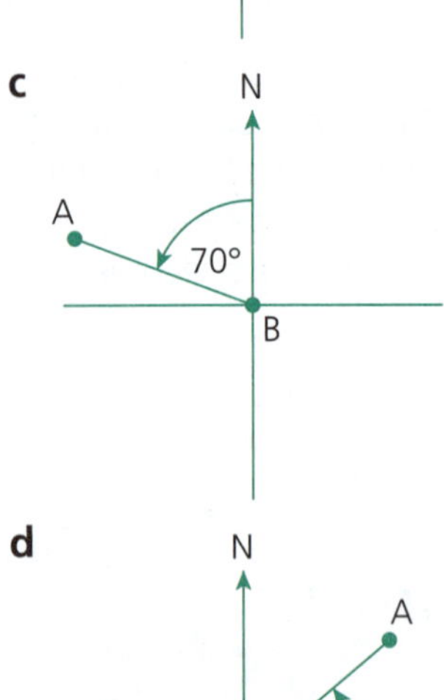

e

f

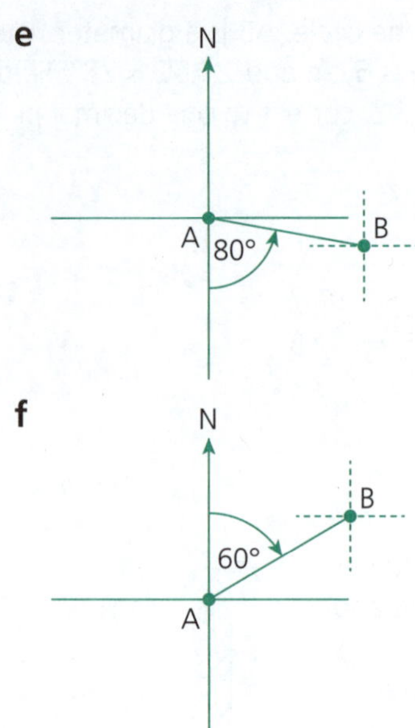

pp. 109–111

40 A bushwalker walks 8 km east and then 12 km south. Find his bearing from his starting point.
pp. 109–111

41 A helicopter travels 20 km on a bearing of 310° from X. How far north of X is the helicopter? Give answer correct to two decimal places.
pp. 109–111

42 A car travels for 2 hours at an average speed of 80 km/h on a bearing of 260°. How far west is the car from its starting point?
pp. 109–111

43 An aircraft travels on a bearing of 160° until it is 300 km south of its starting point. How far has the plane travelled?
pp. 109–111

44 A plane leaves point A and flies on a bearing of 220° for 200 km. How far south of A is the plane? (Answer to nearest km.)
pp. 109–111

45 A large black bear ambles 8 km south and then 14 km east. She then calculates the bearing of the starting point from her new position. Find this bearing to the nearest degree. (See diagram.)

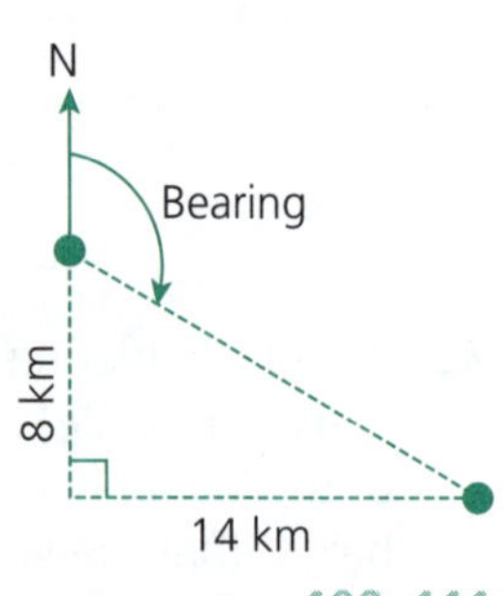

pp. 109–111

Go to p. 219 for **Quick Answers** or to pp. 242–248 for **Worked Solutions**

For a complete understanding of this topic, you must be able to:

✓	Locate the hypotenuse in a right-angled triangle		pp. 94–95
✓	State Pythagoras' Theorem for any right-angled triangle		pp. 94–95
✓	Use Pythagoras' Theorem to calculate the length of the hypotenuse of any right-angled triangle		pp. 95–96
✓	Use Pythagoras' Theorem to calculate the length of a shorter side of any right-angled triangle		pp. 96–97
✓	Use Pythagoras' Theorem to prove that a triangle is right-angled		pp. 97–98
✓	Use Pythagoras' Theorem to show that a set of three numbers is a Pythagorean triad		pp. 98–99
✓	Understand the difference between an exact answer and an approximation		pp. 99–100
✓	Apply Pythagoras' Theorem to problem solving involving right-angled triangles		pp. 100–101
✓	Identify the sides of a right-angled triangle		p. 102
✓	Define the ratios sine, cosine and tangent		pp. 102–103
✓	Use your calculator to find an angle in either degrees, or degrees and minutes, given the value of the ratio		p. 104
✓	Use trigonometry to find the values of both sides and angles of right-angled triangles		pp. 105–108
✓	Solve problems using trigonometry, including those involving angles of elevation and depression, with or without a given diagram.		pp. 108–109
✓	Solve problems using trigonometry, including those involving bearings, with or without a given diagram.		pp. 109–111

Now you are ready to do the tests!

Intermediate Test

Pythagoras and Trigonometry

(45 marks)

1 State Pythagoras' Theorem for the triangle PQR:

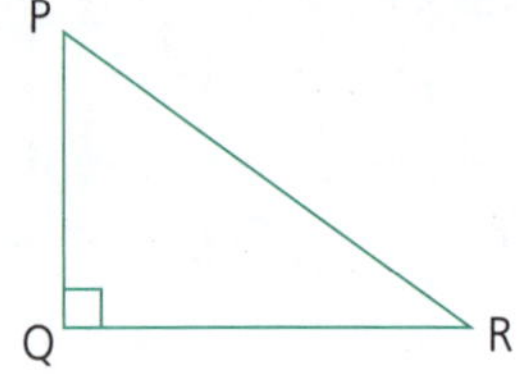

(1 mark)

2 Find the length of the hypotenuse:

a

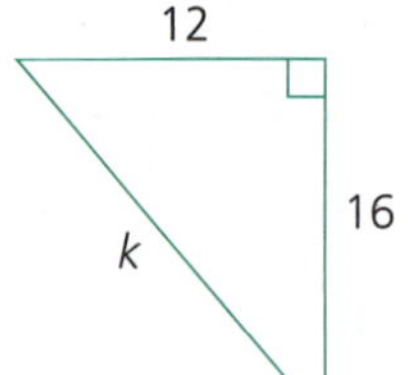

b

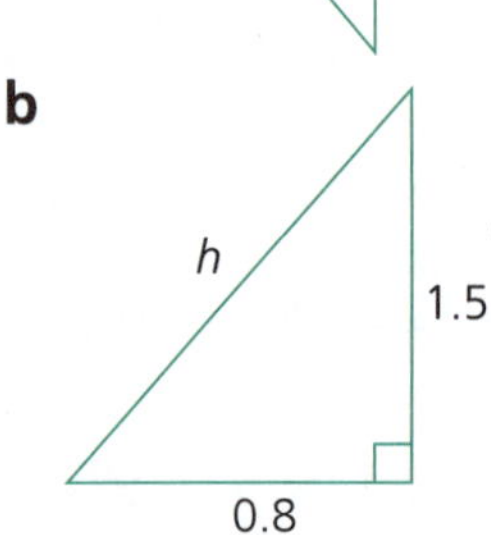

(4 marks)

3 Find the exact lengths of the marked sides:

a

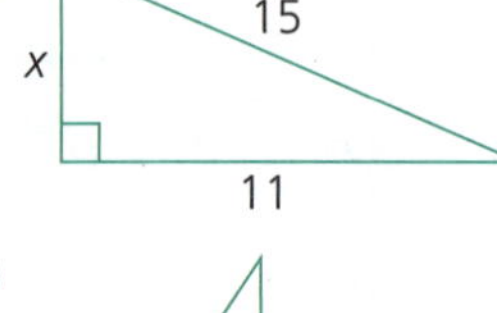

b

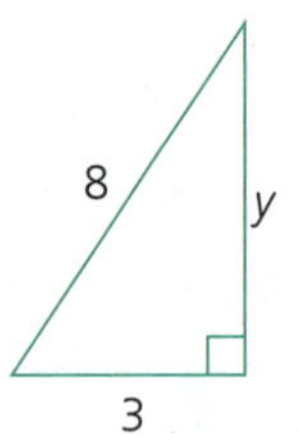

(4 marks)

4 Determine whether these triangles are right-angled:

a

8, 16, 14

b

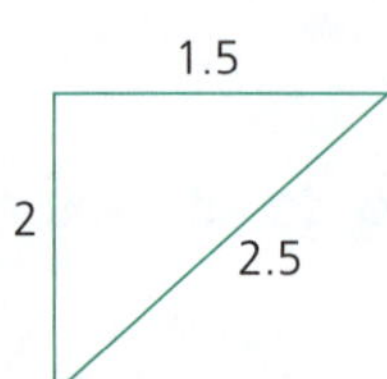

(4 marks)

5 Calculate the length of the diagonal of a square of length 6 cm. Answer correct to one decimal place. (3 marks)

6 Use your calculator to evaluate correct to three decimal places:

a $\cos 83°$ (1 mark)

b $\tan 53° 18'$ (1 mark)

c $25 \sin 52°$ (1 mark)

d $\dfrac{43.8}{\cos 26°10'}$ (1 mark)

7 Find the size of the angle θ correct to the nearest degree if $0° \le \theta \le 360°$:

a $\cos\theta = 0.95$ (1 mark)

b $\tan\theta = \dfrac{6}{8}$ (1 mark)

8 Find the length of the unknown side in these triangles. (Answer correct to one decimal place.)

a

29°, x, 17 cm

(2 marks)

b

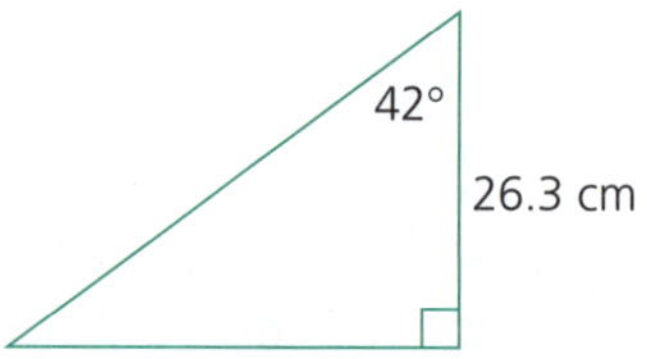

(2 marks)

c

y, 34°19′, 23.2 cm

(2 marks)

d

t, 39°20′, 17.2 cm

(2 marks)

9 Find the size of the marked angle in these diagrams. Answer correct to the nearest degree:

a

(2 marks)

b

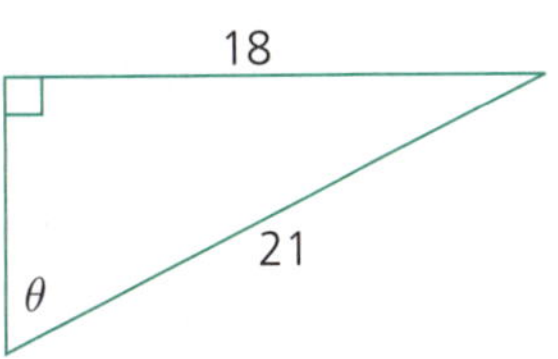

(2 marks)

c

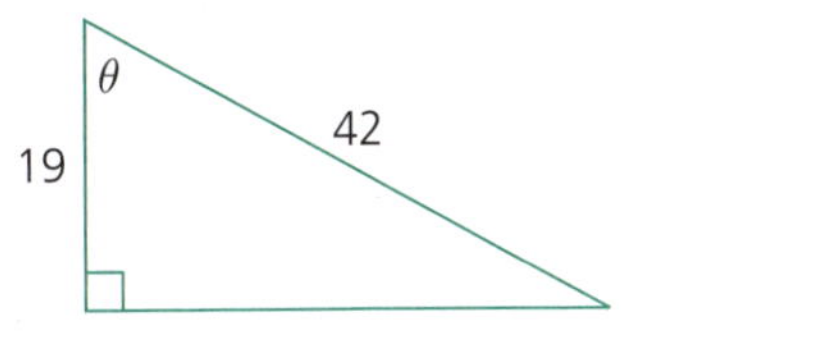

(2 marks)

10 An 8.5-metre ladder reaches 6.7 metres up a wall. Calculate the angle the ladder makes with the wall, correct to the nearest degree.

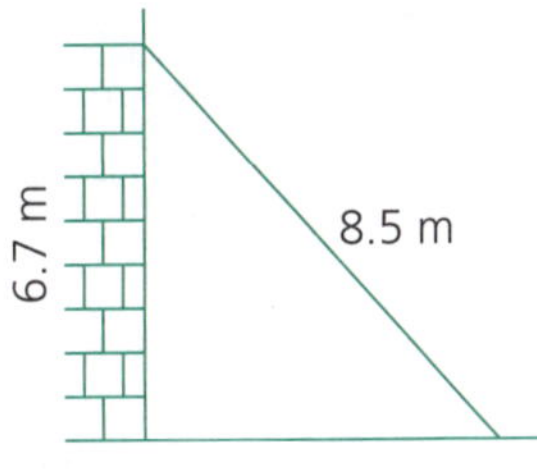

(2 marks)

Hint: redraw the diagram and include all important information.

11 Jeremy measures the angle of elevation of the top of a Banya Tree to be 38°. He knows that he is 50 m from the tree. Calculate the height of the tree correct to the nearest metre.

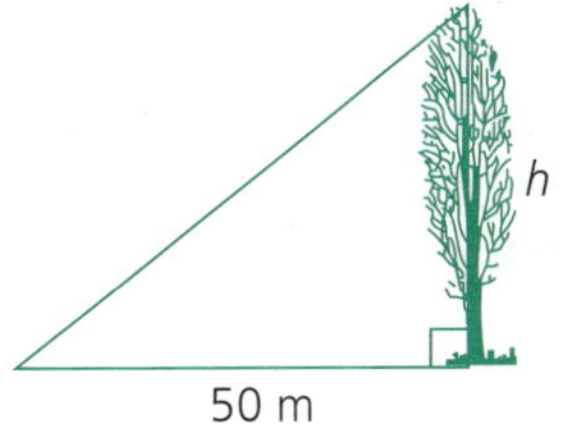

(2 marks)

12 Draw a diagram to illustrate a bearing of:

a 175° (1 mark)

b 230° (1 mark)

13 Mary walks 2.4 km from her sheep at X on a bearing of 132°. Draw a diagram to illustrate this bearing and calculate how far Mary is south of X. (Answer correct to one decimal place.)

(3 marks)

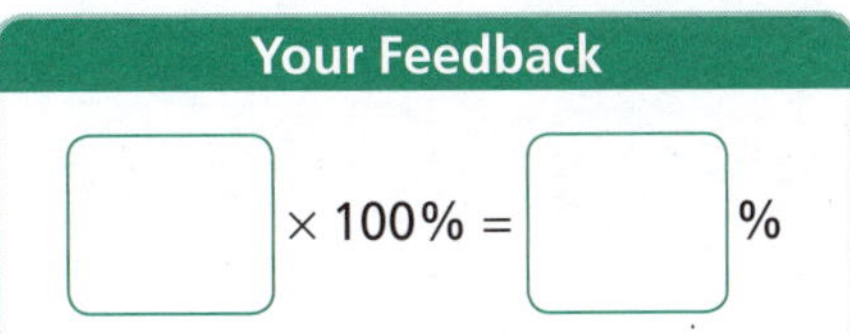

QA PAGE 223

WS PAGE 268

Advanced Test

Pythagoras and Trigonometry

(45 marks)

1 Find the length of the marked sides, correct to one decimal place where necessary:

a

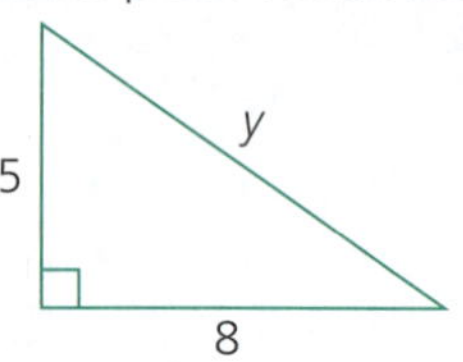

b

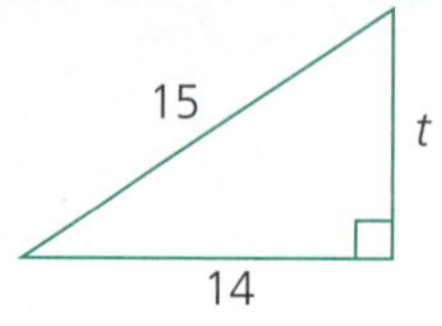

(4 marks)

2 Given that AC = 24 cm, AE = 7 cm, CD = 11 cm, calculate the length of DE:

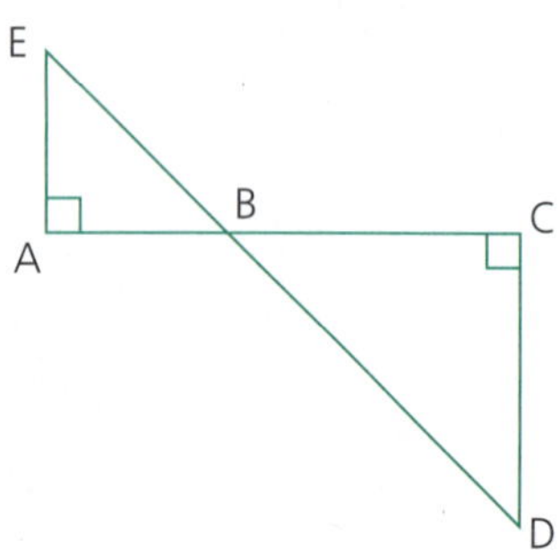

(3 marks)

3 Calculate the perimeter (correct to one decimal place):

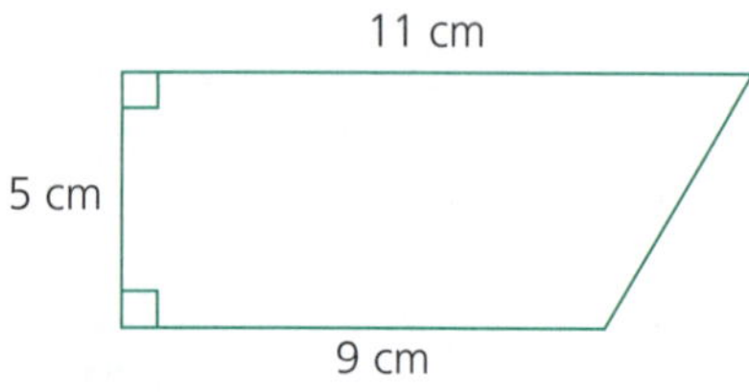

(2 marks)

4 A TV tower 32 m high is to be supported by four wires, fastened 6 m from the top of the mast. Each wire connects to hooks on the ground 15 m from the tower. Calculate the length of wire required. Answer correct to one decimal place. (2 marks)

5

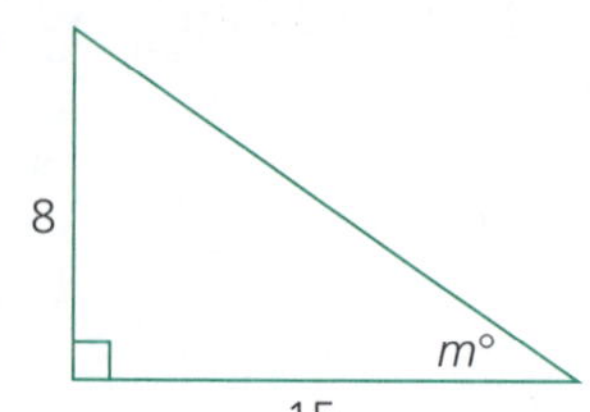

Write down a ratio for:

a $\tan m°$

b $\sin m°$

c $\cos m°$ (4 marks)

6 Find the size of the angle ϕ, correct to the nearest minute for $0° \leq \phi \leq 360°$:

a $\tan \phi = 1.8$

b $\cos \phi = 0.213$

c $\sin \phi = \frac{17}{19}$

d $\tan \phi = \frac{27}{19}$ (4 marks)

7 Given that, in a right-angled triangle, $\cos y° = \frac{12}{13}$, find the simplest expression for $\tan y°$ as a fraction.

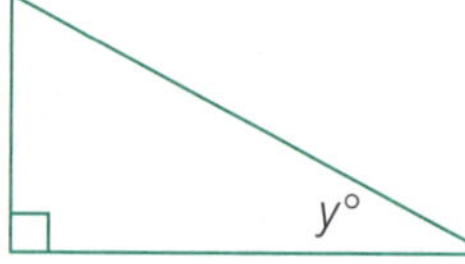

(3 marks)

8 Find the lengths of the unknown side in the following triangles. (Answer correct to one decimal place.)

a

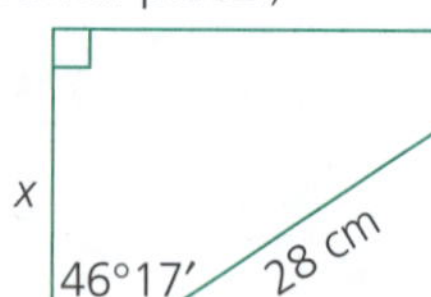

b

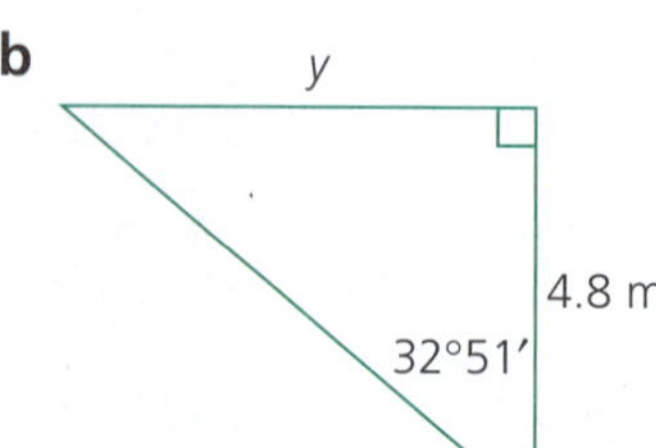

c

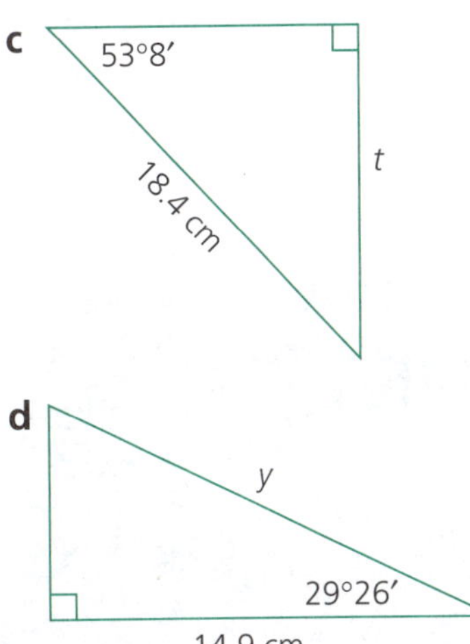

d

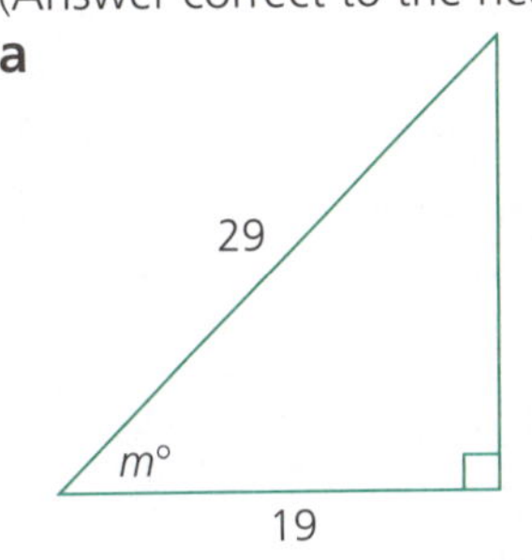

(8 marks)

9 Calculate the size of the marked angles. (Answer correct to the nearest minute.)

a

29

$m°$

19

b

29

19

$m°$

(4 marks)

10 The angle of elevation of the sun is measured to be 38°15′.

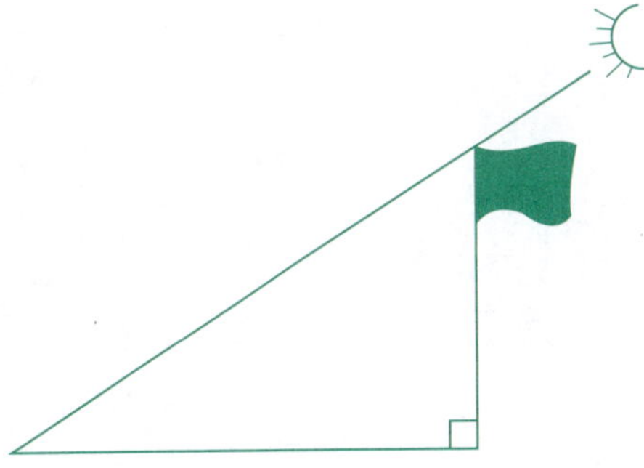

The school flag pole casts a shadow 8.2 metres long. Calculate the height of the flag pole correct to one decimal place. (3 marks)

11 Maria, standing on the roof of the LMC centre, observes Peter at an angle of depression of 11°35′ on the roof of the BAC building. The height of the BAC building is 135 metres and the straight line distance between the two buildings is 300 metres. Draw the diagram to illustrate this situation, showing all important information. Calculate the height of the LMC centre.

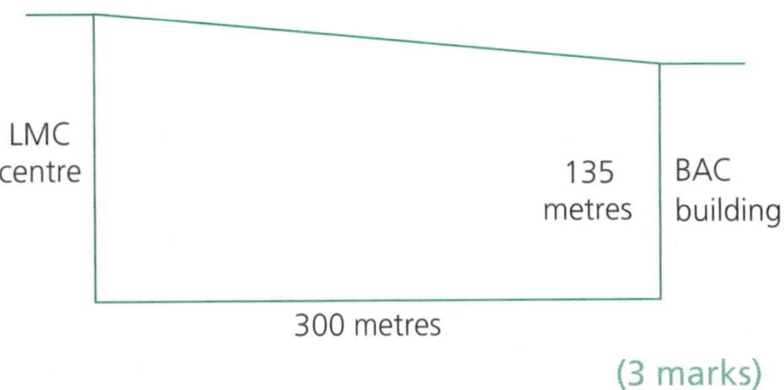

(3 marks)

12 Johannes leaves Kotara (K) on a bearing of 304°18′ for 18.5 km to a point P. Draw a diagram to illustrate this information and hence calculate the distance Johannes is north of Kotara. (3 marks)

13 A plane leaves Rutherford and flies on a bearing of 340°. How far has the plane flown when it is 60 km west of Rutherford? (2 marks)

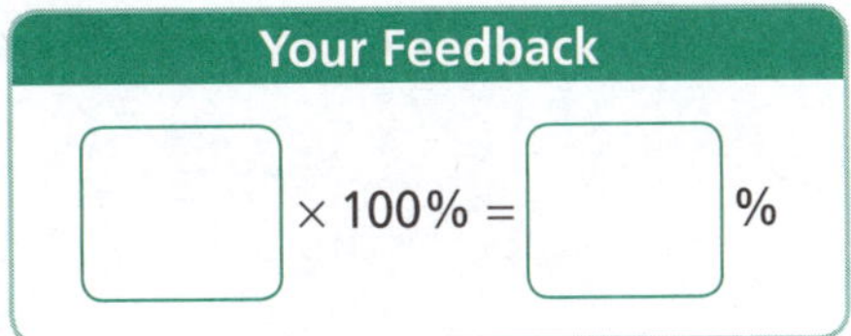

QA PAGE 223

WS PAGE 268

Chapter 9
Measurement

Units of measurement

Length

1000 mm = 1 m
100 cm = 1 m
10 mm = 1 cm
1000 m = 1 km

Mass

1000 mg = 1 g
1000 g = 1 kg
1000 kg = 1 t

Area

$1 \text{ m}^2 = 1000 \text{ mm} \times 1000 \text{ mm}$
$= 1\,000\,000 \text{ mm}^2$
$1 \text{ m}^2 = 100 \text{ cm} \times 100 \text{ cm}$
$= 10\,000 \text{ cm}^2$
$1 \text{ ha} = 100 \text{ m} \times 100 \text{ m}$
$= 10\,000 \text{ m}^2$

Capacity

1000 mL = 1 L
1000 L = 1 kL
1000 kL = 1 ML

Volume and capacity

$1 \text{ cm}^3 = 1 \text{ mL}$
$1000 \text{ cm}^3 = 1 \text{ L}$
$1 \text{ m}^3 = 1000 \text{ L} = 1 \text{ kL}$
Also $1 \text{ cm}^3 = 10 \text{ mm} \times 10 \text{ mm} \times 10 \text{ mm}$
$= 1000 \text{ mm}^3$
$1 \text{ m}^3 = 100 \text{ cm} \times 100 \text{ cm} \times 100 \text{ cm}$
$= 1\,000\,000 \text{ cm}^3$

Keywords

Annulus
Capacity
Composite
Perpendicular height
Surface area

Abbreviations

mm: millimetre
cm: centimetre
m: metre
km: kilometre

mg: milligram
g: gram
kg: kilogram
t: tonne

mL: millilitre
L: litre
kL: kilolitre
ML: megalitre

ha: hectare

Perimeter

The perimeter of a plane figure is the distance around its boundary.

Name	Sketch	Perimeter
Square	s	$P = 4s$
Rectangle	b, l	$P = 2(l + b)$
Circle	r	Circumference (C) $C = 2\pi r$ $= \pi d$

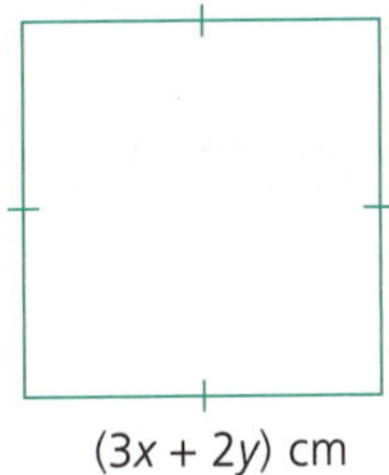

1 Find the perimeter (or circumference) of the following figures:

a

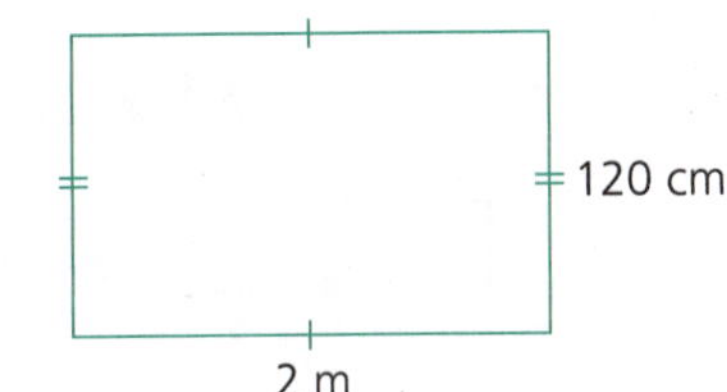

S $P = 4s$
$= 4(3x + 2y)$
$= 12x + 8y$

∴ The perimeter is $(12x + 8y)$ cm.

b

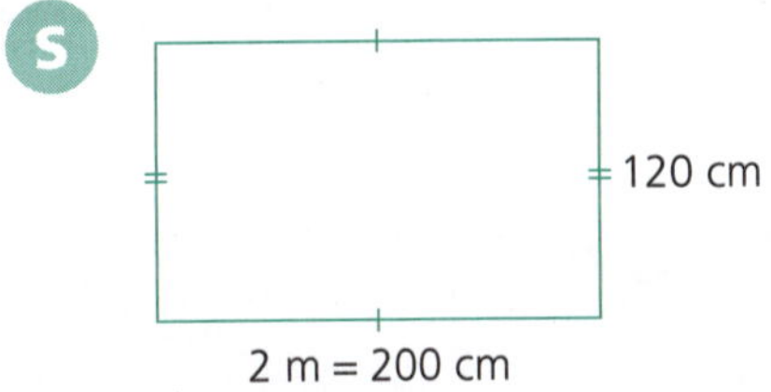

S

$P = 2(l + b)$
$= 2(200 + 120)$
$= 2(320)$
$= 640$

∴ The perimeter is 640 cm.

c

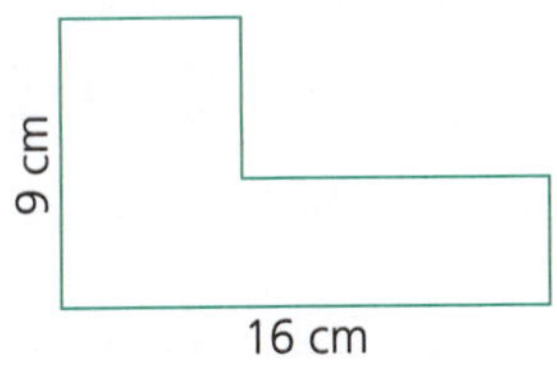

S

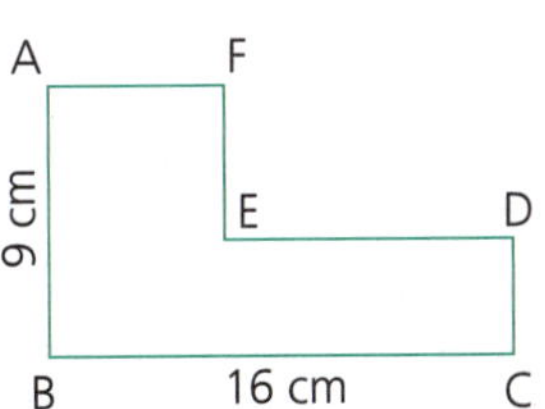

The distance from A to C will be the same, whether using AB + BC or AF + FE + ED + DC.

$P = 2(9 + 16)$
$= 2(25)$
$= 50$

∴ The perimeter is 50 cm.

d

S Find x first, using Pythagoras' Theorem.

$\therefore x^2 = 5^2 + 12^2$
$= 25 + 144$
$= 169$

$x = \sqrt{169}$
$= 13$

$\therefore P = 3 + 12 + 8 + 13$
$= 36$

∴ The perimeter is 36 cm.

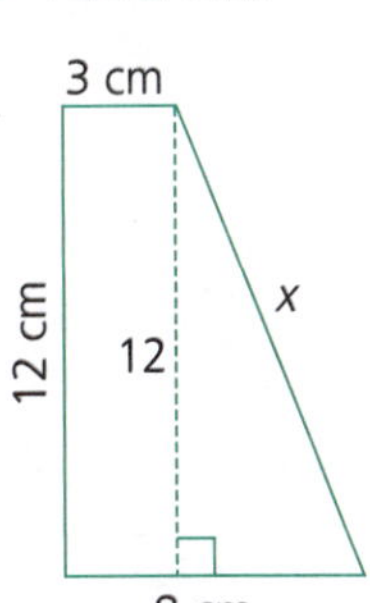

e

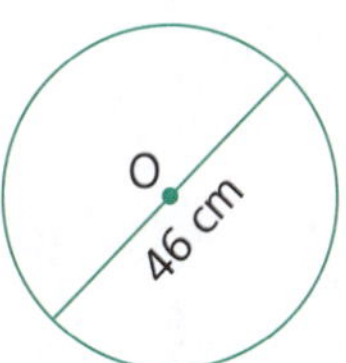

(to 2 decimal places)

S $C = \pi d$
$= \pi \times 46$
$= 144.5132621$
$= 144.51$
(to 2 decimal places)

∴ The circumference is 144.51 cm.

f

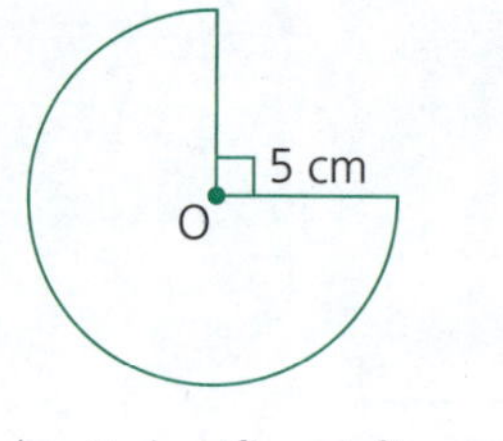

(to 3 significant figures)

S $P = 5 + 5 + \frac{3}{4}$ of circle

$= 5 + 5 + \frac{3}{4} \times 2\pi r$

$= 10 + \frac{3}{4} \times 2 \times \pi \times 5$

$= 33.5619449$

$= 33.6$ (to 3 significant figures)

$\therefore$ The perimeter is 33.6 cm.

Area

The area of a plane figure is the amount of two-dimensional (2D) space it contains.

Name	Sketch	Area
Square	s	$A = s^2$
Rectangle	b, l	$A = lb$
Triangle	h, b	$A = \frac{1}{2}bh$
Circle	r	$A = \pi r^2$
Parallelogram	h, b	$A = bh$
Trapezium	a, h, b	$A = \frac{1}{2}h(a + b)$
Rhombus (and kite)	y, x	$A = \frac{1}{2}$(product of diagonals) $A = \frac{1}{2}xy$

For Example

1 Find the area of the following figures:

a

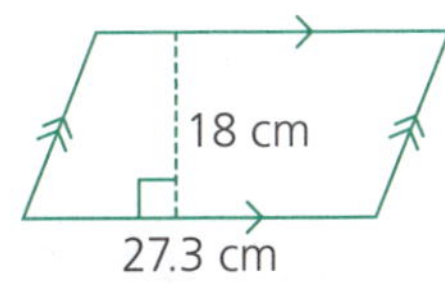

S $A = bh$

$= 27.3 \times 18$

$= 491.4$

$\therefore$ The area is 491.4 cm^2.

b AC = 21 cm
BD = 27 cm

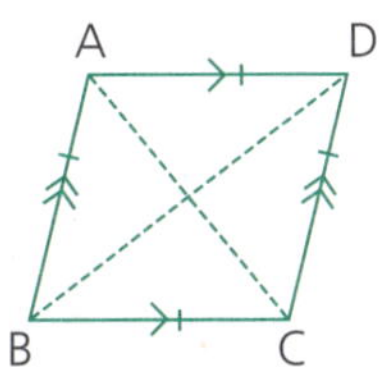

S AC = 21 cm
BD = 27 cm

$A = \frac{1}{2}xy$

$= \frac{1}{2} \times 21 \times 27$

$= 283.5$

$\therefore$ The area is 283.5 cm^2.

c

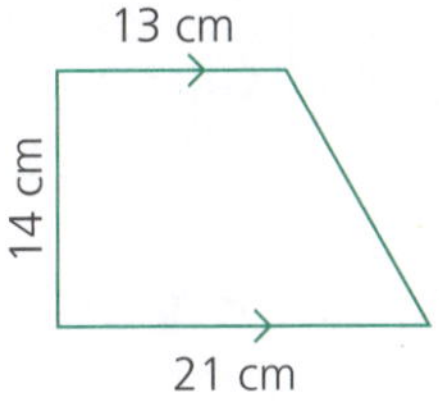

S The figure is a trapezium.

$\therefore A = \frac{1}{2}h(a+b)$

$= \frac{1}{2} \times 14(13+21)$

$= 7(34)$

$= 238$

$\therefore$ The area is 238 cm^2.

d

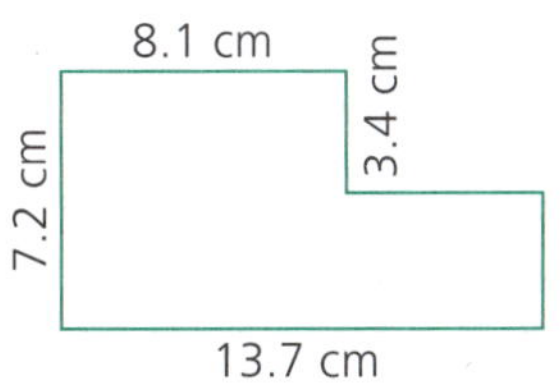

S

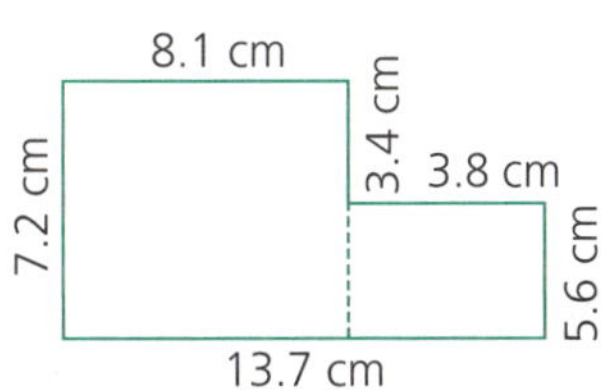

Divide the composite figure into two rectangles and calculate the lengths of the required sides.

$7.2 - 3.4 = 3.8$

and $13.7 - 8.1 = 5.6$

$\therefore A = 8.1 \times 7.2 + 3.8 \times 5.6$

$= 79.6$

$\therefore$ The area is 79.6 cm^2.

e

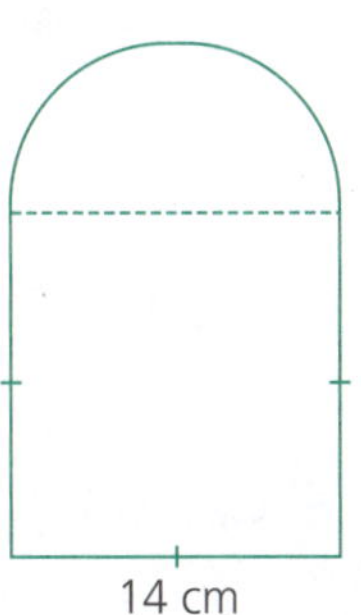

(to 2 decimal places)

S Divide the composite figure into a square and semicircle.

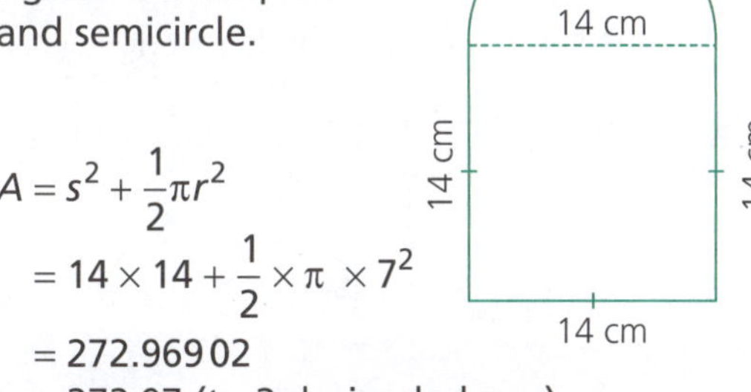

$A = s^2 + \frac{1}{2}\pi r^2$

$= 14 \times 14 + \frac{1}{2} \times \pi \times 7^2$

$= 272.969\,02$

$= 272.97$ (to 2 decimal places)

$\therefore$ The area is 272.97 cm^2.

f

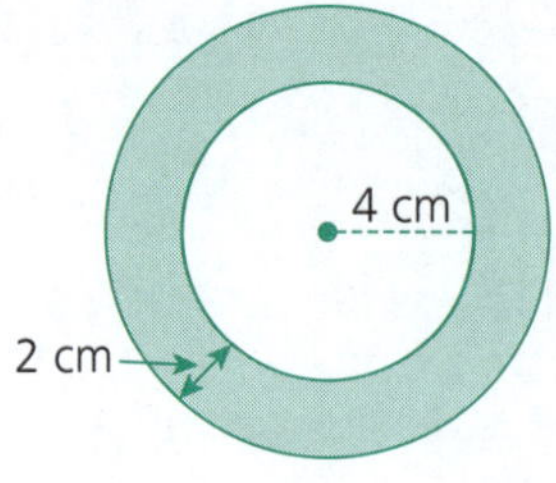

(to 2 decimal places)

S The figure is an annulus. If the radius of the larger circle is R and the radius of the smaller circle is r, then area of annulus:

$\pi R^2 - \pi r^2$

That is:

$$A = \pi \times 6^2 - \pi \times 4^2$$
$$= 62.831\,853\,06$$
$$= 62.83 \text{ (to 2 decimal places)}$$

$\therefore$ The area is 62.83 cm^2.

Surface area

The surface area of a solid is the sum of the areas of its faces.

Name	Sketch	Area
Cube	s	Surface area (SA) = $6s^2$
Rectangle prism	h, b, l	Surface area (SA) = $2lb + 2lh + 2bh$
Right-cylinder	r, h	Total surface area = $2\pi r^2 + 2\pi rh$ Curved surface area = $2\pi rh$

1 Find the surface area of the following solids:

a

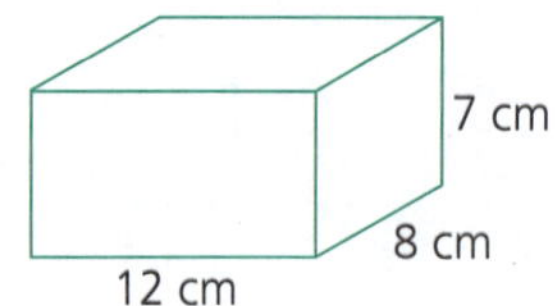

S
$$SA = 2lb + 2lh + 2bh$$
$$= 2 \times 12 \times 8 + 2 \times 12 \times 7 + 2 \times 8 \times 7$$
$$= 472$$

$\therefore$ The surface area is 472 cm^2.

b

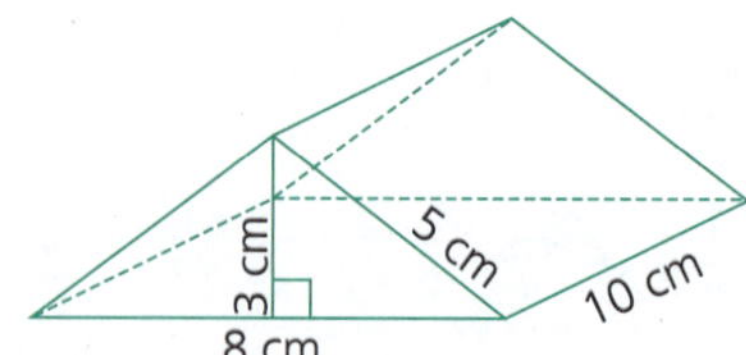

S SA = Sum of five faces
$$= 2 \times \frac{1}{2} \times 8 \times 3 + 2 \times 10 \times 5 + 8 \times 10$$
$$= 24 + 100 + 80$$
$$= 204$$

$\therefore$ Surface area is 204 cm^2.

c

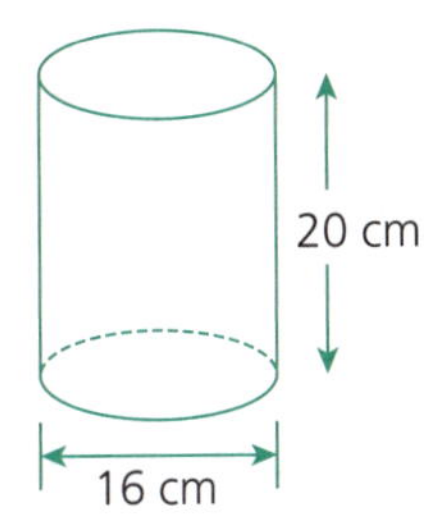

(to 2 decimal places)

S Diameter = 16 cm

$\therefore$ Radius = 8 cm

$$\begin{aligned} SA &= 2\pi rh + 2\pi r^2 \\ &= 2 \times \pi \times 8 \times 20 + 2 \times \pi \times 8^2 \\ &= 1407.433509 \\ &= 1407.43 \text{ (to 2 decimal places)} \end{aligned}$$

$\therefore$ Surface area is 1407.43 cm^2.

Volume

The volume of a solid is a measure of the space contained inside it.
For a prism, the volume is given by V = area of base $\times$ height, that is, $V = Ah$.

Name	Sketch	Area
Cube	s	$V = s^3$
Rectangle prism	h, b, l	$V = lbh$
Cylinder	r, h	$V = \pi r^2 h$

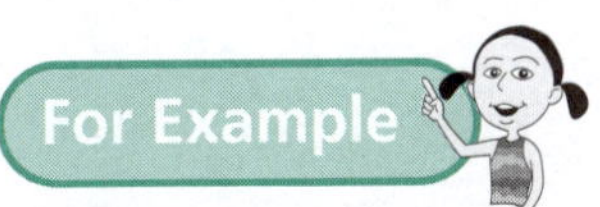

1 Find the volume of the following solids:

a

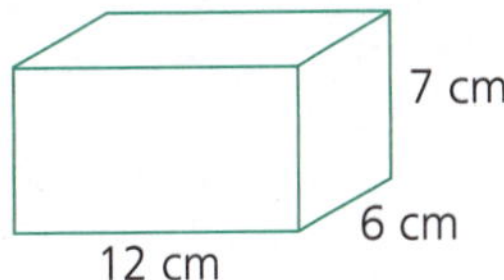

S
$$\begin{aligned} V &= lbh \\ &= 12 \times 6 \times 7 \\ &= 504 \end{aligned}$$

$\therefore$ 504 cm^3

b

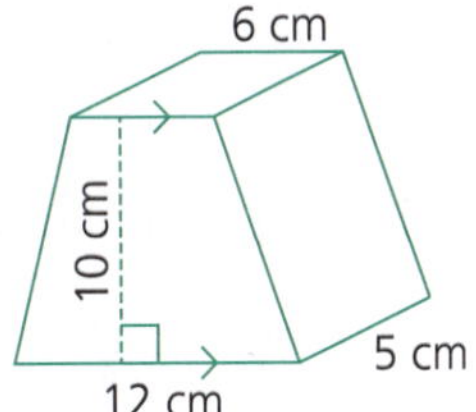

S
$$\begin{aligned} A &= \frac{h}{2}(a + b) \\ &= \frac{10}{2}(6 + 12) \\ &= 90 \end{aligned}$$

$$\begin{aligned} V &= 90 \times 5 \\ &= 450 \end{aligned}$$

$\therefore$ 450 cm^3

c

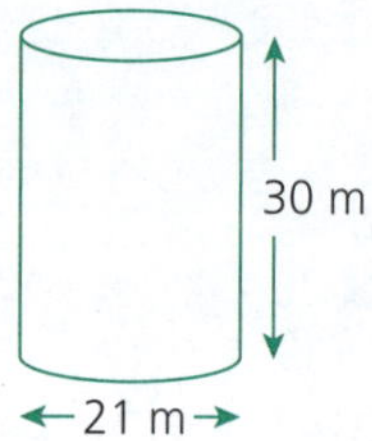

(to the nearest m^3)

S Diameter = 21 m
Radius = 10.5 m

$V = \pi r^2 h$
$= \pi \times 10.5^2 \times 30$
$= 10\,390.8177 \ldots$
$= 10\,391$ (nearest whole)
$\therefore 10\,391 \text{ m}^3$

d

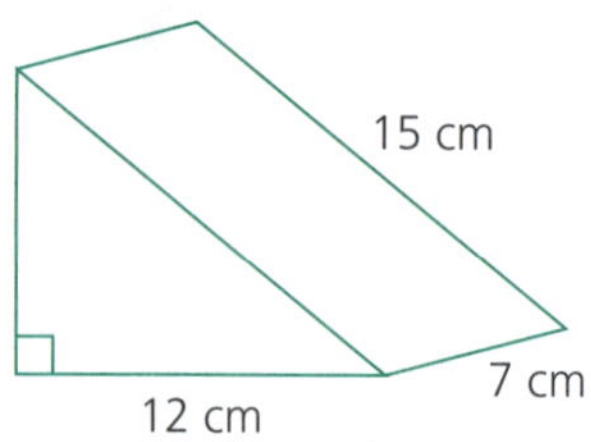

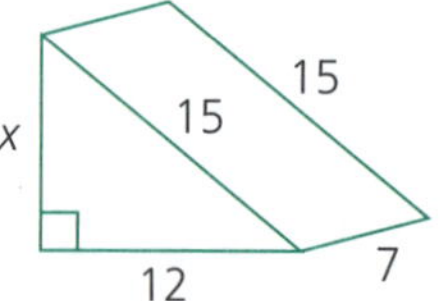

S Using Pythagoras:

$x^2 = 15^2 - 12^2$
$= 225 - 144$
$= 81$
$x = 9$

$V = \frac{1}{2} \times 12 \times 9 \times 7$
$= 378$

$\therefore$ Volume of 378 cm^3

e Find the volume in terms of π.

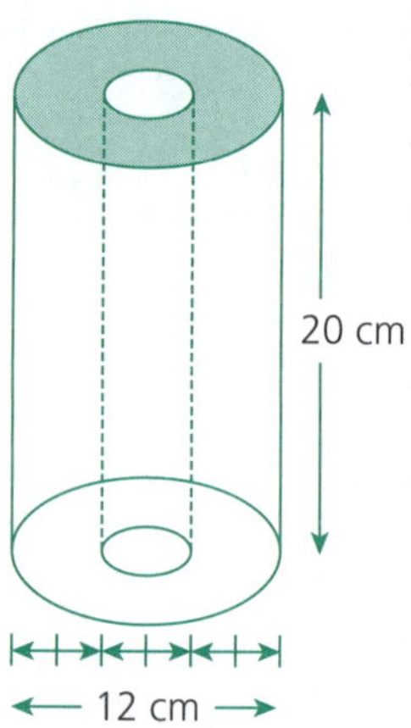

S Outer circle: diameter = 12 cm,
$\therefore$ radius = 6 cm

Inner circle: diameter = 4 cm,
$\therefore$ radius = 2 cm

$V = \pi \times 6^2 \times 20 - \pi \times 2^2 \times 20$
$= 720\pi - 80\pi$
$= 640\pi$

$\therefore$ Volume of $640\pi \text{ cm}^3$

Capacity

Capacity is a measure of the amount of liquid within a container.

$1 \text{ cm}^3 = 1 \text{ mL}$
$1000 \text{ cm}^3 = 1 \text{ L}$
Also, $1 \text{ m}^3 = 1000 \text{ L}$
$= 1 \text{ kL}$

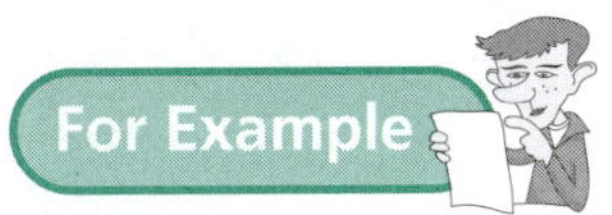

1 Find the capacity of the following solids:

a

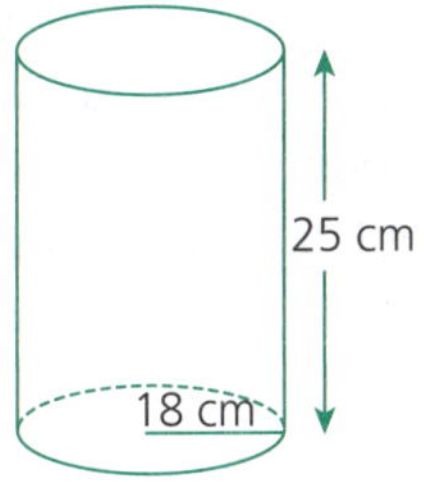

(to the nearest litre)

S $V = \pi r^2 h$

$= \pi \times 18^2 \times 25$

$= 25\,446.900\,49 \text{ cm}^3$

Capacity $= 25\,446.900\,49 \div 1000$

$= 25.446\,900\,49$

$= 25$ (to the nearest litre)

$\therefore$ Capacity is 25 L.

b

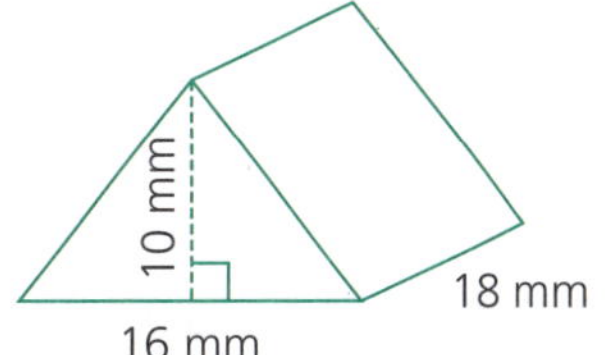

S Change dimensions to cm:

$V = \frac{1}{2} \times 1.6 \times 1 \times 1.8$

$= 1.44$

$\therefore$ Volume of 1.44 cm^3

$\therefore$ Capacity of 1.44 mL

Practise Practise

1 Find the perimeter, correct to three decimal places, of the following figures:

a

b

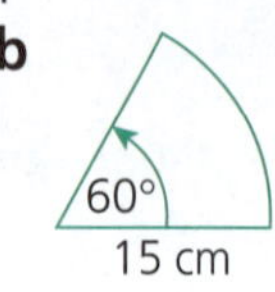

pp. 124–126

2 Find the perimeter of the following figures:

a

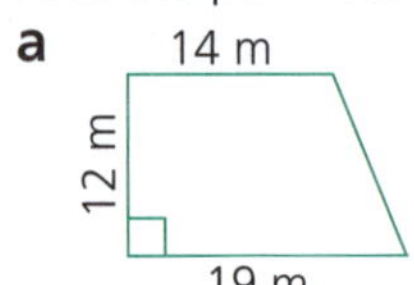

b

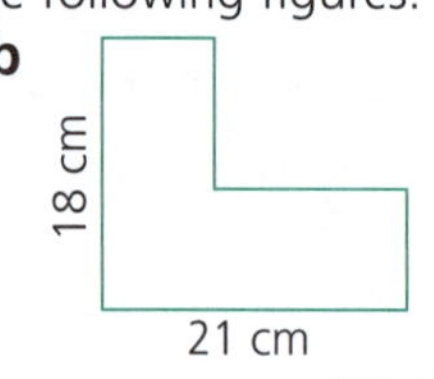

pp. 124–126

3 A rumpus room measuring 4.4 m by 3.6 m has its floor partly covered by a rug so that a border of 50 cm is left between the wall and the edge of the rug on all four sides. Find the perimeter of the rug.
pp. 124–126

4 Syl decides to fit a length of weather stripping completely around his side door, which is in the shape of a semicircle on a rectangle. Calculate the length of weather stripping required, to the nearest cm.

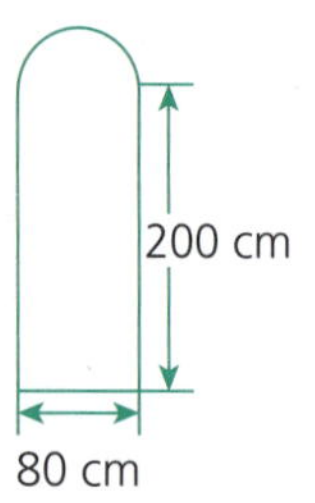

pp. 124–126

5 The lengths of the diagonals of a rhombus are 6 cm and 10 cm. Calculate the area of the rhombus.
pp. 126–128

6 **a** Find the shaded area of the annulus. Leave your answer in terms of π:

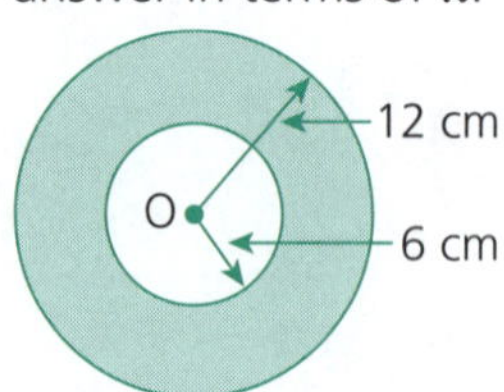

pp. 126–128

b Find the area and perimeter of the rhombus:

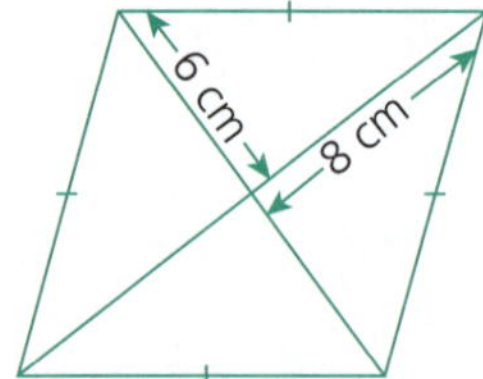

pp. 124–128

7 Find the area of the total surface of a cylinder with a radius of 7 cm and a height of 12 cm. (Answer to 3 significant figures.) pp. 128–129

8 Find the surface area of the triangular prism:

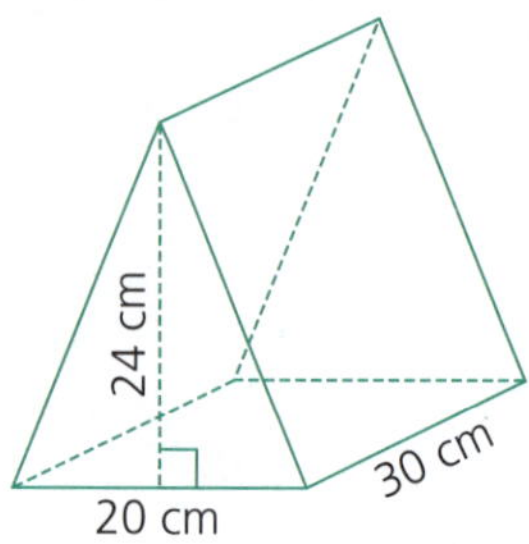

pp. 128–129

9 A rectangular sheet of metal measures 72 cm by 40 cm. From this sheet, circular pieces with diameters of 4 cm are to be stamped out:

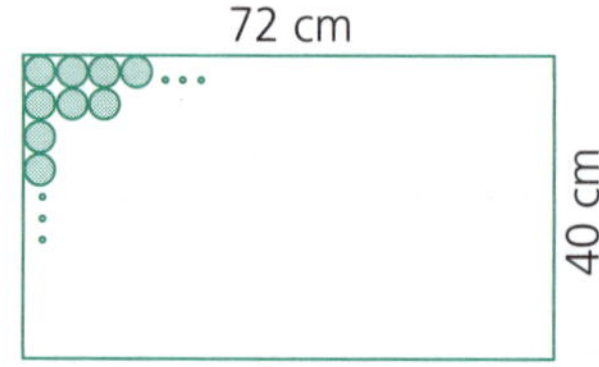

a Find the maximum number of stampings.

b Find the amount of scrap metal, to the nearest cm^2. pp. 126–128

10 The surface area of a cube is known to be 54 cm^2. Find the length of each edge of the cube.
pp. 128–129

11 A tin of fruit salad has a diameter of 7 cm and a height of 10 cm. Find the area of its label, to the nearest cm^2. pp. 128–129

12 Find the volume of a cylinder that has a radius and a height of 7.6 cm. Answer correct to three decimal places. pp. 129–130

13 Find the volume, correct to two decimal places, of the following solids:

a

21 cm
16 cm
0

b

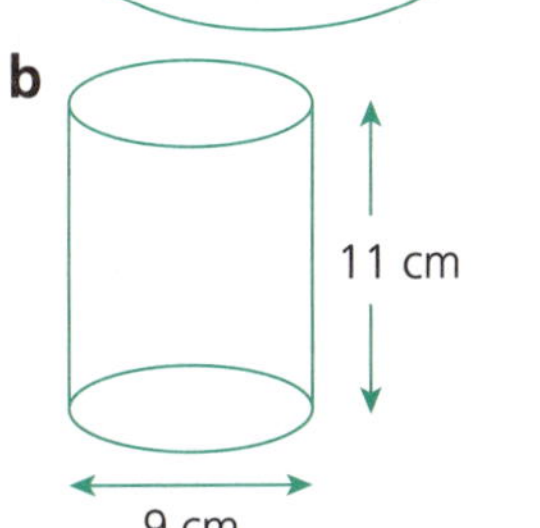

c

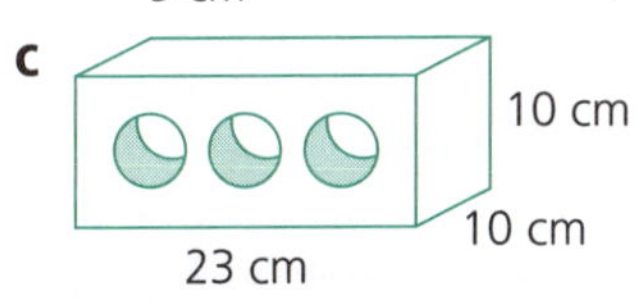

All three holes have diameters of 5 cm.

d

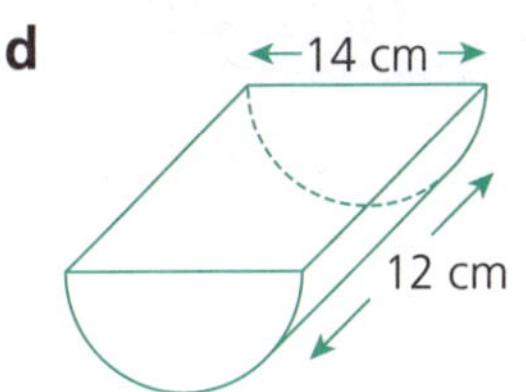

pp. 129–130

14 Find the volume of a cylinder with diameter 6 cm and height 4 cm. Leave your answer in terms of π. pp. 129–130

15 A cylindrical water tank has a radius of 1.8 metres and a height of 2.3 metres. Find its capacity to the nearest litre. pp. 130–131

16 A rectangular prism with base dimensions of 16 cm and 20 cm is filled with a litre of water. How deep is the water? pp. 130–131

17 Ken decided to preserve some of his peaches. The instructions were to fill a cylindrical preserving jar with juice to two-thirds of its total volume. If the radius of the jar was 5 cm and its height was 30 cm, how many litres of juice were required? (Answer to the nearest millilitre.) pp. 130–131

18 A cylindrical oil-storage tank has a height of 12 metres and diameter of 18 metres. Find its capacity to the nearest kilolitre. pp. 130–131

19 A swimming pool has the shape of a trapezoidal prism, as shown in the diagram:

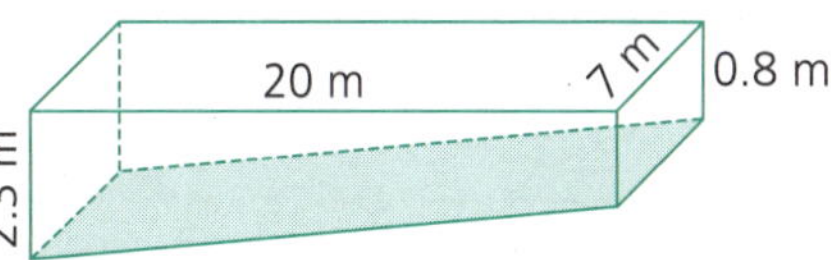

Find the:

a cost of tiling the four walls of the pool, if those tiles cost $42/m^2

b volume of the pool in cubic metres

c volume of water (in kilolitres) required to fill the pool. pp. 130–131

20 A paving brick (paver) is 22.5 cm long and 10 cm wide. Ken purchases a pallet of 1000 pavers to begin to pave his rectangular-shaped pergola area.

a Find the area of the rectangle to be paved.

b What is the area in cm^2?

c How many more pavers will Ken need to complete the job? pp. 126–128

21 A metal cylindrical pipe has an outer diameter of 3 cm and an inner diameter of 2.8 cm. The pipe is 2 metres long. Find the mass of the pipe if it is made of metal of 16.5 g/cm^3. Leave your answer to the nearest kilogram. pp. 129–130

22 A children's swimming pool has a hexagonal shape. It is constructed of a blue vinyl sheet draped over six pipes supported by six poles. The area of the base is 5.2 m^2.

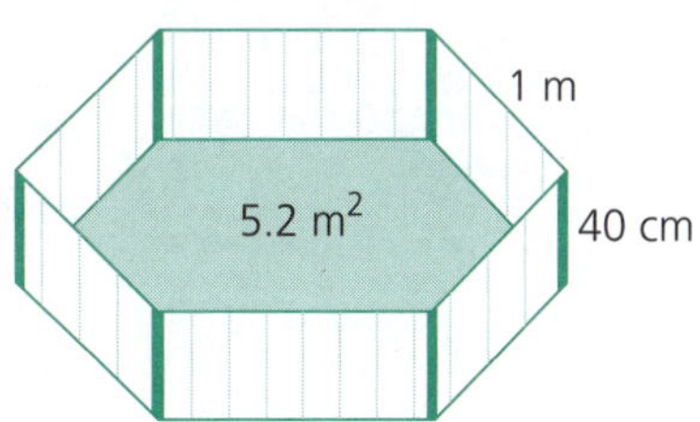

a Find the total surface area of the vinyl sheet.

b Find the capacity of the pool.

c The pool is to be filled to three-quarters of its capacity. How many kilolitres will it contain? pp. 130–131

Go to p. 219 for **Quick Answers** or to pp. 248–250 for **Worked Solutions**

For a complete understanding of this topic, you must be able to:

✓	Convert between units of measurement		p. 124
✓	Find the perimeter of composite figures		pp. 124–126
✓	Find the area of composite figures		pp. 126–128
✓	Find the surface area of composite solids		pp. 128–129
✓	Find the surface area of right-cylinders		p. 129
✓	Find the volume of composite figures		pp. 129–130
✓	Solve problems relating to capacity.		pp. 130–131

Now you are ready to do the tests!

Intermediate Test

Measurement

(30 marks)

1 Find the perimeter, correct to two decimal places:

a

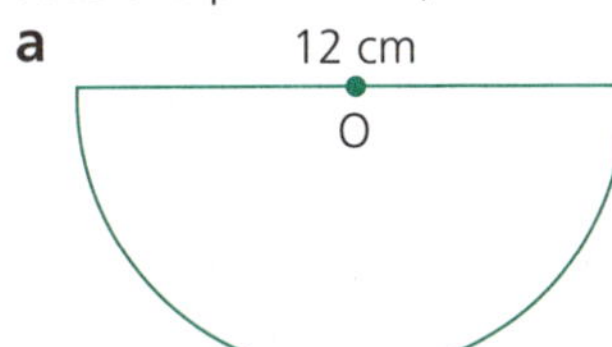

(2 marks)

b

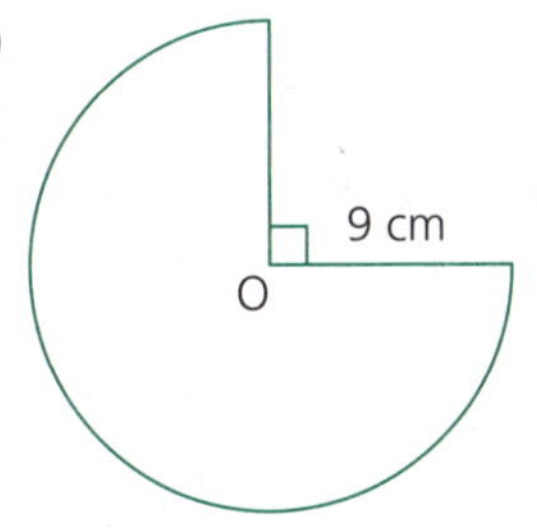

(2 marks)

2 Find the perimeter, correct to three significant figures:

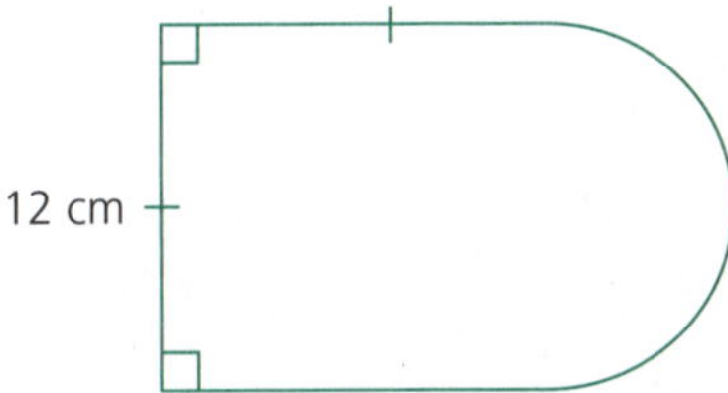

(2 marks)

3 Find the area:

a

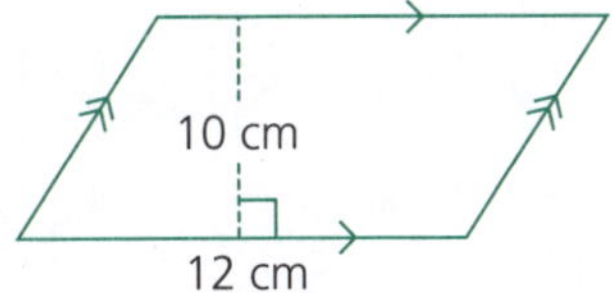

(1 mark)

b

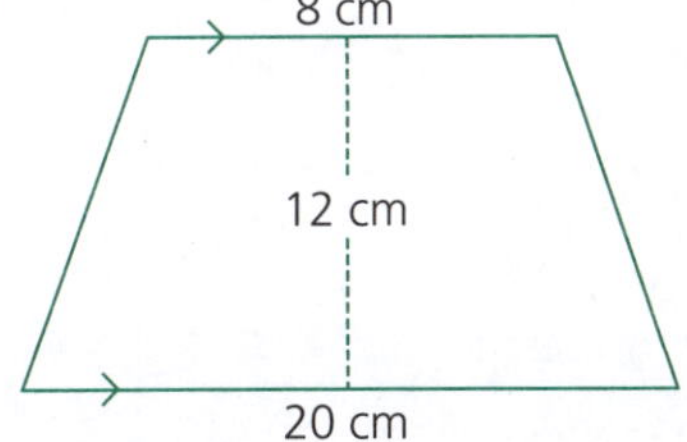

(2 marks)

4 Find area, leaving your answer in terms of π:

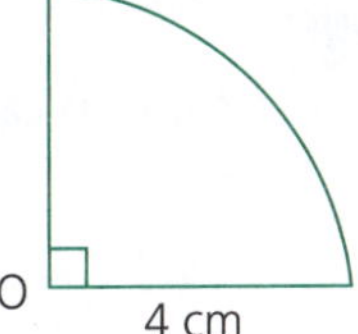

(2 marks)

5 Find the surface area, correct to two decimal places:

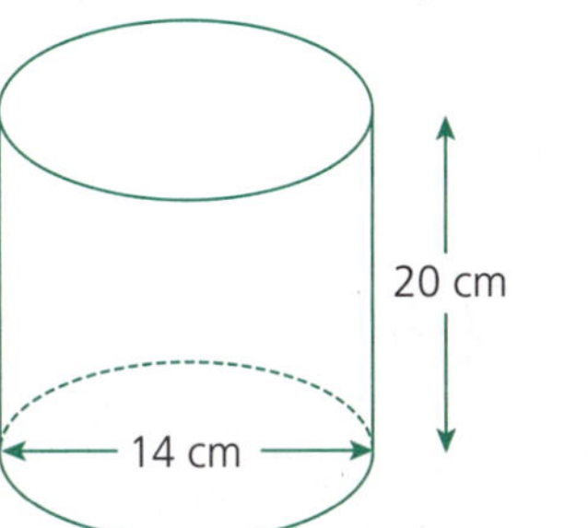

(2 marks)

6 The outside of a pipe with radius 10 cm is to be painted. If the length of the pipe is 14 metres, find the surface area to be painted. Answer to the nearest m^2. (2 marks)

7 Find the surface area:

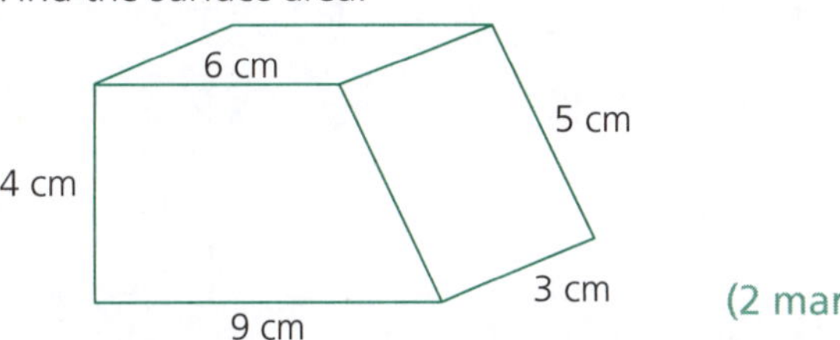

(2 marks)

8 Find the volume, correct to two decimal places:

a

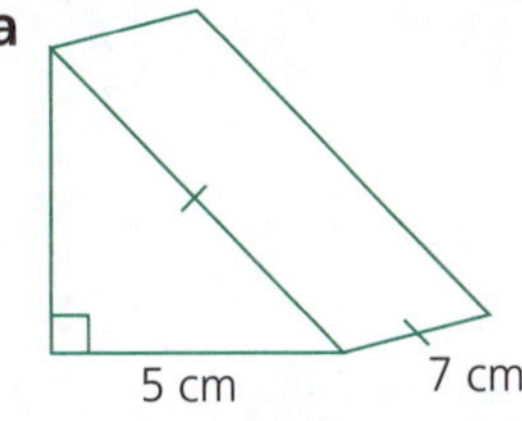

(2 marks)

b

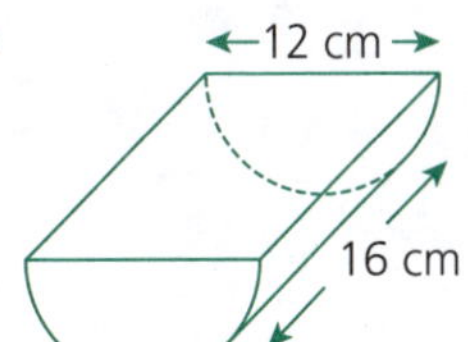

(2 marks)

9 Find the volume of the cylinder in terms of π:

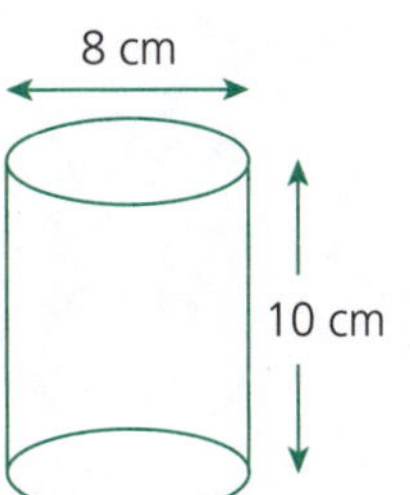

(2 marks)

10 Find the capacity to the nearest kilolitre:

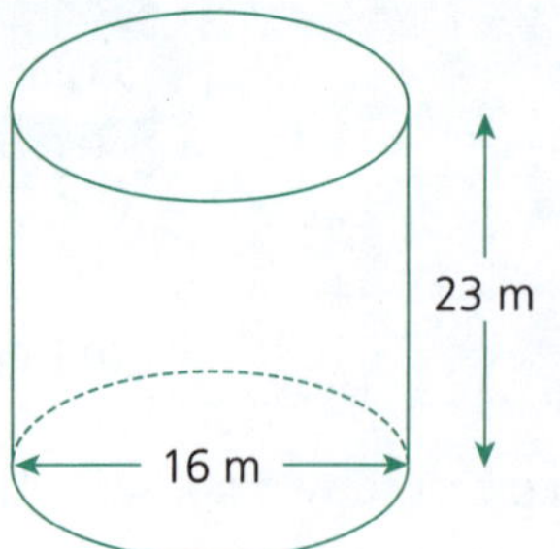

(2 marks)

11 Find the capacity:

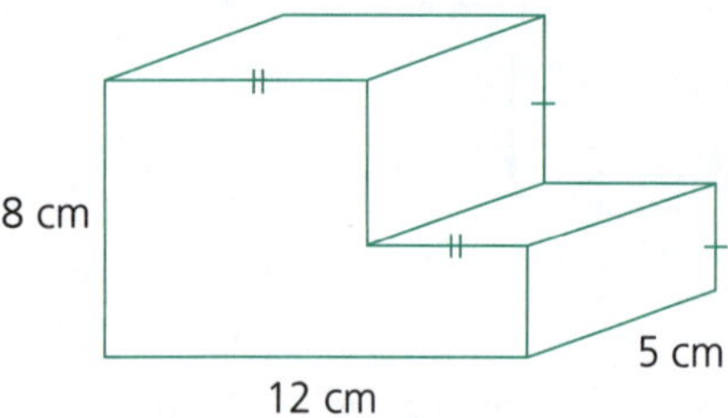

(2 marks)

12

10 cm

5 cm

12 cm

Container A

20 cm

8 cm

Container B

Container A is half-filled with water. Container B has 340 mL of water. If container A is emptied into container B, find the depth of water. (3 marks)

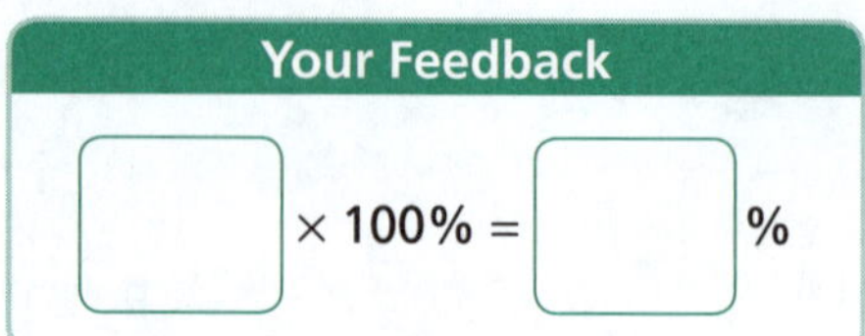

QA PAGE 223

WS PAGE 269

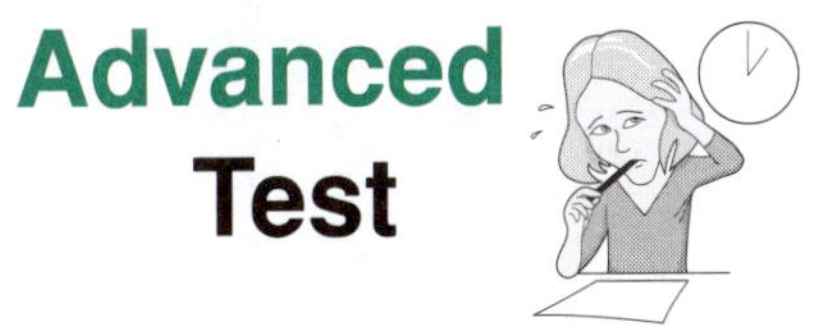

Measurement

(30 marks)

1 Find the perimeter, correct to two decimal places:

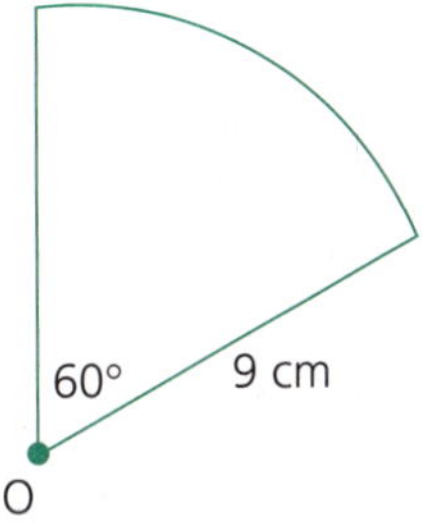

(2 marks)

2 Find the perimeter in terms of π:

a

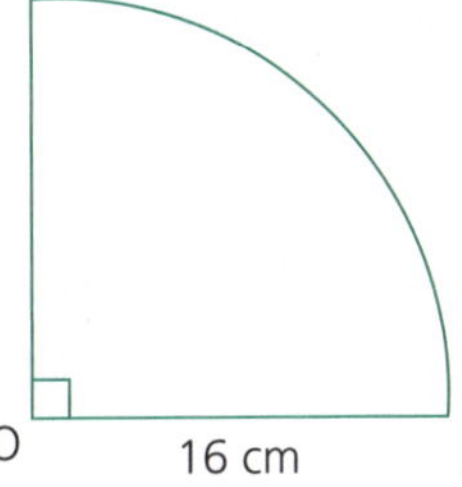

(2 marks)

b

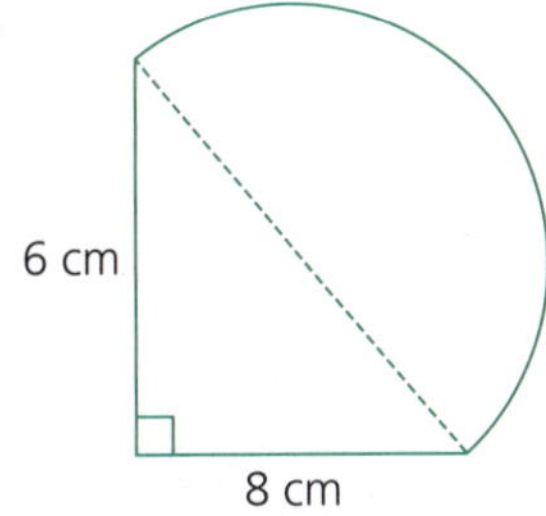

(2 marks)

3 Find the area, correct to three significant figures:

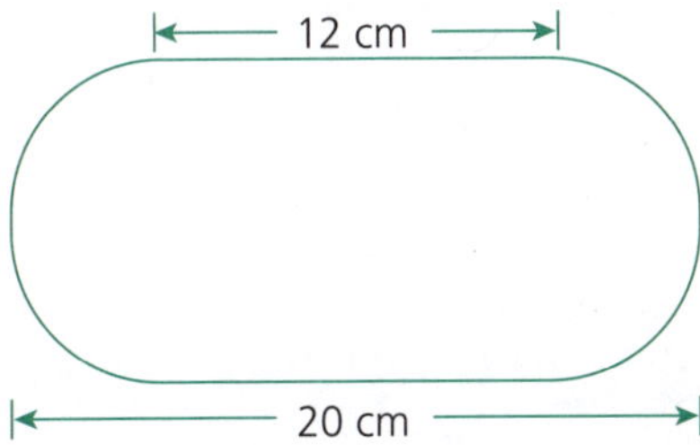

(2 marks)

4 Find the area of sector ABO when OD = $\sqrt{3}$, BD = 1. Leave your answer in terms of π:

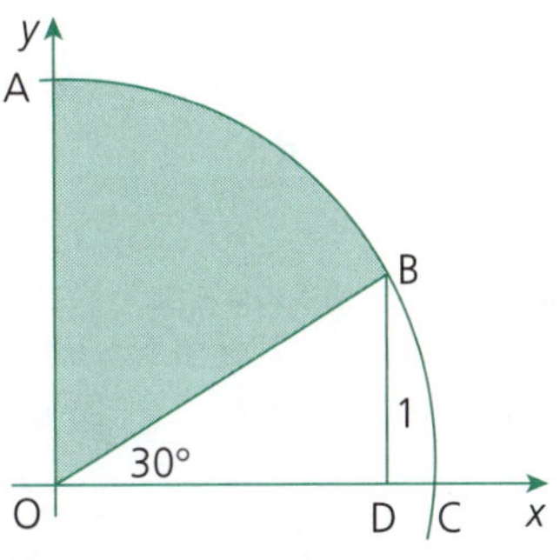

(2 marks)

5 Find the area of the kite ABCD if AC = 12 cm and BD = 9 cm:

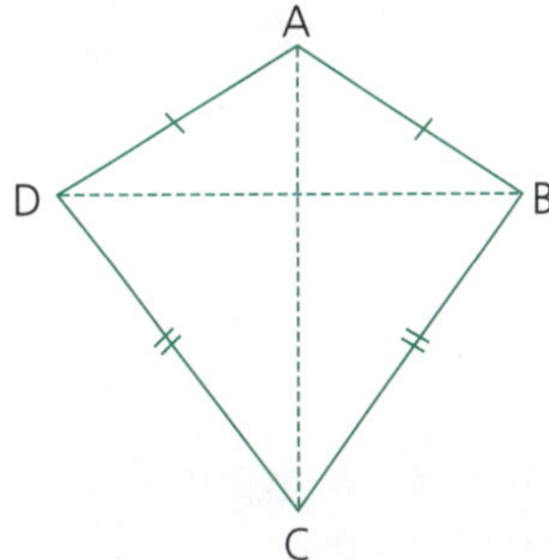

(2 marks)

6 The diagram shows a cylinder with both ends removed. The diameter is 12 cm and the height 20 cm. Find the curved surface area, correct to three decimal places:

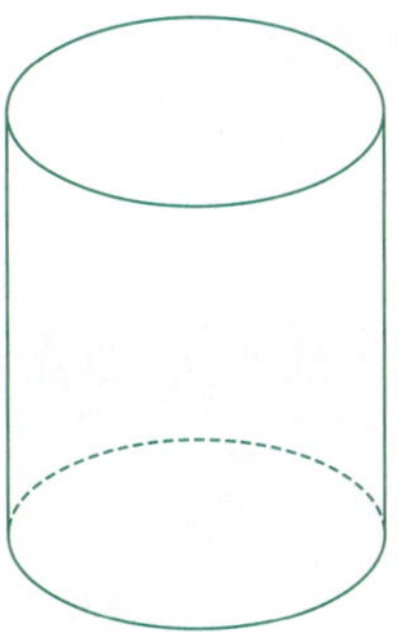

(2 marks)

7 Find the volume:

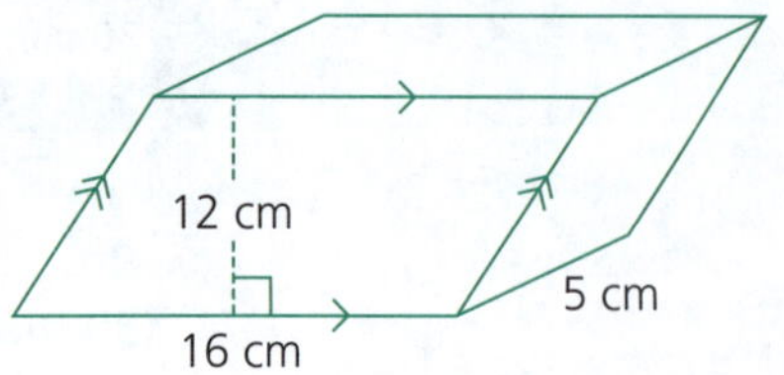

(2 marks)

8

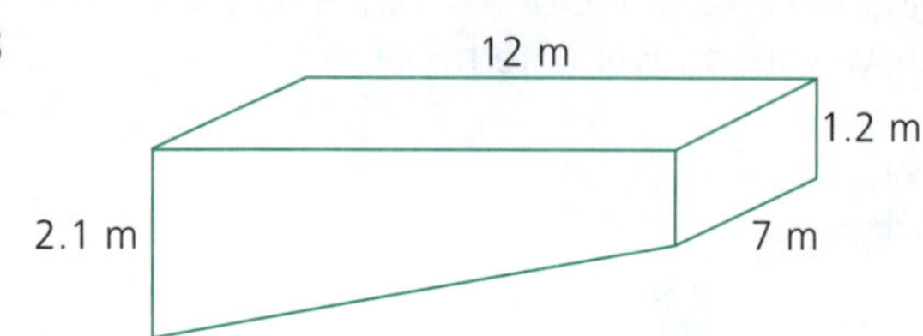

The diagram represents a swimming pool. If the pool is to be filled with water to 20 cm from the top, how much water is required? (2 marks)

9 The tin contains 4 L of paint:

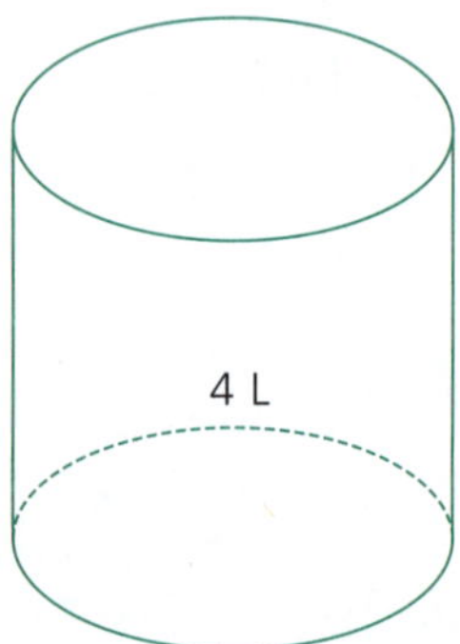

If the diameter is 20 cm, find the depth of the paint. Answer to the nearest cm. (2 marks)

10 The rectangular block is 6 cm thick and is made from an alloy. Five identical holes with diameter 2 cm are drilled through the block. Find the volume of alloy that remains. (Answer correct to two decimal places.)

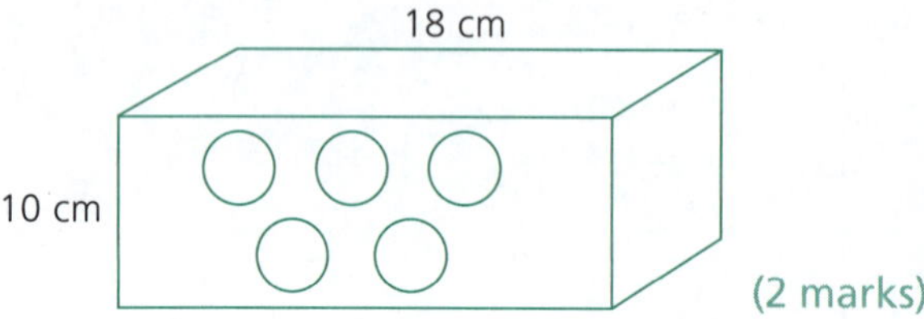

(2 marks)

11 Find the volume:

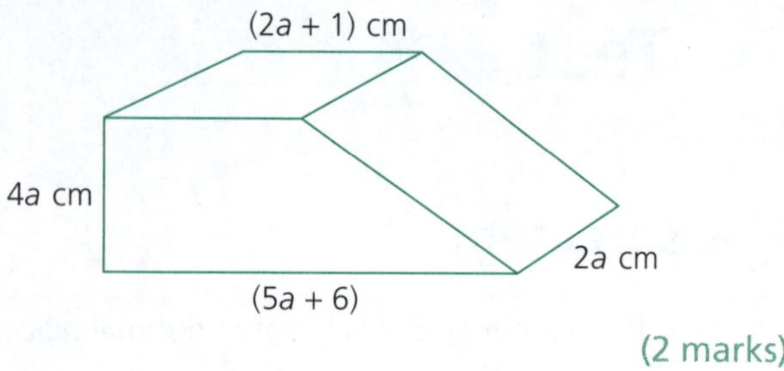

(2 marks)

12 A cube has a surface area of 24 m^2. Water is poured into the cube to a height of 10 cm. How much water in litres is in the cube? (2 marks)

13 Find the volume and surface area, correct to two decimal places:

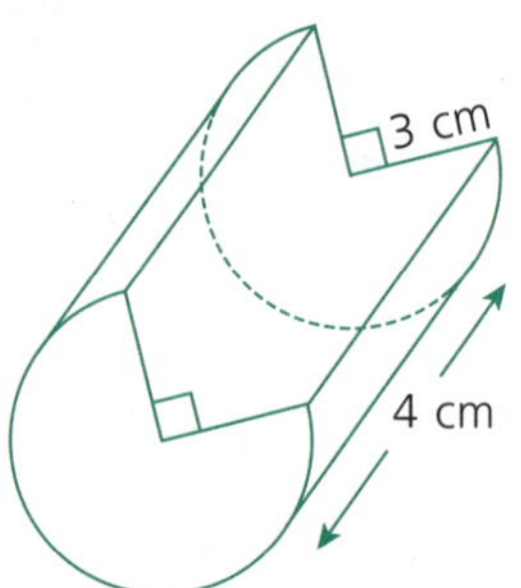

(4 marks)

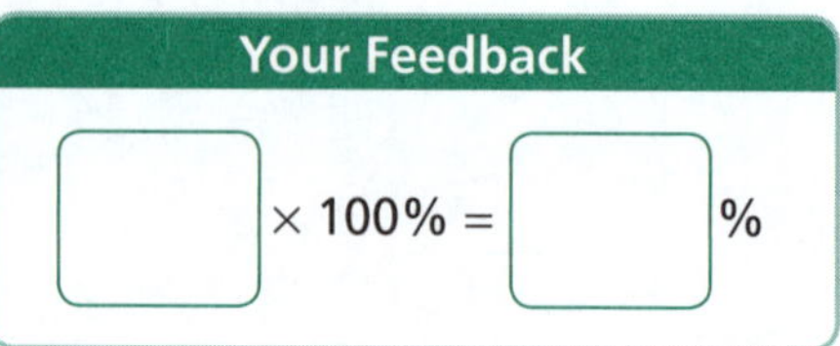

QA PAGE 223

WS PAGE 270

Chapter 10
Geometry

Review

Keywords

Alternate
Bisect
Co-interior
Complementary
Congruent
Corresponding
Decagon
Deduction
Dodecagon
Exterior
Heptagon
Hexagon
Interior
Nonagon
Octagon
Pentagon
Polygon
Produce
Similar
Supplementary
Vertically opposite

Naming angles

When naming angles, we start from arm → vertex → arm.
For example, the angle shown is ∠ABC.

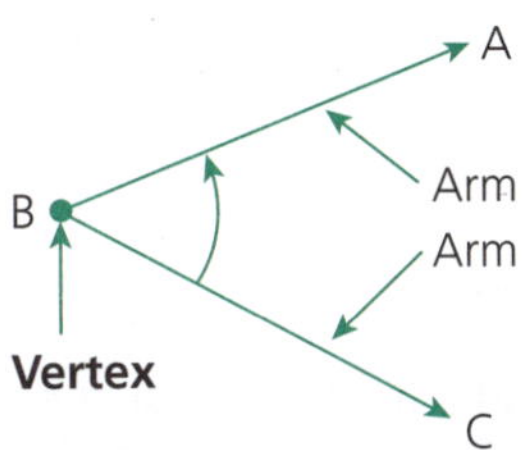

Notation

For the above angles we can use any of the following notations:

- ∠ABC
- $A\hat{B}C$
- $\hat{B}$

Types of angles

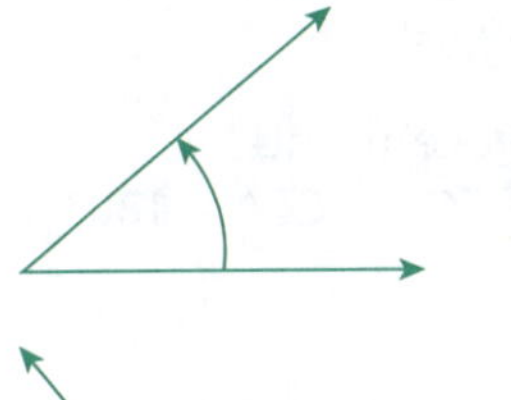

Acute angle—measures between 0° and 90°.

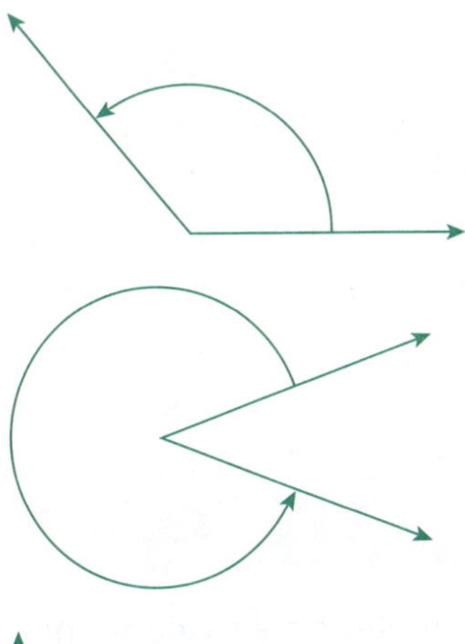

Obtuse angle—measures between 90° and 180°.

Reflex angle—measures between 180° and 360°.

Right-angle—measures 90°.

Straight angle—measures 180°.

Revolution—measures 360°.

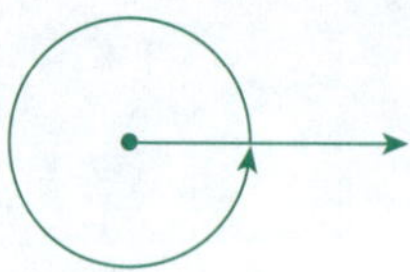

Complementary angles—add up to 90°.

Supplementary angles—add up to 180°.

Adjacent angles—are two angles that have a common arm and common vertex. ∠ABD is adjacent to ∠DBC.

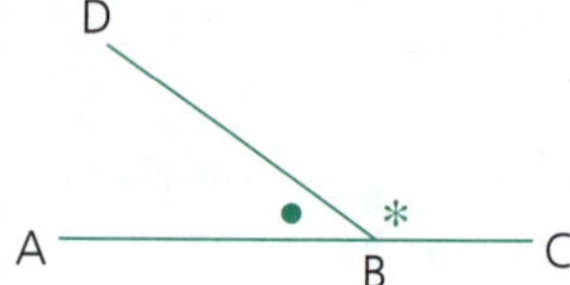

For Example

1

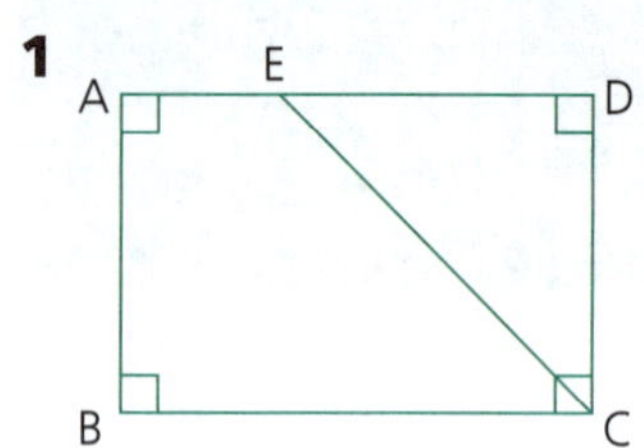

In the figure, ABCD is a rectangle.
Name:

a an acute angle

S ∠DEC is an acute angle (also ∠ECD, ∠EBC).

b an obtuse angle

S ∠CEA is an obtuse angle.

c a right angle

S There are four right angles in the diagram, $\hat{A}$, $\hat{B}$, $\hat{C}$ and $\hat{D}$.

d a straight angle

S ∠DEA is a straight angle.

e a pair of complementary angles

S ∠BCE and ∠ECD are complementary, since ∠BCE + ∠ECD = 90°.

f a pair of supplementary angles.

S ∠DEC and ∠CEA are adjacent and supplementary since ∠DEC + ∠CEA = 180°.

Further angles

Angles on a straight line

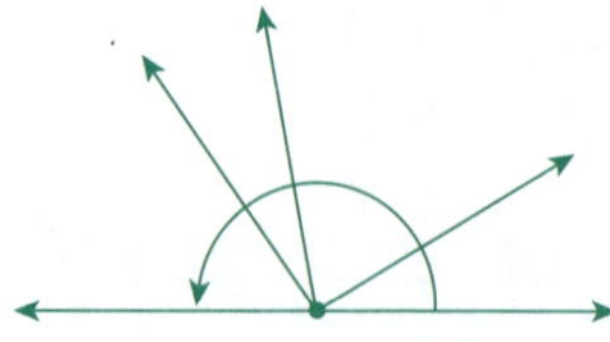

Angles on a straight line add up to 180°.

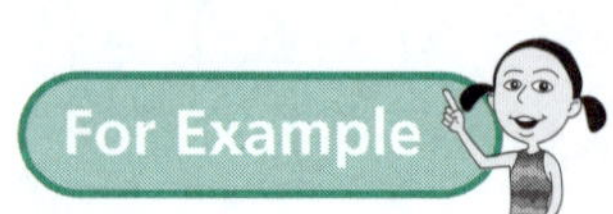

1 Find x.

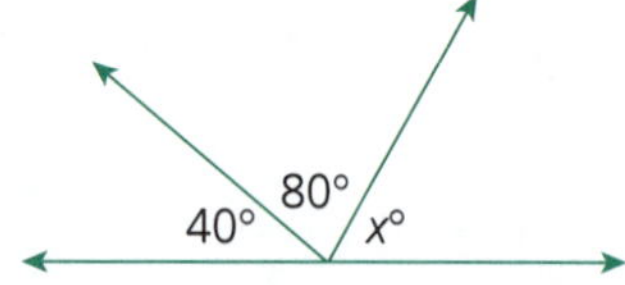

S $x + 80 + 40 = 180$ (angles on a straight line)

$\therefore x = 60$

Vertically opposite angles

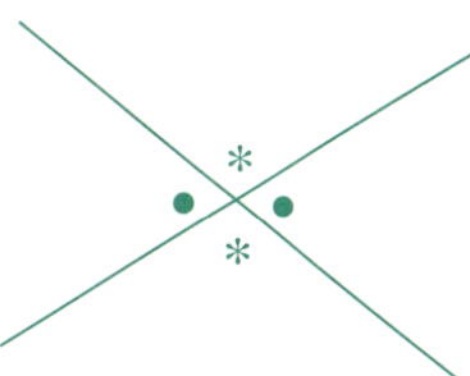

When two lines intersect, two pairs of vertically opposite angles are formed. Vertically opposite angles are equal.

For Example

1 Find x, y and z. (Give reasons.)

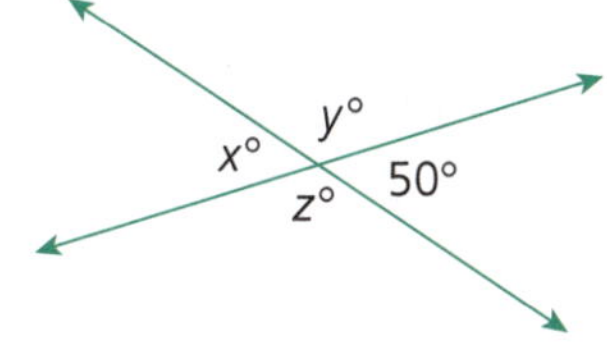

S

$x = 50$ (vertically opposite angles)
$y + 50 = 180$ (supplementary angles)
$\therefore y = 130$
$z = 130$ (vertically opposite angles)

Angles at a point

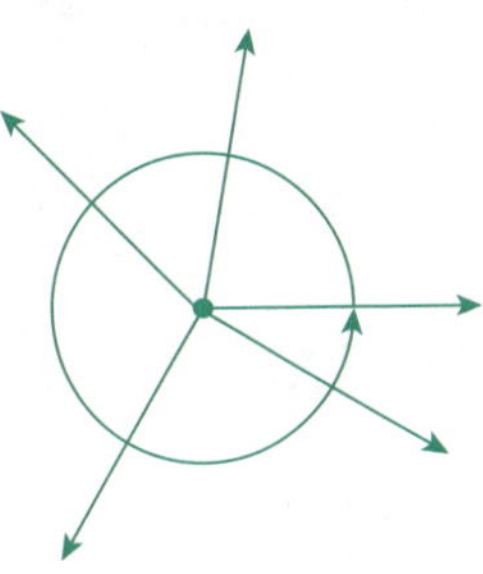

Angles at a point add up to 360°.

For Example

1 Find x, giving reasons.

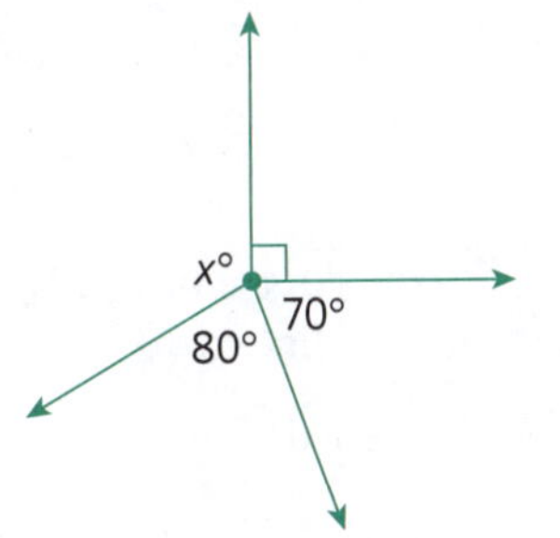

S

$x + 80 + 70 + 90 = 360$
(angles at a point)
$\therefore x + 240 = 360$
$x = 120$

Parallel lines

When a pair of parallel lines is cut by a **transversal**, three types of special angles are formed.

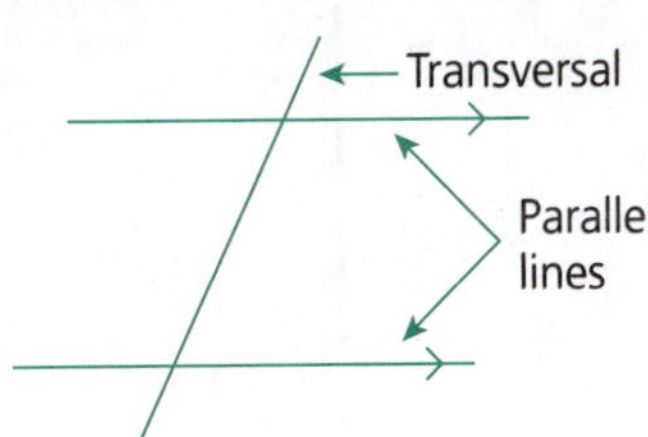

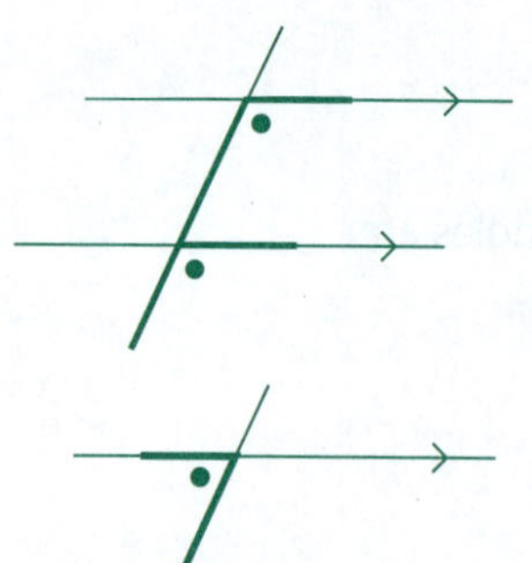

Corresponding angles are equal.

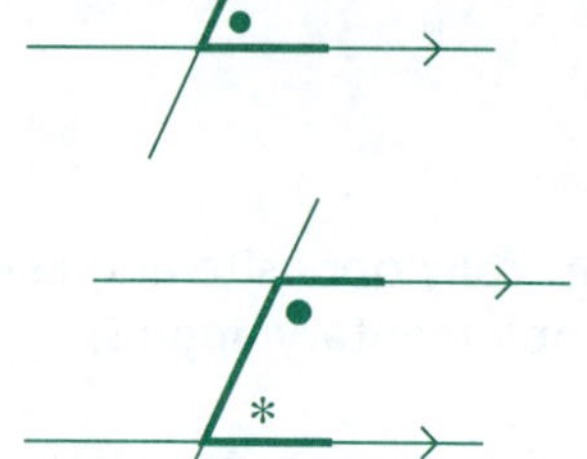

Alternate angles are equal.

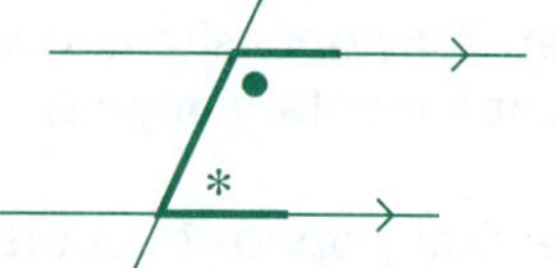

Co-interior angles are supplementary (that is, they add up to 180°).

For Example

1 CD || EF. Find $E\hat{Q}P$.
Give reasons.

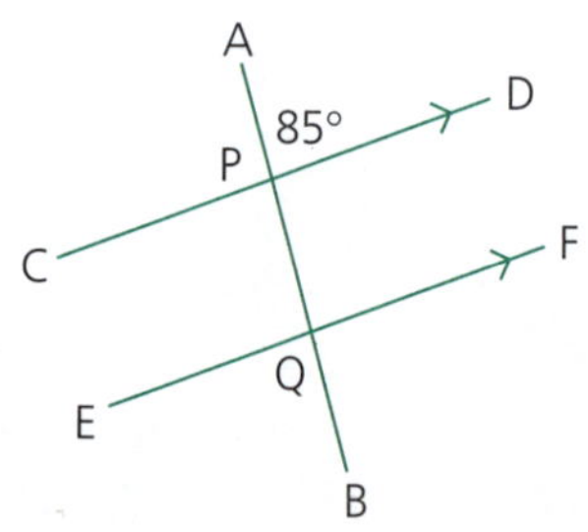

S $Q\hat{P}C = 85°$ (vertically opposite $A\hat{P}D$)
$E\hat{Q}P = 95°$ (co-interior to $Q\hat{P}C$ and CD || EF)

2 AB || CD, $E\hat{G}B = 42°$.
Find $G\hat{H}D$. Give reasons.

S $G\hat{H}D = 42°$ (corresponding to $E\hat{G}B$ and AB || CD)

3 PQ || RS.
Find $R\hat{N}M$.
Give reasons.

S $R\hat{N}M = 50°$ (alternate to $Q\hat{M}N$ and PQ || RS)

4

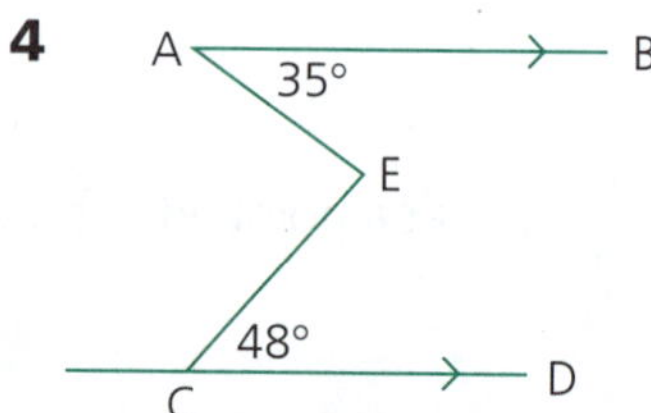

AB || CD. Find ∠CEA. Give reasons.

S

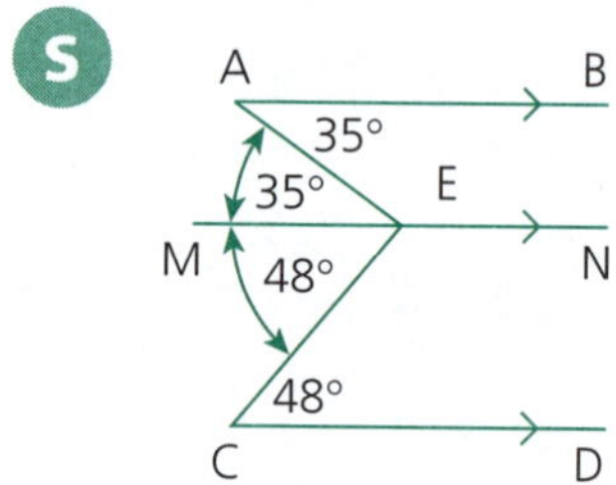

Construct a line MN through E that is parallel to AB.

Since MN || AB and AB || CD, then MN || CD.

$M\hat{E}A = 35°$ (alternate to $B\hat{A}E$ and AB || MN)

$C\hat{E}M = 48°$ (alternate to $E\hat{C}D$ and MN || CD)

$\therefore C\hat{E}A = (35 + 48)° = 83°$

Note: if two lines are parallel to a third line, they are parallel to one another.

Triangles

Angle sum of a triangle

The angle sum of any triangle is 180°.

Isosceles triangle
The base angles of an isosceles triangle are equal.

Equilateral triangle
Any angle in an equilateral triangle is 60°.

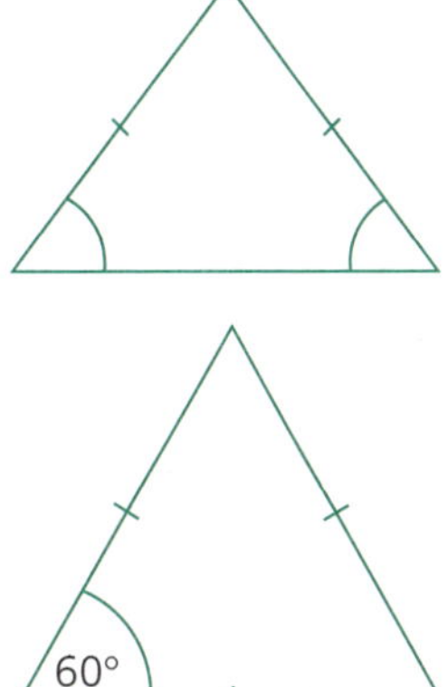

Also, the exterior angle of a triangle equals the sum of the two interior angles, so that $a° = b° + c°$.

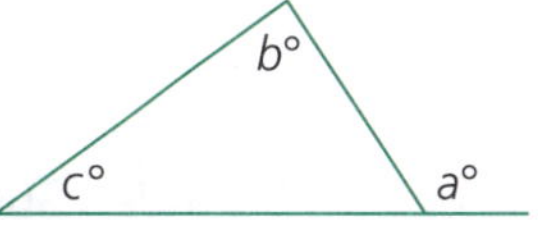

For Example

1 Find x. Given reasons.

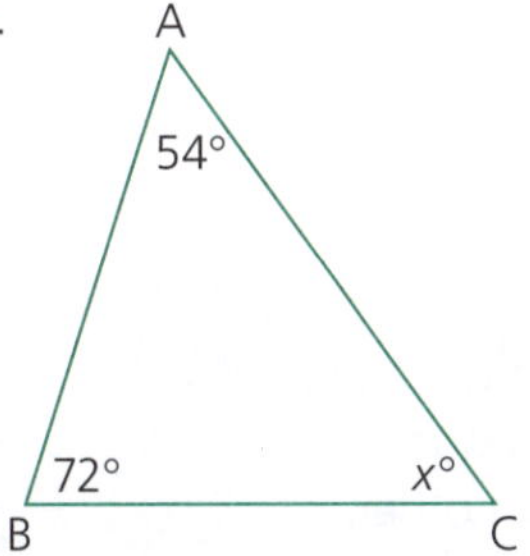

S $x + 72 + 54 = 180$ (angle sum of △ABC)
$x + 126 = 180$
$x = 54$

2 In △FGH, FG = FH and $\text{F}\hat{\text{G}}\text{H} = 50°$.
Find $\text{H}\hat{\text{F}}\text{G}$.
Give reasons.

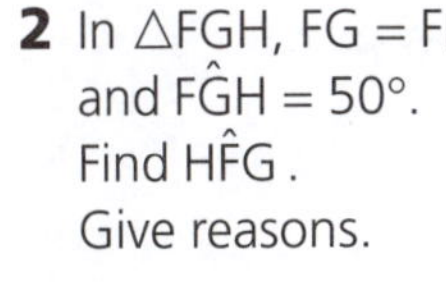

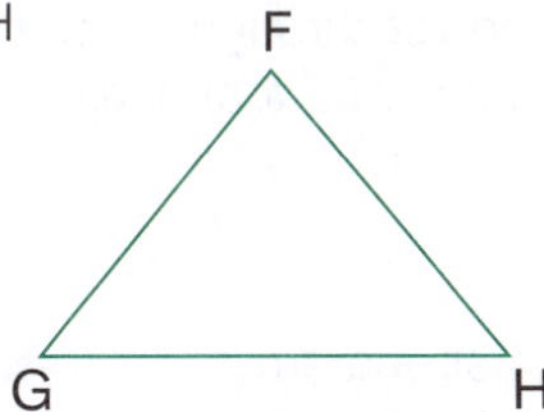

S $\text{G}\hat{\text{H}}\text{F} = 50°$ (base angles of isosceles △FGH)
$\text{H}\hat{\text{F}}\text{G} + 50° + 50° = 180°$ (angle sum of △FGH)
$\therefore \text{H}\hat{\text{F}}\text{G} = 80°$

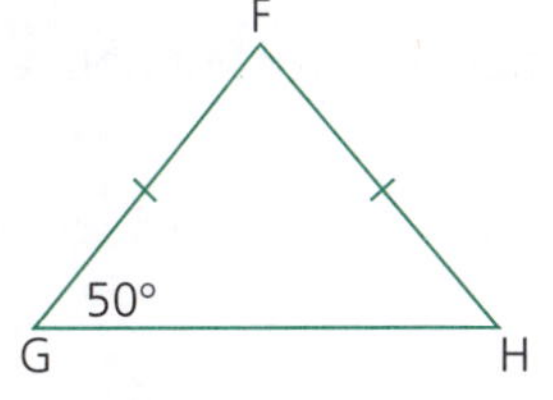

3 In the figure, △ABC is equilateral. Find x, giving reasons.

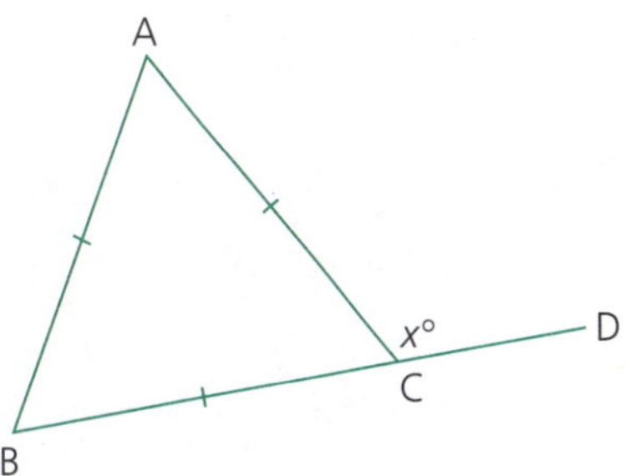

S $\text{B}\hat{\text{C}}\text{A} = 60$ (angle in equilateral △ABC)
$x = 120$ ($\text{A}\hat{\text{C}}\text{D}$ is supplementary to $\text{B}\hat{\text{C}}\text{A}$)

4 In the diagram, FG ∥ BC
$\angle CAB = 68°$
$\angle ABC = 49°$.
Find the value of x, giving reasons.

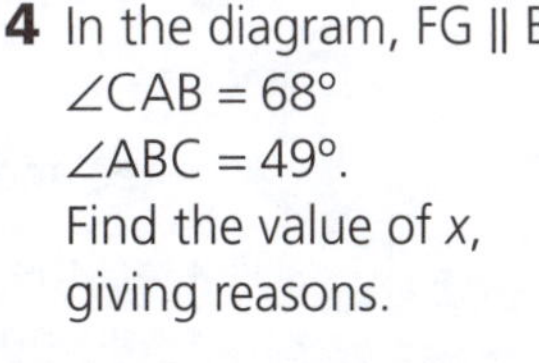

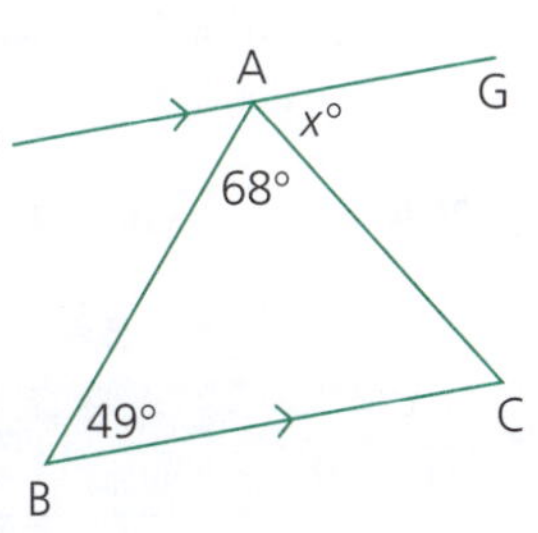

S $\angle BCA + 49° + 68° = 180°$
(angle sum of △ABC)
$\therefore \angle BCA = 63°$
$x = 63$
($\angle GAC$ is alternate to $\angle BCA$ and FG ∥ BC)

Quadrilaterals

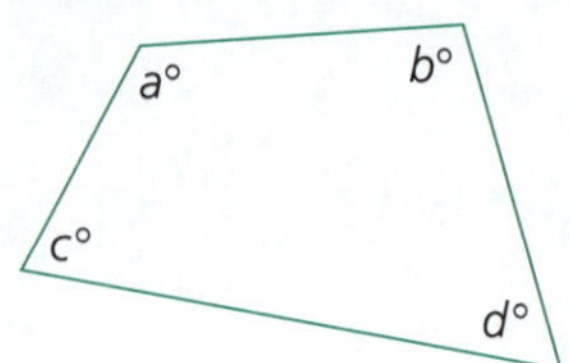

The angle sum of any quadrilateral is 360°, that is, $a + b + c + d = 360$.

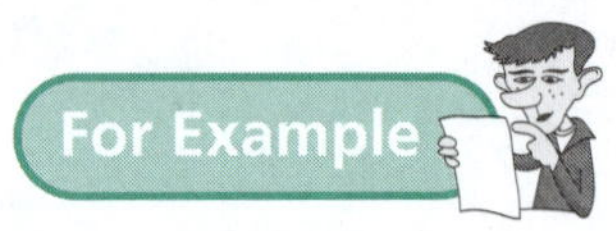

1 Find x, giving reasons.

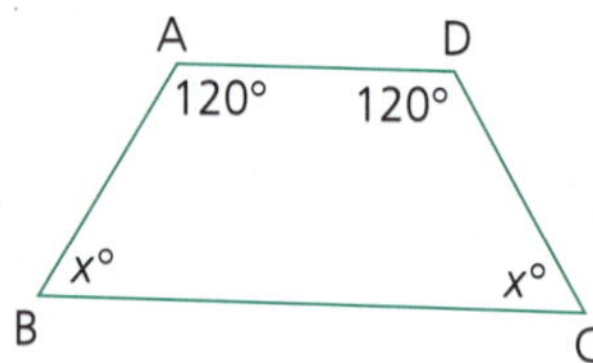

S $x + x + 120 + 120 = 360$
(angle sum of quadrilateral)
$2x + 240 = 360$
$2x = 120$
$x = 60$

2 Find x, giving reasons.

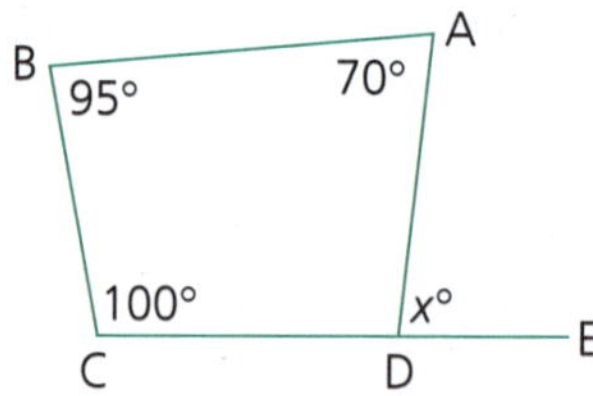

S $\angle CDA + 70° + 95° + 100° = 360°$
(angle sum of quadrilateral ABCD)
$\therefore \angle CDA = 95°$
$x = 85$ ($\angle ADE$ is supplementary to $\angle CDA$)

Properties of special quadrilaterals

Parallelogram

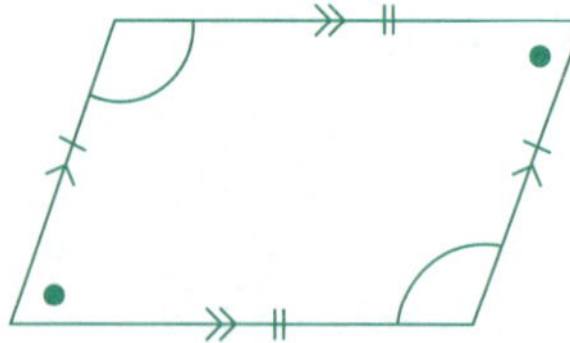

In a parallelogram:

- opposite sides are equal and parallel
- opposite angles are equal
- diagonals bisect each other.

Rhombus

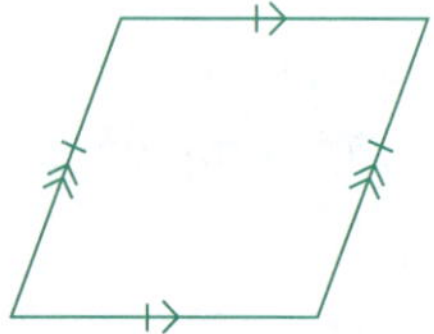

A rhombus has:

- all the properties of a parallelogram
- all sides equal
- diagonals bisecting the angles through which they pass
- diagonals bisecting each other at right angles.

Rectangle

A rectangle has:

- all the properties of a parallelogram
- all angles right angles
- the diagonals equal in length.

Square

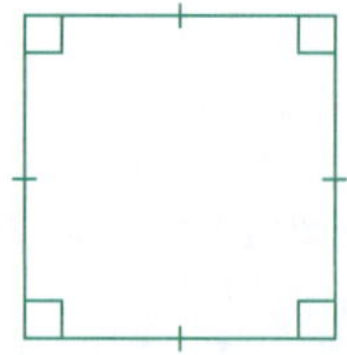

A square has all the properties of both a rhombus and a rectangle.

1

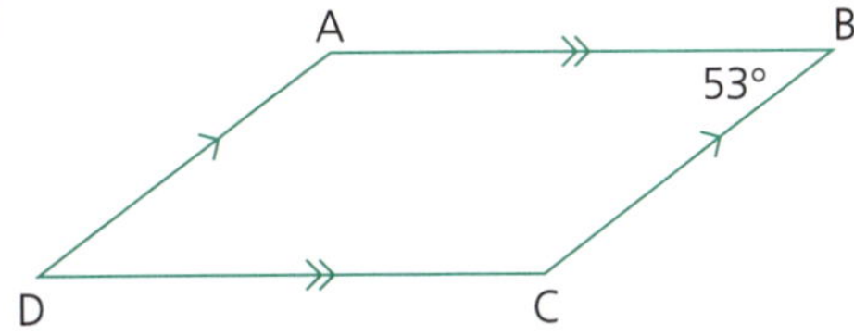

ABCD is a parallelogram. $C\hat{B}A = 53°$.

a Find $A\hat{D}C$, giving reasons.

S **$A\hat{D}C = 53°$ (opposite angles in a parallelogram are equal)**

b Find $B\hat{A}D$, giving reasons.

S **$B\hat{A}D = 127°$ (co-interior to $C\hat{B}A$ and $AD \parallel BC$)**

2

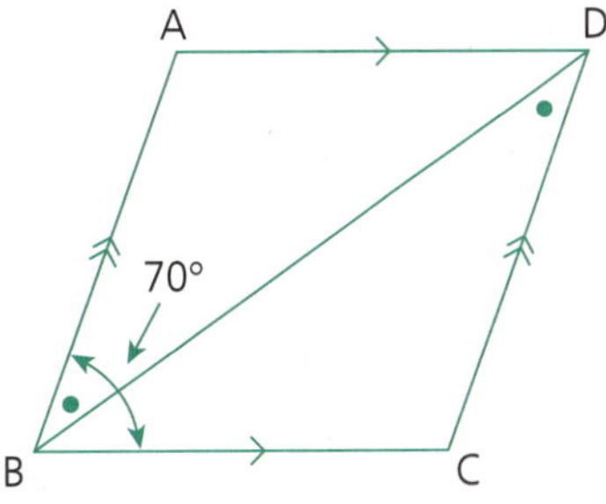

ABCD is a rhombus, $A\hat{B}C = 70°$.

Find $C\hat{D}B$, giving reasons.

S **$A\hat{B}D = 35°$ (diagonal BD of rhombus bisects $A\hat{B}C$)**
$C\hat{D}B = 35°$ (alternate to $A\hat{B}D$ and $AB \parallel DC$)

3

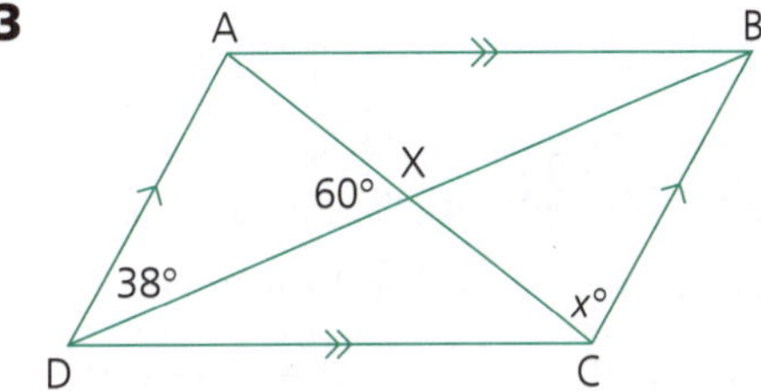

ABCD is a parallelogram. Find x, giving reasons.

S
$$
\begin{aligned}
C\hat{B}D &= 38° \\
&\quad \text{(alternate to } A\hat{D}B \text{ and } AD \parallel BC) \\
B\hat{X}C &= 60° \text{ (vertically opposite } D\hat{X}A) \\
x + 60 + 38 &= 180 \text{ (angle sum of } \triangle BXC) \\
x &= 82
\end{aligned}
$$

Polygons

A *regular* polygon has all angles and sides equal.

Interior and exterior angles

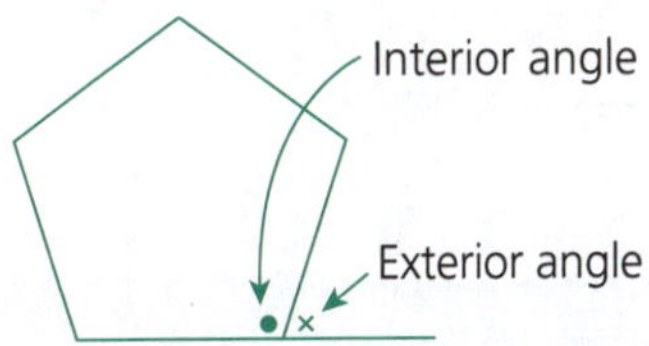

Angle sum of an *n*-sided polygon

The angle sum of an n-sided polygon is $(n - 2) \times 180°$, where n is the number of sides of the polygon.

Interior angle of regular polygon

In a regular n-sided polygon, the interior angle can be calculated by dividing the angle sum by the number of sides.

Therefore, interior angle $= \dfrac{(n-2)\times 180°}{n}$.

1 Find the size of the interior angle of a regular decagon.

Decagon $\Rightarrow n = 10$

Angle sum $= (10 - 2) \times 180°$
$= 8 \times 180°$
$= 1440°$

Size of each interior angle $= 144°$

2 Find the angle sum of a pentagon and hence find the size of each interior angle of a regular pentagon.

Note: regular polygons have all angles and sides equal.

A pentagon has five sides, that is, $n = 5$.

Angle sum of pentagon $= (n - 2) \times 180°$
$= (5 - 2) \times 180°$
$= 3 \times 180°$
$= 540°$

A pentagon has five angles.

Therefore each interior angle $= 540° \div 5$
$= 108°$

3 In the figure, ABCDEF is a regular hexagon. Find the size of $F\hat{E}G$. Give reasons.

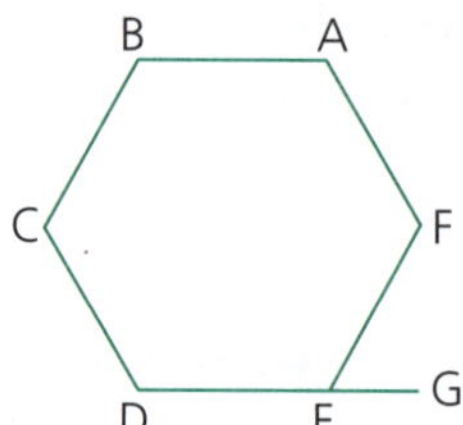

S A hexagon has six sides, that is, $n = 6$.

Angle sum of hexagon $= (n - 2) \times 180°$
$= (6 - 2) \times 180°$
$= 4 \times 180°$
$= 720°$

$D\hat{E}F = \dfrac{720°}{6}$ (angle in a regular hexagon)
$= 120°$

$F\hat{E}G = 60°$ (supplementary to $D\hat{E}F$)

Exterior angles of polygons

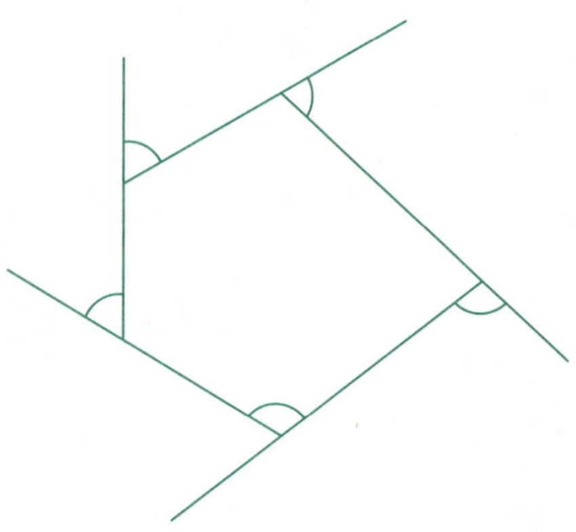

In a regular polygon with n sides.

An n-sided polygon has n exterior angles. They are formed by producing (extending) each side in order around the polygon.

In the diagram, there are five sides so there are five exterior angles.

The sum of the **exterior** angles of any polygon is 360°.

Note: the exterior and interior angles of any polygon are supplementary (that is, they add up to 180°).

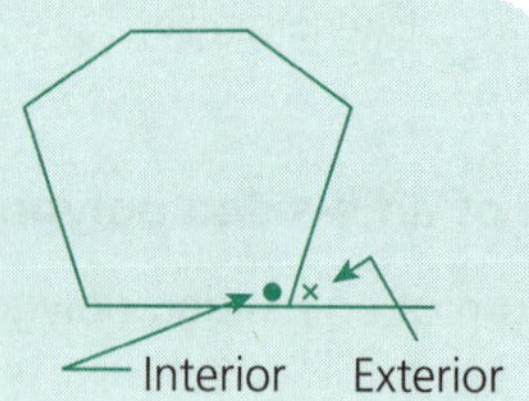

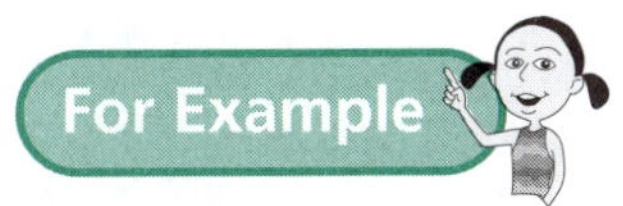

1 Find the exterior angle of a regular hexagon.

Exterior angle of regular polygon $= \dfrac{360^\circ}{6}$
For hexagon, $n = 6$

$$\Rightarrow \text{Each exterior angle} = \frac{360^\circ}{6} = 60^\circ$$

2 Find the size of the interior angle of a regular decagon.

The interior angle can be found by first finding the exterior angle.

$$\text{Exterior angle} = \frac{360^\circ}{10} = 36^\circ$$

Interior angle + Exterior angle = 180°

$\therefore$ Interior angle of regular decagon
$= (180 - 36)^\circ$
$= 144^\circ$

Congruent triangles

Congruent triangles are equal in all respects.

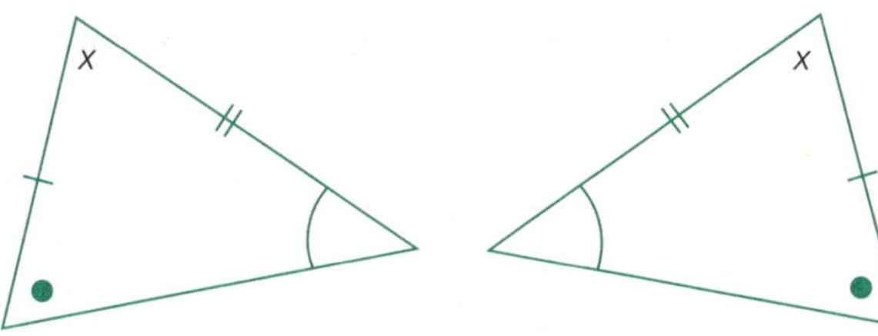

Note:
- Three sides of one equal to three sides of the other.
- Three angles of one equal to three angles of the other.
- Their areas are equal.
- Their perimeters are equal.

Tests for congruent triangles

Two triangles are congruent if any one of the following four tests (conditions) is true.

(SSS) test

Three sides of one triangle are respectively equal to the three sides of the other (SSS).

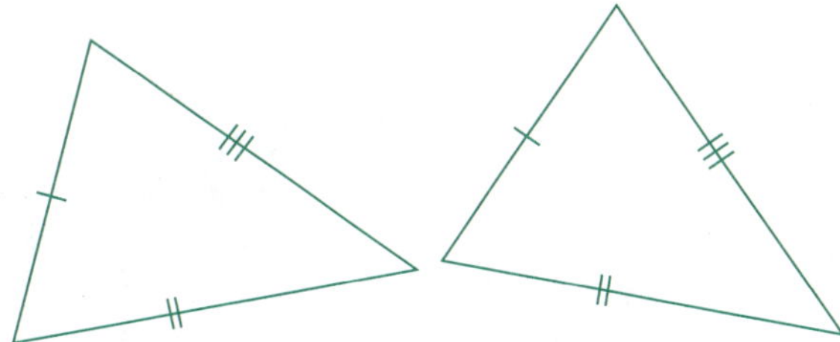

(SAS) test

Two sides of one triangle and the included angle are respectively equal to two sides and the included angle of the other (SAS).

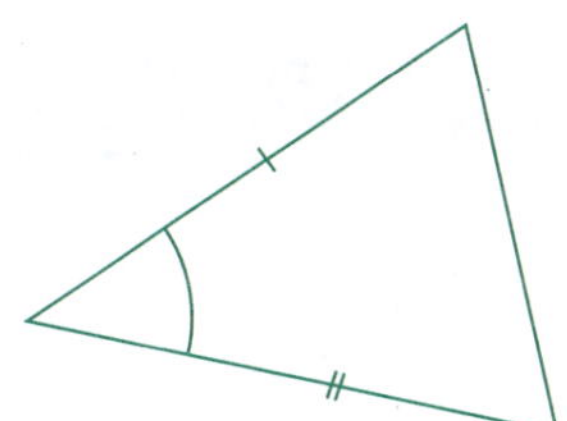

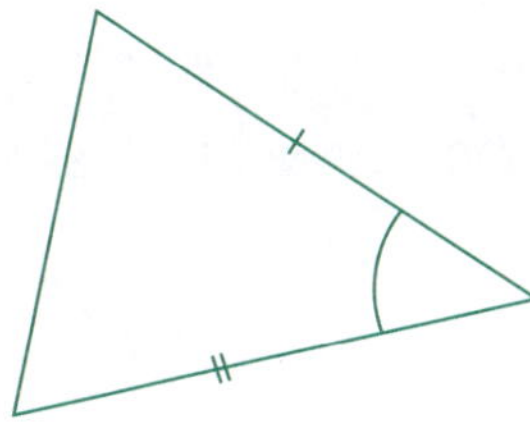

(AAS) test

Two angles and one side of one triangle are respectively equal to two angles and the corresponding side of the other (AAS).

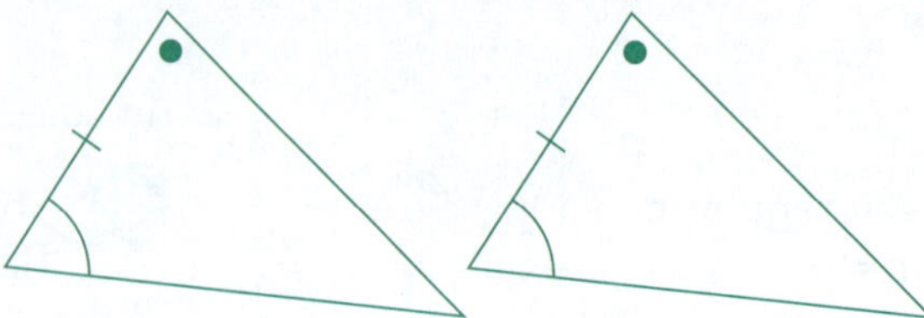

(RHS) test

Two right-angled triangles are congruent if the hypotenuse and one side of one triangle are respectively equal to the hypotenuse and one side of the other (RHS).

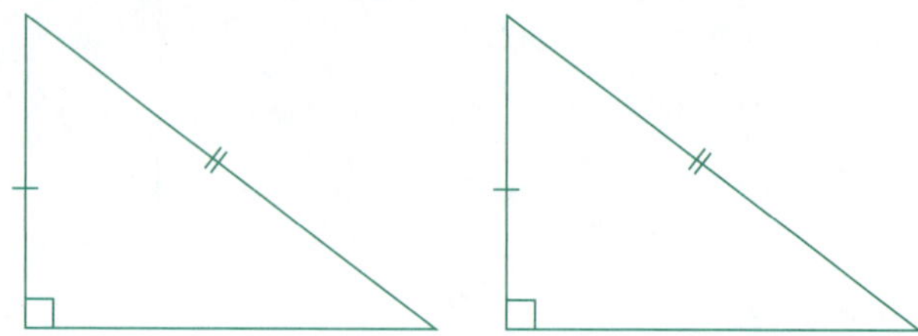

1 In each case below, state which of the congruence tests would be used in proving each pair of triangles congruent.

a

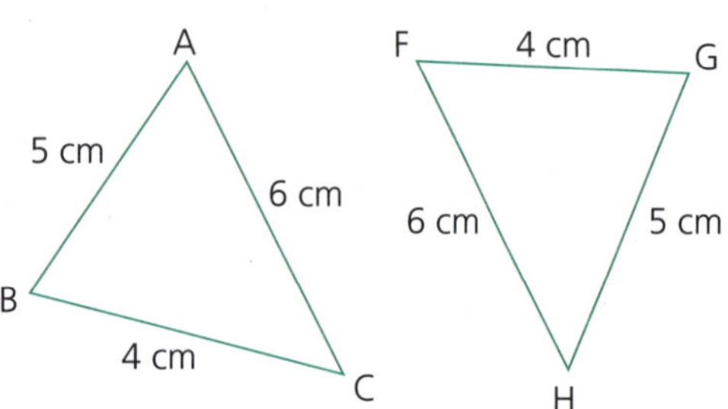

S △ABC ≡ △HGF (SSS)

Note: ≡ means 'is congruent to'.

b

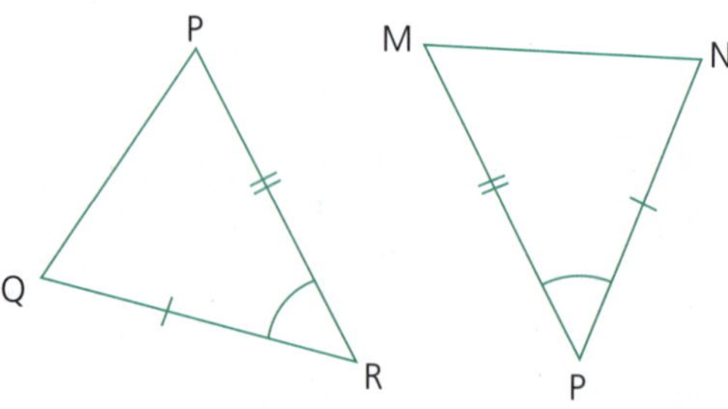

S △QRP ≡ △NPM (SAS)

c

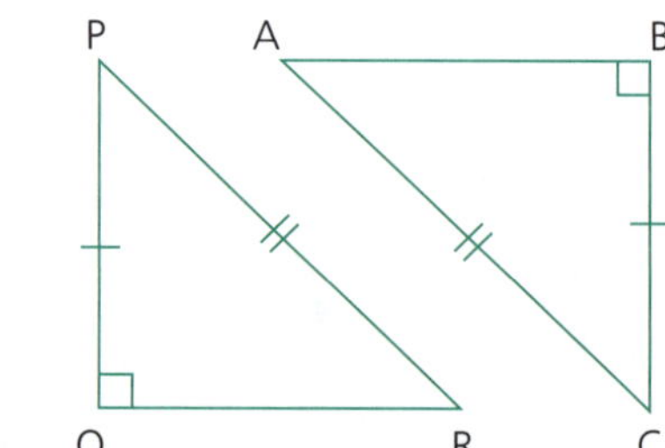

S △QRP ≡ △BAC (RHS)

d

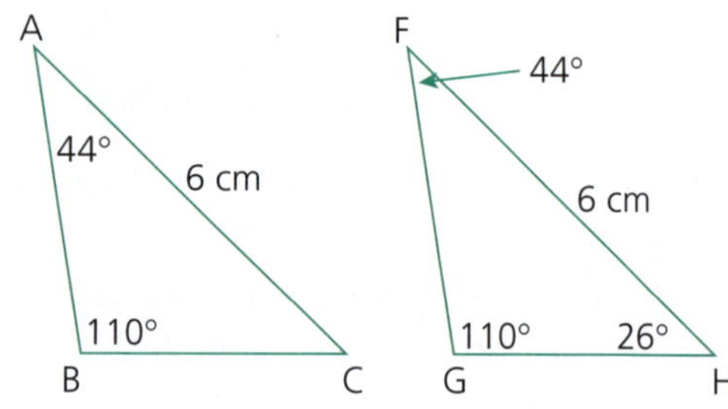

S △ABC ≡ △FGH (AAS)

Formal proofs for congruent triangles

When proving congruent triangles:

- mark all information given (data) on the diagram and then mark any other angles or sides that you can prove are equal
- remember that a congruence proof consists of three equality statements
- always give reasons.

1 Given that AC = EC and BC = CD, prove triangles ABC and EDC are congruent.

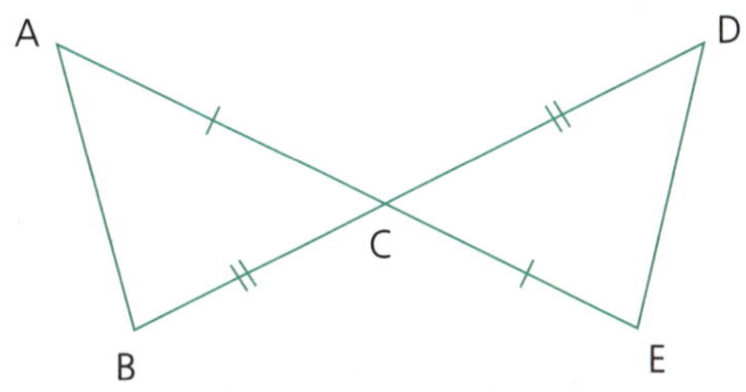

Note: the included angle is the one contained between the sides proved equal.

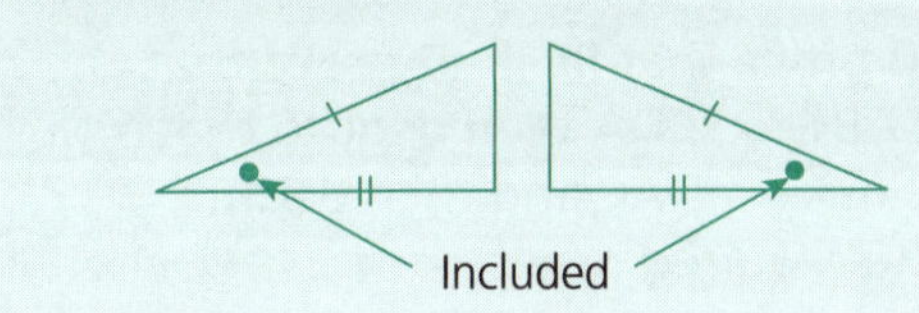

S

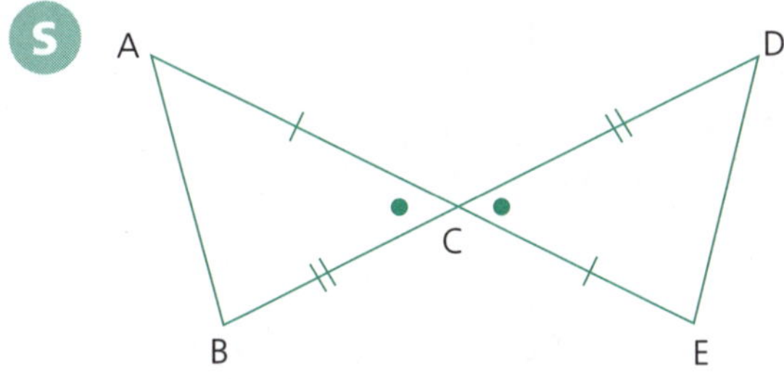

In the triangles ABC and EDC:

AC = EC	(S)	(data)
BC = DC	(S)	(data)
∠BCA = ∠DCE	(A)	
(vertically opposite angles)		
∴ △ABC ≡ △EDC		(SAS)

2 Given that ∠CDA = ∠ABC, and also ∠ACD and ∠CAB are both right-angles, prove that △ACD ≡ △CAB.

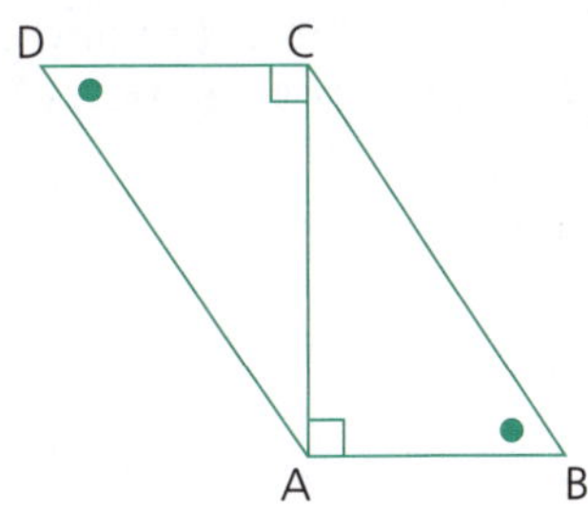

S

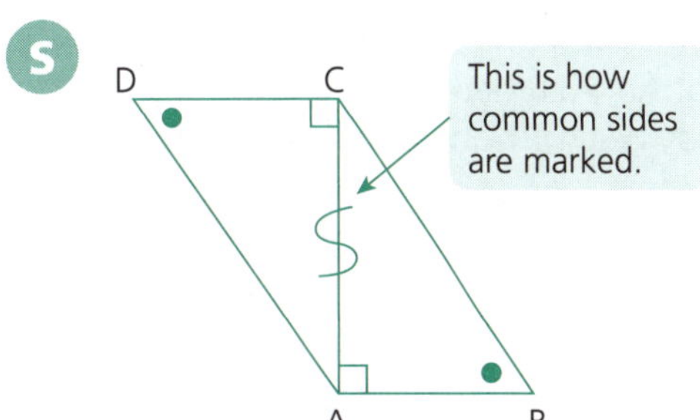

In △s ACD and CAB

∠CDA = ∠ABC	(A)	(data)
∠ACD = ∠CAB	(A)	(data, both 90°)
AC = CA	(S)	(common)
∴ △ACD ≡ △CAB	(AAS)	

3 Given that AB ⊥ DC (∠CDB = 90°) and AC = BC, prove that △ADC ≡ △BDC.

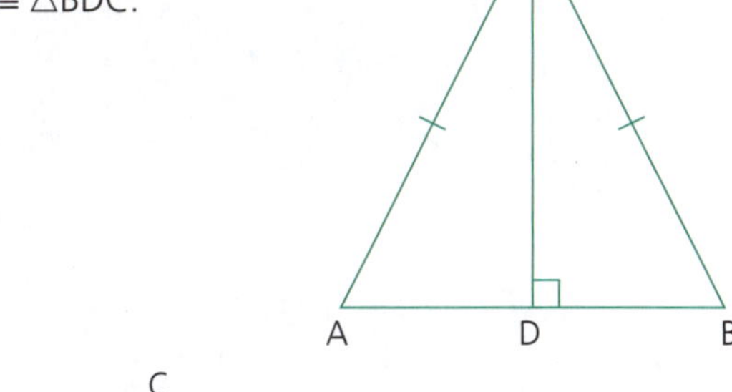

S

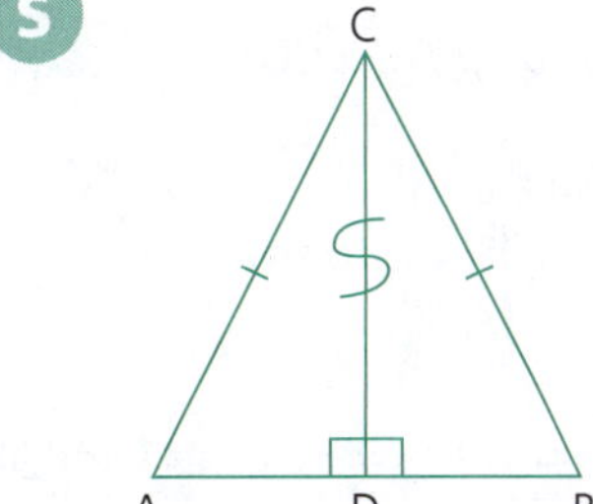

In △s ACD and BDC

∠ADC = ∠BDC = 90°	(R)	(AB ⊥ DC)
AC = BC	(H)	(data)
CD = CD	(S)	(common side)
∴ △ADC ≡ △BDC	(RHS)	

Using congruent triangles to deduce other geometrical results

If two triangles are proved to be congruent, then all other corresponding sides and angles of the triangles are known to be equal. If △ABC ≡ △EDC:

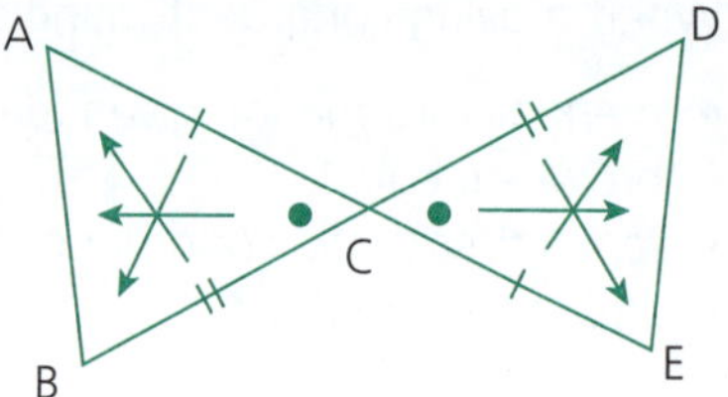

Then AB = ED (corresponding sides in congruent △s)
∠ABC = ∠EDC (corresponding angles in congruent △s)
∠CAB = ∠CED (corresponding angles in congruent △s)

1 Given that AB = CD and AB || CD, prove that BX = CX by firstly proving that △ABX ≡ △DCX.

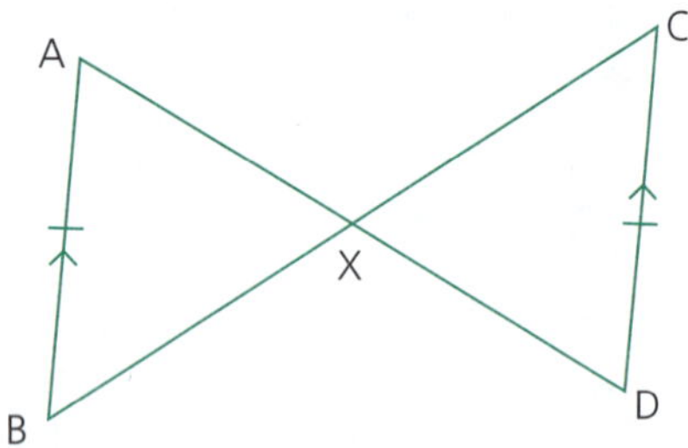

S To prove that BX = CX, prove that △ABX ≡ △DCX and then deduce the required result.

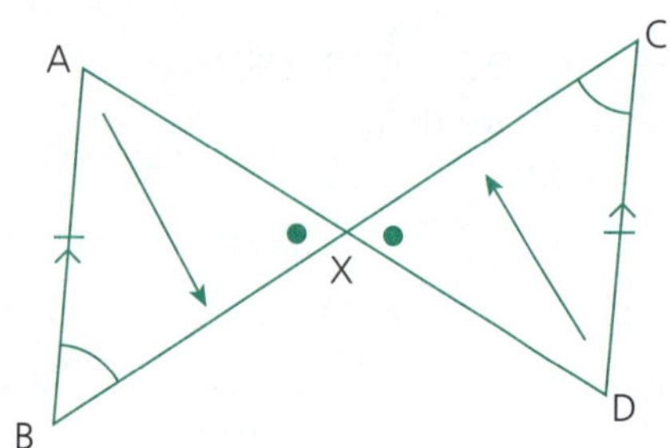

In △s ABX and DCX
∠ABX = ∠DCX (A)
(alternate angles and AB || CD)
∠BXA = ∠CXD (A)
(vertically opposite angles)
AB = DC (S) (data)
△ABX ≡ △DCX (AAS)
∴ BX = CX
(corresponding sides in congruent triangles)

Both BX and CX are opposite the unmarked angle in each △.

2 Given AB = CD and ∠CAB = ∠ACD (= 90°), prove that BC || AD.

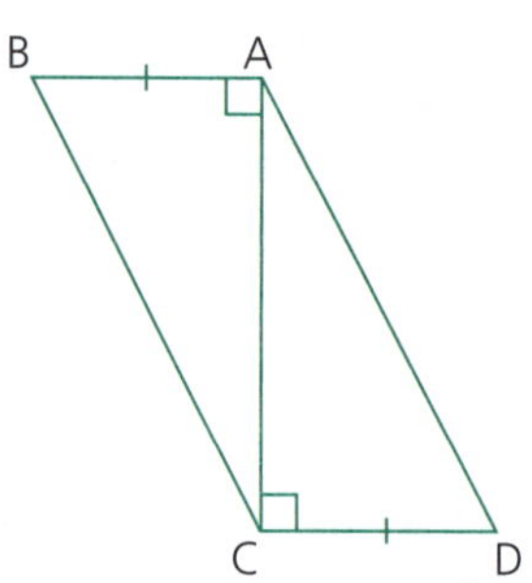

S To prove that BC || AD prove that △ABC ≡ △CDA then deduce that ∠BCA = ∠DAC and then deduce the required result.

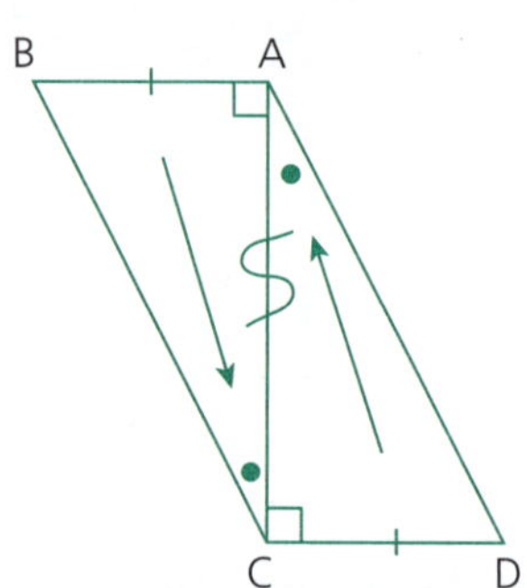

In △s ABC and CDA
AB = CD (S) (data)
∠CAB = ∠ACD
= 90° (A) (data)
AC = CA (S)
(common side)
△ABC ≡ △CDA (SAS)
∠BCA = ∠DAC (corresponding angles in congruent triangles)
∴ BC || AD (a pair of alternate angles BCA and DAC are equal)

Note: || means '**is parallel to**'.

Similarity

Similar figures have their corresponding angles equal (equiangular) and their corresponding sides are in the same proportion (ratio).

For example, figure ABCD is similar to figure EFGH. **The symbol ||| means 'is similar to'**.

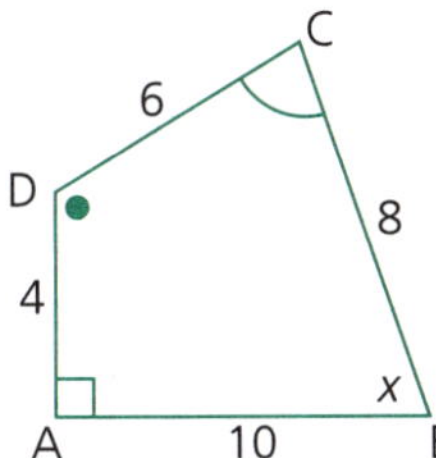

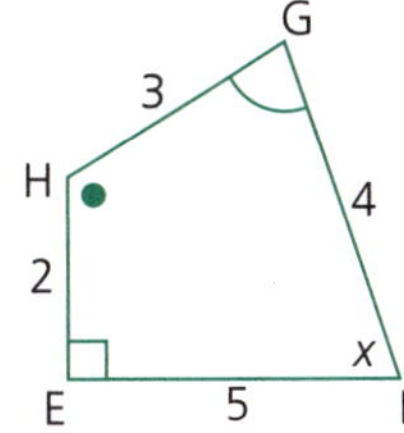

Note:

1 The angles of figure ABCD are equal to the corresponding angles of figure EFGH. That is: $\angle A = \angle E$, $\angle B = \angle F$, $\angle C = \angle G$ and $\angle D = \angle H$.

2 The corresponding sides of the two similar figures ABCD and EFGH are in proportion (the same ratio). That is:

$$\frac{AB}{EF} = \frac{BC}{FG} = \frac{CD}{GH} = \frac{DA}{HE} = \frac{2}{1} \quad \text{or} \quad \frac{10}{5} = \frac{8}{4} = \frac{6}{3} = \frac{4}{2} = 2$$

For Example

1 Find the values of the pronumerals in these pairs of similar figures:

a

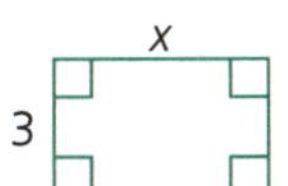

S The two rectangles are similar, therefore

$\frac{x}{10} = \frac{3}{6}$ (corresponding sides are in the same ratio)

$x = \frac{3}{6} \times 5$

$x = 5$

b

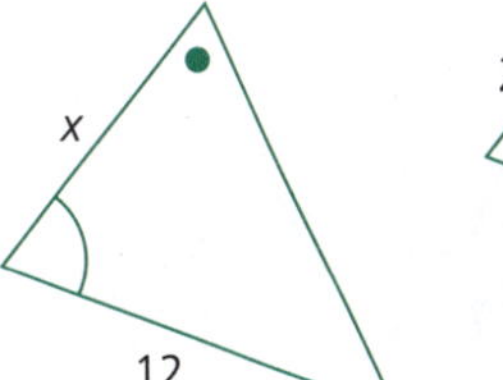

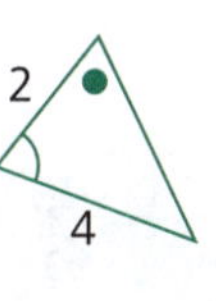

S The two triangles are similar, therefore

$\frac{x}{2} = \frac{12}{4}$ (corresponding sides are in proportion)

$x = \frac{12}{4} \times 2$

$x = 6$

Similar triangles

Two triangles are similar if:

a two angles of one triangle are equal to two angles of the other, or
b the lengths of the sides of the two triangles are in proportion, or
c the lengths of two sides are in proportion and their included angles are equal.

The following figures illustrate the three tests:

a

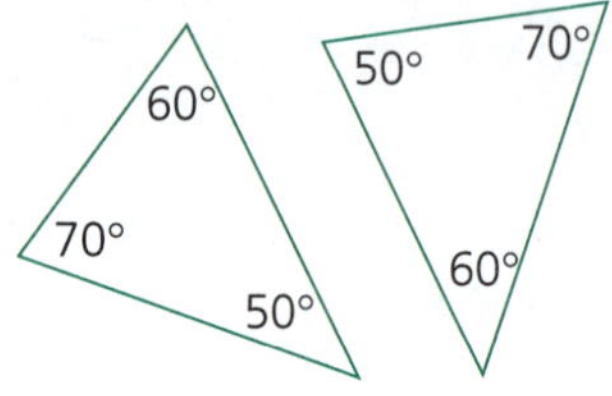

Triangles are similar because they are **equiangular**.

b

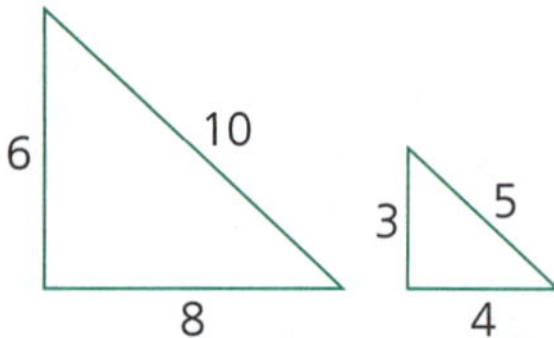

$$\frac{10}{2} = \frac{6}{3} = \frac{8}{4} = 2$$

Triangles are similar because their **corresponding sides are in proportion**.

c

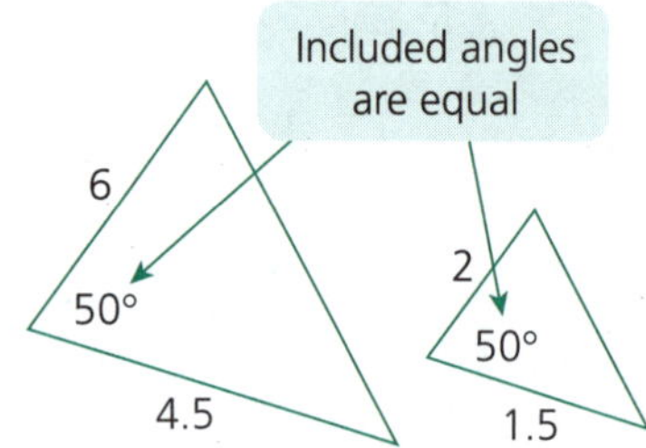

$$\frac{6}{2} = \frac{4.5}{1.5} = 3$$

∴ sides are in proportion. Also included angles are both 50°.

Triangles are similar because the lengths of **two sides are in proportion and their included angles are equal**.

Included angles are ones between proportional sides.

For Example

1 For the following pairs of triangles, state the reasons why they are similar and find the value of each pronumeral.

Note 1: to prove two △s equiangular, only two angles of one need to be shown to be equal to two angles of the other.

The third angles are automatically equal also as all △s have the same angle sum of 180°.

Note 2: corresponding sides in similar triangles are opposite the equal angles.

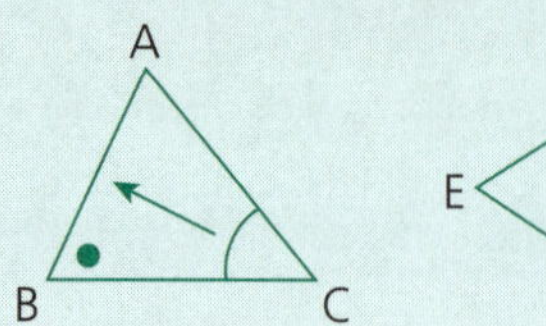

In △ABC and EFG:
AB corresponds to EF
BC corresponds to DF
AC corresponds to ED

a

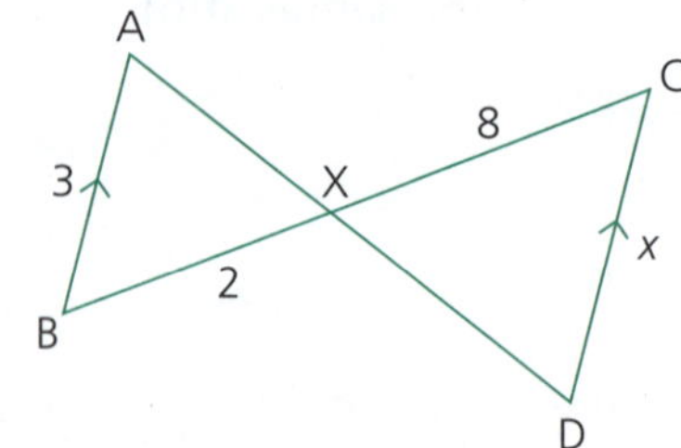

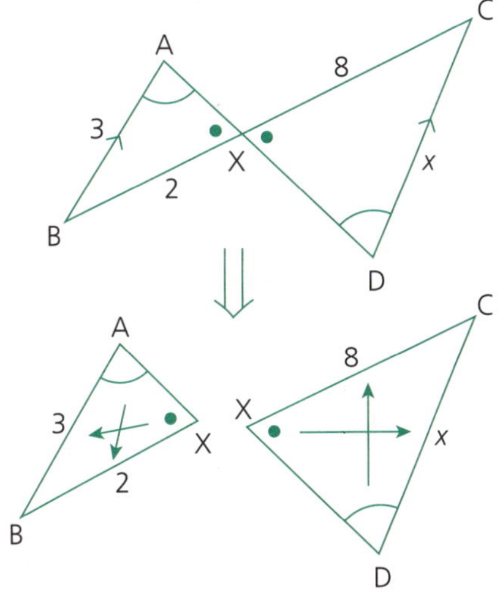

In △s ABX and DCX
∠BXA = ∠CXD (vertically opposite angles)
∠XAB = ∠XDC (alternate angles and AB ∥ CD)
∴ △ABX ||| △DCX (equiangular)

Using ratios of corresponding sides of similar triangles

$$\frac{x}{3} = \frac{8}{2}$$

Solving the equation

$$x = \frac{8}{2} \times 3$$
$$x = 12$$

b

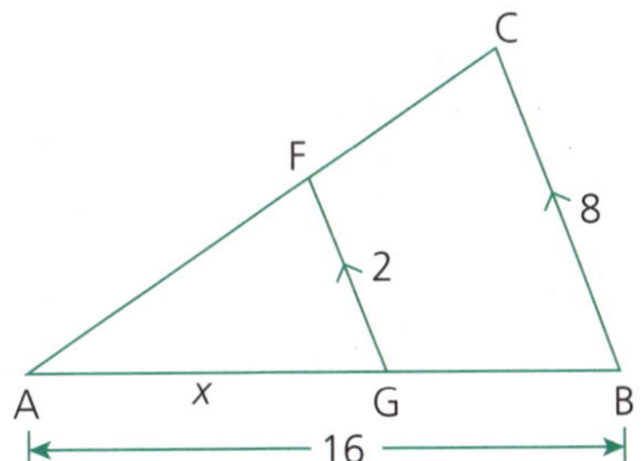

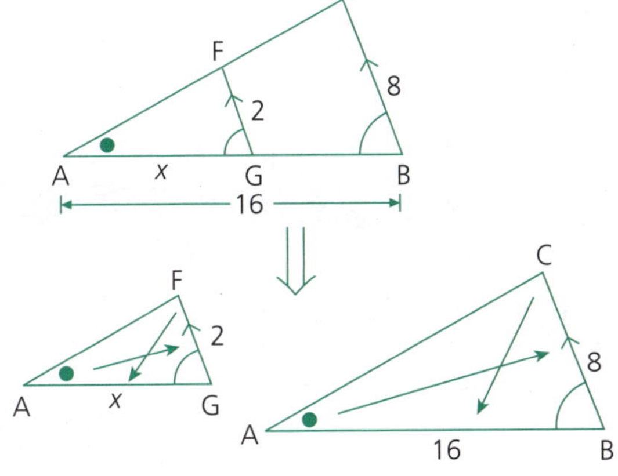

In △s ABC and AGF
∠CAB = ∠FAG (common to both △s)
∠ABC = ∠AGF (corresponding angles and FG ∥ CB)
∴ △ABC ||| △AGF (equiangular)

Since the triangles are similar then corresponding sides must be in the same ratio. That is:

$$\frac{x}{16} = \frac{2}{8}$$
$$x = \frac{2}{8} \times 16$$
$$x = 4$$

c

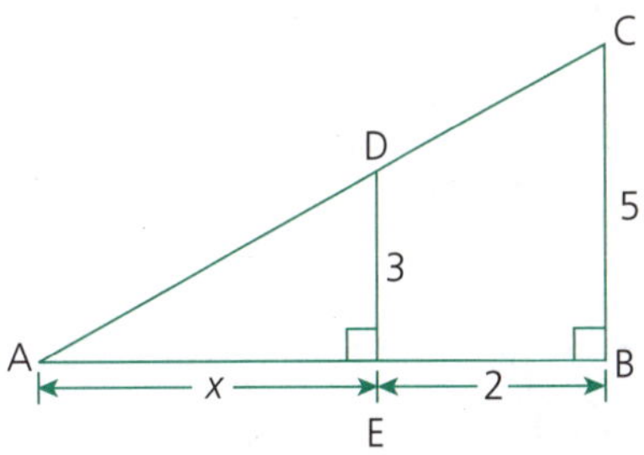

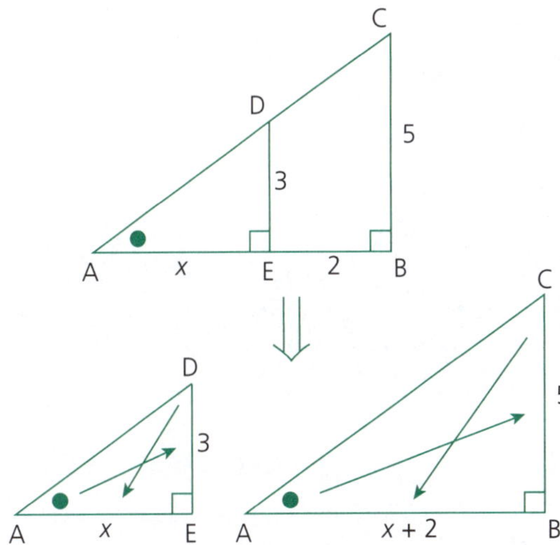

In △s ABC and AED
∠CAB = ∠DAE (common angle)
∠ABC = ∠AED = 90° (given)
∴ △ABC ||| △AED (equiangular)

Using ratios of corresponding sides:

$$\frac{x}{x+2} = \frac{3}{5}$$

Cross-multiply

Solving the equation

$$5x = 3(x + 2)$$
$$5x = 3x + 6$$
$$\therefore 5x - 3x = 6$$
$$2x = 6$$
$$x = 3$$

Practise Practise

1 In each of the following cases, evaluate the pronumeral, giving a brief reason:

a

A D 72° $x°$ B C

b

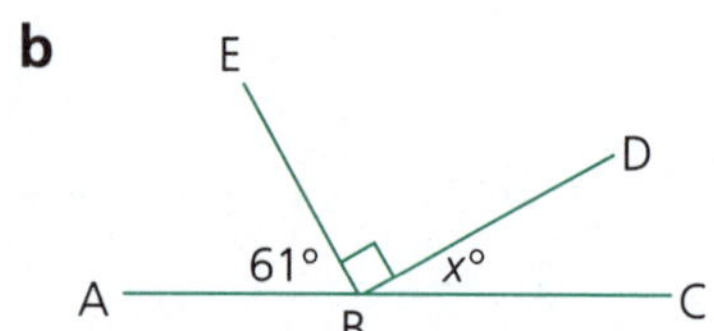

c

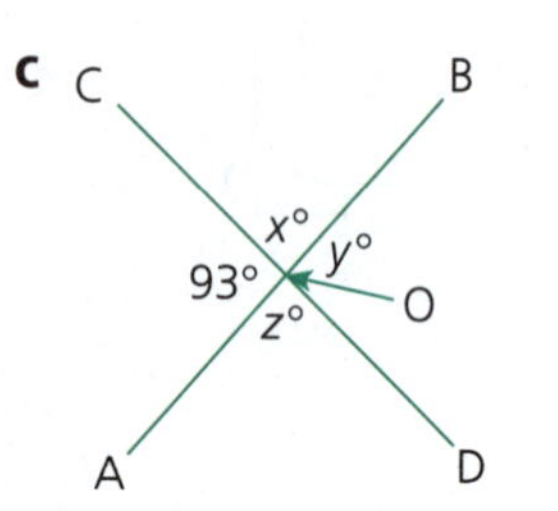

d

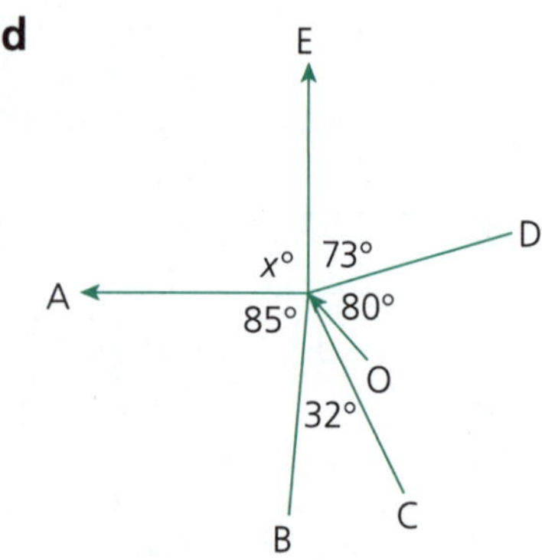

e

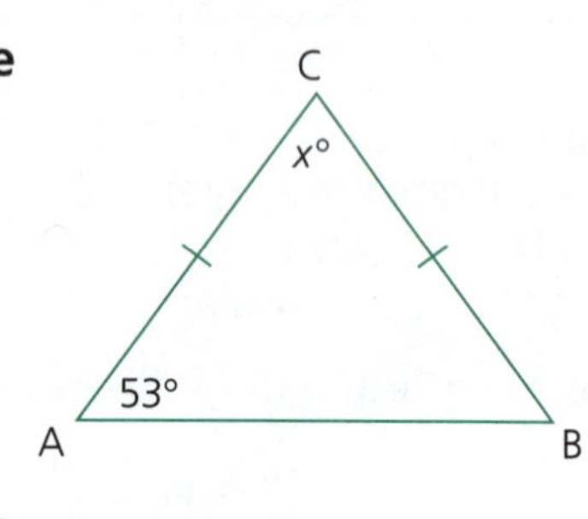

f

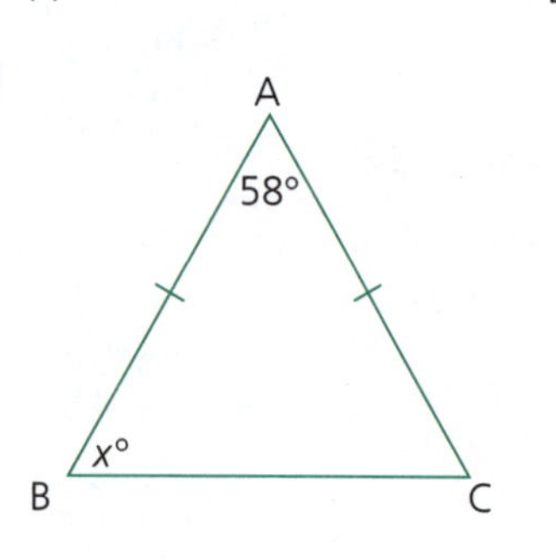

g

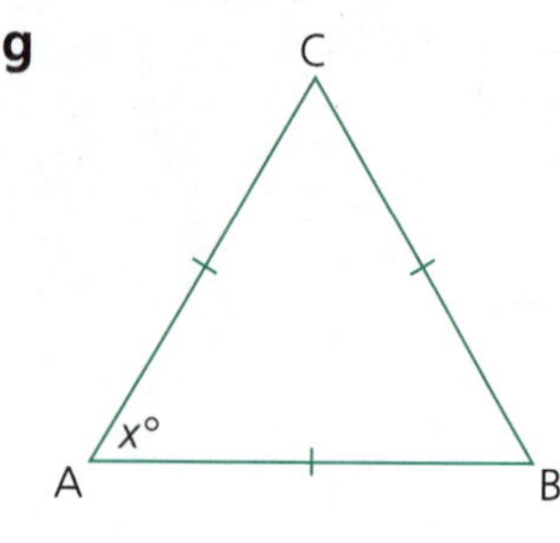

h

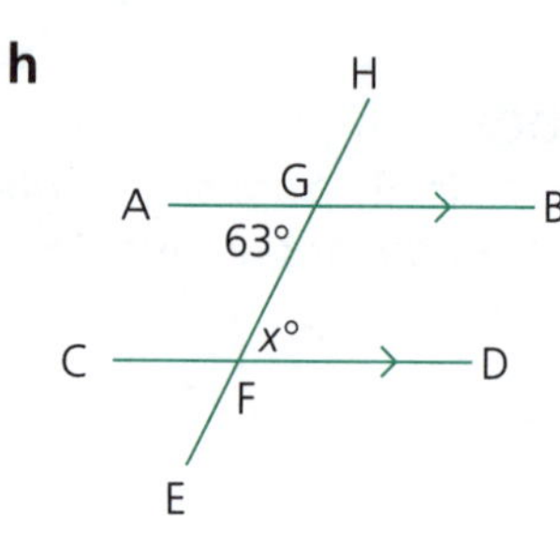

i

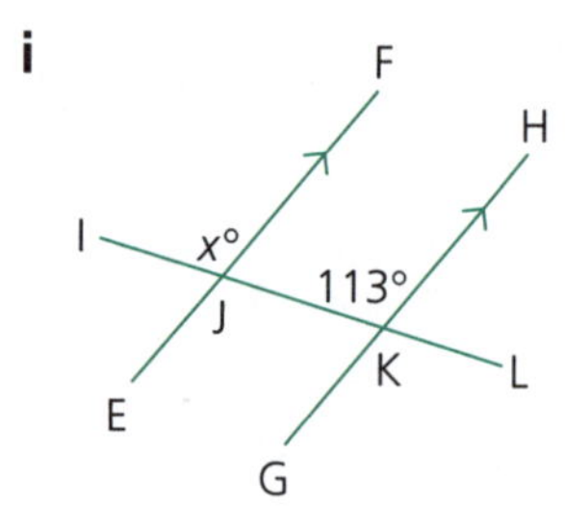

j

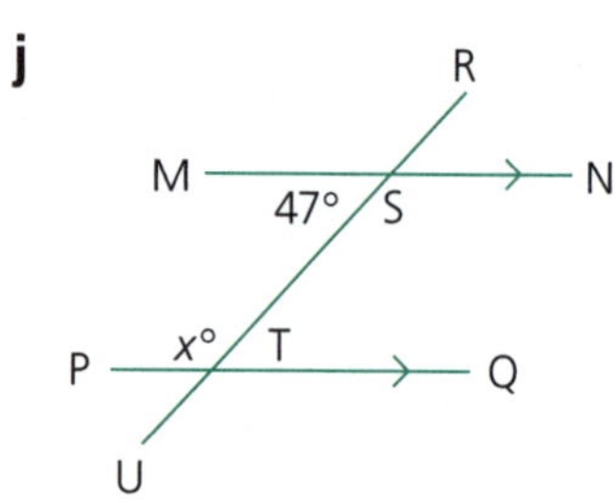

k

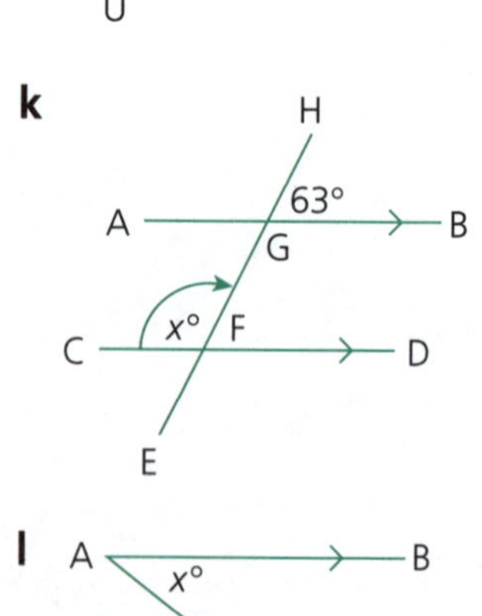

l

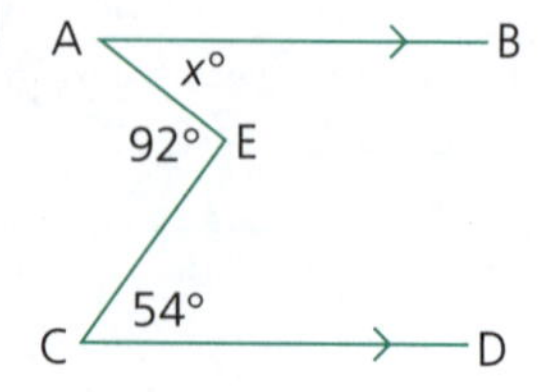

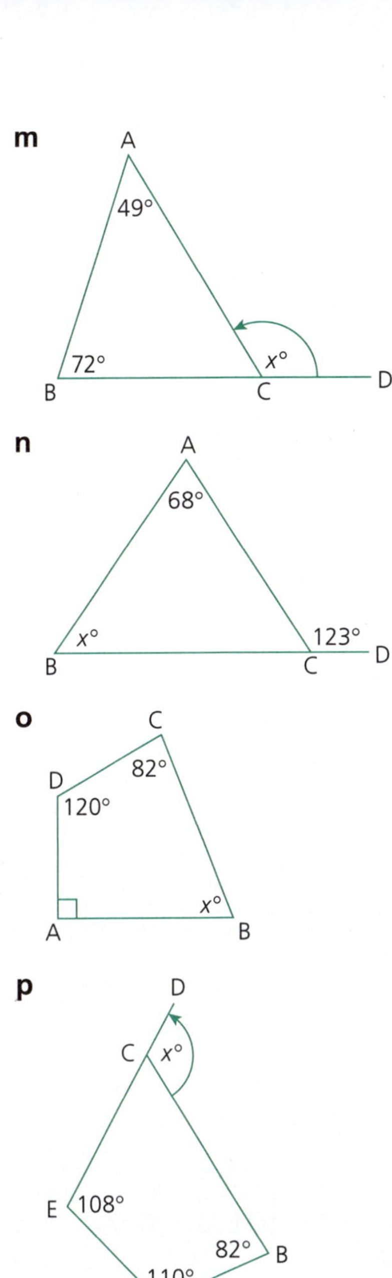

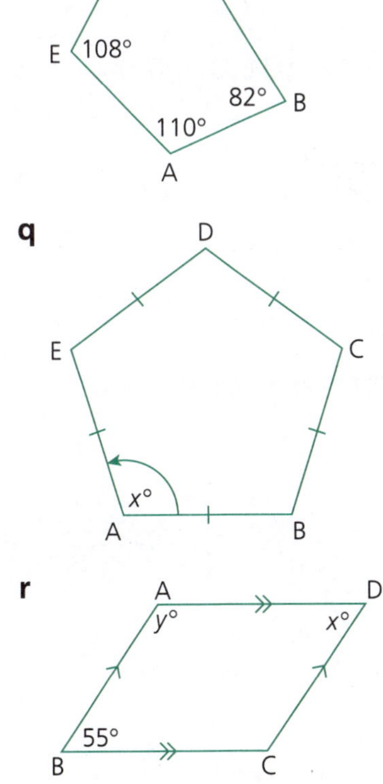

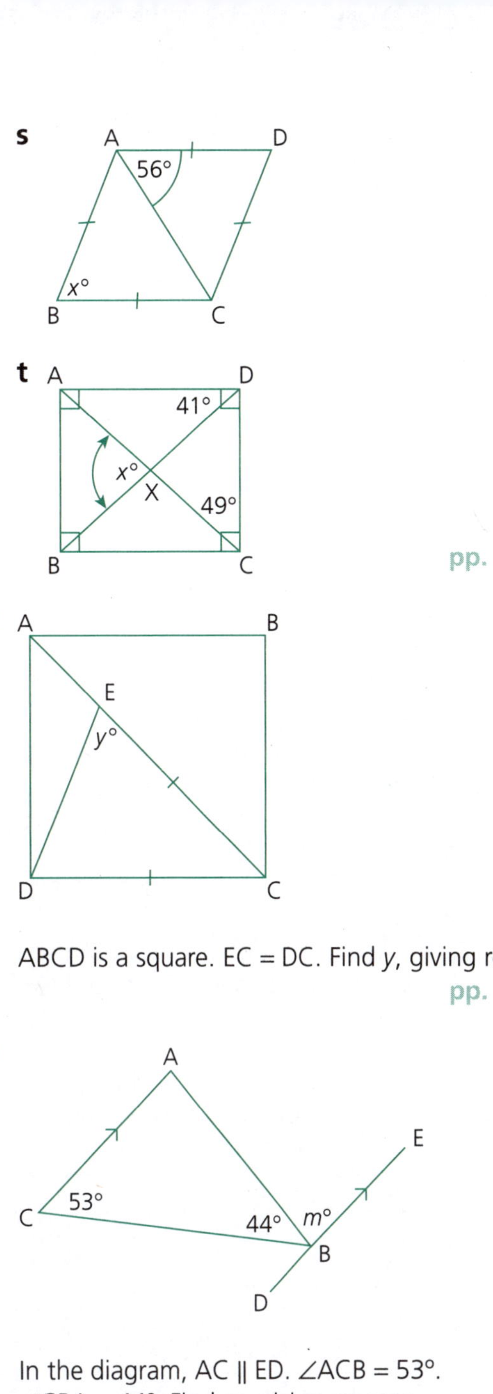

pp. 139–144

ABCD is a square. EC = DC. Find y, giving reasons.

pp. 139–144

In the diagram, AC || ED. ∠ACB = 53°. ∠CBA = 44°. Find m, giving reasons. pp. 141–142

4

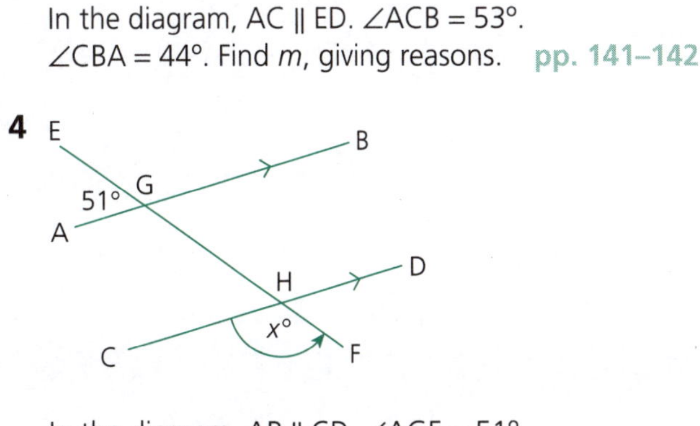

In the diagram, AB || CD. ∠AGE = 51°. Find x, giving reasons. pp. 141–142

5

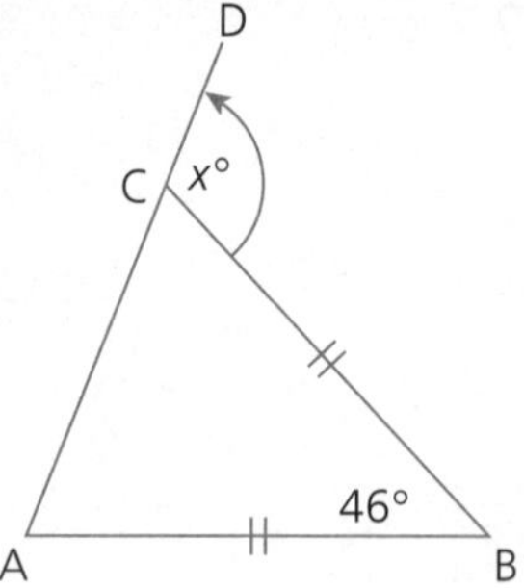

In the figure AB = BC. ∠ABC = 46°.
Find x, giving reasons.

p. 143

6 **a** Find the size of each interior angle of a:
- **i** regular pentagon
- **ii** regular heptagon
- **iii** regular dodecagon
- **iv** regular 20-sided polygon.

b Find the size of each exterior angle of a:
- **i** regular octagon
- **ii** regular nonagon (9 sides).

c
- **i** Each interior angle of a regular polygon is 156°. Find the number of sides of the polygon.
- **ii** Each exterior angle of a regular polygon is 20°. Find the numbers of sides of the polygon.

pp. 145–147

7 Name the test you would use in proving each of the following pairs of triangles congruent:

a

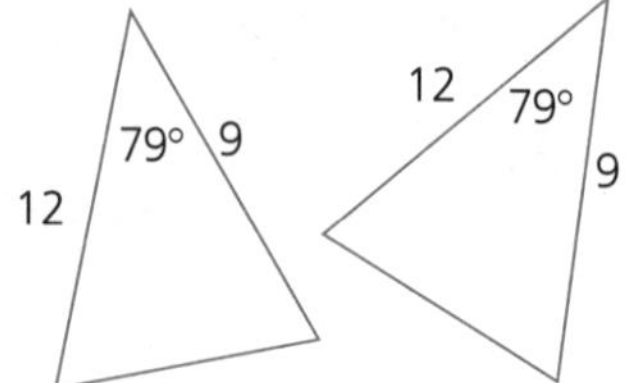

b

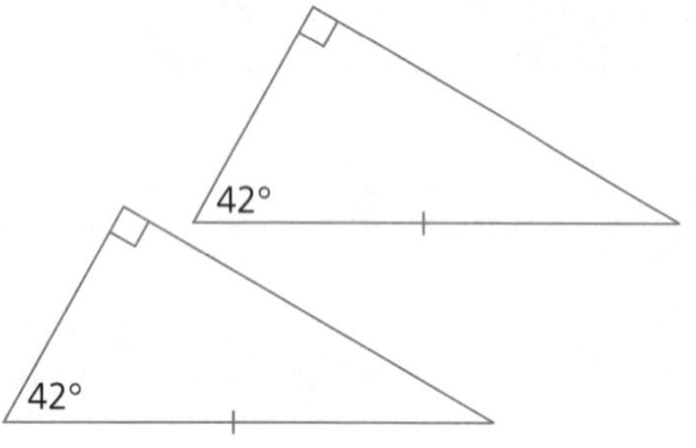

c

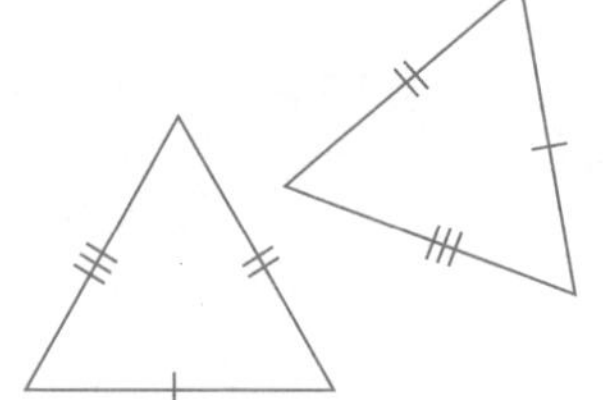

d

e

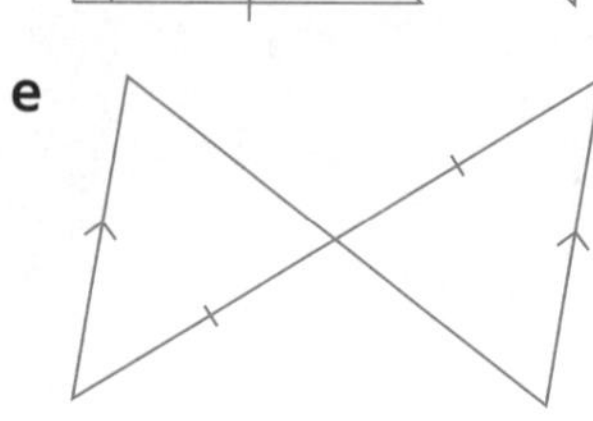

pp. 147–148

8 **a**

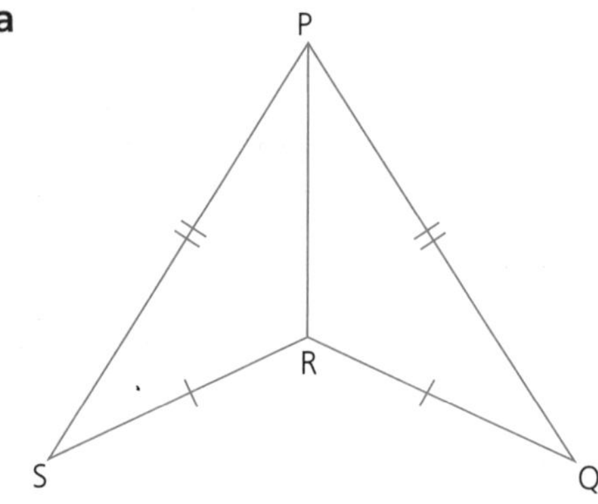

Given PS = PQ and SR = QR, prove that △PRS ≡ △PRQ.

b

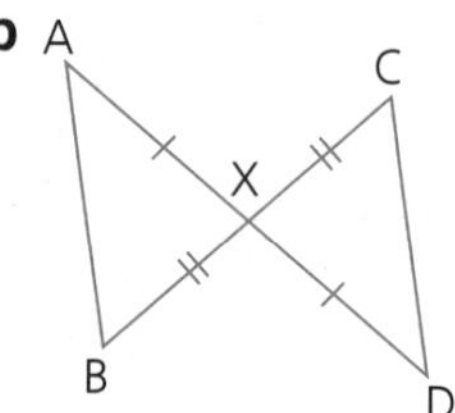

Given that AX = DX and BX = CX, prove that △ABX ≡ △DCX.

c

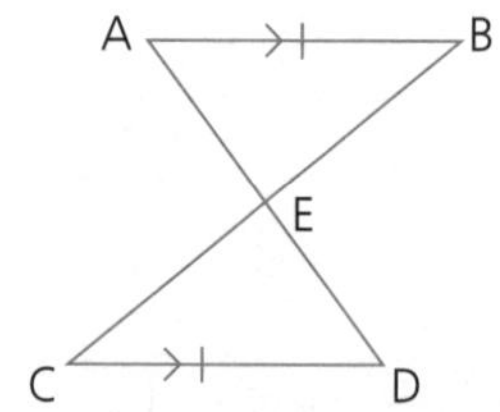

Given that AB || CD and AB = CD, prove that △BAE ≡ △CDE.

d

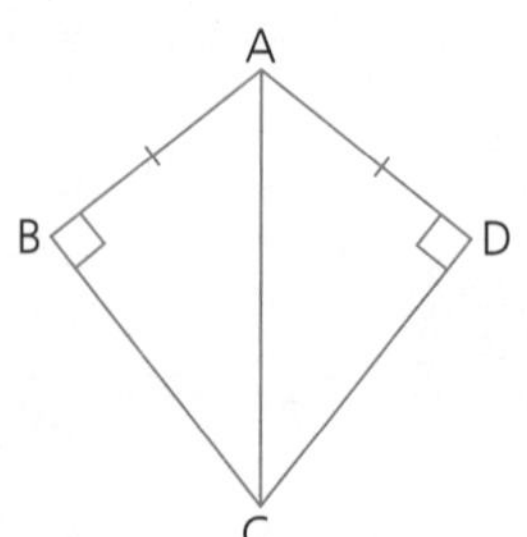

In the diagram AB = AD and
△ABC = △CDA = 90°.
Prove that △ABC ≡ △ADC.

p. 149

9

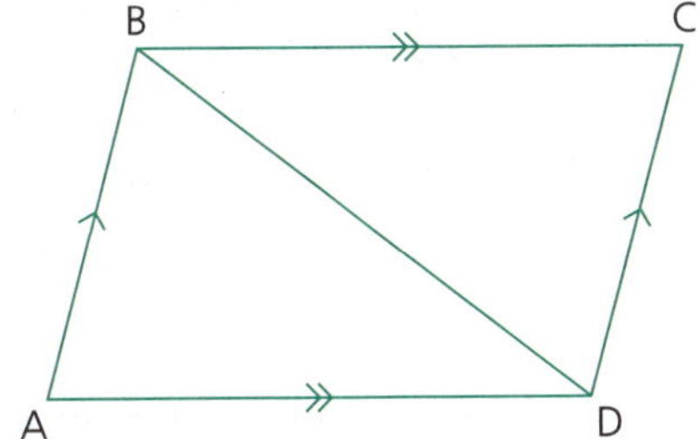

In the diagram, AB || DC and AD || BC.

a Prove that △BAD ≡ △DCB.

b Hence prove that:

i AD = BC

ii ∠BAD = ∠DCB.

pp. 149–151

10

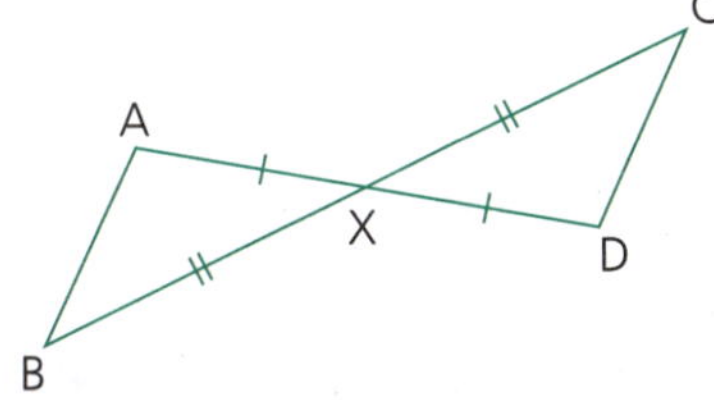

In the figure, AX = DX and BX = CX.

Prove that:

a △ABX ≡ △DCX, and hence that

b AB || CD.

pp. 149–151

11

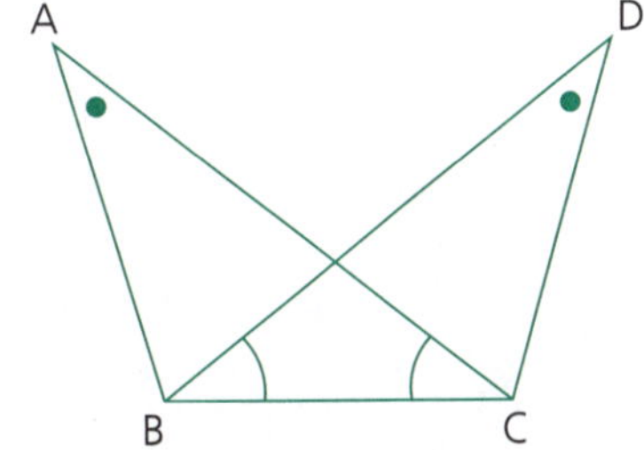

In the diagram, ∠CAB = ∠BDC and ∠BCA = ∠CBD.

a Prove that △ABC ≡ △DCB.

b Show that AB = DC.

pp. 149–151

12

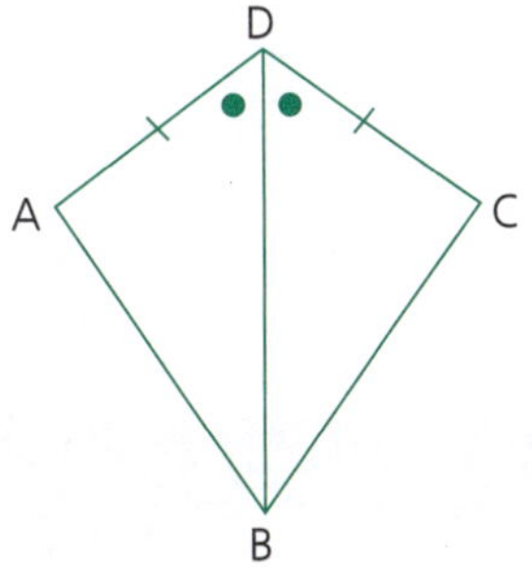

In the figure AD = CD and DB bisects ∠CDA.

Prove that:

a △ABD ≡ △CBD

b AB = CB

c DB bisects ∠ABC (that is, ∠ABD = ∠CBD).

pp. 149–151

13 Find the values of the pronumerals in the pairs of similar figures below:

a

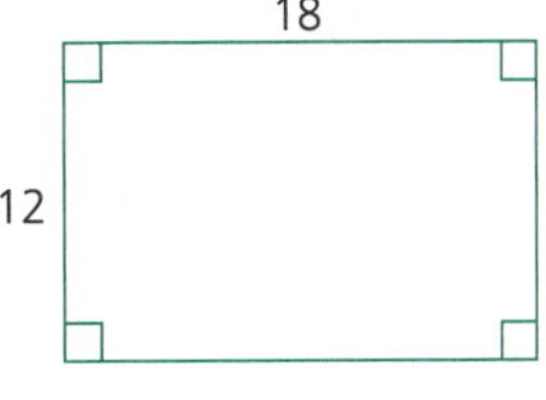

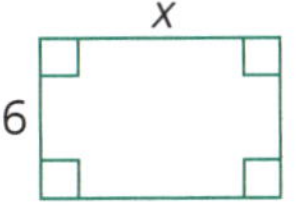

b

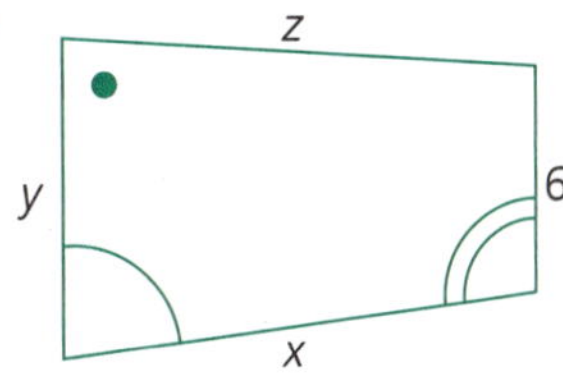

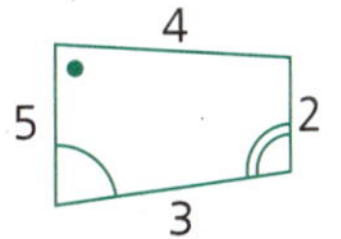

c

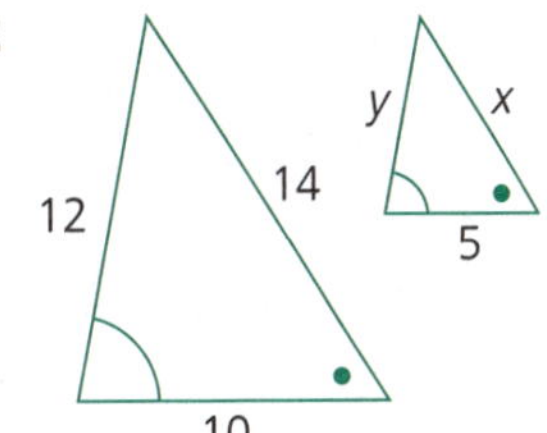

d

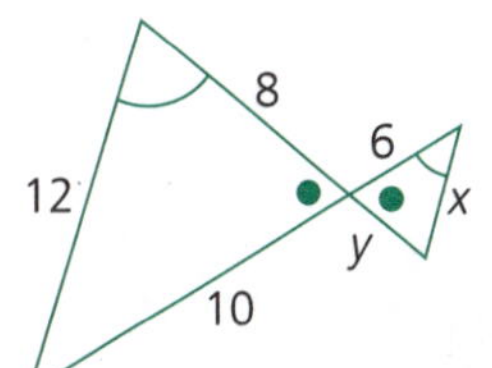

pp. 151–153

14 For the following pairs of triangles, prove that they are similar and find the value of each pronumeral:

a

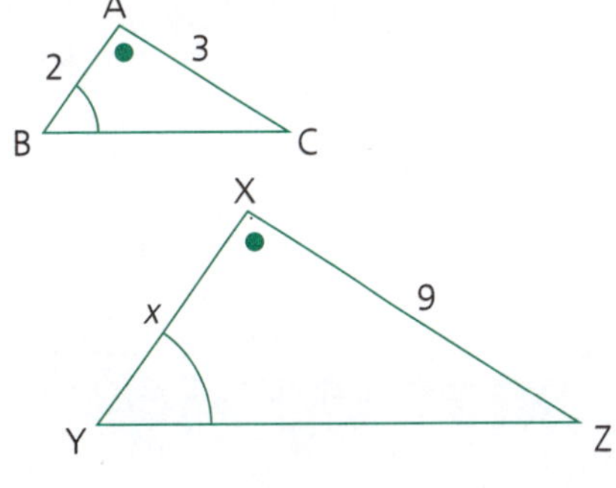

b

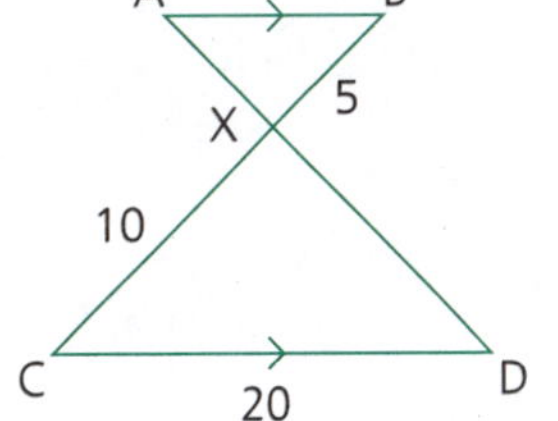

c

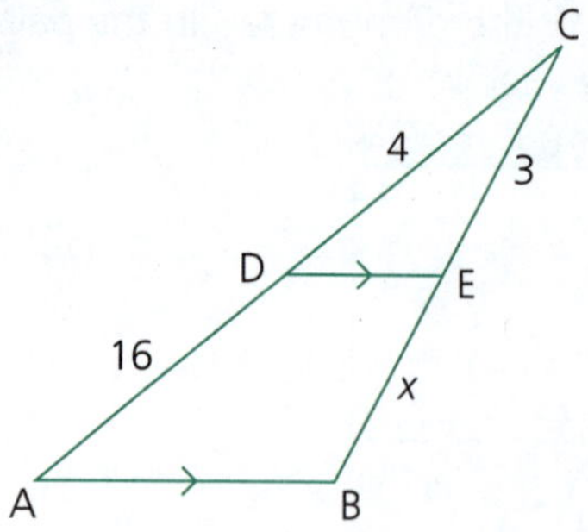

d

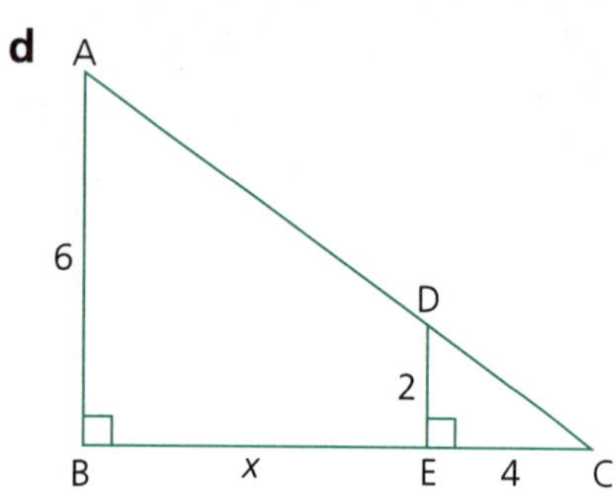

e

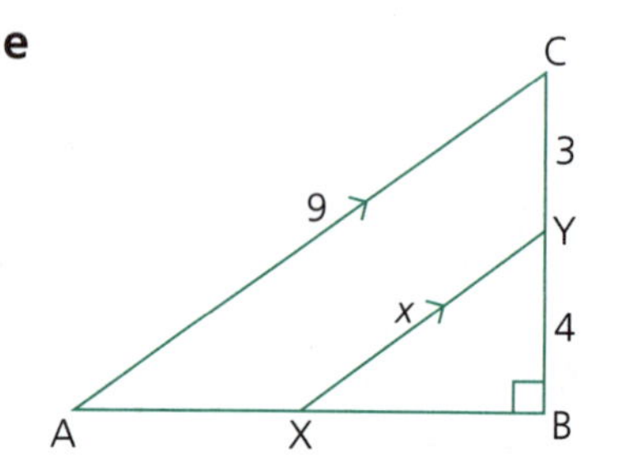

f

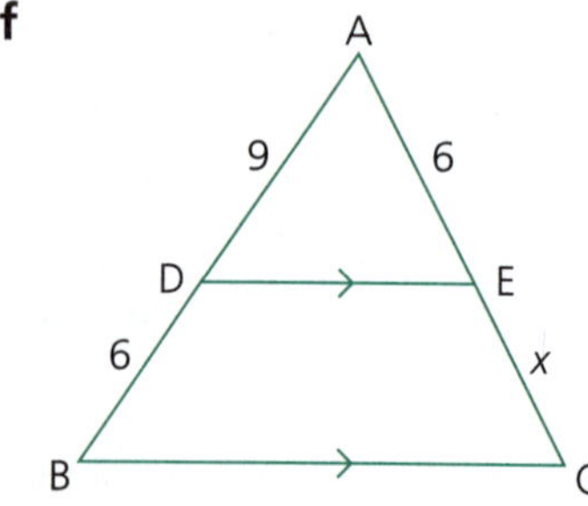

pp. 152–153

15 A metre ruler casts a shadow 0.5 metres long. At the same time, a building casts a shadow 10 metres long.

a Draw a diagram to illustrate this information.

b Use similar triangles to calculate the height of the building. pp. 152–153

16 A pole 5 metres high casts a shadow 2 metres long. At the same time a tree casts a shadow 12 metres long. How high is the tree? pp. 152–153

Go to p. 220 for **Quick Answers** or to pp. 250–253 for **Worked Solutions**

For a complete understanding of this topic, you must be able to:

✓	Name angles		p. 138
✓	Use geometrical facts to solve problems with angles, triangles and parallel lines		pp. 138–143
✓	Use properties of special quadrilaterals		pp. 144–145
✓	Use angle properties of polygons		pp. 145–147
✓	Recognise congruent triangles		p. 147
✓	Prove that two triangles are congruent		pp. 147–148
✓	Use congruent triangles to deduce geometrical results		pp. 149–150
✓	Recognise similar figures		p. 151
✓	Prove that two triangles are similar		pp. 152–153
✓	Use similarity to find unknown sides and angles of triangles.		pp. 152–153

Now you are ready to do the tests!

Intermediate Test

Geometry

(40 marks)

1 Find the value of x:

a

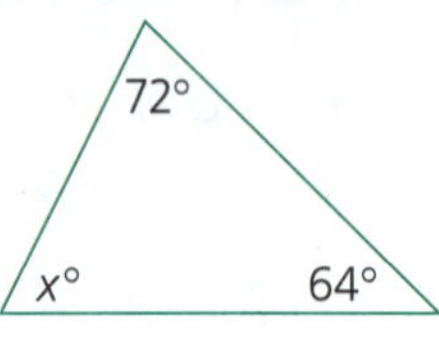

(1 mark)

b

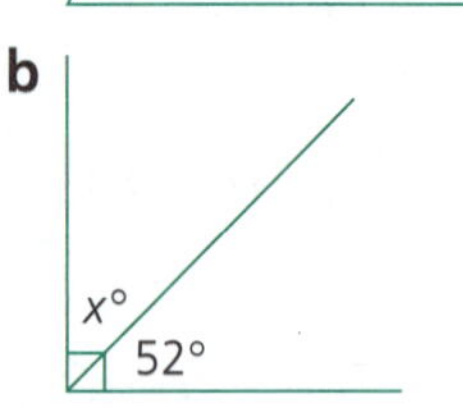

(1 mark)

c

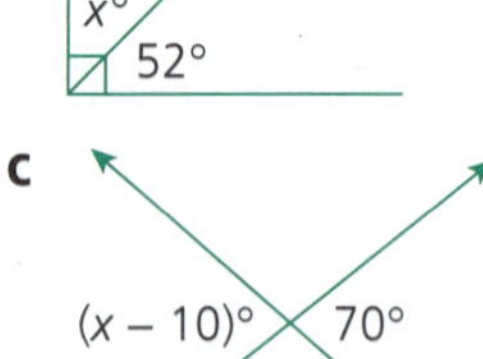

(1 mark)

d

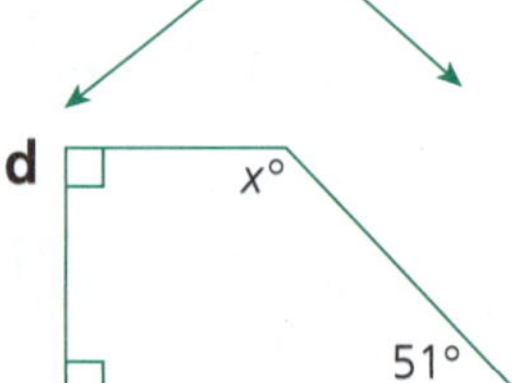

(1 mark)

e

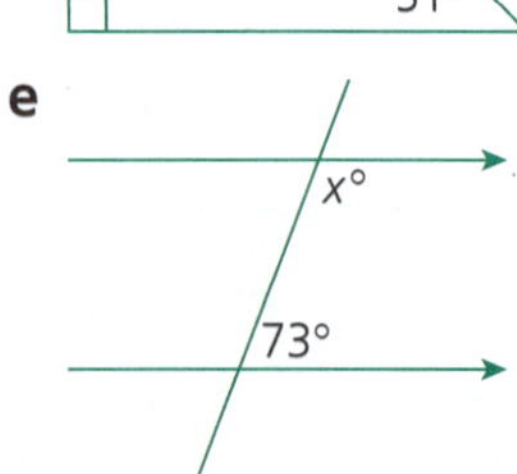

(1 mark)

f

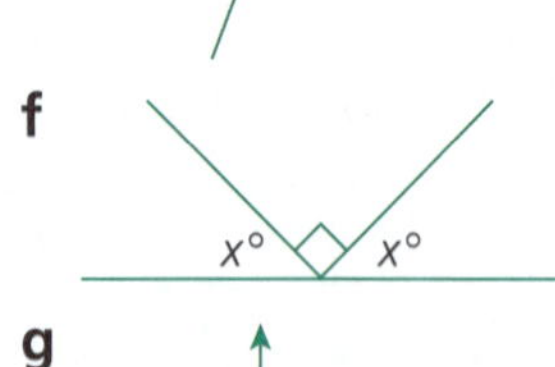

(1 mark)

g

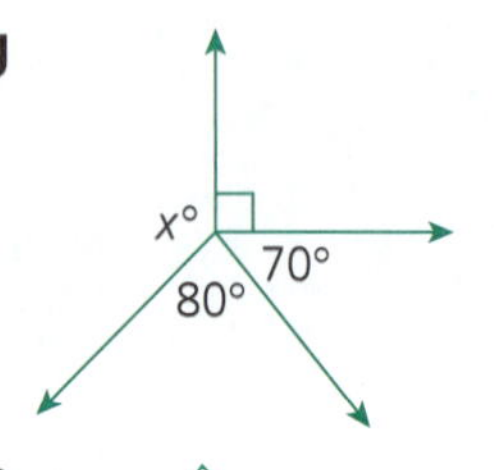

(1 mark)

h

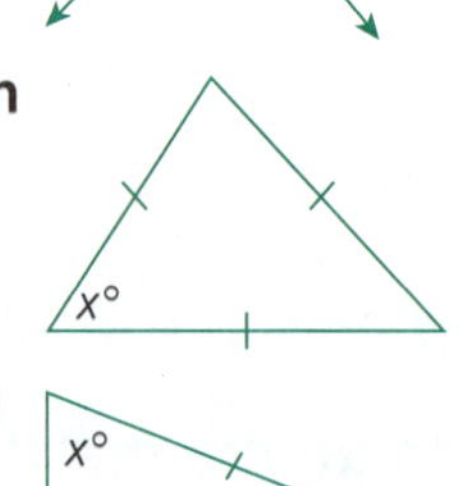

(1 mark)

i

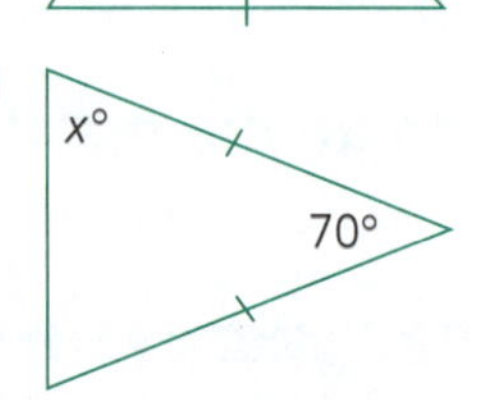

(1 mark)

2 Which of the following is a property of all parallelograms?
A The diagonals are equal.
B All angles are equal.
C Angles add up to 180°.
D Opposite sides are parallel. (1 mark)

3 Complete the following statements:
a In __________ triangles the corresponding sides and corresponding angles are equal.
b In __________ triangles the corresponding angles are equal and corresponding sides are in the same ratio. (2 marks)

4 Eleni makes the following statements concerning a rhombus:
I The diagonals bisect each other.
II The diagonals bisect at right angles.
III The diagonals are equal.
She is correct in:
A I, II and III
B II and III only
C I and II only
D I and III only. (1 mark)

5 A hexagon has six angles and six sides. Is the statement true or false? (1 mark)

6 Triangle ABC is congruent to triangle ZYX.

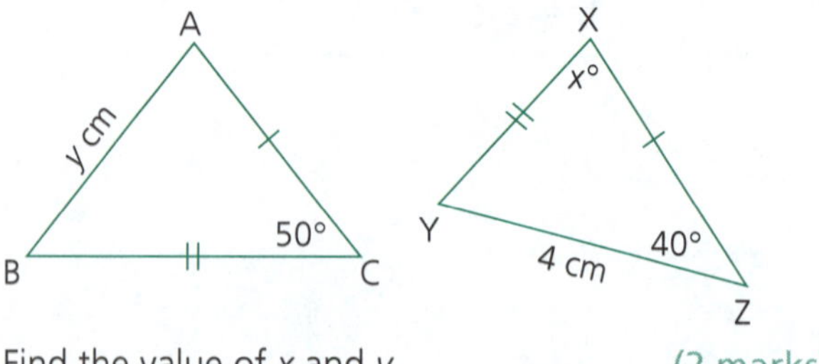

Find the value of x and y. (2 marks)

7 In the diagram below △ABC is similar to △XYZ.

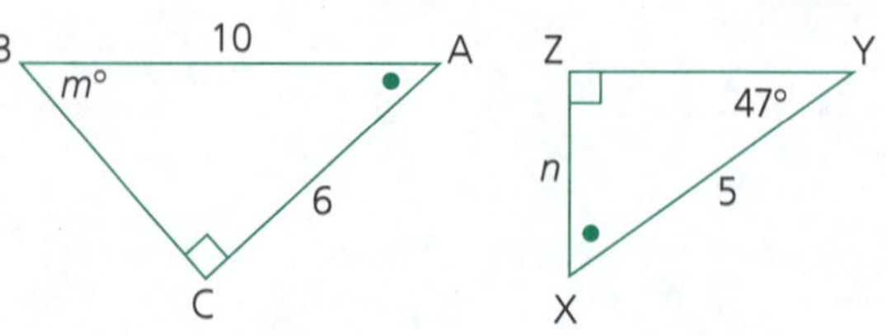

Find the value of m and n. (2 marks)

8 **a** Find the angle sum of a 15-sided polygon. (2 marks)

b Find the interior angle of a regular 15-sided polygon. (1 mark)

c What is the size of the exterior angle of a regular 15-sided polygon? (1 mark)

9 A regular polygon has an angle sum of 1080°. How many sides does it have? (2 marks)

10 Name the test you would use in proving each of the following pairs of triangles congruent:

a

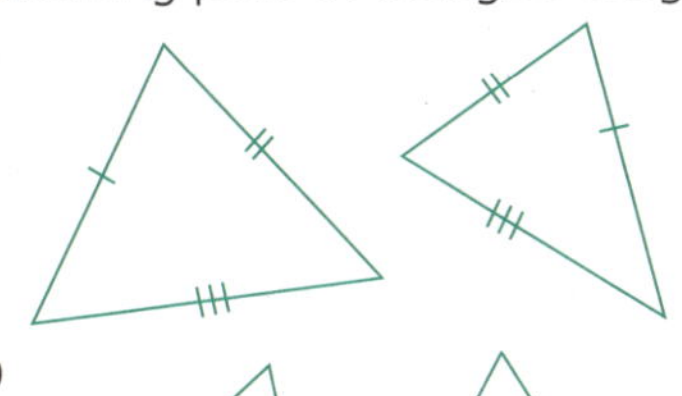

b

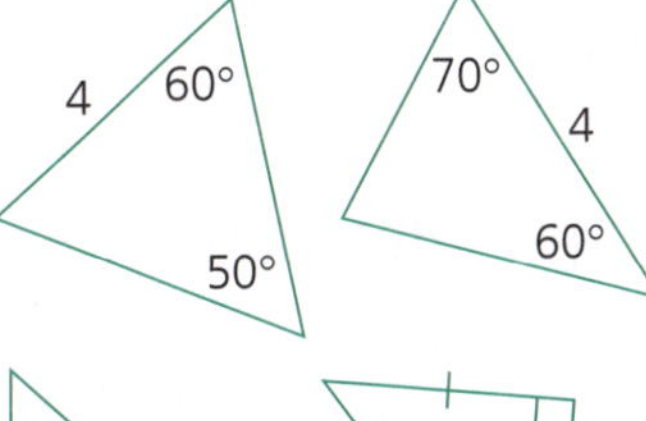

c

(3 marks)

11 In the diagram below, ABCD is similar to NMQP. Find the value of x, y and z.

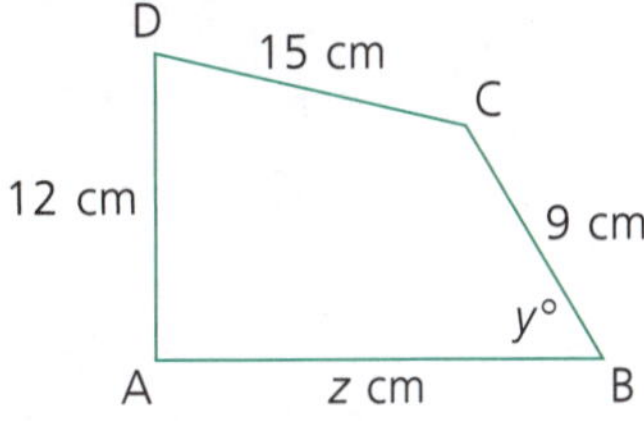

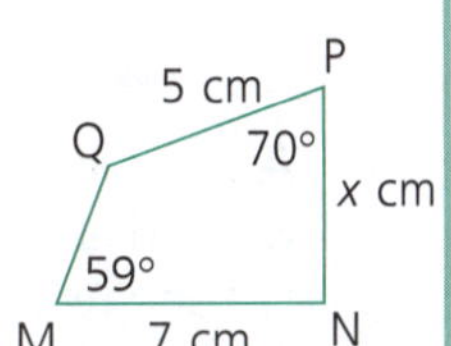

(3 marks)

12

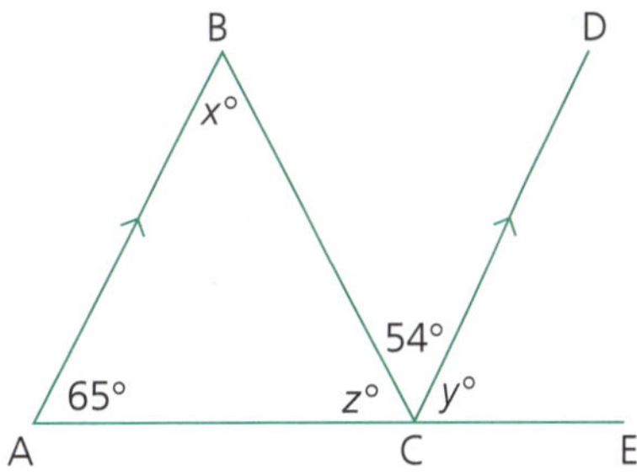

In the figure, AB || CD
$\angle BAC = 65°$
$\angle BCD = 54°$.

Find the value of:

a y, giving reasons

b x, giving reasons

c z, giving reasons. (3 marks)

13

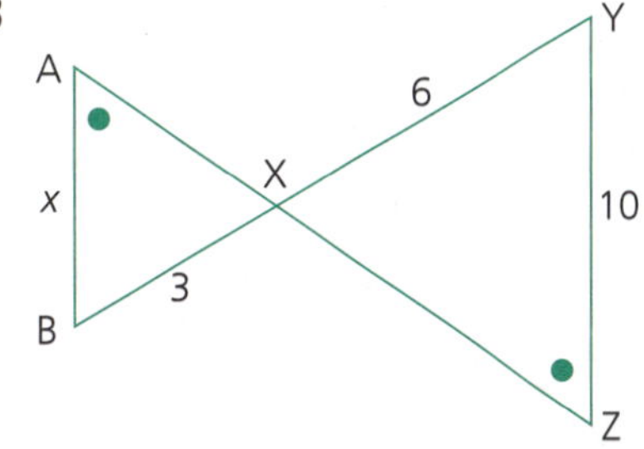

a Prove that $\triangle ABX$ is similar to $\triangle ZYX$. (2 marks)

b Hence, find the value of x. (1 mark)

14

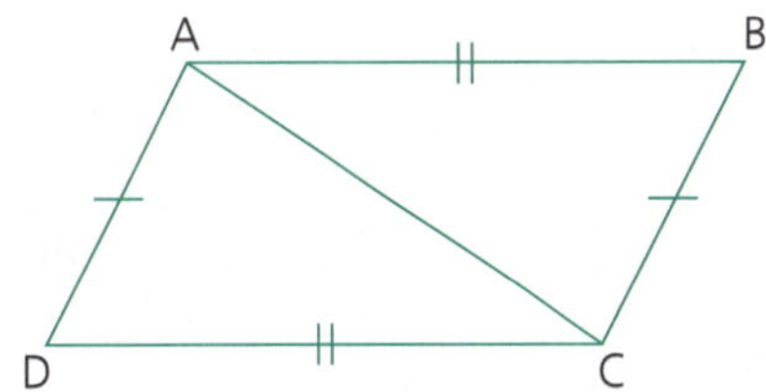

a Prove that $\triangle ADC$ is congruent to $\triangle CBA$. (3 marks)

b Show that $\angle ADC = \angle CBA$. (1 mark)

QA PAGE 224

WS PAGE 271

Advanced Test

Geometry

(40 marks)

1 Find the value of x, giving reasons.

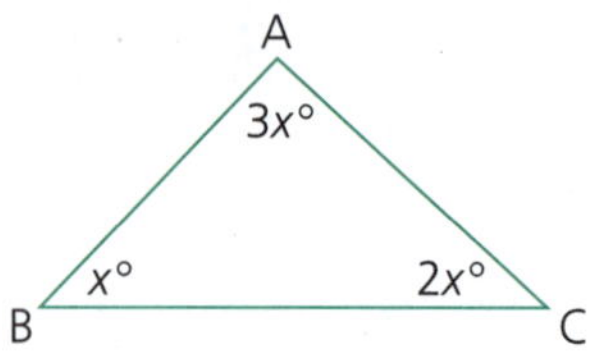

(2 marks)

2 Find x, giving reasons.

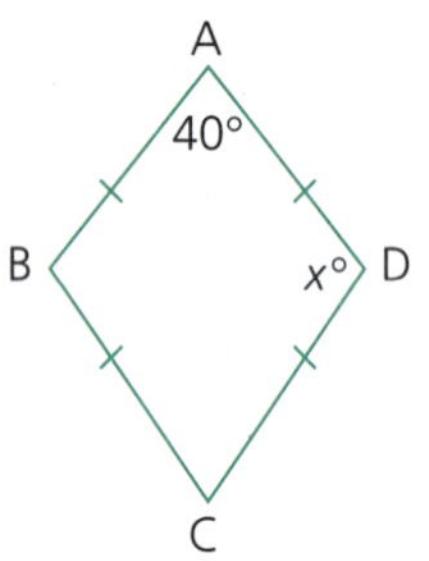

(2 marks)

3 Find x, giving a reason.

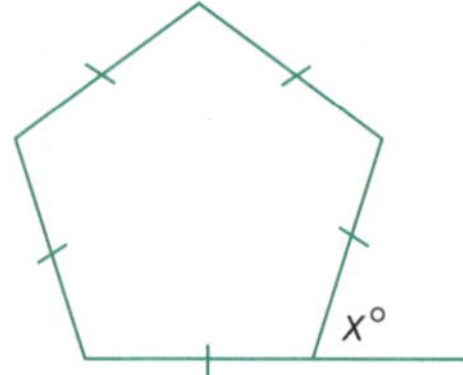

(2 marks)

4 Find the size of ∠CAB. Give reasons.

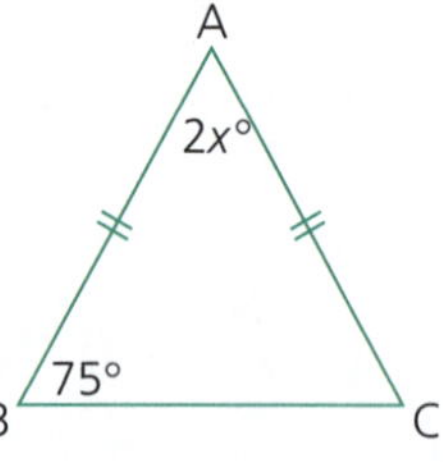

(2 marks)

5 BC ∥ DE. Find x, giving reasons.

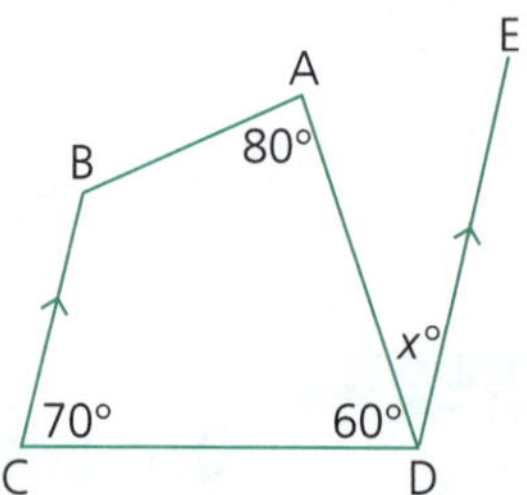

(2 marks)

6

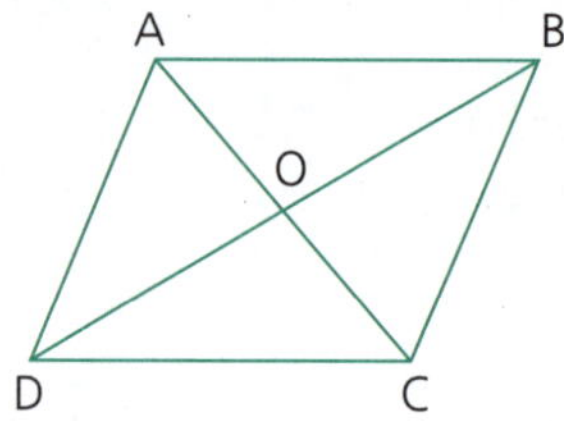

ABCD is a rhombus.

Which statement is not true?

A AB = BC = CD = AD
B AC = BD
C OA = OC
D ∠ADB = ∠BDC

(1 mark)

7 Which two rectangles are similar?

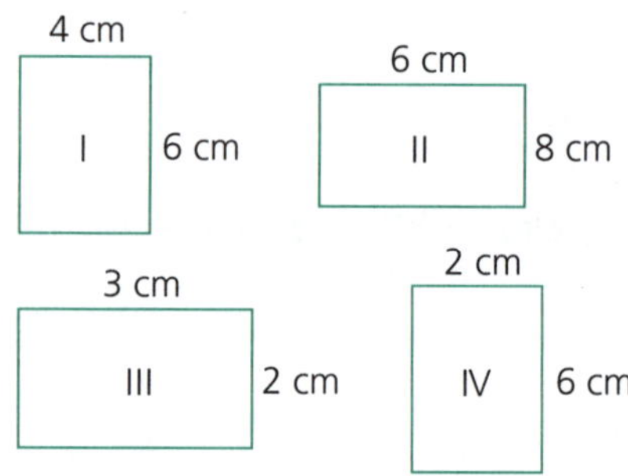

A I and II
B I and III
C I and IV
D III and IV

(1 mark)

8 Which triangles are congruent?

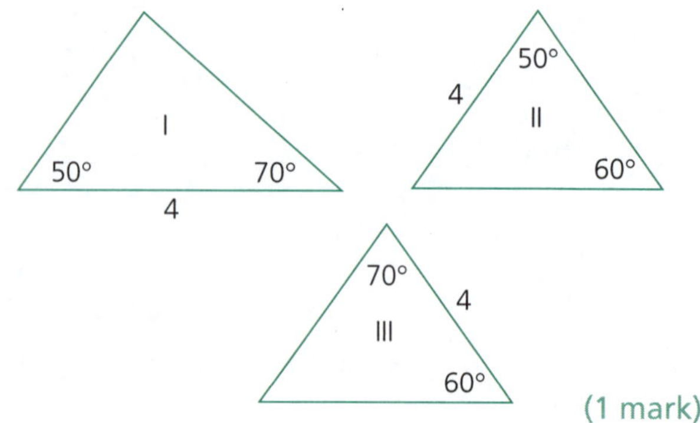

(1 mark)

9 Find x:

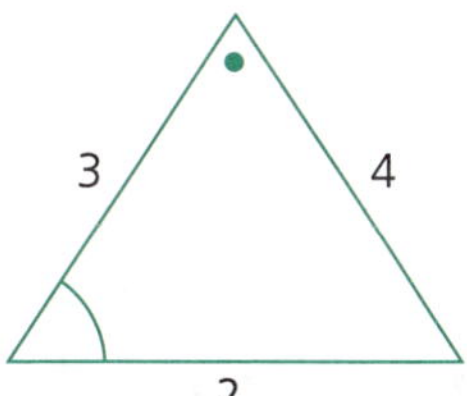

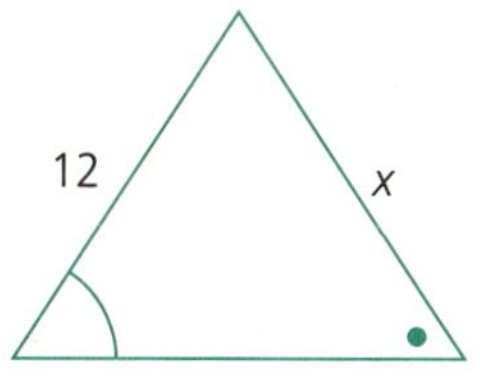

(2 marks)

10 In the figure below, triangle A is reflected about line m, to form triangle B. Triangle B is translated then rotated to form triangle C.

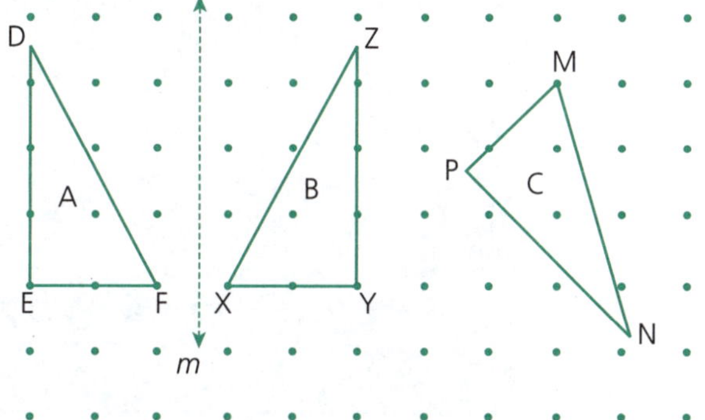

a Is triangle A congruent to triangle C? (1 mark)

b True or false: 'Triangle B is similar to triangle C.' (1 mark)

c Which side of triangle C corresponds to side DE of triangle A? (1 mark)

d Which angle of triangle A corresponds to angle NMP of triangle C? (1 mark)

e True or false: 'The area and perimeter of triangle A is the same as the area and perimeter of triangle C.' (1 mark)

11 In the following figure, $\triangle ABC \parallel\!\mid \triangle XYZ$, $BC \neq YZ$.

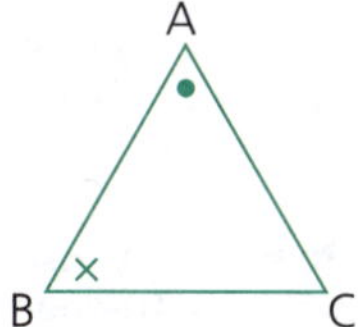

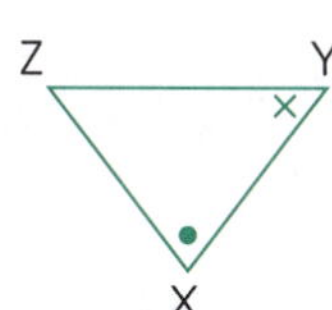

Write true or false for the following statements:

a $\angle BCA = \angle YZX$

b $AB = XY$

c Area of $\triangle ABC$ = Area of $\triangle XYZ$

d $\dfrac{AB}{XY} = \dfrac{AC}{XZ}$ (4 marks)

12 **a** Find the size of an interior angle of a regular 11-sided polygon (hendecagon). (2 marks)

b Find the size of an exterior angle of a regular dodecagon. (2 marks)

c The size of an exterior angle of a regular polygon is 9°. How many sides does the polygon have? (2 marks)

13

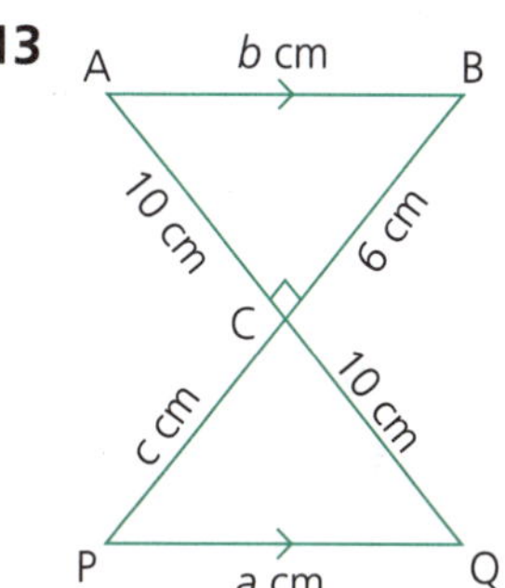

In the diagram $AB \parallel PQ$ and ACQ is a straight line, with $AC = QC$.

a Prove that $\triangle ABC \equiv \triangle QPC$. (3 marks)

b Find the value of a, b and c. (3 marks)

14

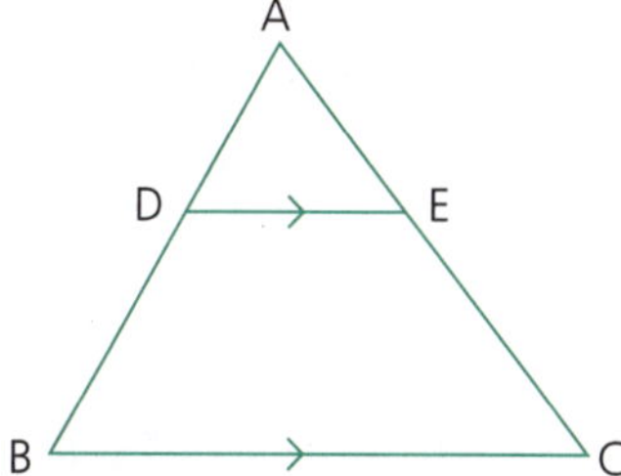

In the figure, $DE \parallel BC$, $DE = 2$ cm, $BC = 6$ cm, $AD = 3$ cm.

a Prove $\triangle ABC \parallel\!\mid \triangle ADE$. (2 marks)

b Hence, find the length of DB. (2 marks)

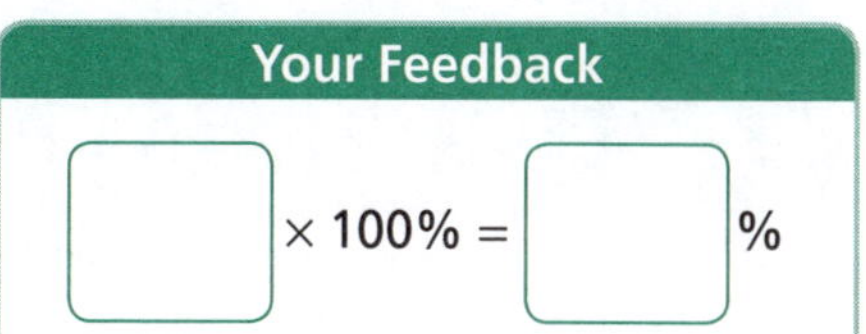

QA PAGE 224

WS PAGE 271

Chapter 11
Statistics

Single variable (univariate) data analysis

Some definitions and formulae

Range = Highest score – Lowest score

Mode = Most common score or score with the highest frequency

Median = Middle score when scores are arranged in *ascending* order

$$\textbf{Mean} = \text{Average} = \frac{\text{Total of the scores}}{\text{Number of scores}} = \frac{\Sigma xf}{\Sigma f}$$

A note on notation

x = score

f = frequency

c.f. = cumulative frequency

xf = (score) × (frequency)

Σ = sum

Σf = sum of frequency column

Σxf = sum of xf column

$\bar{x}$ = mean

Keywords

Box plot, Bi-modal, Bivariate, Cumulative, Dependent, Distribution, Dot plot, Frequency, Frequency polygon, Histogram, Independent, Interquartile range, Lower quartile, Mean, Median, Mode, Normal, Outcome, Prediction, Range, Relative, Relationship, Scatterplot, Skewed, Stem-and-leaf, Trend, Univariate, Upper quartile

For Example

1 The results for 25 Year 10 students in a weekly spelling test are as follows:

7	8	9	(10)	4	7	6
4	5	(3)	8	5	6	7
9	9	8	7	6	5	7
8	6	4	9			

a Construct a frequency distribution table to answer the following questions.

S Frequency distribution table:

Score (x)	Tally	Frequency (f)	Cumulative frequency (c.f.)	Score × frequency (xf)
3	I	1	1	3 ⇐ (3 × 1)
4	III	3	4	12 ⇐ (4 × 3)
5	III	3	7	15 ⇐ (5 × 3)
6	IIII	4	11	24
7	~~IIII~~	5	16	35
8	IIII	4	20	32
9	IIII	4	24	36
10	I	1	25	10
	Σf	25	Σxf	167

b How many students scored:

i 8 marks?

ii 8 or less marks?

iii less than 8 marks?

iv more than 8 marks?

S

i 4 (from frequency column)

ii 20 (from c.f. column)

iii 16 [same as 7 or less]

iv 25 – 20 = 5 [total minus 8 or less]

c What percentage of students scored:

i 8 marks?

ii 8 or less marks?

iii less than 8 marks?

iv more than 8 marks?

S

i Percentage $= \frac{4}{25} \times 100\% = 16\%$

ii Percentage $= \frac{20}{25} \times 100\% = 80\%$

iii Percentage $= \frac{16}{25} \times 100\% = 64\%$

iv Percentage $= \frac{5}{25} \times 100\% = 20\%$

d Find the range and the mode.

S

Range = Highest score – Lowest score
= 10 – 3 = 7

Mode = Score with highest frequency
= 7 (frequency is 5)

e Calculate the mean.

S

$$\text{Mean} = \frac{\Sigma xf}{\Sigma f} = \frac{167}{25} = 6.68$$

f Find the median score.

S

To find the median, we need the middle score. This is the 13th score. From the c.f. column, the 13th score is a 7. (The 12th, 13th, 14th, 15th and 16th scores are all 7.)

When we have 25 scores the 13th is the middle score, that is, 12 before it and 12 after it.

g Construct a:

i frequency histogram

ii cumulative frequency histogram.

S

i Histogram

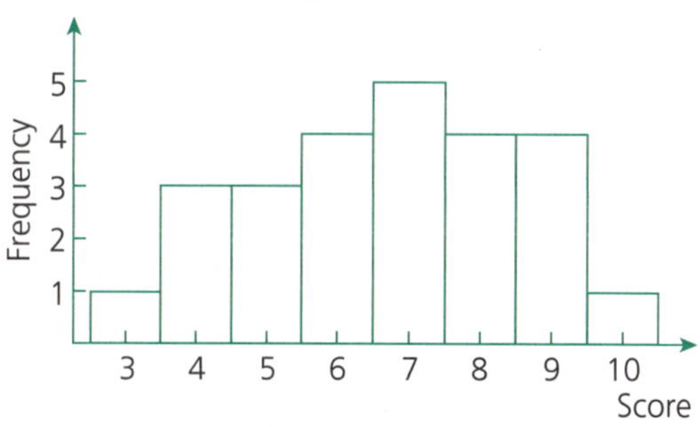

ii Cumulative frequency histogram

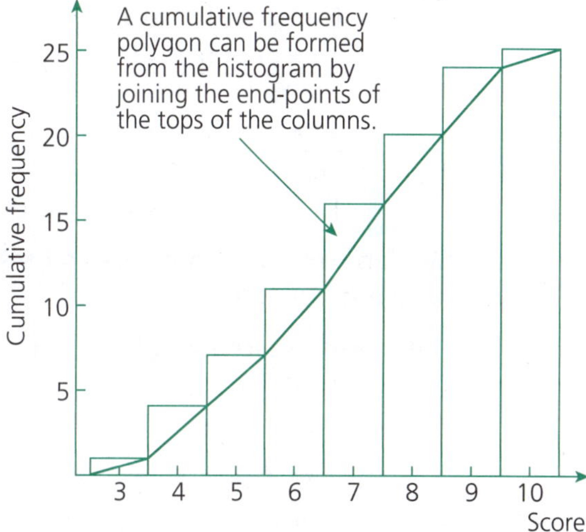

Note that if a frequency polygon had been required, the graph would be as follows:

Frequency polygon

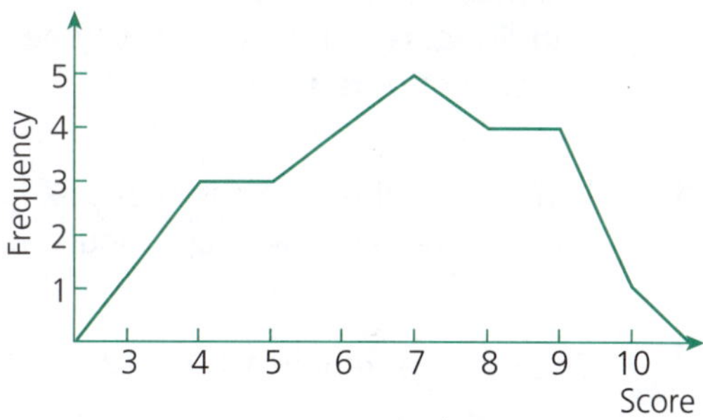

This can also be achieved by joining the midpoints of the tops of the columns in the histogram.

2 Gregory, over his first six innings, managed to score 18, 32, 12, 12, 20 and 12.

a Find the range of these scores.

S Range = Highest score − Lowest score
= 32 − 12
= 20

b Write down the modal score.

S Mode = Most common score
= 12

c Calculate the mean and the median.

S $\text{Mean} = \dfrac{\text{Total of scores}}{\text{Number of scores}}$

$= \dfrac{18 + 32 + 12 + 12 + 20 + 12}{6}$

$= \dfrac{106}{6}$

= 17.7 (to one decimal place)

Median = middle score when scores are in ascending order

The scores in ascending order are:
12, 12, 12, 18, 20, 32

Middle scores

$\text{Median} = \dfrac{12 + 18}{2}$

= 15

When there is an even number of scores, the median will be the average of the two middle scores, that is, half-way between the two middle scores.

d If in his next three innings he scores 17, 15 and 20, find the new range, mode, mean and median.

S The scores are now 18, 32, 12, 12, 20, 12, 17, 15 and 20.

Range = 32 − 12 = 20 unchanged

Mode = 12 unchanged

Mean

$= \dfrac{18 + 32 + 12 + 12 + 20 + 12 + 17 + 15 + 20}{9}$

$= \dfrac{158}{9}$ = 17.6 (to 1 decimal place)

The scores in ascending order are:

12, 12, 12, 15, 17, 18, 20, 20, 32.

Middle score

When there is an odd number of scores there is a single middle score.

The mean has dropped marginally while the median has risen.

e How many runs must he score in his tenth innings to have an overall average of 18?

S For Gregory to average 18 over ten innings, his total for ten innings would be (10 × 18) runs, that is, 180 runs. Over nine innings, he has scored 158 runs, thus he would have to score (180 − 158) runs, that is, 22 runs in his tenth innings.

3 Naomi averaged 17 points per game over the first five games of the basketball season. Over the following seven games her average was 23 points per game. Calculate Naomi's average for the season.

S Over the first five games Naomi scored (5 × 17) = 85 points.

Over the next seven games Naomi scored (7 × 23) = 161 points.

Overall score = 85 + 161
= 246 points

$\text{Season average} = \dfrac{246}{12}$

= 20.5

$\dfrac{\text{Total points}}{\text{Number of games}}$

Naomi's season average was 20.5 points.

Dot plots

A dot plot is an alternative method of displaying information. It is simpler to draw and simpler to read than a histogram.

1 The type of cars passing along Macquarie Street over a 10-minute period was noted and the results recorded in the following table:

Brand	Frequency
Ford	5
Holden	7
Toyota	8
Hyundai	4
Other	6
Total	30

Record the data in a dot plot.

A dot diagram was constructed from the information in the table.

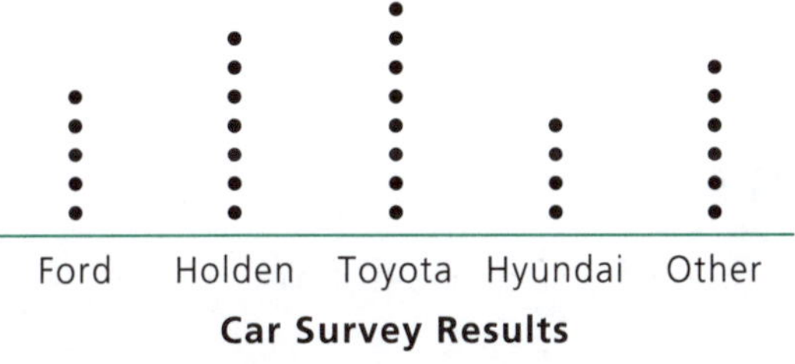

2 Consider the results of 25 Year 10 students in a weekly spelling test:

7	8	9	10	4
7	6	4	5	3
8	5	6	7	9
9	8	7	6	5
7	8	6	4	9

Record the data in a dot plot and a frequency table.

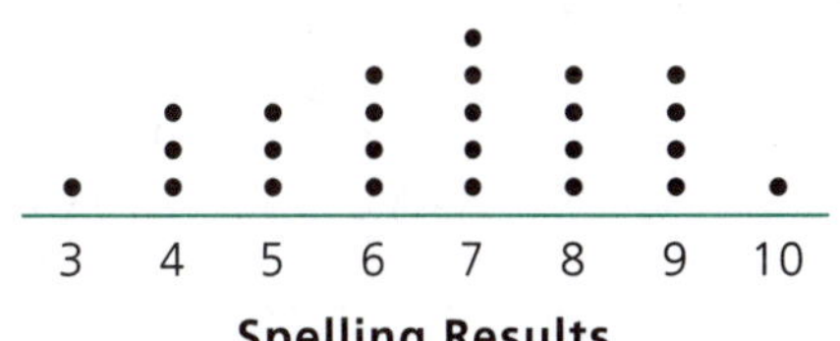

Score	Frequency
3	1
4	3
5	3
6	4
7	5
8	4
9	4
10	1
	25

Stem-and-leaf plots

Stem-and-leaf plots are another method of displaying numerical data. The leaf is the final digit of a number, while the stem is the first digit or digits. The leaf is always a single digit. The stem can contain any number of digits. For example, if 68, 647 and 24 317 are organised into stem-and-leaf then:

68	6 stem	8 leaf
647	64 stem	7 leaf
24 317	2431 stem	7 leaf

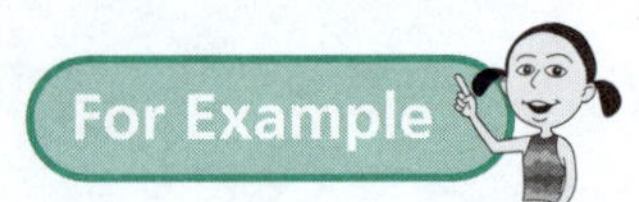

1 The results in a mathematics test out of 50 were:

19	48	36	40	31	22
18	27	18	20	18	36
49	50	45	13	9	17
22	31	39	26	28	30
44	8	23	19	46	33

Draw a stem-and-leaf plot for this data.

S **It is helpful to use an initial plot:**

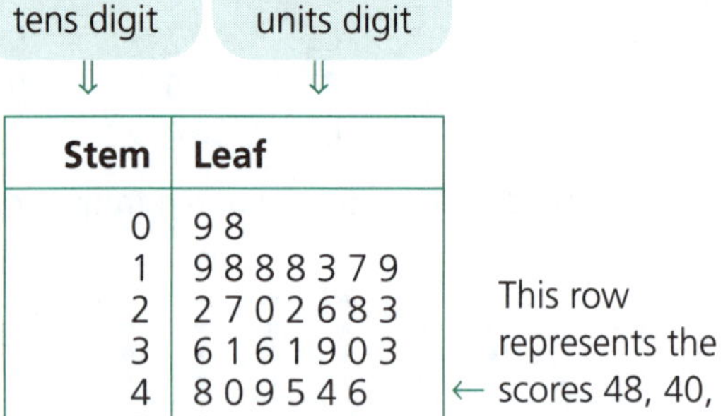

Stem	Leaf
0	9 8
1	9 8 8 8 3 7 9
2	2 7 0 2 6 8 3
3	6 1 6 1 9 0 3
4	8 0 9 5 4 6
5	0

← This row represents the scores 48, 40, 49, 45, 44, 46

Note: in the initial plot the numbers are entered as they occur.

Refined plot:

Stem	Leaf
0	8 9
1	3 7 8 8 8 9 9
2	0 2 2 3 6 7 8
3	0 1 1 3 6 6 9
4	0 4 5 6 8 9
5	0

The digits in the leaf are here arranged in increasing order in order to enable greater understanding of the scores.

A stem-and-leaf plot allows the smallest and largest numbers to be seen, as well as the size of the distribution and any clustering of data. It gives an indication of the groupings to be used in a grouped table and makes the tallying easier.

Outlying scores stand out.

Back-to-back stem-and-leaf plots

Two sets of data can be compared easily when arranged as a back-to-back stem-and-leaf plot.

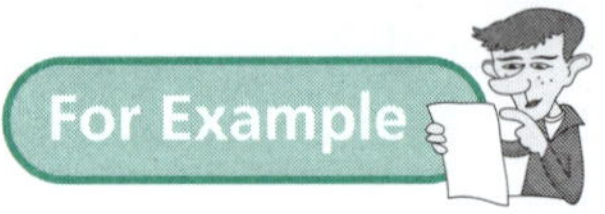

1 The results of a history test for 9A and 9B were recorded:

9A		9B
	4	1
2	5	0 3 4
9 7 4 1 1 0	6	3 6 6 7 8 8 9
9 7 6 6 6 5 3	7	0 0 2 4 4 4 8
9 8 6 5 5 0	8	0 2 4 5 5 6
6 5 4 2 0	9	2

Comment on the results of the students in the classes regarding the:

a number of students

S **Each class has 25 students.**

b outliers.

An outlier is a score considerably higher or lower than other scores.

S **The outlier in 9B scored 41.**

Using the stem-and-leaf plot to find the range, mode and median

A stem-and-leaf plot can be used to find statistical measures.

1 From the stem-and-leaf plot (below) showing the results of 30 students in a maths test, find the range, mode and median.

Stem	Leaf
0	8 9
1	3 7 8 8 8 9 9
2	0 2 2 3 6 7 8
3	0 1 1 3 6 6 9
4	0 4 5 6 8 9
5	0

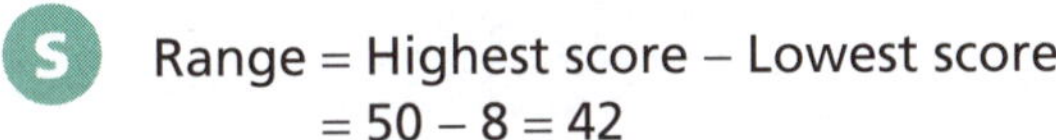

S Range = Highest score − Lowest score
= 50 − 8 = 42

Mode = Score(s) with highest frequency
= 18

Look for the number occuring the most in each stem.

Median = middle score when scores are arranged in order.

The stem-and-leaf plot has organised the scores in the correct order.

Begin crossing a number from the start and from the end one at a time until you reach the middle number (odd number of scores) or middle numbers (even number of scores).

Start here ... ⇓

Stem	Leaf
0	8 9
1	3 7 8 8 8 9 9
2	0 2 2 3 6 (7 8)
3	0 1 1 3 6 6 9
4	0 4 5 6 8 9
5	0

⇑ then here.

In this case

median = average of 27 and 28

$= \dfrac{27 + 28}{2}$

$= 27.5$

Then range = 42
mode = 18
median = 27.5

2 Twenty boys and 16 girls from Year 9 had their height measured (in centimetres).

Boys: 180, 186, 164, 166, 158, 170, 172, 175, 182, 154, 151, 158, 167, 169, 170, 174, 179, 177, 181, 171

Girls: 147, 142, 165, 168, 174, 142, 155, 159, 160, 179, 181, 176, 166, 159, 154, 162

a Display the two sets of data back-to-back on a stem-and-leaf plot.

S Initial plot:

Leaf (G)	Stem	Leaf (B)
2 2 7	14	
4 9 9 5	15	8 4 1 8
2 6 0 8 5	16	4 6 7 9
6 9 4	17	0 2 5 0 4 9 7 1
1	18	0 6 2 1

Final plot:

Leaf (G)	Stem	Leaf (B)
7 2 2	14	
9 9 5 4	15	1 4 8 8
8 6 5 2 0	16	4 6 7 9
9 6 4	17	0 0 1 2 4 5 7 9
1	18	0 1 2 6

b Calculate for each plot the range and median.

S Boys

15	1 4 8 8
16	4 6 7 9
17	0 0 1 2 4 5 7 9
18	0 1 2 6

Range = 186 − 151 = 35 cm

$$\text{Median} = \frac{170 + 171}{2} = 170.5 \text{ cm}$$

Girls

7 2 2	14
9 9 5 4	15
8 6 5 2 0	16
9 6 4	17
1	18

Range = 181 − 142 = 39 cm

$$\text{Median} = \frac{160 + 162}{2} = 161 \text{ cm}$$

c Comment on the two sets of data.

S
- There is a greater spread of heights for the girls.
- The median is higher for boys.
- Clustering for girls is in the 160s; for boys in 170s.
- From this data it is appears that boys in Year 9 are taller than girls on average, but there is a greater variation among girls.

Interquartile range

When scores are arranged in ascending order:

The **median** is the middle score of the distribution.

The **upper quartile** is the middle score between the median and the highest score.

The **lower quartile** is the middle score between the median and the lowest score.

The **interquartile range** is equal to the upper quartile minus the lower quartile.

Note: 50% of the scores are represented by the interquartile range.

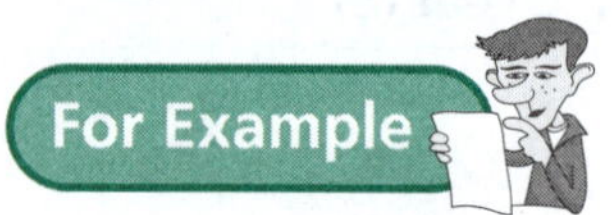

1 Consider the following scores for **a** and **b**:

a 3, 5, 4, 6, 4, 4, 8, 7, 6, 8
Find the:
- **i** median
- **ii** upper quartile
- **iii** lower quartile
- **iv** interquartile range.

S Arrange in ascending order:

⇓
3, 4, 4, 4, 5, 6, 6, 7, 8, 8
⇑

i Median = Average of 5th and 6th scores

$$= \frac{5 + 6}{2}$$
$$= 5.5$$

Even number of scores ∴ two middle numbers.

ii For the upper quartile, use only the scores above the median:

6, 6, 7, 8, 8
⇑

Upper quartile
= Middle number of scores above median
= 7

Odd number of scores ∴ one middle number.

iii For the lower quartile, use only the scores below the median:

3, 4, 4, 4, 5
⇑

Lower quartile
= Middle number of scores below median
= 4

iv Interquartile range
= Upper quartile − Lower quartile
= 7 − 4
= 3

b 18, 20, 16, 16, 18, 14, 18, 21, 19

Find the:

i median

ii upper quartile

iii lower quartile

iv interquartile range.

S Arrange in ascending order
⇓

14, 16, 16, 18, 18, 18, 19, 20, 21
⇑

i Median = middle score = 18

One middle score

ii For the upper quartile, use only the scores above the median:

18, 19, 20, 21
⇑

Upper quartile
= Average of 2nd and 3rd scores

$= \frac{19 + 20}{2} = 19.5$

iii For the lower quartile, use only the scores below the median:

14, 16, 16, 18
⇑

Lower quartile
= Average of 2nd and 3rd scores
= 16

iv Interquartile range
= Upper quartile − Lower quartile
= 19.5 − 16
= 3.5

Interquartile range from stem-and-leaf plots

A stem-and-leaf plot can be used to find both the median and the quartiles.

1 Consider the stem-and-leaf plot from before. The results of 30 students in a mathematics test out of 50 have been tabulated. Find the interquartile range.

Stem	Leaf
0	8 9
1	3 7 8 8 8 9 9
2	0 2 2 3 6 (7 8)
3	0 1 1 3 6 6 9
4	0 4 5 6 8 9
5	0

← There are 30 scores so that the median is the average of the 15th and 16th scores.

That is, median $= \frac{27 + 28}{2}$

$= 27.5$

For the upper quartile, consider only numbers above the median (28, 30, …). There are 15 of these so the upper quartile will be the 8th number above the median.

Cross off 7 numbers
⇓

2	8
3	0 1 1 3 6 6 (9)
4	0 4 5 6 8 9
5	0

15 numbers

Upper quartile = 39.

For the lower quartile consider only numbers below the median (8, 9, …, 27). There are 15 of these so the lower quartile will be the 8th number above the smallest score. Cross off the first seven numbers as follows:

Stem	Leaf
0	8 9
1	3 7 8 8 8 9 9
2	0 2 2 3 6 7

15 numbers

The lower quartile = 19
Interquartile range = 39 − 19 = 20

Box plots

A box plot uses five selected pieces of information—lowest score (lower extreme), highest score (upper extreme), median, and upper and lower quartiles—to give an alternative display for a set of scores.

These five numbers are also called the **five point summary**.

1 Consider a set of data with lowest score 8, highest score 50, median 27.5, upper quartile 39 and lower quartile 19. Draw a box plot for the data.

This is sometimes called a box-and-whiskers plot. The lines at each end are the whiskers, while the centre rectangle is the box. The box contains 50% of the scores.

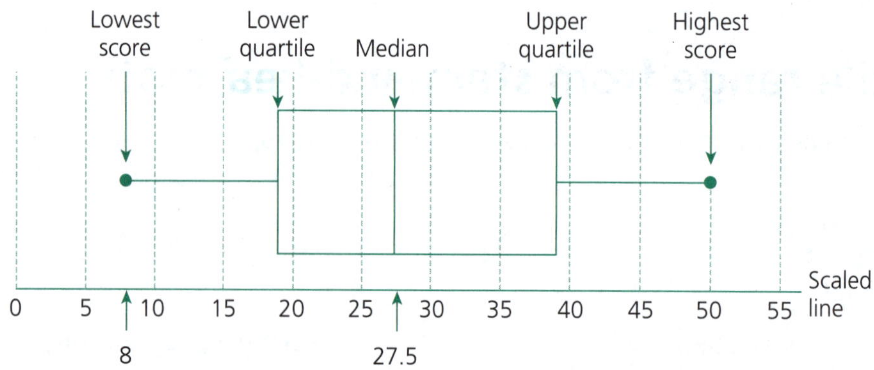

A box plot can be drawn to illustrate any numerical data regardless of how the quartiles and the median are calculated.

1 For the scores 3, 5, 4, 6, 4, 4, 8, 7, 6, 8 calculate the median, upper and lower quartiles and draw a box plot.

Rearrange scores:
3, 4, 4, 4, 5, 6, 6, 7, 8, 8
⇑

Median = $\frac{5+6}{2}$ = 5.5

For lower quartile: 3, 4, 4, 4, 5
⇑
Lower quartile = 4

For upper quartile: 6, 6, 7, 8, 8
⇑
Upper quartile = 7

Interquartile range = 7 − 4 = 3

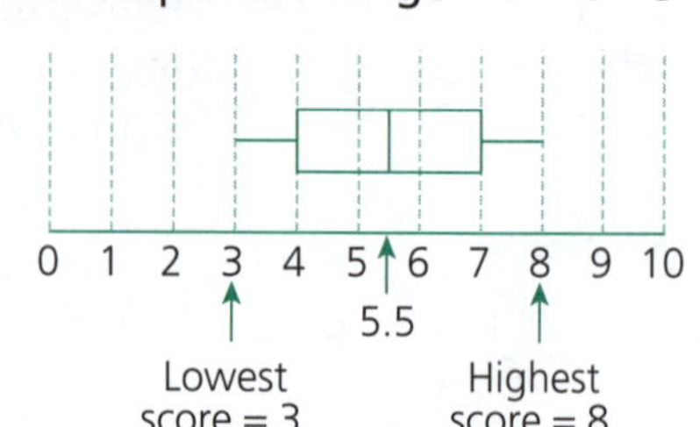

2 A stem-and leaf plot has been constructed to represent the scores of 25 students in a spelling test out of 30. Construct a box plot from the information.

Stem	Leaf
0	7 7 8 9 9
1	0 1 1 4 4 6 6 9
2	1 1 2 2 5 8 9 9 9
3	0 0 0

← Median

Middle of lower scores.

Middle of upper scores.

S There are 25 scores,
∴ the median is the 13th score.

Median = 19

For lower quartile consider only lower 12 scores. The two middle scores (6th and 7th) are 10 and 11:

$$\text{Lower quartile} = \frac{10 + 11}{2} = 10.5$$

For upper quartile consider only upper 12 scores. The two middle scores are 28 and 29.

$$\text{Upper quartile} = \frac{28 + 29}{2} = 28.5$$

Summary:

This is called the 'five points summary'.

Lowest score = 7
Highest score = 30
Lower quartile = 10.5
Median = 19
Upper quartile = 28.5

Box plot:

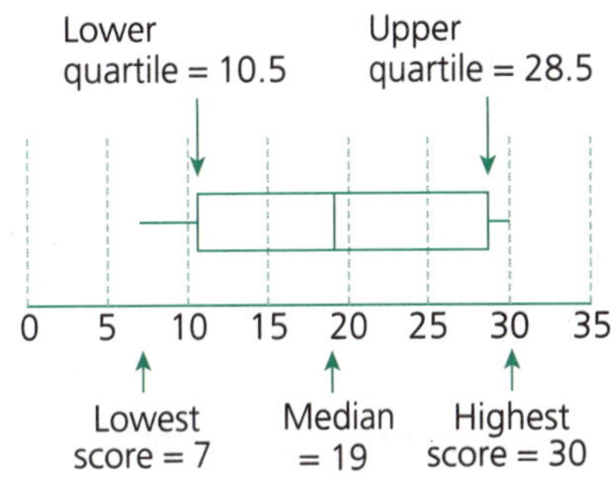

Comparing data sets

Comparing data sets using box plots

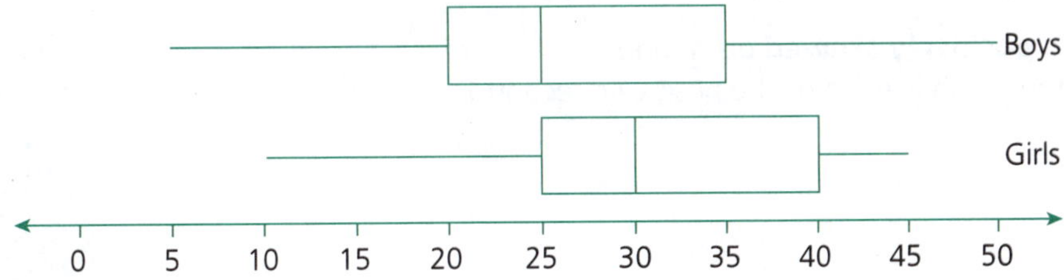

Box plots can be used to compare different data sets:

- Both highest and lowest scores were boys (50, 5).
- Both have the same interquartile range (15). Clustering of scores is thus the same for the middle 50%, but boys have a greater variation in overall scores.
- 75% of girls scored 25 or more; 50% of boys scored 25 or more.
- Girls have, as a whole, performed better than the boys.

Comparing data sets using back-to-back stem-and-leaf plots

Test marks in Class 10T out of 70:

Boys		Girls
7	1	3
5	2	4
6 3	3	2 2 5
6 1 1	4	0 1 5
8 7 5 2	5	3 3 3 9
9 4 1	6	2 4 8
0	7	

From the stem-and-leaf plot:

Median boys = 52
Median girls = 45

Lower quartile boys = 36
Lower quartile girls = 32

Upper quartile boys = 61
Upper quartile girls = 59

Interquartile range boys = 61 – 36 = 25

Interquartile range girls = 59 – 32 = 27

Although the boy's median is significantly higher, the interquartile ranges are similar.

Skew

When data sets are compared, the distribution of the data can be considered.

This histogram is **symmetrical** and the mode, mean and median are all equal. The majority of scores are **clustered** around the middle and this is known as a **normal distribution**.

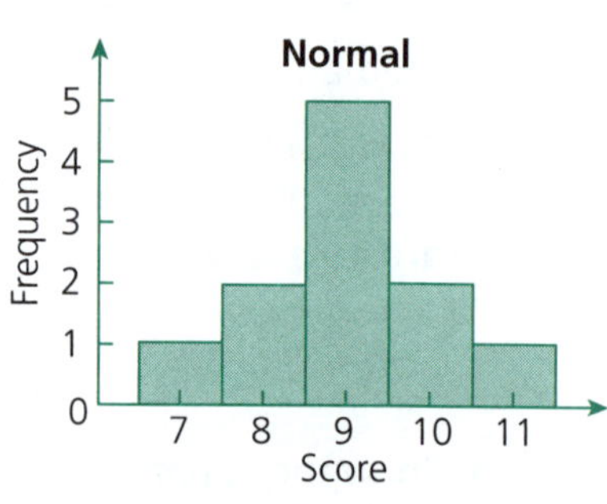

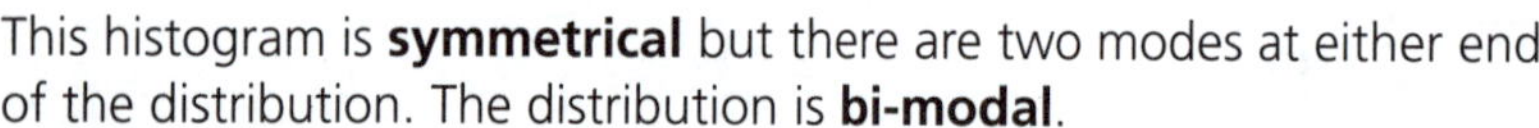

This histogram is **symmetrical** but there are two modes at either end of the distribution. The distribution is **bi-modal**.

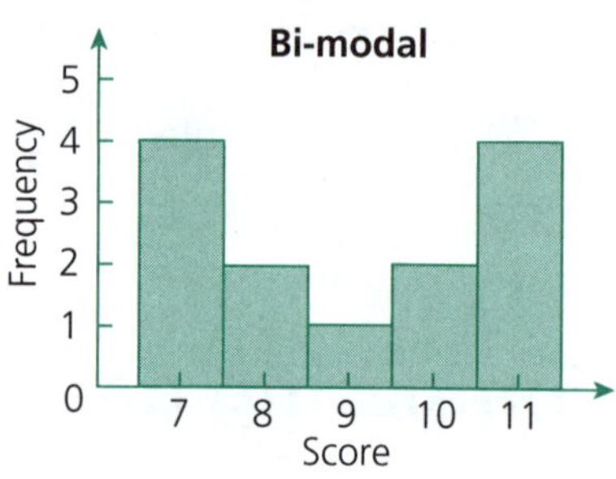

The histograms below are **not symmetrical** and the scores are gathered at the lower end, or the upper end, of the distribution. The distributions are said to be **skewed**.

This is an example of a **positively skewed** distribution.
(Note that the tail of the graph is towards the positive direction.)

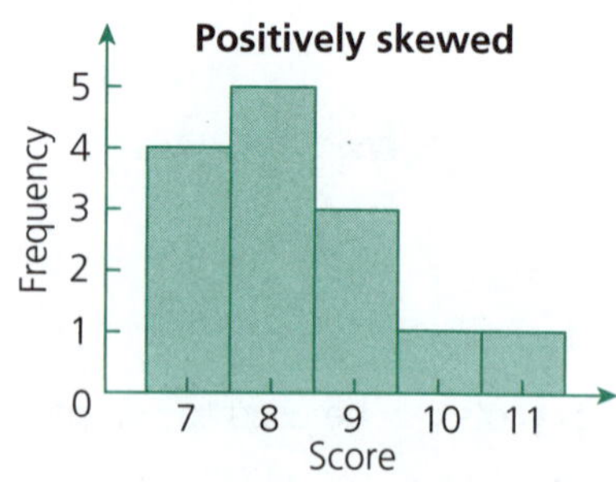

This is an example of a **negatively skewed** distribution.
(Note that the tail of the graph is towards the negative direction.)

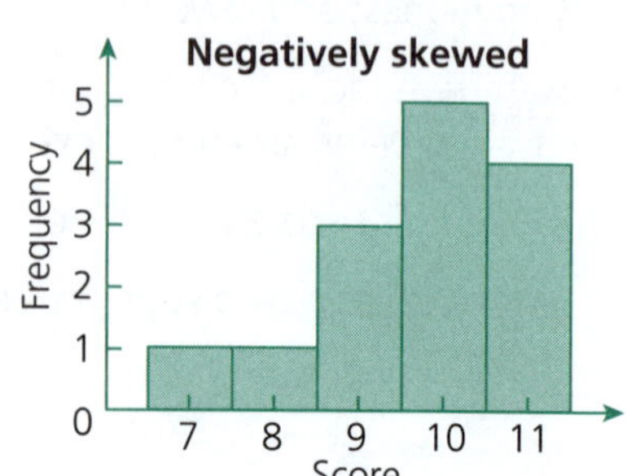

For Example

1

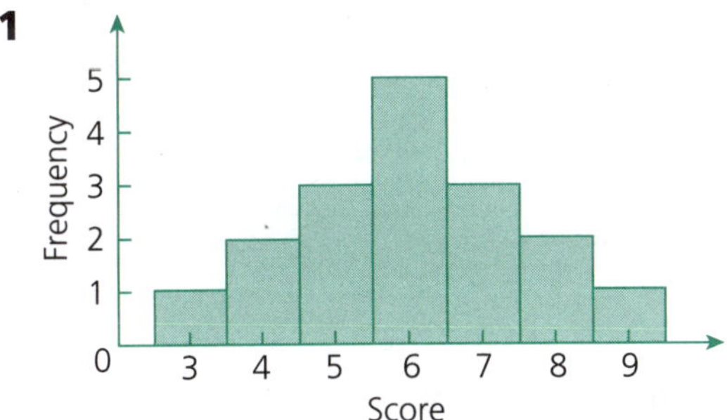

Refer to the above histogram to complete the following.

a Draw a frequency table.

S

Score (x)	Freq (f)	$f \times x$
3	1	3
4	2	8
5	3	15
6	5	30
7	3	21
8	2	16
9	1	9
Total	17	102

b Find the mean, mode and median.

S Mean $= \frac{102}{17}$

$= 6$

Mode = 6

Median: as there are 17 scores, then the 9th score is the middle.

∴ Median is 6.

c Determine whether the data is normally distributed.

S Yes, normal distribution.

2 The diagrams below show the frequency polygons of two distributions. Comment on the skew of each data set.

a

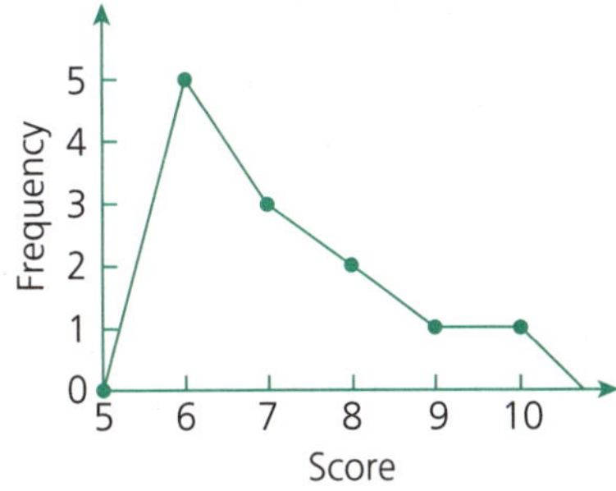

S Positively skewed.

b

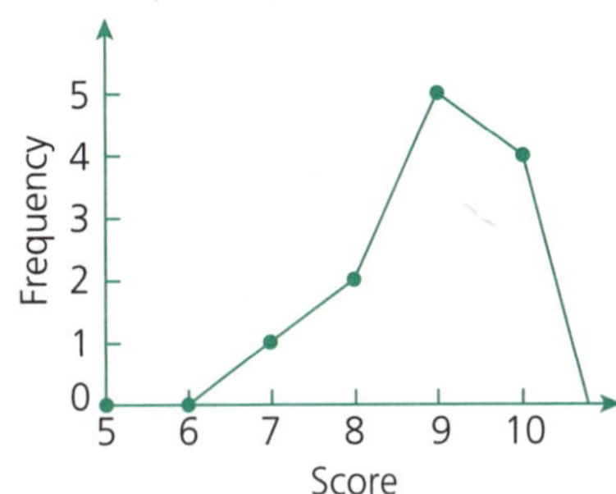

S Negatively skewed.

3 The dot plot shows the results in a history test. Comment on the skew of the distribution.

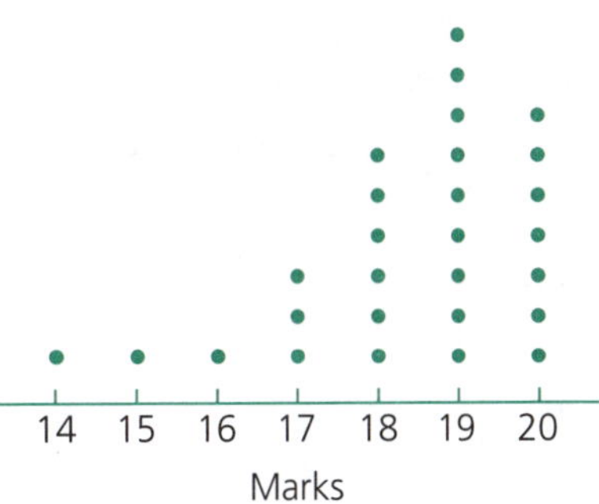

S The data is negatively skewed.

4 A teacher gave two tests to a class. As the class started a new topic they sat a pre-test, and then a post-test when the topic was completed. The back-to-back stem-and-leaf plot shows the results for both tests:

Pre-test		Post-test
9 9 9 8 7 5 5 3	0	2
9 8 8 6 3 1	1	4 5
8 5 4	2	0 1 8 8
1	3	2 4 5 7 8 8
	4	6 7 7 8 9

Comment on the skew of the distributions.

S **The pre-test was positively skewed and the post-test was negatively skewed.**

5 Draw a box plot for each and comment on the skew of each distribution:

a 2, 4, 4, 6, 6, 6, 8, 8, 10

S Lowest score = 2
Lower quartile = 4
Median = 6
Upper quartile = 8
Highest score = 10

Normal distribution

b 2, 4, 6, 7, 7, 8, 8, 8, 9, 9, 9, 9, 10

S Lowest score = 2
Lower quartile = 6.5
Median = 8
Uper quartile = 9
Highest score = 10

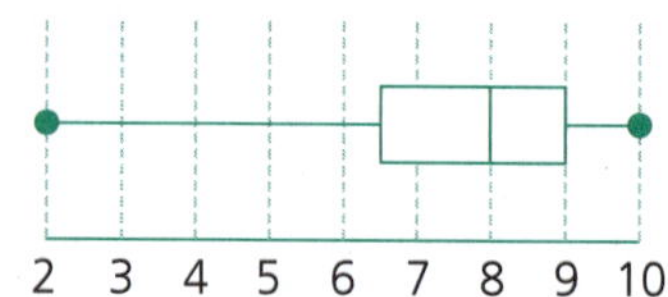

Negatively skewed

Bivariate data analysis

Introduction

In this chapter we have concentrated on single variable data such as marks in a test, or height of students. Now we consider pairs of variables such as height and weight and any **relationship** that may exist between the two variables. One variable is the **independent** variable (e.g. time) and the other is the **dependent** variable. When data is written in a table the independent variable is usually the top row, or the left-hand column. When graphed, the independent variable is placed on the horizontal axis.

The simplest way to graph bivariate data is to use a **scatterplot**. The relationship between the two variables can be described as:

- **Positive or negative**

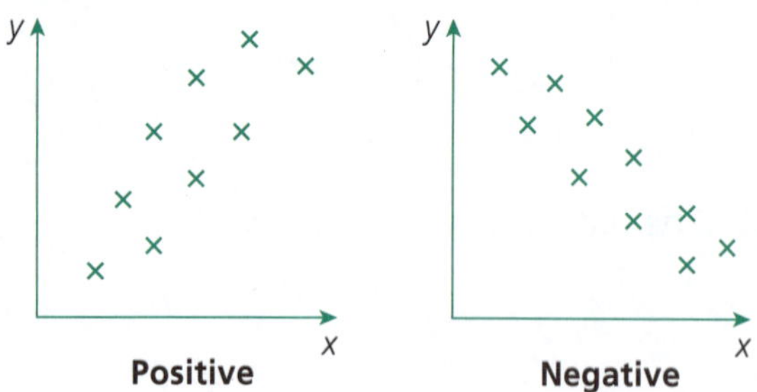

- **Strong or weak**

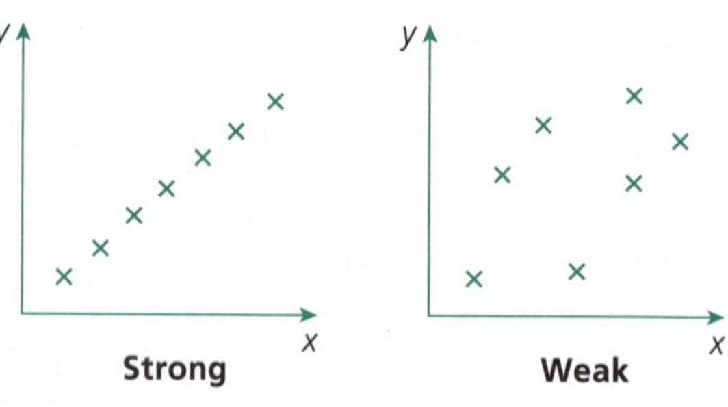

The scatterplot can be used to identify **trends** and make **predictions**.

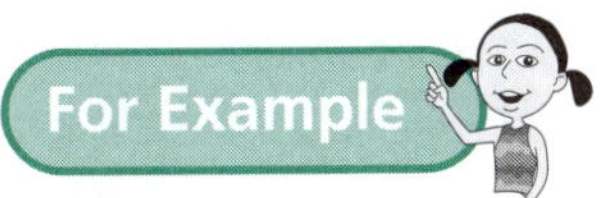

1 State the independent and dependent variables in each of the following pairs of data:

a Age and height of a primary school student

S **Independent: age; dependent: height**

b Speed of a car and the distance travelled

S **Independent: speed; dependent: distance**

c Wage earned and number of hours worked

S **Independent: hours worked; dependent: wage**

2 State the type of relationship between x and y for each of the following scatterplots:

a

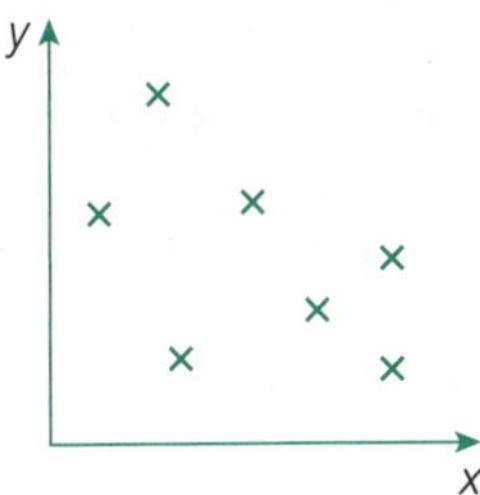

S **Weak negative relationship.**

b

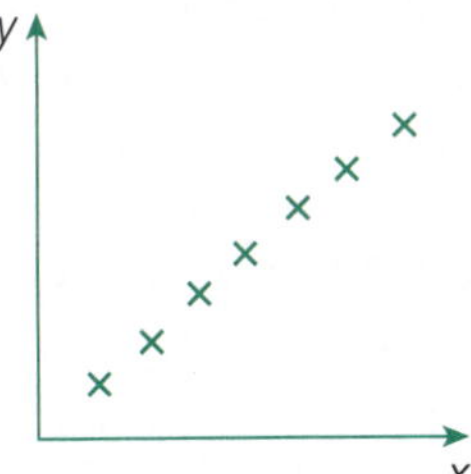

S **Strong positive relationship.**

3 A survey was conducted to find the time (t, in weeks) people had spent on a particular diet and the corresponding mass loss (m, in kg). The results of the survey are shown in the table:

t	6	2	8	4	5	2	4	7	2	3
m	5	1	6	3	6	4	2	6	2	3

a Draw a scatterplot.

S

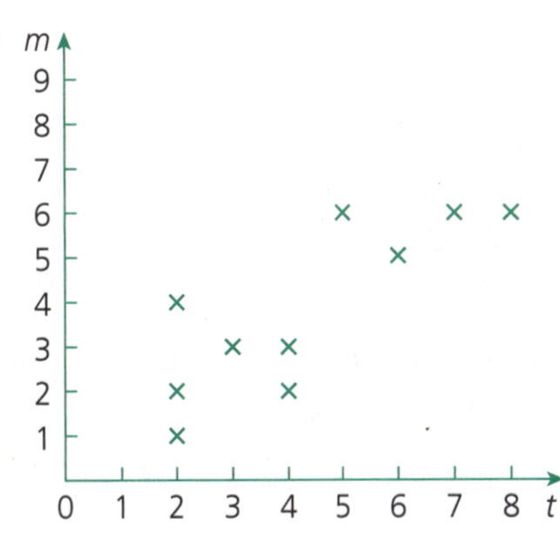

b Describe the relationship between t and m.

S **Weak positive relationship.**

c What is the trend suggested by the scatterplot?

S **The longer a person stays on the diet the more mass will be lost.**

Practise Practise

1 Johann Bark surveyed the number of notes used in each line of a sheet of music. His results were:

5 3 6 3 4 2 5 3 4 7
2 2 4 4 5 5 4 3 2 5

From this information, draw up a frequency distribution table and answer the following questions:

a How many lines contained:
  i 4 notes?
  ii 4 or fewer notes?
  iii more than 4 notes?

b Find the range and mode.

c What percentage of the lines contained:
  i 4 notes?
  ii 4 or fewer notes?

d Calculate the mean and median.

e Draw a frequency polygon and a cumulative frequency histogram.

f What is the probability that a line selected at random would contain 4 notes? pp. 164–166

2 Goldilocks surveyed 25 bears in the forest to find out the number of lumps each had in their porridge. She wrote down the following set of numbers:

7 4 0 2 3 6 4 5 3 2 1 3 7
6 6 4 5 2 4 5 5 4 6 7 6

Draw up a frequency distribution table for these data (about lumps) and answer the following questions:

a Find the range and the mode of the number of lumps.

b How many bears found:
  i 4 lumps?
  ii 4 or less lumps?
  iii less than 4 lumps?

c What percentage of bears had:
  i 4 lumps?
  ii at least 4 lumps?

d If a bear is chosen at random, what is the probability that his porridge contained:
  i 4 lumps?
  ii less than 4 lumps?

e Calculate the mean number of lumps and the median.

f Draw a histogram.

g Construct a cumulative frequency polygon and mark the median on the diagram.

pp. 164–166

3 Gai has been playing soccer for six years and has recorded her goal-scoring record. Over these years Gai scored 8, 7, 4, 8, 6 and 3 goals.

a **i** Calculate Gai's mean number of goals.
  ii Find the range and the mode.
  iii Find the median.

b In her seventh year of soccer Gai struck brilliant form to score 13 goals. How will this affect Gai's mean and median?

c After eight years, Gai's mean number of goals slumped to 6.5. How many goals did she score in her eighth year? pp. 164–166

4 For the following sets of scores, find the mean and median (to one decimal place if necessary):

a 7, 4, 7, 3, 5, 6, 4, 8, 4

b 2, 4, 6, 3, 5, 7, 9, 11, 7, 4 pp. 164–166

5 Zoltan needs to average 75 in his yearly examinations to be allowed to drive the family car. His average for his first four exams is 70. What mark must he score in his fifth and final exam to average 75? pp. 164–166

6 Twenty-five students were surveyed. They were asked the question: 'What colour is your favourite cap?'

The results of the survey were: green 8, red 7, blue 3, black 2, yellow 1, other colour 4. Draw a dot plot to illustrate this information. p. 167

7 The shoe sizes of 22 students from a Year 10 class were collected:

$6\frac{1}{2}$ 5 8 $7\frac{1}{2}$ 7 5 5 $5\frac{1}{2}$
7 6 $4\frac{1}{2}$ 4 7 6 $5\frac{1}{2}$ 6
5 4 5 7 7 $6\frac{1}{2}$

By first making a dot plot, draw up a frequency distribution table to illustrate this information.

p. 167

8 The height of 20 Year 9 boys was measured and recorded. The heights in cm were:

148 150 151 155 149 154 163
159 152 162 165 167 147 158
168 154 149 150 160 152

Construct a stem-and-leaf plot to illustrate this information. p. 168

9 The results of students in an English test out of 70 were as follows:

26	47	39	64	48	50	52	33	61	55
47	62	46	52	60	39	28	54	39	62
49	53	57	44	64	43	40	33	47	50

a Construct a stem-and-leaf plot to illustrate this information.

b Write down the range, mode and find the median. p. 168

10 For the following sets of scores find the median:

a 7, 9, 8, 6, 6, 5, 4, 8

b 9, 8, 9, 6, 7, 6, 6, 5, 9, 8, 7

c 18, 14, 17, 12, 19, 16, 16, 14, 18, 15

d 11, 7, 9, 12, 12, 7, 8, 11, 12, 8, 7, 14, 13, 10, 9

pp. 164–166

11 From the following stem-and-leaf plots find the median:

a

Stem	Leaf
1	0 7 8
2	1 2 4 6 9
3	7 7 7 8 8 9
4	0 0 6 7 8 8 9
5	1 4

b

Stem	Leaf
10	5 8
11	3 3 6 6 9
12	4 4 7
13	5 5 6 8 8
14	7

pp. 169–170

12 For the following sets of scores, find the median, the upper and lower quartiles and the interquartile range and draw a box plot:

a 7, 9, 8, 6, 6, 5, 4, 8

b 9, 8, 9, 6, 7, 6, 6, 5, 9, 8, 7

c 18, 14, 17, 12, 19, 16, 16, 14, 18, 15

d 11, 7, 9, 12, 12, 7, 8, 11, 12, 8, 7, 14, 13,10, 9

pp. 170–173

13 From the following box plots:

a

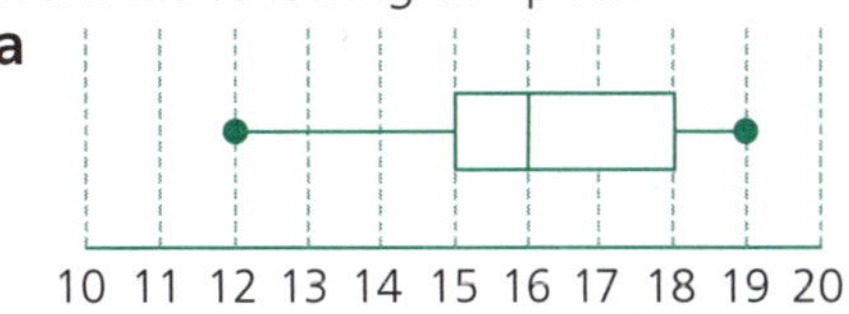

b

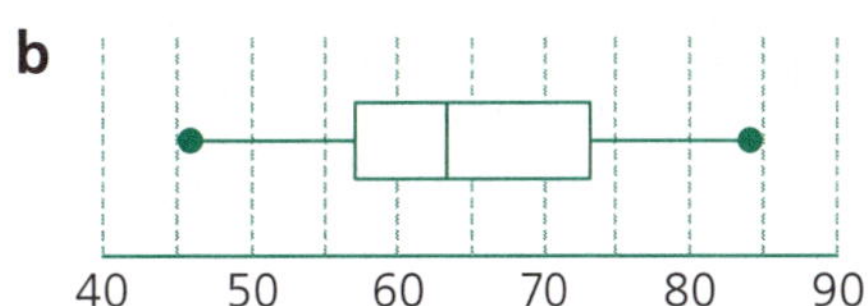

i write down the median

ii calculate the interquartile range.

pp. 172–173

14 From the following stem-and-leaf plots find the median, the upper and lower quartiles and construct a box plot:

a

Stem	Leaf
1	0 7 8
2	1 2 4 6 9
3	7 7 7 8 8 9
4	0 0 6 7 8 8 9
5	1 4

b

Stem	Leaf
10	5 8
11	3 3 6 6 9
12	4 4 7
13	5 5 6 8 8
14	7

pp. 171–173

15 The heights of 30 Year 9 boys were measured and tabulated. The results were as follows:

161	149	150	161	167	169
172	180	182	150	174	167
168	170	175	182	181	168
154	166	162	160	159	169
166	158	164	174	172	170

a Construct a stem-and-leaf plot from this information.

b From the stem-and-leaf plot find the median, the upper and lower quartiles and the interquartile range.

c Construct a box plot to illustrate this information.

pp. 171–173

16 Comment on the skew of the following data sets:

a

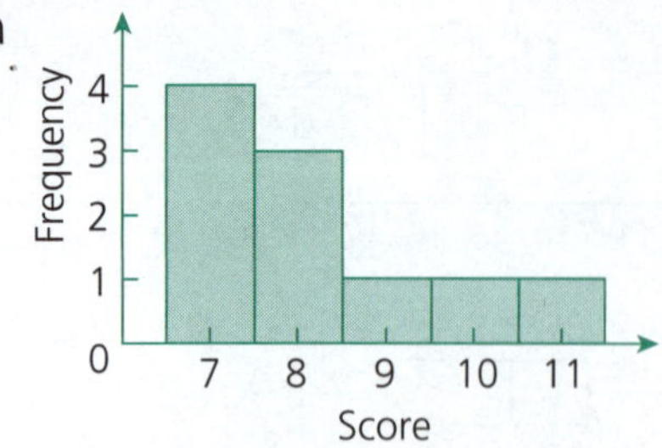

b

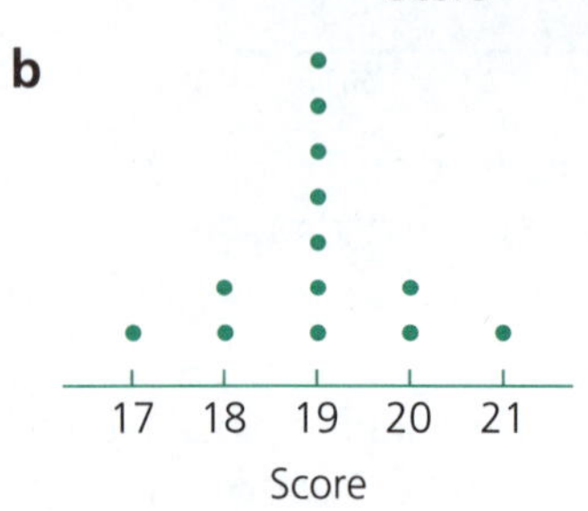

c

Stem	Leaf
1	4 7 7 8 9
2	3 5 6
3	1
4	4

d

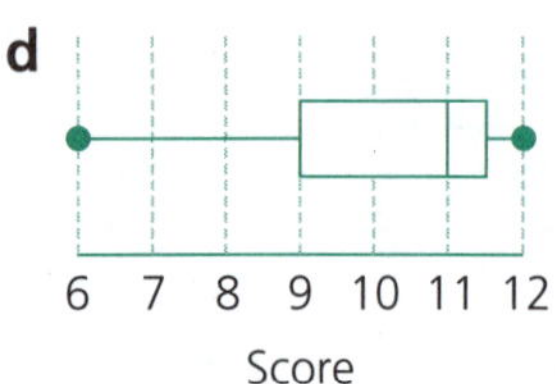

e

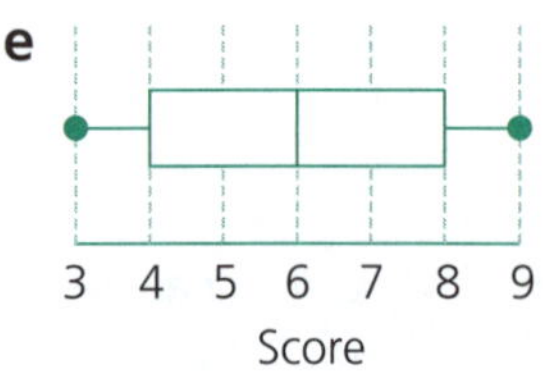

pp. 174–176

17

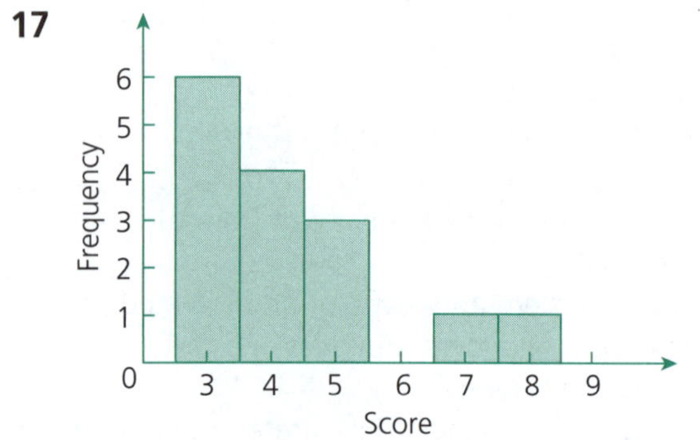

Use the histogram to:

a draw a frequency table

b find the mean, mode and median

c comment on the skew of the data. pp. 174–176

18 The diagrams below show the frequency polygons of two distributions. Comment on the skew of each data set:

a

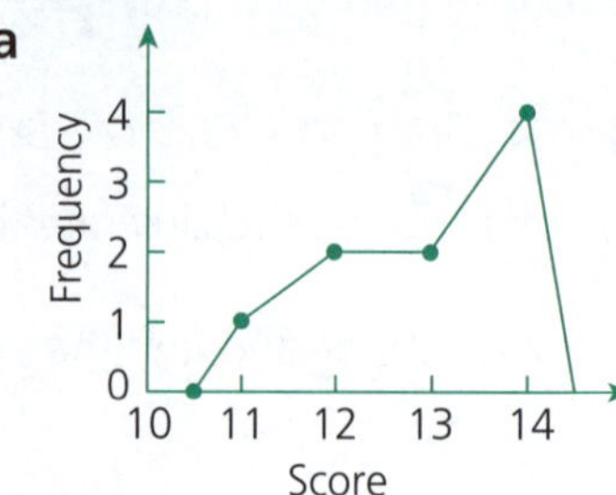

b

pp. 174–176

19 A group of students chose ten-pin bowling for sport. The dot plot shows the shoe sizes that were used by the group:

a Comment on the skew.

b Find the median, mode and mean of the data.

pp. 174–176

20 The results of two maths tests are recorded below in the back-to-back stem-and-leaf plot.

Comment on the skew of both tests:

Results of tests		
Trigonometry		**Equations**
9 9 6	0	
8 6 5 4 2 0	1	8 9
7 4 3	2	3 5 7 8 9
5 4	3	2 4 6 7 7 8 9 9
0	4	0

p. 174

21 For the following data sets, draw a box plot and comment on the skew of each distribution:

a 9, 13, 15, 16, 16, 17, 17, 17, 18
b 8, 9, 10, 10, 10, 11, 18
c 10, 12, 12, 14, 14, 14, 16, 16, 18 pp. 172–176

22 Determine the independent variable in each of the following pairs of data:

a rainfall (r in mm) and crop yield (A in kg)
b height of plant (h in cm) and age (n in weeks)
c car's engine size (s in L) and fuel economy (E in L/100 km). pp. 176–177

23 The graphs show a relationship between x and y. Comment on the type of relationship, using terms such as weak/strong and positive/negative.

a

b

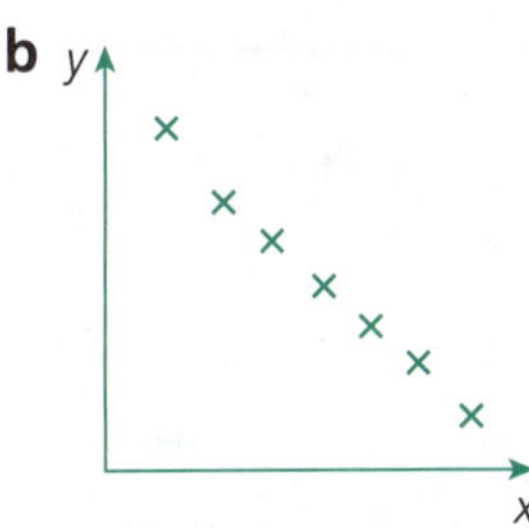

pp. 176–177

24 Eight students measured their heights (h) and arm spans (s) and the results are recorded in the table below. All lengths are in cm.

h	161	165	158	157	159	162	160	158
s	163	167	159	159	162	165	162	160

a Draw a scatterplot.
b Describe the relationship between h and s.
c Use the scatterplot to predict the arm span of a student with a height of 164 cm.
d What is the trend suggested by the scatterplot?

pp. 176–177

25 The same eight students then compared their heights (h) with the distance from the top of their heads to the middle of their chests (d). The results are recorded in the table below. All lengths are in cm.

h	161	165	158	157	159	162	160	158
d	41	42	40	39	40	41	40	39

a Draw a scatterplot.
b Use the scatterplot to predict the arm span of a student with a height of 164 cm.

pp. 176–177

26 Consider the data set:

x	4	11	9	5	2	7	1	8	4	6
y	7	0	2	5	8	3	9	3	8	4

a Draw a scatterplot for the data.
b Describe the relationship between x and y as positive or negative.
c Describe the relationship between x and y as strong or weak. pp. 176–177

Go to p. 220 for **Quick Answers** or to pp. 253–256 for **Worked Solutions**

For a complete understanding of this topic, you must be able to:

✓	Recall and use basic statistics concepts		pp. 164–166
✓	Construct dot plots and stem-and-leaf plots and use them to find range, mode and median		pp. 167–170
✓	Calculate upper and lower quartiles, interquartile range and construct box plots		pp. 170–173
✓	Compare data sets using stem-and-leaf and box plots		pp. 173–174
✓	Recognise normal and skewed distributions		pp. 174–176
✓	Recognise the independent and dependent variables in bivariate data displays		pp. 176–177
✓	Display bivariate data in a scatterplot		pp. 176–177
✓	Recognise relationships of bivariate data as positive/negative and strong/weak.		pp. 176–177

Now you are ready to do the tests!

Intermediate Test

Statistics

(30 marks)

1 Which of the distributions shown below is bi-modal? (1 mark)

a 2, 2, 3, 3, 3, 4, 5 **b** 14, 16, 16
c 1, 2, 3, 3, 4, 4 **d** 0, 0, 1, 2

2 Grace's netball team scored 26, 32, 28, 39 and 30 points in their first five games of the season.

a Find the mean number of points scored in the first five games. (1 mark)

b How many points does her team need to score in the next game to maintain the same mean score? (1 mark)

c If Grace's team scores 37 points in the sixth game, what is their overall mean score for the six games? (1 mark)

d After the seventh game the mean score was calculated and found that it had dropped by two points. If the opposing team scored 28 points, by how many points did Grace's team lose? (2 marks)

3 Each day for a week Thomas ate two mandarins at lunchtime and counted the number of seeds he found in each mandarin. The number of seeds were:

4 6 3 5 4 2 6
4 5 3 5 3 1 5

a Calculate the mean, mode and range. (3 marks)

b Find the median and the interquartile range. (2 marks)

c Draw a box plot. (2 marks)

4 A box plot is drawn to show the distribution of marks for a class of 28 students.

The test was marked out of 25.

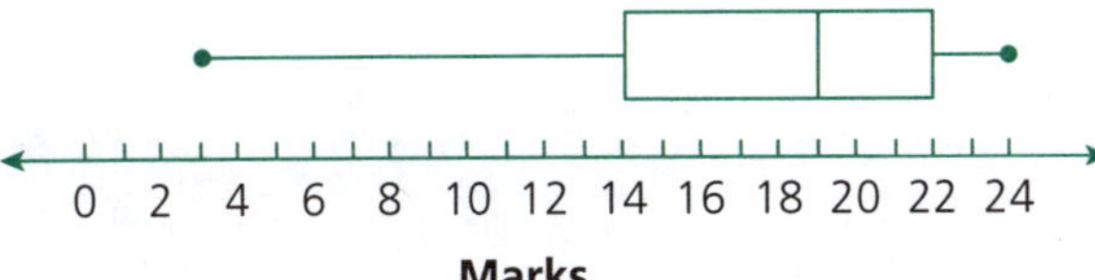

a Find the highest and lowest marks for the test. (2 marks)

b Approximately what percentage of students scored:

i 19 or less? (1 mark)

ii 14 or more? (1 mark)

iii 88% or more? (1 mark)

c How many students scored 14 or less? (1 mark)

d If you scored 72%, would you be placed in the bottom half or top half of the class? (1 mark)

e Comment on the skew of the distribution. (1 mark)

5 The Wildfires and the Diamonds play in the same basketball competition and the number of points scored in each game of the season is recorded in the back-to-back stem-and-leaf plot:

Wildfires		Diamonds
Leaf	**Stem**	**Leaf**
9 8	0	
9 9 7 6 4	1	9
2	2	1
4 2 0	3	2 6 6 7
2 1	4	0 2 4 7 8 9
1	5	6 7

a Comment on the team's:

i range of scores (1 mark)

ii mode (1 mark)

iii median. (1 mark)

b Comment on the skew of each team's points. (1 mark)

6 Consider the bivariate data set:

x	2	5	3	6	5	2	1	6	4	4
y	4	9	7	10	10	5	3	11	8	7

a Draw a scatterplot for the data. (2 marks)

b Which is the dependent variable? (1 mark)

c Is the relationship positive or negative? (1 mark)

d Is the relationship weak or strong? (1 mark)

Your Feedback

☐ × 100% = ☐ %

QA PAGE 224

WS 

Advanced Test

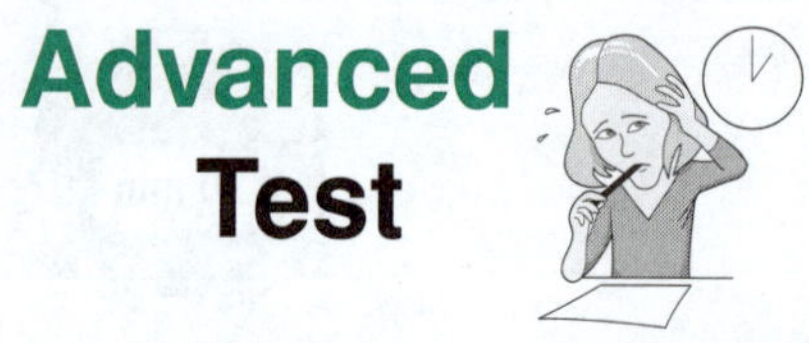

Statistics

(30 marks)

1 Match the terms negatively skewed, positively skewed, bi-modal and normally distributed to each of the following histograms: (4 marks)

a

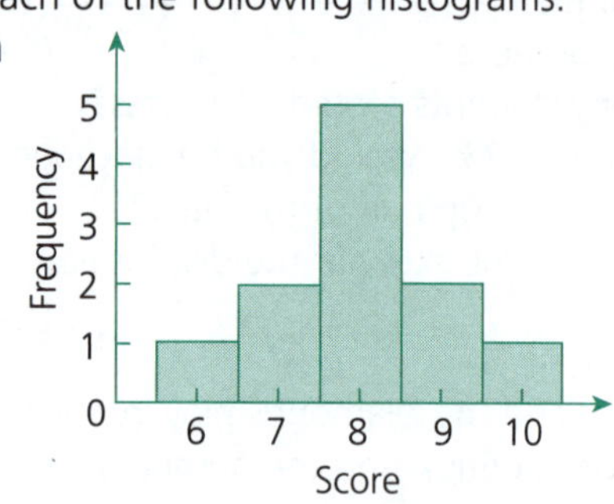

b

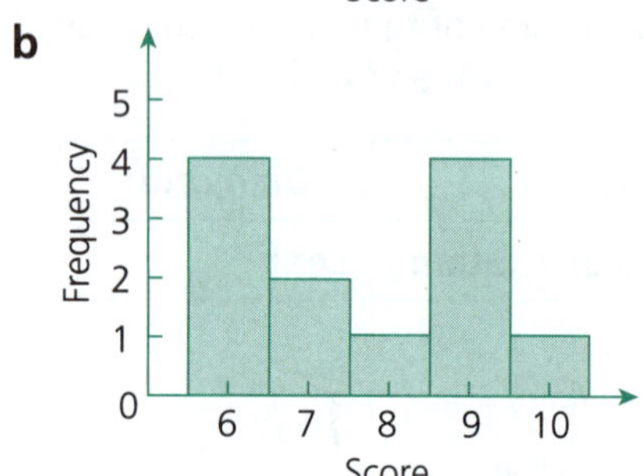

c

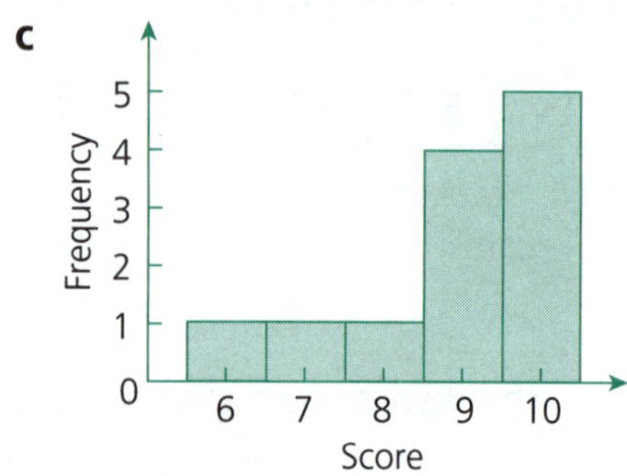

d

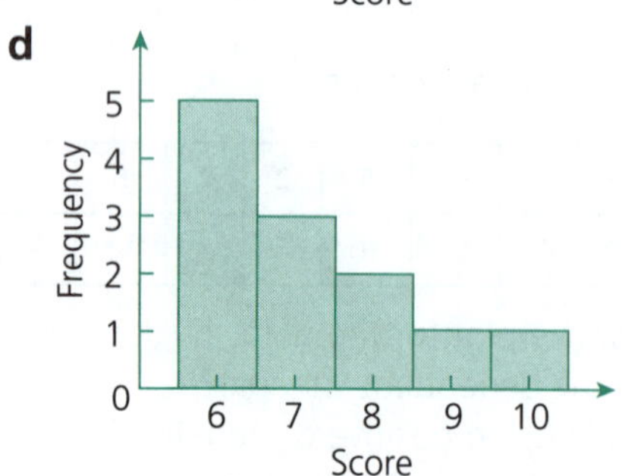

2 A set of scores is displayed in a stem-and-leaf plot:

Stem	Leaf
0	7
1	0 0 2 5 8
2	6 7
3	1 6
4	2

a Find the mode and range. (2 marks)
b Find the median. (1 mark)
c Find the interquartile range. (2 marks)
d Draw a box plot. (2 marks)

3 The table shows the results of a spelling test for a particular class of Year 10 students.

Score	Frequency
6	1
7	3
8	7
9	12
10	7

a How many students were in the class? (1 mark)
b The data set is skewed. Is it positive or negative? (1 mark)
c The mean result in all Year 10 classes was 8.2. How does this class compare? (2 marks)

4 The prices of books in two shops were recorded and displayed in the box plot below:

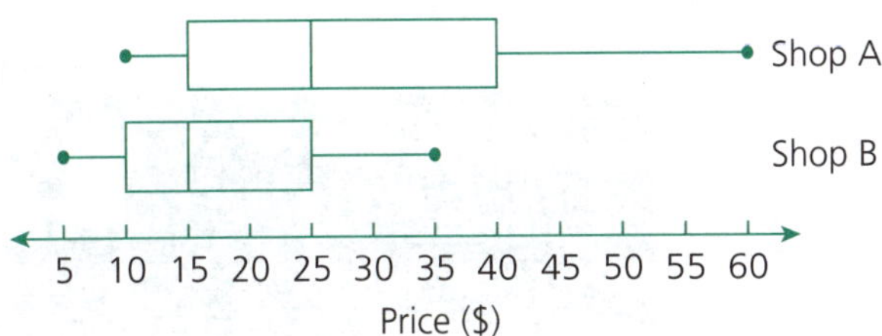

a What was the price of the cheapest book in Shop A? (1 mark)

b What percentage of books in Shop B were priced at $25 or less? (1 mark)

c If Shop A had 400 books for sale, how many were priced at $40 or more? (1 mark)

d Compare the two shops' interquartile range. (2 marks)

5 The number of people (n) at a beach at midday compared with the temperature (T) over a 10-day period is recorded in the table below:

T	27	29	31	20	23	26	31	36	41	38
n	86	102	105	11	21	46	68	121	85	92

a Draw a scatterplot. (2 marks)

b Is the relationship positive or negative? (1 mark)

c Comment on the strength of the relationship and give a reason. (2 marks)

6 Find the interquartile range for a data set if 75% of the scores are 7.4 or above and 25% of the scores are 10.1 or above. (1 mark)

7 Ten students sat a history test and a science test and the results are presented on the scatterplot below:

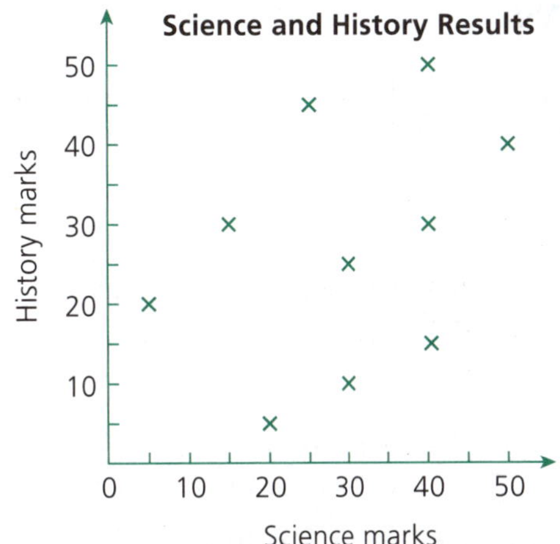

a Is the relationship between the pair of results weak or strong? (1 mark)

b What was the highest science mark? (1 mark)

c The results from the scatterplot are recorded in the table below:

Science	25	30	40	40	50	30	b	5	20	15
History	a	10	15	50	40	25	30	20	5	30

Find the missing values of a and b. (2 marks)

Your Feedback

☐ × 100% = ☐ %

QA PAGE 224

WS PAGE 273

Chapter 12 Probability

Review

A coin is tossed once. There is one chance in two that it will land with a head facing up. We say that the probability of throwing a head is one out of two, or:

$P(\text{Head}) = \frac{1}{2}$

As each outcome is equally likely, the probability of throwing a tail is also $\frac{1}{2}$, that is:

$P(\text{Tail}) = \frac{1}{2}$

Keywords
Favourable
Outcomes
Tree Diagram
Complementary event
Conditional

When throwing a die, there are six faces and each is equally likely to turn up. It cannot be predicted which number will be face up. The probability of throwing a four is one chance out of six, that is, $P(4) = \frac{1}{6}$.

Definition of probability

The probability of an event occurring is defined as follows:

$$\text{Probability (event)} = \frac{\text{Number of favourable outcomes}}{\text{Number of possible outcomes}}$$ provided each outcome is equally likely to occur.

Consider the die. The number of favourable outcomes was one (throwing a six), while the number of possible outcomes was six (each face):

$$\therefore P(6) = \frac{\text{Number of favourable outcomes}}{\text{Number of possible outcomes}} = \frac{1}{6}$$

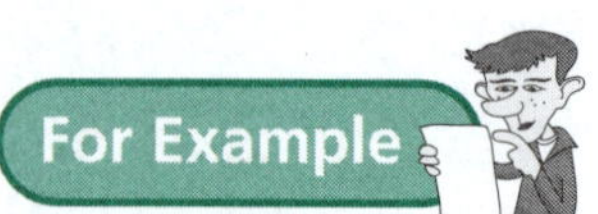

1 A fair die is thrown. Find the probability of rolling:

a a 3

S $P(3) = \frac{1}{6}$

b a 5

S $P(5) = \frac{1}{6}$

c an odd number

S There are three odd numbers (1, 3, 5):

$\therefore P(\text{odd}) = \frac{3}{6} = \frac{1}{2}$

d an even number

S $P(\text{even}) = \frac{3}{6} = \frac{1}{2}$

e a number larger than 4.

S There are two numbers larger than 4:5 and 6.

$P(>4) = \frac{2}{6} = \frac{1}{3}$

Range of probability

A certainty has a probability of 1. An impossibility has a probability of 0.

The probability of any specific event occurring has a value between 0 and 1. A value outside this range is meaningless. Thus, if there are two possibilities, the probability of the event (say, A) not occurring is $1 - P(A)$.

These are called **complementary events**.

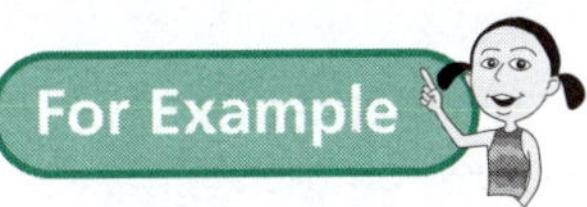

1 Michelle's school bag contains two black socks and three grey socks. If one sock is drawn out at random, what is the probability that the sock is:

a black

S $P(\text{black}) = \dfrac{\text{Number of favourable outcomes}}{\text{Number of possible outcomes}}$

$= \frac{2}{5}$

Total possible outcomes = Total number of socks.

b grey

S $P(\text{grey}) = 1 - P(\text{black})$

$= 1 - \frac{2}{5}$

$= \frac{3}{5}$

Relative frequency and probability

The relative frequency column can be used to find experimental probability.

Nicholas throws an unbiased die 180 times. The results are below.

The probability of throwing a one is theoretically $\frac{1}{6}$, or 0.167 (to three decimal places). Nicholas has thrown 28 ones out of 180 throws.

Score	Frequency	Relative frequency
1	28	0.156
2	30	0.167
3	31	0.172
4	29	0.161
5	32	0.178
6	30	0.167
Σ	180	1.000

Experimental probability of throwing a 1

$= \frac{28}{180} = 0.156$ (to three decimal places)

← Remember, r.f. $= \dfrac{f}{\text{Total } f}$

[Total of relative frequency is 1, even though rounding here means a total of 1.001.]

Venn Diagrams

Venn Diagrams provide a means of representing outcomes visually. Usually consisting of circles and a rectangle, the overlapping sections identify common outcomes.

For Example

1 A survey found the number of students who enjoyed Science and History and the results detailed on the Venn Diagram.

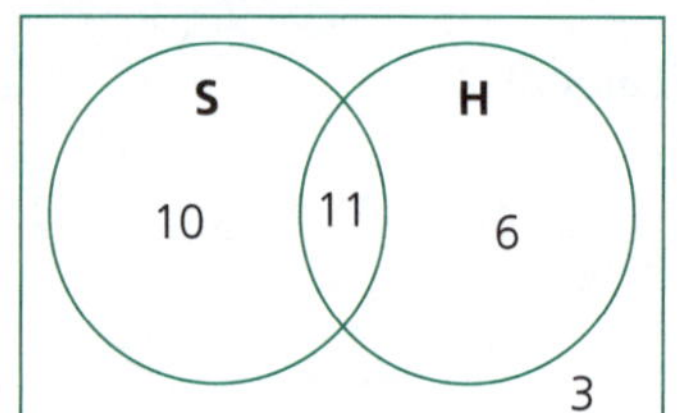

S: Science
H: History

a How many students were surveyed?

S 30

b If a student was chosen at random, what is the probability that the student:

i enjoys both subjects?

S $\therefore P(\text{both}) = \frac{11}{30}$

ii enjoys History but not Science?

S $\therefore P(\text{History, not Science}) = \frac{6}{30} = \frac{1}{5}$

iii does not enjoy either subject?

S $\therefore P(\text{neither}) = \frac{3}{30} = \frac{1}{10}$

2 On a particular day, the training patterns of a squad of female triathletes were recorded on the Venn Diagram:

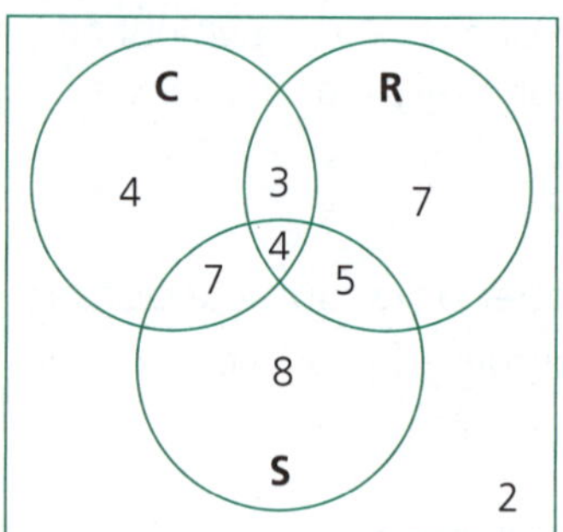

C: Cycling
R: Running
S: Swimming

a If a triathlete was chosen at random, what is the probability that she:

i swam and cycled?

S $\therefore P(\text{swam and cycled}) = \frac{11}{40}$

ii ran, cycled and swam?

S $\therefore P(\text{all three}) = \frac{4}{40} = \frac{1}{10}$

b If a triathlete who had cycled that day was chosen at random, find the probability that she:

i ran

S $\therefore P(\text{cyclist also ran}) = \frac{7}{18}$

ii did not swim

S $\therefore P(\text{cyclist did not swim}) = \frac{7}{18}$

Multiple events using tables

Multiple events are sometimes called multistage events and involve more than one event, such as flipping a coin twice, or choosing two balls from a bag. If replacement is allowed then the outcomes from each selection can be repeated. If replacement is not allowed then the outcomes cannot be repeated.

For Example

1 A bag contains three balls numbered 1, 2 and 3. The first ball chosen is replaced before the second ball is selected.

a Complete a table listing the possible outcomes.

S

		1st choice		
		1	2	3
2nd choice	1	(1, 1)	(2, 1)	(3, 1)
	2	(1, 2)	(2, 2)	(3, 2)
	3	(1, 3)	(2, 3)	(3, 3)

b What is the probability of:

i selecting a 3 followed by a 1?

S $\therefore P(3, 1) = \frac{1}{9}$

ii selecting a double?

S $\therefore P([1,1] \text{ or } [2, 2] \text{ or } [3, 3]) = \frac{3}{9} = \frac{1}{3}$

iii selecting a total of 5?

S $\therefore P(\text{total of } 5) = \frac{2}{9}$

iv selecting a total less than 4?

S $\therefore P(\text{less than } 4) = \frac{3}{9} = \frac{1}{3}$

2 Another bag contains three balls numbered 4, 5 and 6. The first ball chosen is **not** replaced before the second ball is selected.

a Complete a table listing the possible outcomes.

S

		1st choice		
		4	5	6
2nd choice	4	×	(5, 4)	(6, 4)
	5	(4, 5)	×	(6, 5)
	6	(4, 6)	(5, 6)	×

b What is the probability of:

i selecting a total of 9?

S $\therefore P(\text{total of } 9) = \frac{2}{6} = \frac{1}{3}$

ii selecting a total more than 10?

S $\therefore P(\text{more than } 10) = \frac{2}{6} = \frac{1}{3}$

3 Mitchell throws two dice and adds the two face-up numbers. Draw up a table to show all possibilities when the scores are added.

S **From the table, count the total number of possible outcomes—36.**
So, all answers will be out of 36.

First Die

Second Die	+	1	2	3	4	5	6
	1	2	3	4	5	6	7
	2	3	4	5	6	7	8
	3	4	5	6	7	8	9
	4	5	6	7	8	9	10
	5	6	7	8	9	10	11
	6	7	8	9	10	11	12

Find the probability that the sum of numbers face up will be:

a 7

S $P(7) = \frac{6}{36} = \frac{1}{6}$ Highlighted in table.

b 8

S $P(8) = \frac{5}{36}$

c 12

S $P(12) = \frac{1}{36}$

d greater than 7

S $P(>7) = \frac{15}{36} = \frac{5}{12}$

e 7 or more

S $P(\geq 7) = \frac{21}{36} = \frac{7}{12}$

f 1

S $P(1) = 0$, as impossible with two dice

g even or less than 7

S P(even or less than 7)
$= \frac{24}{36}$
$= \frac{2}{3}$

h even and less than 7

S P(even and less than 7)
$= \frac{9}{36}$
$= \frac{1}{4}$

4 Scott throws two dice and then calculates the product of the two numbers face up.

S The first step is to draw up the table.

×	1	2	3	4	5	6
1	1	2	3	4	5	6
2	2	4	6	8	10	12
3	3	6	9	12	15	18
4	4	8	12	16	20	24
5	5	10	15	20	25	30
6	6	12	18	24	30	36

Now count the total number of possible outcomes—again it is 36. This then becomes the denominator in all the probabilities.

Find the probability that the product of numbers face up will be:

a 6

S $P(6) = \frac{4}{36} = \frac{1}{9}$ Highlighted in table.

b 36

S $P(36) = \frac{1}{36}$

c less than 24

S $P(<24) = \frac{30}{36} = \frac{5}{6}$

d even

S $P(\text{even}) = \frac{27}{36} = \frac{3}{4}$

e even and less than 24

S $P(\text{even and less than 24}) = \frac{22}{36} = \frac{11}{18}$

f even or less than 24

S $P(\text{even or less than 24}) = \frac{35}{36}$

g is not 36.

S $P(\text{not } 36) = 1 - P(36) = 1 - \frac{1}{36} = \frac{35}{36}$

5 The table represents the dominant hand of boys and girls in a class of 30 students:

Predisposition	Boy	Girl	Totals
Left-handed	4	2	6
Right-handed	10	14	24
Totals	14	16	30

a If a student is chosen at random, what is the probability that the person is a right-handed boy?

S Ten right-handed boys out of 30 students:
$\therefore P(\text{right-handed boy}) = \frac{10}{30}$
$= \frac{1}{3}$

b If a girl is chosen at random what is the probability that she is left-handed?

S Two left-handed girls out of 16 students:
$\therefore P(\text{left handed}) = \frac{2}{16}$
$= \frac{1}{8}$

Multiple events using tree diagrams (or probability trees)

A tree diagram is a branching diagram that is useful to list the possible outcomes.

1 Lee tosses two coins.

S Draw the tree diagram.
Total number of outcomes = 4

Diagram	Outcomes
First coin H → Second coin H	H H
First coin H → Second coin T	H T
First coin T → Second coin H	T H
First coin T → Second coin T	T T

Four outcomes

Find the probability that Lee throws:

a two heads

S $P(\text{HH}) = \frac{1}{4}$

b two tails

S $P(\text{TT}) = \frac{1}{4}$

c a head and a tail.

S $P(\text{H and T}) = \frac{2}{4} = \frac{1}{2}$

Could be H, T or T, H. The order is not specified.

2 Susie has three red and two blue socks in her drawer. She picks out two socks at random.

S Total number of outcomes = 20

Diagram	Outcomes
R → R, R, B, B	RR, RR, RB, RB
R → R, R, B, B	RR, RR, RB, RB
R → R, R, B, B	RR, RR, RB, RB
B → R, R, R, B	BR, BR, BR, BB
B → R, R, R, B	BR, BR, BR, BB

Find the probability that Susie has chosen:

a two red

S $P(\text{RR}) = \frac{6}{20} = \frac{3}{10}$

b two blue

S $P(\text{BB}) = \frac{2}{20} = \frac{1}{10}$

c one of each colour

S $P(\text{one of each}) = \frac{12}{20} = \frac{3}{5}$

d a pair of the same colour.

S $P(\text{pair}) = \frac{8}{20} = \frac{2}{5}$ [RR or BB]

Note: $P(\text{not a pair}) + P(\text{pair}) = 1$.
If the probabilities for all possible outcomes are totalled, the result will always be 1.

3 The letters A, B, C are written on identical cards and placed in a bag. Three cards are drawn from the bag without replacement to make a three-letter word. Use a tree diagram to list the possible outcomes and find the probability that the word is BAC:

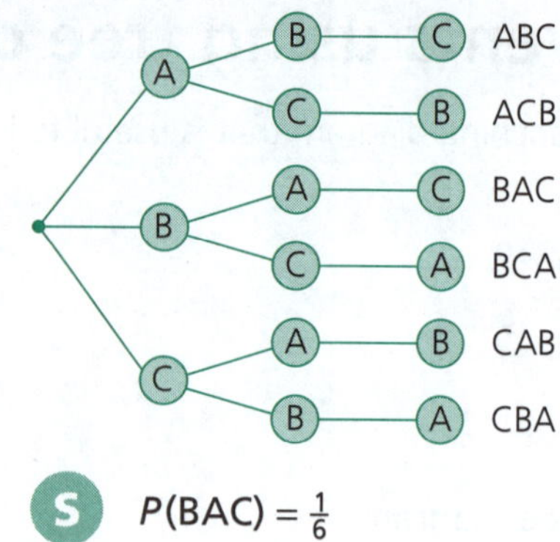

S $P(\text{BAC}) = \frac{1}{6}$

The tree diagram can be simplified by writing probabilities corresponding to a branch. The probability of each outcome is found by multiplying the probabilities on each branch.

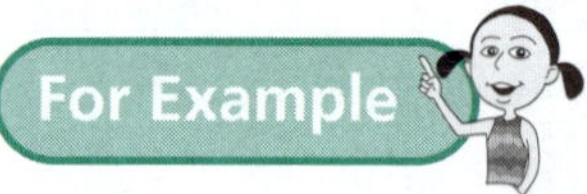

1 René's top drawer contains three grey headbands and two red ones. Two headbands are taken from the drawer, one after the other, and the colours noted. Find the probability that the headbands are either both red, or both grey, or different colours.

Draw up a tree as before but mark the probabilities for each possible outcome along the branch.

Diagram	Outcome	Probability
First pick / Second pick		Multiply along each branch
$\frac{2}{5}$ R, $\frac{1}{4}$ R	R R	$P(\text{RR}) = \frac{2}{5} \times \frac{1}{4} = \frac{1}{10}$
$\frac{2}{5}$ R, $\frac{3}{4}$ G	R G	$P(\text{RG}) = \frac{2}{5} \times \frac{3}{4} = \frac{3}{10}$
$\frac{3}{5}$ G, $\frac{2}{4}$ R	G R	$P(\text{GR}) = \frac{3}{5} \times \frac{2}{4} = \frac{3}{10}$
$\frac{3}{5}$ G, $\frac{2}{4}$ G	G G	$P(\text{GG}) = \frac{3}{5} \times \frac{2}{4} = \frac{3}{10}$

Grey selected first, so still 2 reds out of 4.

Grey selected first, so only 2 left out of 4.

Prob. (grey first) = $\frac{3}{5}$

Prob. (red first) = $\frac{2}{5}$

Find the probability the headbands may be:

a both red

S $P(\text{RR}) = \frac{2}{5} \times \frac{1}{4} = \frac{1}{10}$

b both grey

S $P(\text{GG}) = \frac{3}{5} \times \frac{2}{4} = \frac{3}{10}$

c different colours.

S $P(\text{different}) = P(\text{RG}) + P(\text{GR})$

$= \frac{3}{10} + \frac{3}{10}$ from diagram

$= \frac{3}{5}$

$\therefore$ 3 grey, 2 red = 5 altogether

Note: to find the probability of two or more branches combined we add the probabilities of the branches.

2 The names of eight boys and four girls are placed in one hat, while the names of six boys and four girls are placed in a second hat. A name is drawn from each hat. Find the probability that the names drawn are either both girls, or both boys, or one boy, or at least one girl.

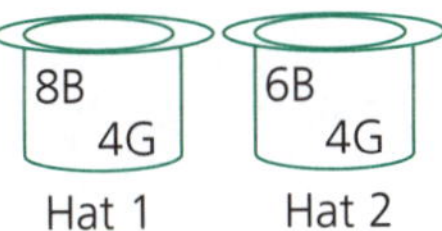

Diagram	Outcomes	Probability
First hat, Second hat		
$\frac{8}{12}$ B, $\frac{6}{10}$ B	BB	$P(BB) = \frac{8}{12} \times \frac{6}{10} = \frac{2}{5}$
$\frac{8}{12}$ B, $\frac{4}{10}$ G	BG	$P(BG) = \frac{8}{12} \times \frac{4}{10} = \frac{4}{15}$
$\frac{4}{12}$ G, $\frac{6}{10}$ B	GB	$P(GB) = \frac{4}{12} \times \frac{6}{10} = \frac{1}{5}$
$\frac{4}{12}$ G, $\frac{4}{10}$ G	GG	$P(GG) = \frac{4}{12} \times \frac{4}{10} = \frac{2}{15}$

Find the probability that the names may be:

a both girls

S $P(GG) = \frac{2}{15}$

From diagram.

b both boys

S $P(BB) = \frac{2}{5}$

From diagram.

c one boy

S $P(\text{one boy}) = P(BG) + P(GB)$

$= \frac{4}{15} + \frac{1}{5}$

$= \frac{7}{15}$

d at least one girl.

At least one girl means 1 girl, or 2 girls, but not 2 boys, that is, everything except 2 boys.

↓

S $P(\text{at least one girl}) = 1 - P(BB)$

$= 1 - \frac{2}{5}$

$= \frac{3}{5}$

3 A box contains five red balls and three blue balls. Brae chooses three balls at random, without replacement.

a Draw a tree diagram.

S

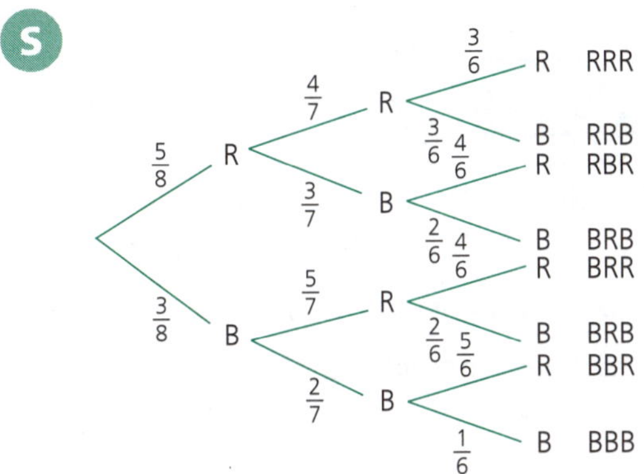

b What is the probability that Brae chose:

i three red balls?

S $P(RRR) = \frac{5}{8} \times \frac{4}{7} \times \frac{3}{6}$

$= \frac{5}{28}$

ii balls of the same colour?

S $P(RRR \text{ or } BBB) = \frac{5}{8} \times \frac{4}{7} \times \frac{3}{6} + \frac{3}{8} \times \frac{2}{7} \times \frac{1}{6}$

$= \frac{11}{56}$

iii two blue balls and a red ball?

S $P(RBB \text{ or } BRB \text{ or } BBR)$

$= \frac{5}{8} \times \frac{3}{7} \times \frac{2}{6} + \frac{3}{8} \times \frac{5}{7} \times \frac{2}{6} + \frac{3}{8} \times \frac{2}{7} \times \frac{5}{6}$

$= \frac{15}{56}$

Conditional probability

Conditional probability is when the chance of something occurring depends on (is conditional on) another event happening first. The probability that B occurs, given that A occurs, can be written as $P(B \mid A)$.

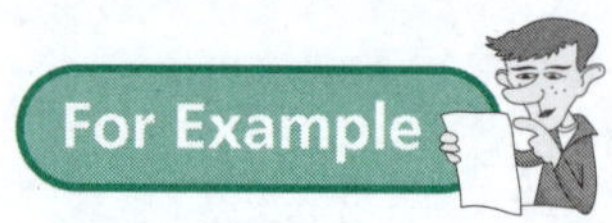

1 A die is rolled. Given that the number is odd, what is the probability that it is less than 4?

S **Odd numbers are 1, 3, 5**

$\therefore P(\text{less than } 4) = \frac{2}{3}$

2 A survey of 40 students found the number of students who use mobile phones and tablets for study. The results are recorded in the Venn Diagram.

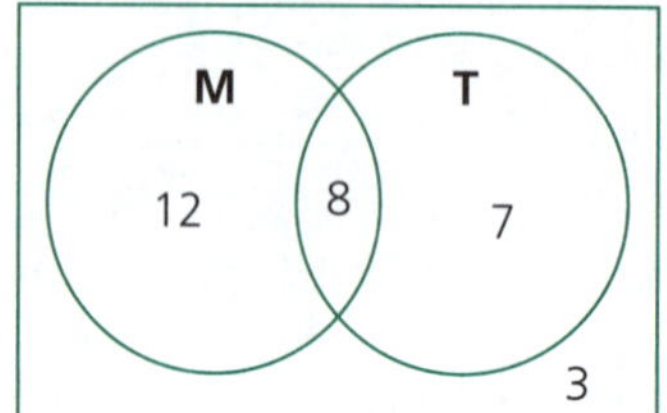

M: Mobile phone
T: Tablet

A student is chosen at random. What is the probability that they use a:

a mobile phone, given that they use a tablet?

S **8 + 7 = 15 students use a tablet; 8 of these students also use a mobile phone:**

$\therefore P(\text{mobile} \mid \text{tablet}) = \frac{8}{15}$

b tablet, given that they do not use a mobile phone?

S **7 + 3 = 10 students do not use a mobile phone; 7 of these students use a tablet:**

$\therefore P(\text{tablet} \mid \text{not mobile}) = \frac{7}{10}$

3 In the music class of 24 students, 16 play the piano and 12 play the guitar. There are three students who play neither the piano nor the guitar.

a Draw a Venn Diagram to summarise this information.

S

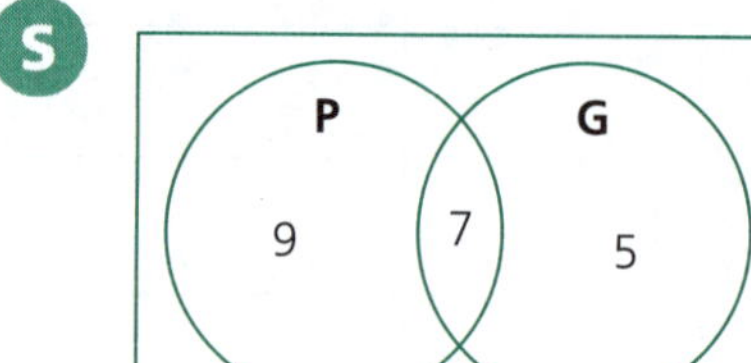

P: Piano
G: Guitar

b What is the probability that a student chosen at random plays the piano, given that they do not play the guitar?

S **9 + 3 = 12 students do not play guitar; 9 of these students play the piano:**

$\therefore P(\text{piano} \mid \text{not guitar}) = \frac{9}{12} = \frac{3}{4}$

Practise Practise

1 A pack has 52 cards divided into four suits: diamonds, hearts, clubs and spades. Each suit contains a 2, 3, 4, 5, … , 10, J(ack), Q(ueen), K(ing), A(ce). If one card is drawn at random from the pack, find the probability that it is:

a an A
b a K
c a heart
d a spade
e a black suit
f a picture card (J, Q, K)
g not a picture card
h a number less than 7
i the six of hearts
j red card or a 6
k a red 6. pp. 186–187

2 The probability of rain at the weekend is assessed as $\frac{7}{10}$. What is the probability that it will not rain at the weekend? pp. 186–187

3 A jar contains seven red, five black and three white jelly babies. If one jelly baby is drawn at random from the jar, find the probability that it is:

a red
b not red
c black
d red or white
e green. pp. 186–187

4 In a supermarket promotion, 10 gold tokens numbered 1 to 10 are hidden. Melville found one of the tokens. Find the probability that Melville found the token with number:

a 7
b even
c less than 7
d more than 7. pp. 186–187

5 Robyn keeps loose change in her car console. Last Saturday she noted that there were four 50-cent coins and six 20-cent coins. She selected one coin from the console to pay for a 45c iceblock. What is the probability that the coin selected is sufficient? pp. 186–187

6 The table below summarises the birth place of a group of students:

Continent	Number
Africa	14
America	18
Asia	36
Australia	112
Europe	20

If a student is chosen at random, what is the probability they will be:

a born in Asia?
b born in Europe or Africa?
c not born in Australia? pp. 186–187

7 The graph below represents the results of a survey of the people living in the households in Mooney Street:

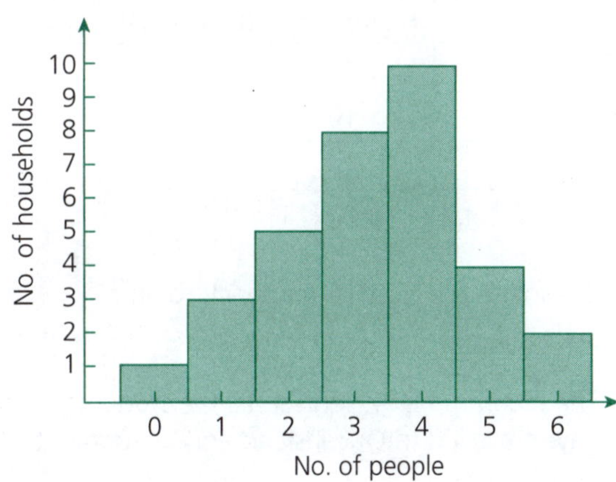

a How many households in Mooney Street?
b If a household was chosen at random what is the probability that it contains:
  i three people?
  ii less than two people?
  iii at least four people? pp. 186–187

8

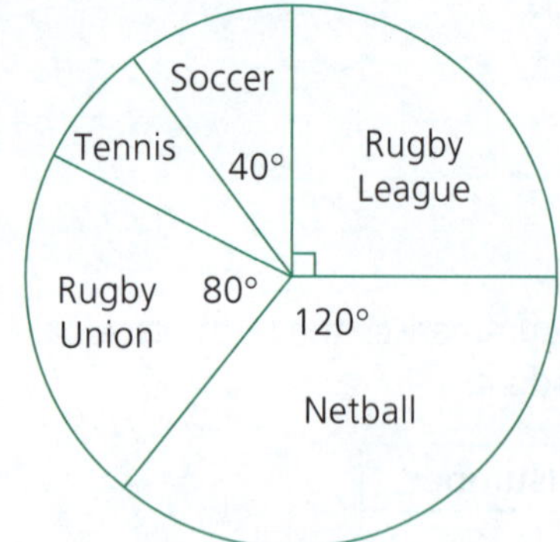

The sector graph represents the favourite sports of a number of students. If a student is chosen at random, what is the probability that the student favours:

a Netball?
b Tennis?
c Soccer or Rugby League?

pp. 186–187

9 A bag contains red balls, green balls and blue balls. A ball is chosen from the bag, its colour noted and then returned to the bag. This occurs 100 times. The results are recorded in the table below:

Colour	Frequency	Relative frequency
Red	48	
Green	19	
Blue	33	
Total	100	

Complete the relative frequency column.

pp. 186–187

10 The diagram below represents the students who have travelled to Queensland and Victoria in the past 5 years:

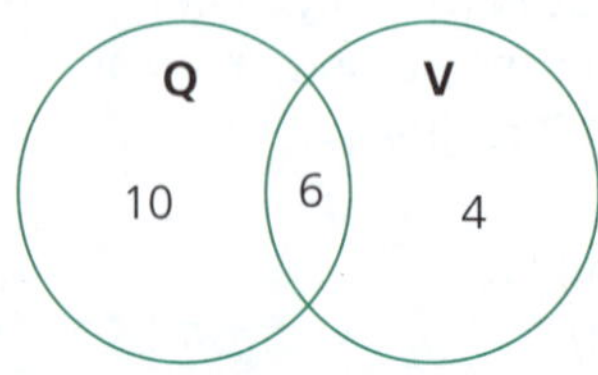

a How many students are represented in the diagram?
b If a student is chosen at random, find the probability that this student visited:
  i Queensland
  ii Victoria but not Queensland
  iii Victoria and Queensland.

p. 188

11

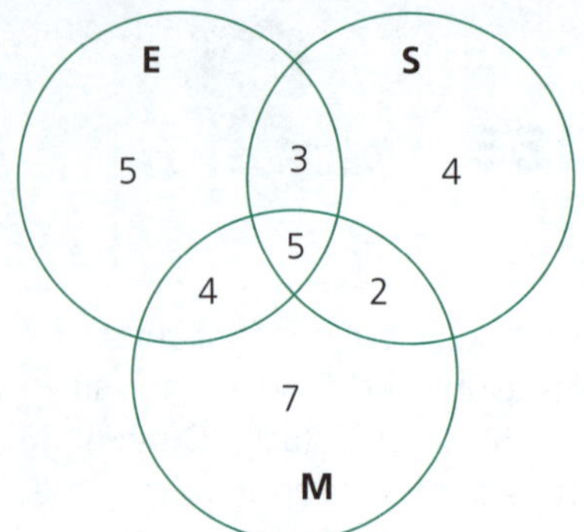

The diagram above represents the students who enjoy English (E), Science (S) and Mathematics (M).

a How many students are represented on the diagram?
b If a student is chosen at random, what is the probability that the student likes:
  i Mathematics?
  ii English and Mathematics?
  iii all three subjects?
  iv English or Science but not Mathematics?

p. 188

12 A coin is tossed and a die is thrown.

a Use a table to list all the possible outcomes.
b What is the probability of getting
  i a head and a 4?
  ii a tail and an odd number?

pp. 189–190

13 Two dice are thrown (faces numbered 1 to 6). The sum of the two uppermost faces is noted. Find the probability that the sum is:

a 9
b 5
c odd
d greater than 9
e 9 or less
f not 9
g odd or less than 5
h odd and less than 5.

pp. 189–190

14 If, instead of adding the numbers on the dice, the product is calculated, find the probability that this product is:

a 24
b odd
c less than 12
d 12 or less
e divisible by 3
f odd and divisible by 3
g odd or divisible by 3.

pp. 189–190

15 Two dice (faces numbered 1 to 6) are thrown and the highest number from the two faces is noted. For example, if a 3 and 5 are thrown, only the 5 is noted. Draw up a table to illustrate all possible outcomes. Find the probability that the number noted is:

a 6
b 1
c not 6
d not 6 or 1
e even
f a prime
g an even prime
h a prime or even
i divisible by 3
j greater than 6. pp. 189–190

16 A sample of students was conducted and the results shown in the following table:

	Born in Australia	**Born overseas**	Totals
Female	74	46	
Male	56	24	
Totals			

a Complete the table.
b If a student is chosen at random, what is the probability that the student is a:
  i male born in Australia?
  ii female born overseas?
c If a male is chosen at random, find the probability that he was born:
  i overseas
  ii in Australia. pp. 189–190

17 Mia has two tickets in a raffle that has two prizes. Draw a tree diagram to list the possible outcomes using W (wins) and L (loses). pp. 191–193

18 There are three children in a family. By drawing a tree diagram, list all possible outcomes using M (male) and F (female). pp. 191–193

19 A die is thrown twice. Complete the tree diagram, listing the possible ways of tossing a six (S) and not tossing a six (N):

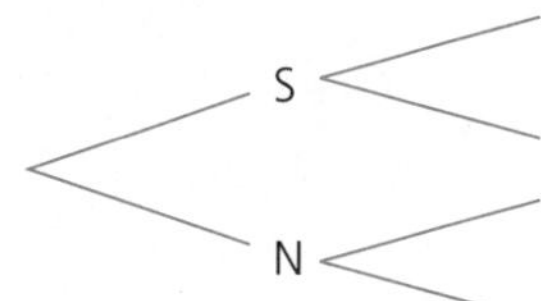

pp. 191–193

20 Three coins are tossed and the uppermost faces noted. Draw a tree diagram to illustrate the possible outcomes. Find the probability that the toss results in:

a three heads
b two heads
c one head
d no heads
e at least one head. pp. 191–193

21 A bag of Smarties contains three red and two blue ones. Two are drawn at random from the bag, the colour noted and then eaten. Find the probability that they are:

a both red
b both blue
c different colours
d the same colour. pp. 191–193

22 A biscuit barrel contains five choc-chip and four walnut biscuits. Two biscuits are drawn in succession and consumed. Find the probability that of the two biscuits:

a both are choc-chip
b one is a choc-chip
c only the first is a choc-chip
d neither is a choc-chip
e at least one is a choc-chip. pp. 191–193

23 One hundred tickets are sold in a Chook Raffle. Gregory John purchases two tickets, hoping to win both first and second prizes. Find the probability that he wins:

a first prize
b both prizes
c only second prize
d a prize
e no prize
f at least one prize. pp. 191–193

24 A die is rolled. Given that the number is even, what is the probability that it is:

a a 4
b less than 5? pp. 191–193

25 The letters of the alphabet are written on 26 identical cards and placed in a box. One card is chosen at random.

a What is the probability that the card is a vowel?
b Given that the card is a vowel, what is the probability that it is an E?
c Given that the card is a consonant, what is the probability that it is a P? pp. 193–194

26 The gender and age of bus travellers were recorded and the results shown in the table below:

Survey of Bus Travellers

		Age		
		12 and under	**13 to 17**	**18 and over**
Gender	**Female**	5	12	10
	Male	8	11	4

A person is chosen at random.

a What is the probability that the person is female and aged 13 to 17?

b Given that the person is female, what is the probability that she is aged 18 and over?

c Given that the person is aged 12 and under, what is the probability that the person is male?

pp. 193–194

27 Data on the cholesterol level of a group of people was collected and summarised in the table below:

	Low	**Normal**	**High**	Totals
Men	16		26	
Women		22	14	60
Totals		50		

a Complete the table.

b If a person is chosen at random, what is the probability that they are:

i male with a normal cholesterol level?

ii female with a low cholesterol level?

c If a women is chosen at random, what is the probability that she:

i has a normal cholesterol level?

ii does not have a low cholesterol level?

d If a person with a high cholesterol level is chosen what is the probability that the person is female?

pp. 193–194

28 A card is chosen from a deck of cards.

a If a red card is chosen, what is the probability that is a diamond?

b If a 7 is chosen, what is the probability that it is black?

c If a red queen is chosen, what is the probability that it is the queen of hearts?

pp. 193–194

29 The Venn Diagram shows what was consumed by people at a morning tea, using T = tea, C = coffee and P = pastry:

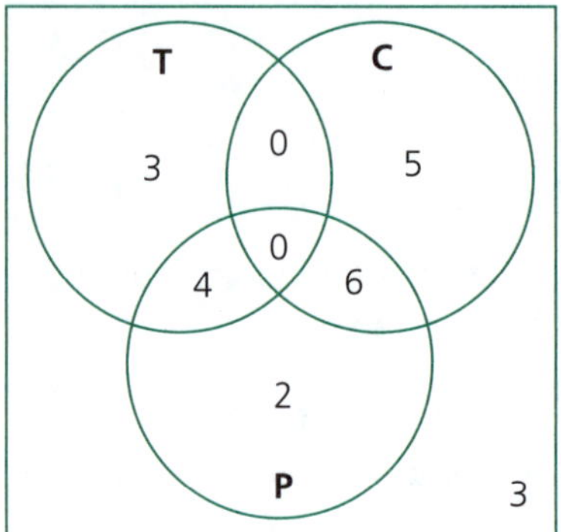

What is the probability that a randomly chosen person had a:

a coffee?

b tea and a pastry?

c pastry and a cup of coffee?

pp. 193–194

30 Of a group of 30 supporters at a football game, 17 were wearing a team jersey and 12 were wearing a team cap. Nine fans were wearing both.

a Draw a Venn Diagram.

b What fraction of those wearing a jersey were also wearing a cap?

c What fraction of those who were not wearing a cap, were wearing a jersey?

pp. 193–194

Go to p. 220 for **Quick Answers** or to pp. 256–258 for **Worked Solutions**

For a complete understanding of this topic, you must be able to:

✓	Recall basic probability concepts		pp. 186–187
✓	Calculate probabilities of events contained in Venn Diagrams		p. 188
✓	Calculate probabilities of compound events contained in tables		pp. 189–190
✓	Find the probability of compound events using tree diagrams		pp. 191–193
✓	Calculate the conditional probability of an event.		pp. 193–194

Now you are ready to do the tests!

Intermediate Test

Probability

(30 marks)

1 A coin is tossed twice. What is the probability that:
a two heads were tossed? (1 mark)
b at least one head is tossed? (1 mark)

2 A bag contains three red and five black balls. Two balls are selected at random, the first replaced before the second is drawn.
a Complete the probability tree: (1 mark)

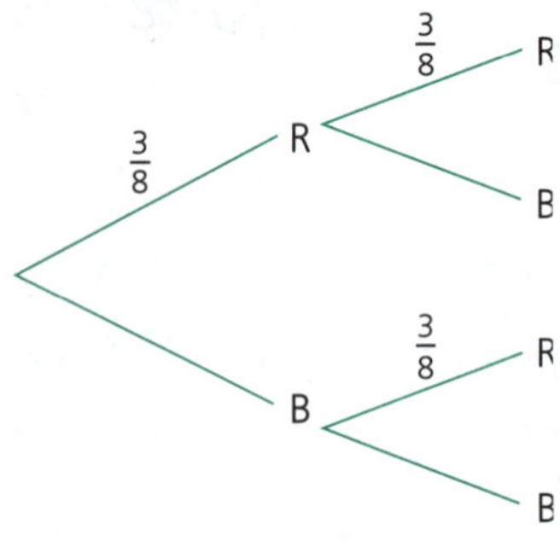

b What is the probability of choosing:
i two black balls? (1 mark)
ii balls of different colours? (1 mark)

3 A box has four balls marked 1, 2, 3, 4. A ball is selected at random from the box and the pointer, shown below, is rotated on the circular disc:

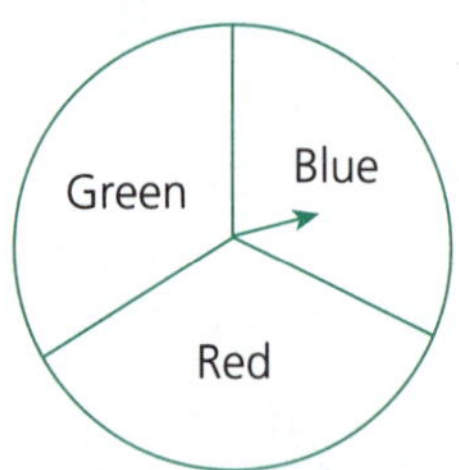

What is the probability of:
a a 3 and a green? (1 mark)
b an even number and not blue? (1 mark)

4 A survey was conducted to find the smoking habits of 200 people at a suburban shopping centre. The results of the survey are summarised:

	Males	**Females**	Total
Smoker	27	36	63
Non-smoker	73	64	137
Total	100	100	200

a A male is chosen at random. What is the probability that he is a smoker? (1 mark)
b A female is selected at random. What is the probability that she does not smoke? (1 mark)
c A person is chosen at random. What is the probability that this person is a non-smoker? (1 mark)

5 City commuters were surveyed to find whether they had used a bus (B) or a train (T) in the previous week. The results are summarised below:

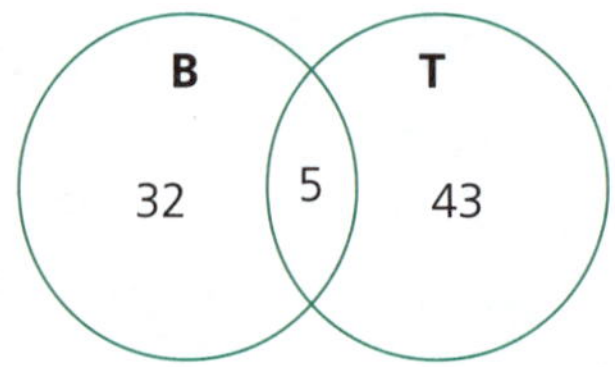

A commuter is chosen at random. What is the probability that this person travelled by:
a bus? (1 mark)
b train only? (1 mark)

6 A coin is flipped and a die is rolled. What is the probability of getting a:
a head and a 5? (1 mark)
b tail and an odd number? (1 mark)
c head and a number less than 3? (1 mark)

7 A bag contains three cards with the letters A, B and C. A card is chosen at random from the bag, and then returned to the bag before a second card is chosen.

a Draw a tree diagram to list all possible outcomes. (2 marks)

b What is the probability of choosing:

i BB? (1 mark)

ii BA? (1 mark)

iii at least one B? (1 mark)

iv AA or BB or CC? (1 mark)

v different letters? (1 mark)

8 A box contains three white discs and four green discs:

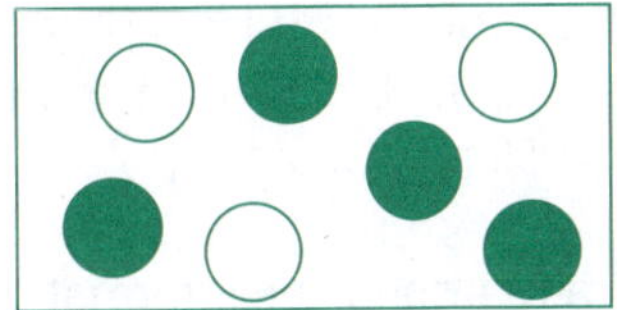

Two discs are chosen from the box. The first is not replaced before the second is chosen.

a Complete the tree diagram: (2 marks)

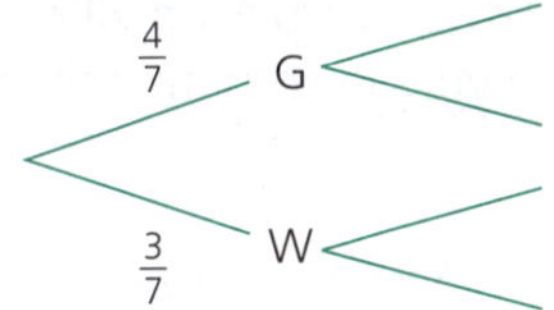

b What is the probability of selecting:

i GW? (1 mark)

ii WG? (1 mark)

iii WW? (1 mark)

9 Cards with the numbers from 1 to 20 are placed in a bag. A card is chosen at random from the bag.

a Find the probability that the number is less than 6. (1 mark)

b If the number is more than 10, what is the probability that it is even? (1 mark)

c The number is a multiple of 4. What is the probability that it is a factor of 24? (1 mark)

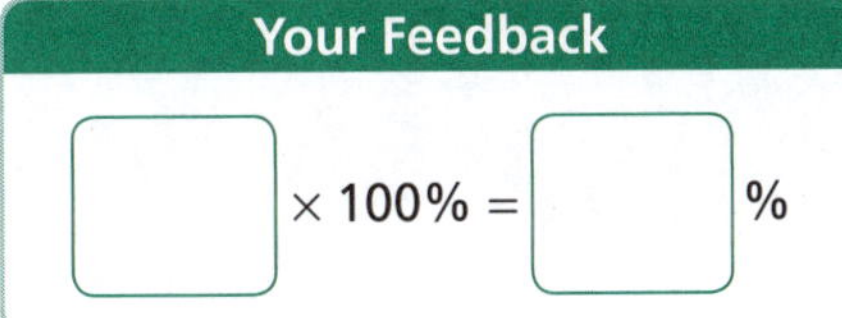

QA PAGE 224

WS PAGE 273

Advanced Test

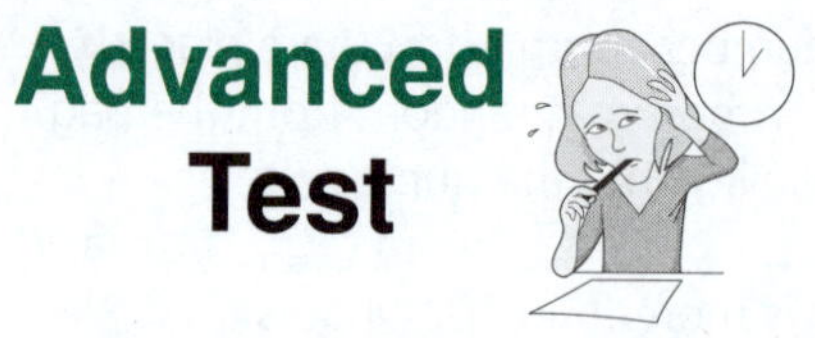

Probability

(30 marks)

1 A bag contains four green and six yellow balls. Two balls are selected at random, the first not replaced before the second is drawn.

a Draw a probability tree. (1 mark)

b What is the probability of choosing:

i different-coloured balls? (1 mark)

ii at least a yellow ball? (1 mark)

2 The table shows the results of a survey of voters at a recent state election:

	Males	Females
Lib–NP	35	45
ALP	45	40
Other	20	15

a A woman is chosen at random. What is the probability that she voted Lib–NP? (1 mark)

b A person is chosen at random. What is the probability that this person voted ALP? (1 mark)

3 A survey of 100 Year 10 students was conducted to find the subjects that they enjoyed. (Mathematics was not an option, as everyone enjoyed it.) The three subjects were English (E), Science (S) and PDHPE (P). The results of the survey are shown:

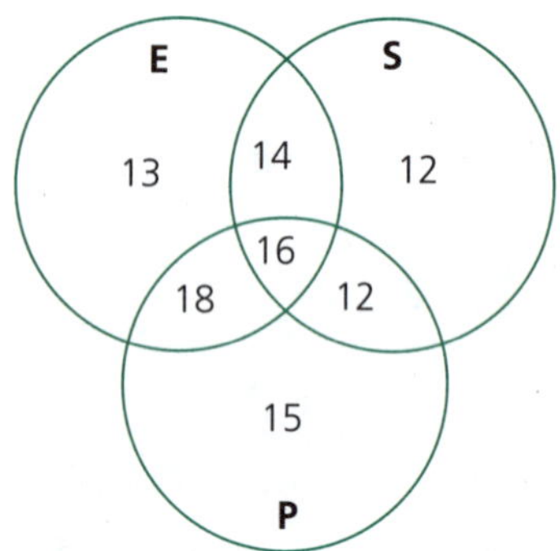

a A student is chosen at random. What is the probability that the student:

i likes PDHPE? (1 mark)

ii likes English, but not Science? (1 mark)

b A student who likes Science was chosen at random. What is the probability that this person:

i likes English? (1 mark)

ii likes all three subjects? (1 mark)

4 A survey of 100 students was conducted at a local school. Forty-five per cent of the students play team sport on a weekend sport, 35% play team sport during the week and 40% play do not play a team sport.

a Use a Venn Diagram to display the information. (1 mark)

b If a student is chosen at random, what is the probability that they play during the week as well as the weekend? (1 mark)

c What is the probability that a person who does not play on the weekend, plays during the week? (1 mark)

5 A group of cricket bowlers were surveyed and the results shown in the table below:

		Bowl	
		Left-handed	Right-handed
Bat	Left-handed	6	4
	Right-handed	5	15

a If a player was chosen at random, what is the probability that they bat and bowl left-handed? (1 mark)

b If a left-handed batman was chosen at random, what is the probability that they bowl right-handed? (1 mark)

6 A coin is tossed three times.

a Draw a tree diagram to list the possible outcomes. (2 marks)

b Find the probability of tossing:

i three tails (1 mark)

ii a head and two tails (1 mark)

iii at least one head. (1 mark)

7 Bag A contains three red and four blue balls. Bag B contains five red and six blue balls. A ball is chosen from each bag.

a Complete the tree diagram. (1 mark)

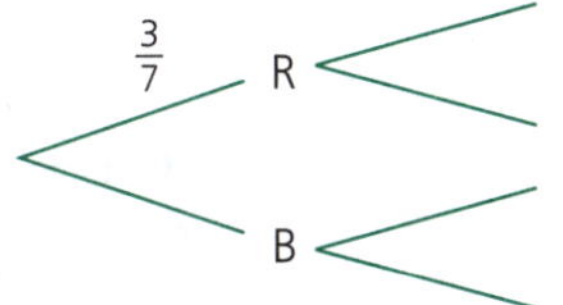

b Find the probability of choosing:

i two reds (1 mark)

ii the same colour (1 mark)

iii different colours (1 mark)

iv at least one blue. (1 mark)

8 William has 0.7 chance of passing his Science test, a 0.8 chance of passing his English test and a 0.9 chance of passing his Mathematics test.

a Complete the tree diagram: (3 marks)

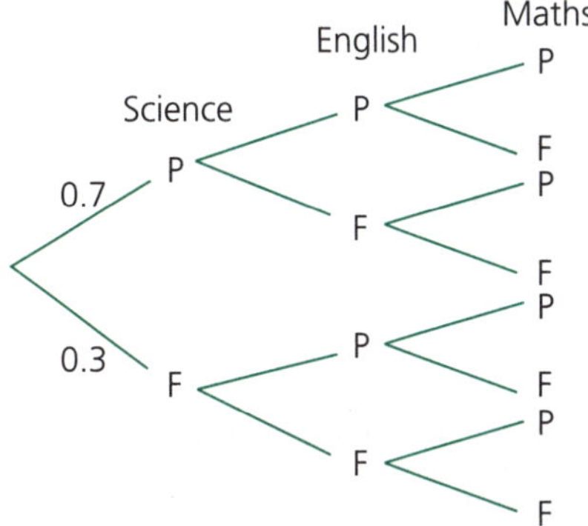

b Find the probability that William:

i passed all three tests (1 mark)

ii passed Maths but failed English and Science (1 mark)

iii failed only one test. (1 mark)

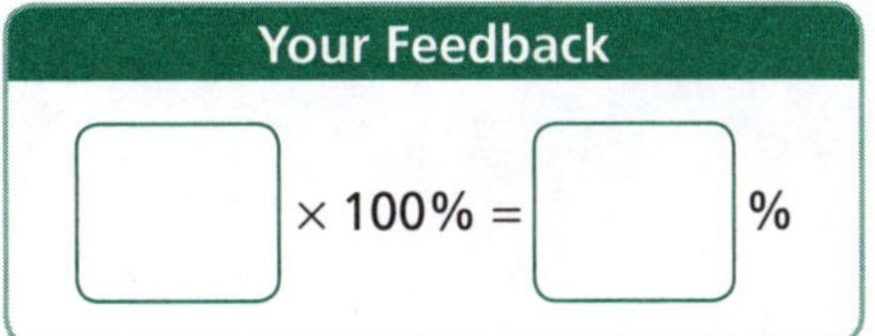

QA PAGE 224

WS PAGE 274

Sample Examination 1

(90 marks)

Financial Mathematics

1 a Delta's normal pay rate is $13.40 per hour. In one week she works 30 hours at normal pay, 6 hours at time and a half, and 5 hours at double time. What is her total pay for the week? (1 mark)

b Find the simple interest on $6800 at $7\frac{1}{4}$% per annum for 8 years. (1 mark)

c

Taxable income	Tax on this income
0–$18 200	Nil
$18 201–$37 000	19c for each $1 over $18 200
$37 001–$80 000	$3572 plus 32.5c for each $1 over $37 000
$80 001–$180 000	$17 547 plus 37c for each $1 over $80 000
$180 001 and over	$54 547 plus 45c for each $1 over $180 000

Rhett's taxable income is $62 420. Use the table to calculate the tax payable. (2 marks)

d Mitchell purchased a pair of jeans that were originally marked at $80 but were discounted by 20%.

i How much will Mitchell pay for the jeans? (1 mark)

ii Laura works at the shop and is given a further 10% discount on a similar priced pair of jeans. What is Laura's total discount? (1 mark)

e A hardware store reduced the price of a drill by 15%. It later dropped the price by another 20% to match a competitor's price. If the original price was $76, find the final price. (1 mark)

f A DVD recorder was purchased 2 years ago for $680. If it has depreciated in value at 18% p.a., what is its present value, to the nearest dollar? (1 mark)

g Leong invests $25 000 for 4 years at 6% p.a. compounded monthly. Find the amount of interest earned by Leong. (2 marks)

Algebraic Techniques and Indices

2 a Expand and simplify:

i $2 - (x + 3)$ (1 mark)

ii $4(2x - 5) - 3(4x - 2)$ (2 marks)

b Factorise:

i $12ab - 4a$ (1 mark)

ii $x^{12} + x^6$ (1 mark)

c Simplify:

i $\frac{2x}{3} + \frac{x}{4}$ (1 mark)

ii $\frac{6x^2y}{3ab} \div \frac{xy^2}{2a}$ (2 marks)

d Simplify:

$4x^{-3} \div 2x^{-4}$ (1 mark)

e Expand and simplify: $(3x - 4)(2x + 1)$ (1 mark)

Equations

3 **a** Solve $4 + 2x \le 6$ and graph your solution on a number line. (2 marks)

b Solve: $\dfrac{4y - 5}{3} = \dfrac{3y + 5}{4}$ (2 marks)

c If $S = \dfrac{n}{2}[2a + (n - 1)d]$, find S when $n = 12$, $a = 4$, $d = -2$. (2 marks)

d Solve $3y^2 - 75 = 0$. (1 mark)

e Solve $x^2 - 3x + 2 = 0$. (1 mark)

f Three burgers and two drinks cost \$15.80, while two burgers and a drink cost \$9.80. By solving simultaneous equations, find the cost of a burger. (2 marks)

Linear and Non-Linear Relationships

4 **a** For the points P(–2, 4) and Q(3, –1), find:

i distance PQ (1 mark)

ii midpoint PQ (1 mark)

iii gradient PQ. (1 mark)

b Graph the line $y = 2x - 3$. (2 marks)

c If the point $(5, p)$ lies on the line $2x - y + 3 = 0$, find the value of p. (2 marks)

d Find the equation of the line: (2 marks)

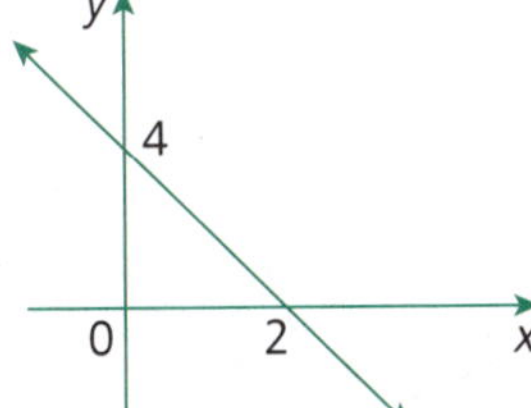

e The equation of the curve graphed on the number plane could be:

A $y = 4 - 2x^2$

B $y = 4x$

C $y = x^2 + 4$

D $y = 4 - x^2$ (1 mark)

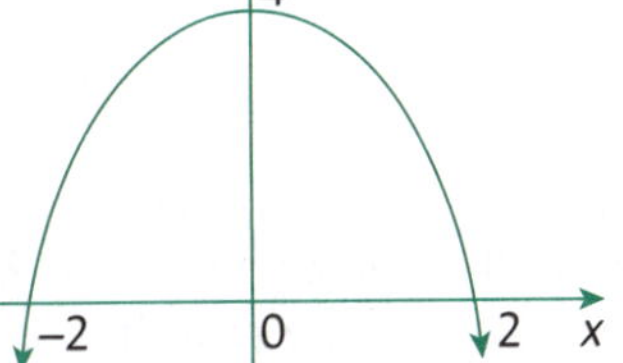

Rates and Proportion

5 **a** Convert 30 m/s to km/h. (1 mark)

b A hose is used to fill an empty container with water. It takes 8 minutes to fill the container with 160 L of water.

i At what rate is the water filling the container, in L/min? (1 mark)

ii Write a rule linking V (volume in litres) and t (time in minutes). (1 mark)

iii Draw a graph of volume against time for $t \le 100$. (1 mark)

iv How much water is in the container after an hour? (1 mark)

v How long would it take to fill a tank with a capacity of 6000 litres? (1 mark)

c It is known that $y \propto x$ (y is proportional to x) and if $y = 36$ then $x = 9$.

i Write an equation that links y and x using the constant of proportionality k. (1 mark)

ii Find the value of k. (1 mark)

iii Find the value of y when $x = 15$. (1 mark)

d Charlotte's wage is directly proportional to the number of hours she works. When she works 24 hours she is paid \$444. Let n = number of hours worked and W = wages in dollars. By finding the constant of proportionality, write an equation that links W and n. (1 mark)

Pythagoras and Trigonometry

6 a Find the value of x: (1 mark)

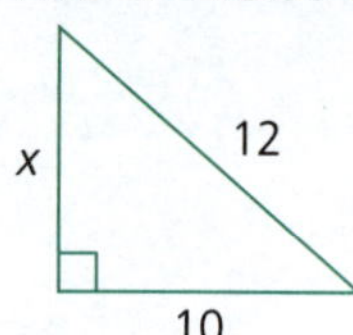

b Find the value of x, correct to two decimal places:

i (2 marks)

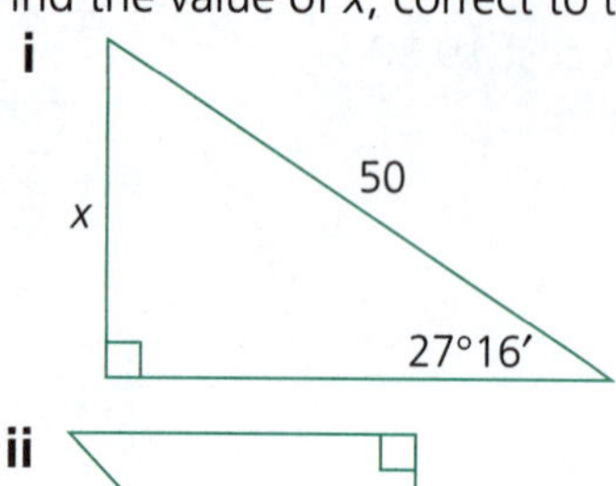

ii (2 marks)

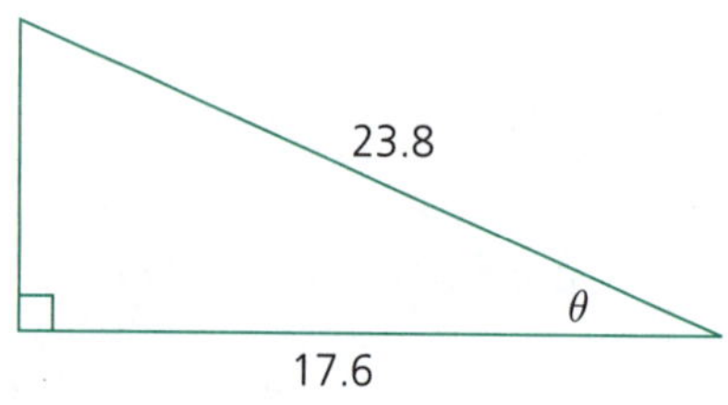

c Find θ, to the nearest degree: (1 mark)

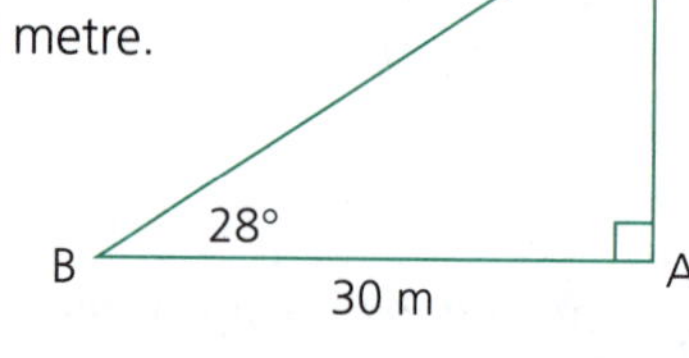

d The angle of elevation of the top of a tower TA from B, 30 m from the base of the tower, is 28°.
Find the height of the tower to the nearest metre. (2 marks)

e A boat leaves A and sails on a bearing of 140° for 30 km to B.
How far south of A is B? (Answer correct to one decimal place.) (2 marks)

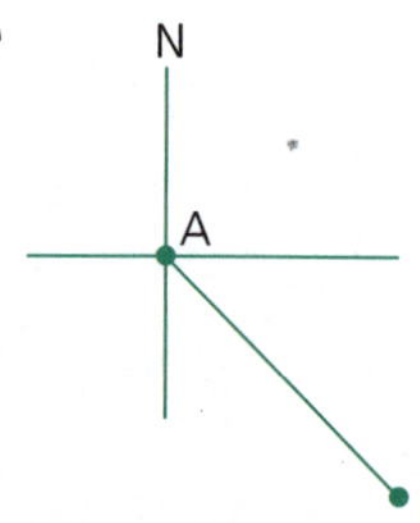

Measurement

7 a Find the perimeter, correct to two decimal places: (2 marks)

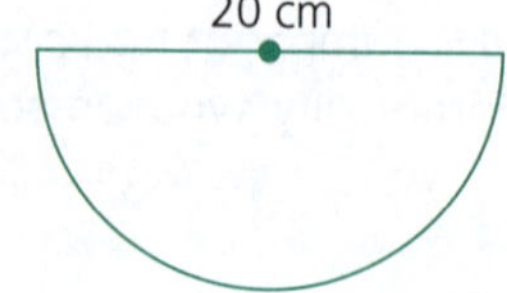

b Find the area, correct to two decimal places: (2 marks)

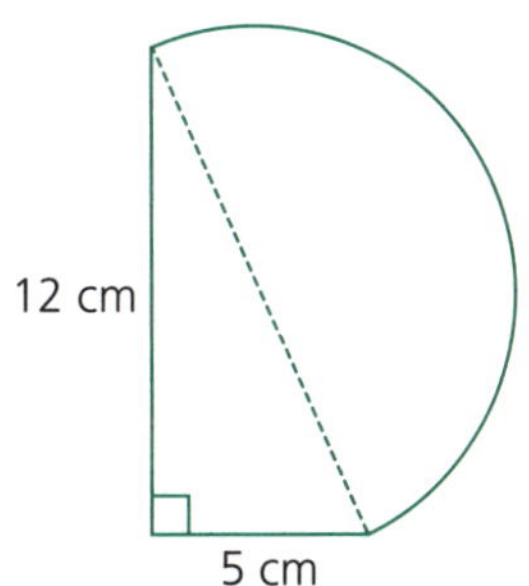

c Find the surface area of the closed cylinder: (2 marks)

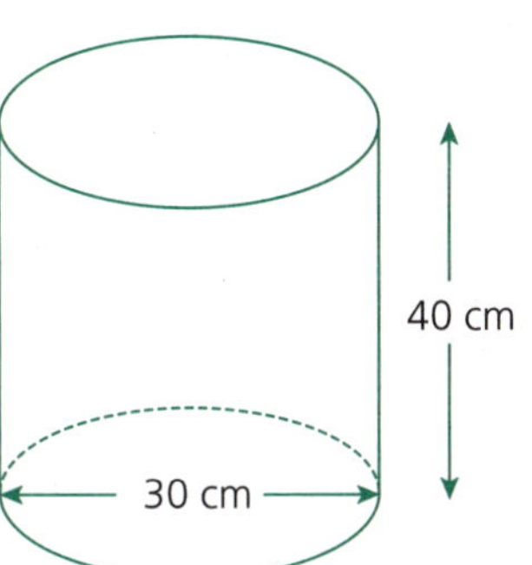

d Find the volume: (2 marks)

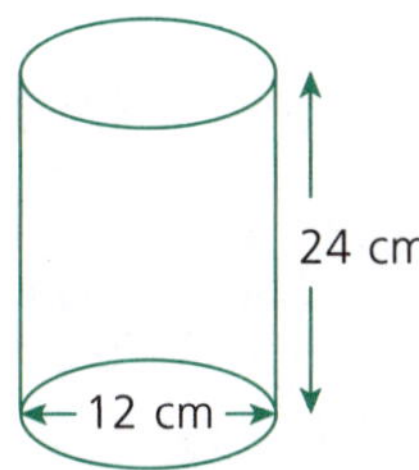

e Find the capacity of a water trough in the shape of half a cylinder with radius 80 cm and length 3 metres. Give your answer to the nearest litre. (2 marks)

Geometry

8 a Find the value of the pronumerals:

i (1 mark)

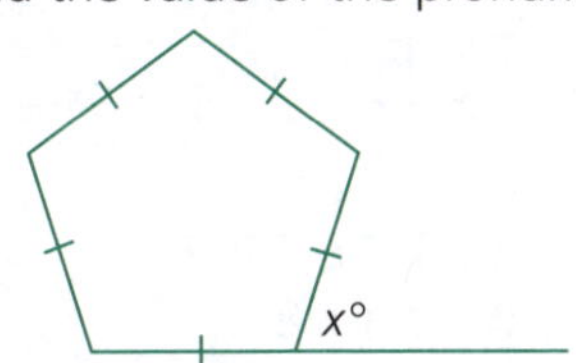

ii PQRSTUVW is a regular octagon. ST || YX. Find x. (1 mark)

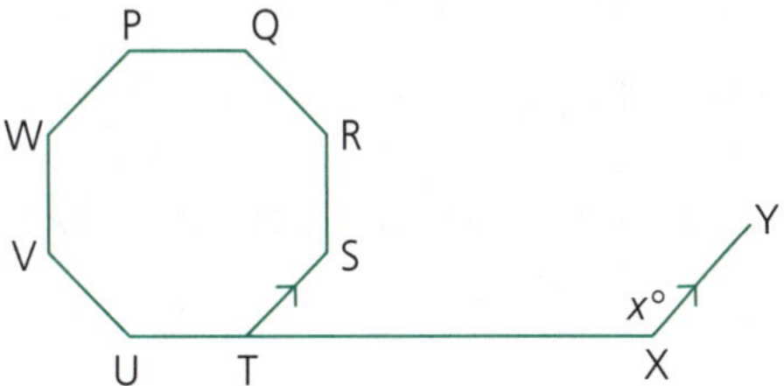

b Find the size of an exterior angle of a regular decagon. 

(1 mark)

c Which test is used to prove $\triangle ACB \equiv \triangle EDF$? 

(1 mark)

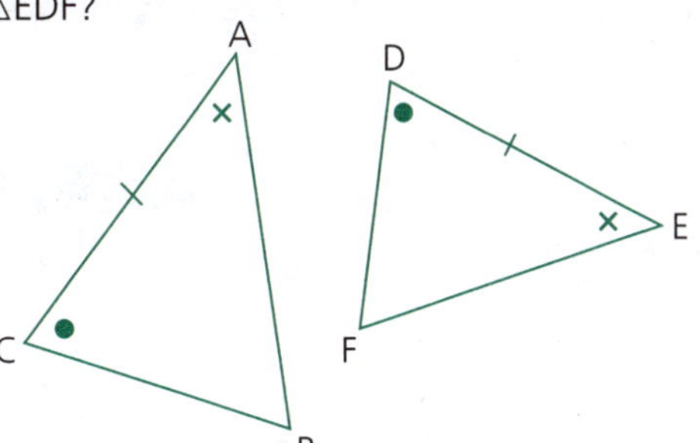

d Which test proves $\triangle ACE \mathbin{|||} \triangle BCD$? (1 mark)

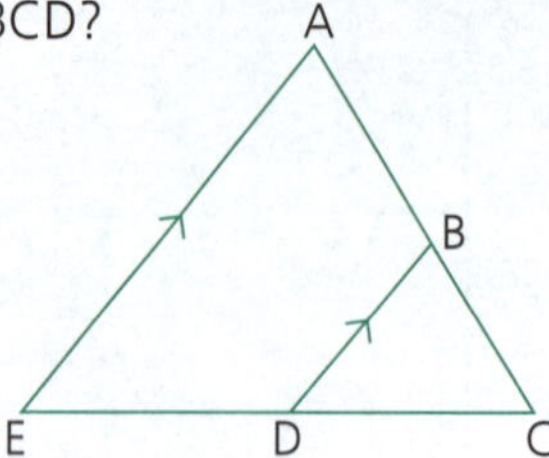

e Find the value of the pronumerals:

i (2 marks)

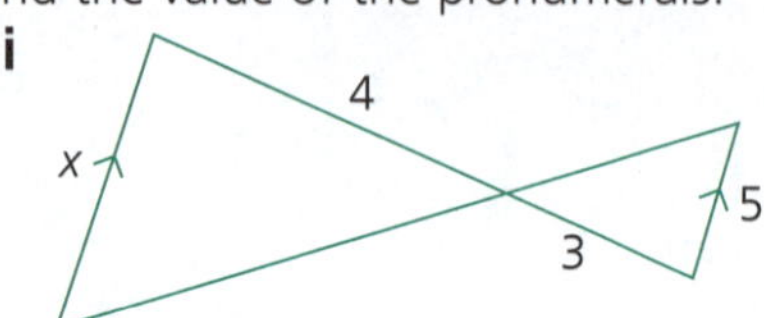

ii (2 marks)

x
6
4
10

f ABCD is a rectangle. $\angle ADB = 60°$. Find the size of $\angle BCE$. (1 mark)

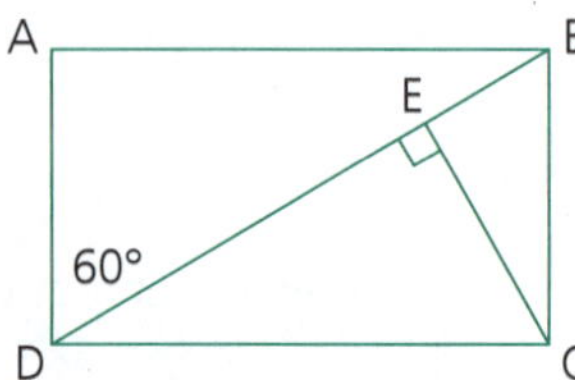

Statistics and Probability

9 a The classes 10M3 and 10M4 sat for the same mathematics test. The results are summarised in the back-to-back stem-and-leaf plot.

i Write down the:

α median of 10M3 scores (1 mark)

β mode of 10M4 scores. (1 mark)

ii For 10M4, write down a five point summary and then draw a box plot for 10M4's results. (3 marks)

10M3		10M4
9	4	7 9 9
9 8 7 7 6 1	5	5 6 8 9
8 8 4 3 0	6	0 1 7 8 8 9
9 9 8 6 4 2	7	2 5 7 7 8 9 9 9
6 4 4 4 3 0	8	0 3 3 4 7 8
9 8 5 3 2 0	9	0 4 6

b A bag contains identical cards numbered 1 to 16. A card is selected from the bag at random. It is known that the card is a multiple of 3. What is the probability that it is even? (1 mark)

c Two six-sided dice are thrown and the product of their upper-most faces is calculated. Find the probability of scoring:

i 1 (1 mark)

ii 7 (1 mark)

iii more than 20. (1 mark)

d A survey found the number of students who had consumed water and juice for breakfast. Twelve drank water, six juice, and three neither. If 20 students were surveyed, how many drank both water and juice? (1 mark)

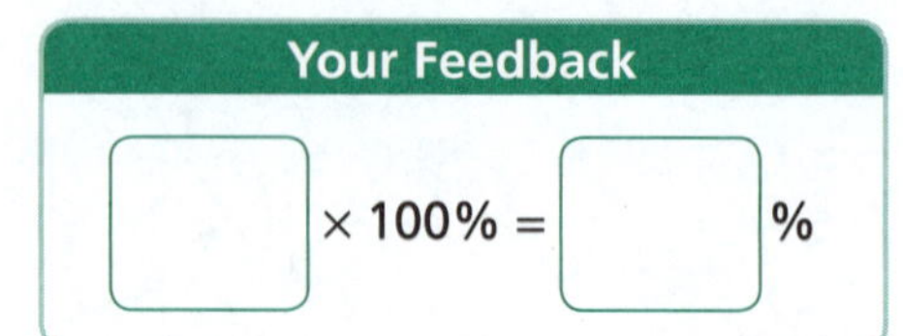

QA PAGE 225

WS PAGE 275

Sample Examination 2

Part A (Multiple-Choice Questions) (45 marks)

Read each question and choose an answer A, B, C or D.

1 Use the calculator to evaluate $\frac{8.23}{2.7 \times 6.3}$ correct to two decimal places.

A 19.20 **B** 3.53 **C** 0.48 **D** 1.48 (1 mark)

2 Which of the following points lies on the parabola $y = x^2 - x$?

A (1, −1) **B** (−1, 1) **C** (0, 2) **D** (2, 2) (1 mark)

3 A maths quiz was marked out of 20 and the results shown in the box plot below:

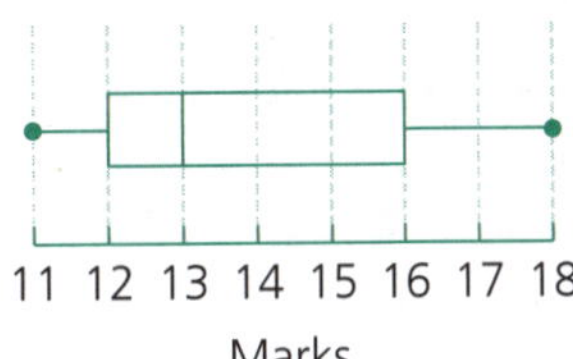

If 40 students sat the quiz, approximately how many students scored between 12 and 18?

A 6 **B** 24 **C** 30 **D** 75 (1 mark)

4 What is the radius of the circle with equation $x^2 + y^2 = 16$?

A 1 **B** 4 **C** 8 **D** 16 (1 mark)

5 If $y = kx$ and $y = 6$ when $x = 3$, find the value of k.

A $k = 2$ **B** $k = \frac{1}{2}$ **C** $k = 18$ **D** $k = \frac{1}{18}$ (1 mark)

6 Write 219 000 in scientific notation.

A 2.19×10^3 **B** 2.19×10^5 **C** 21.9×10^4 **D** 2.2×10^4 (1 mark)

7 This graph represents:

−2 −1 0 1 2 3

A $x < 2$ **B** $x = 2$ **C** $x \leq 2$ **D** $x > 2$ (1 mark)

8 Write $12\frac{1}{2}\%$ as a decimal.

A 12.5 **B** 0.125 **C** $\frac{1}{8}$ **D** $\frac{12\frac{1}{2}}{100}$ (1 mark)

9 How many full 375 mL glasses of orange juice can be poured from a 2 L bottle?

A 6 **B** 5.3 **C** 50 **D** 5 (1 mark)

10 Simplify $\frac{5x}{4} + \frac{x}{3}$.

A $\frac{6x}{7}$ **B** $\frac{6x}{12}$ **C** $\frac{19x}{12}$ **D** $19x$ (1 mark)

11 The distance of AB is:

A 5 **B** 25

C $-\frac{3}{4}$ **D** 4

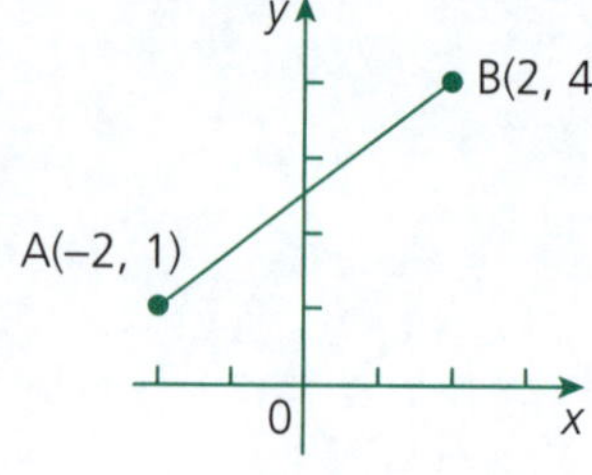

(1 mark)

12 An adult's brain is approximately 2.5% of the body mass. If the mass of a woman is 63 kg, her brain has a mass of how many kilograms?

A 15.75 kg **B** 1.575 kg **C** 0.1575 kg **D** 3.968 kg (1 mark)

13 An employee is paid at the casual rate of $496 for a 40-hour week. What will he receive for a week where he works 37 hours?

A $458.80 **B** $462.20 **C** $465.80 **D** $453.40 (1 mark)

14 The volume of a cylinder with diameter 4 cm and height 2 cm is:

A 4π cm^3 **B** 8π cm^3 **C** 16π cm^3 **D** 32π cm^3 (1 mark)

15 Use Pythagoras' Theorem to calculate the length of BD to the nearest cm.

A 13 **B** 4

C 10 **D** 9

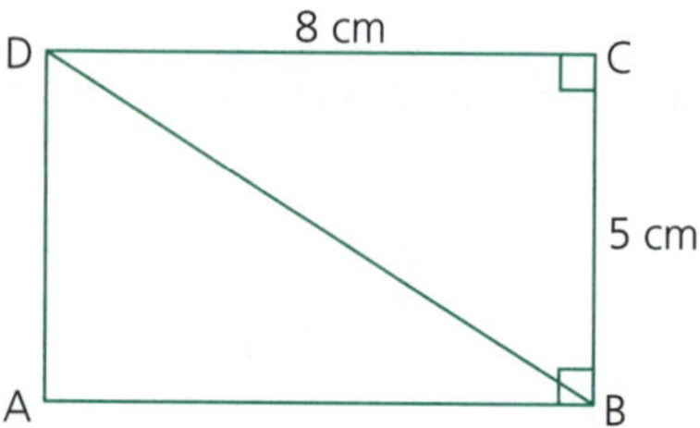

(1 mark)

16 Which of the following lines is perpendicular to $y = 2x - 1$?

A $y = x^2$ **B** $y = x - 2$ **C** $y = \frac{-1}{2}x$ **D** $y = \frac{2}{x}$ (1 mark)

17 A coin is tossed twice. What is the probability of tossing two tails?

A $\frac{1}{2}$ **B** $\frac{1}{4}$ **C** $\frac{1}{3}$ **D** $\frac{1}{8}$ (1 mark)

18 If $a = 3$, what does $2a^2$ equal?

A 36 **B** 25 **C** 18 **D** 12 (1 mark)

19 $3a^4 \times 2a^2 \times a = ?$

A $6a^7$ **B** $6a^6$ **C** $6a^8$ **D** $5a^8$ (1 mark)

20 The diagram shows a prism with cross-sectional area 15 cm^2 and length 6 cm. The volume of the prism in cm^3 is:

A 21 **B** 45

C 1350 **D** 90

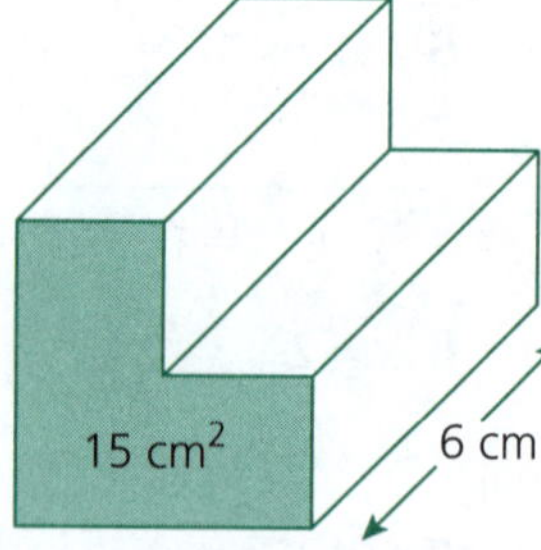

(1 mark)

21 The interquartile range of the scores 4, 6, 8, 8, 12, 16, 20, is:

A 6 **B** 16 **C** 4 **D** 10 (1 mark)

22 In the diagram, $\sin\theta = ?$

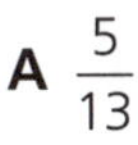

A $\frac{5}{13}$ **B** $\frac{12}{5}$

C $\frac{12}{13}$ **D** $\frac{5}{12}$

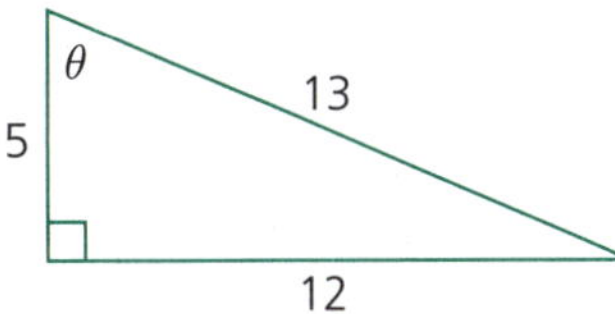

(1 mark)

23 Simplify $3(a + b) - (a - b)$.

A $2a$ **B** $2a + 2b$ **C** $2a + 4b$ **D** $4a + 2b$ (1 mark)

24 Solve $(x - 3)(x + 1) = 0$.

A $x = -3, 1$ **B** $x = 3, -1$ **C** $x = -3, -1$ **D** $x = 3, 1$ (1 mark)

25 Factorise $x^2 - 16$.

A $(x - 4)(x + 4)$ **B** $(x - 4)(x - 4)$ **C** $(x - 8)(x - 8)$ **D** $(x + 8)(x - 8)$ (1 mark)

26 Find the value of x in this diagram:

A $5\frac{1}{3}$ **B** 3

C 48 **D** 15

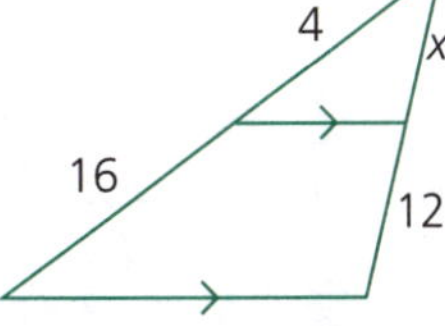

(1 mark)

27 Which of the following data sets is negatively skewed?

A

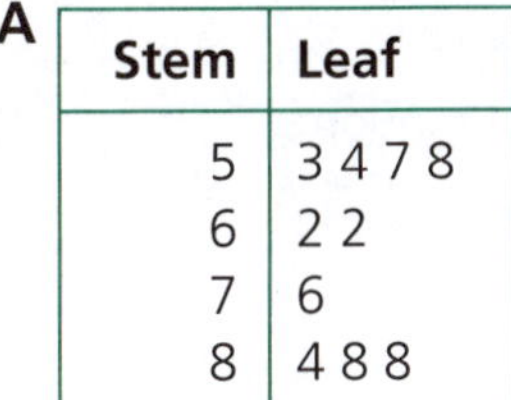

Stem	Leaf
5	3 4 7 8
6	2 2
7	6
8	4 8 8

B

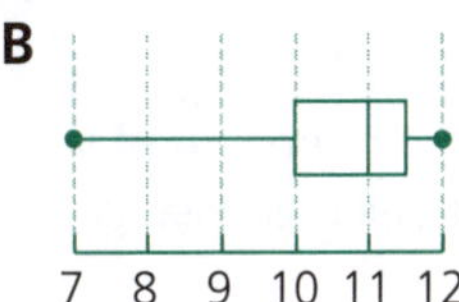

C

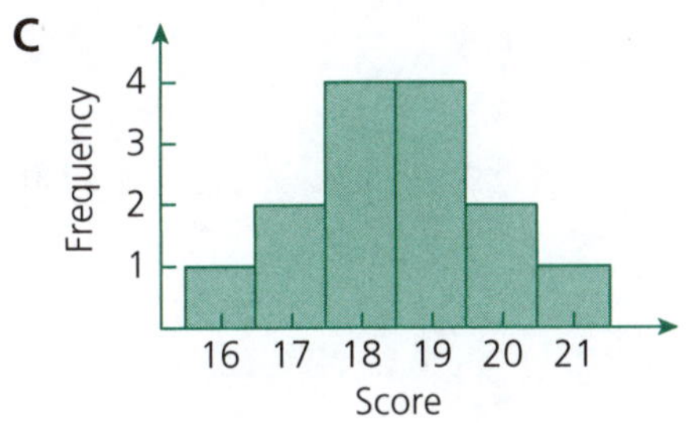

D

(1 mark)

28 The value of x is:

A 40 **B** 70

C 110 **D** 140

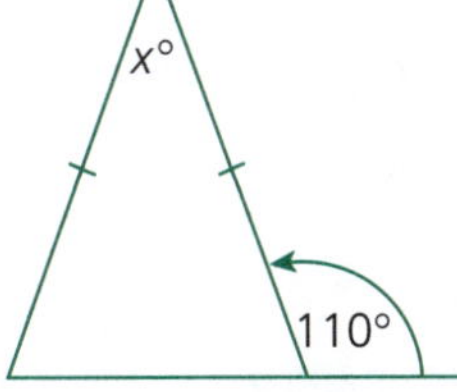

(1 mark)

29 A car travelled from a point A to a point B with a constant speed of 70 km/h. If the journey took 2 hours 15 minutes, what is the distance from A to B? (Answer to the nearest kilometre.)

A 157 **B** 158 **C** 151 **D** 150 (1 mark)

30 The shape shown is a semicircle. The closest approximation to the perimeter is:

A 56 cm **B** 42 cm

C 77 cm **D** 36 cm

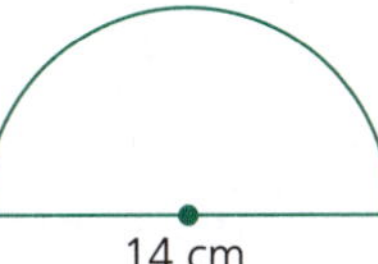

(1 mark)

31 Which of the following triangles are congruent?

I

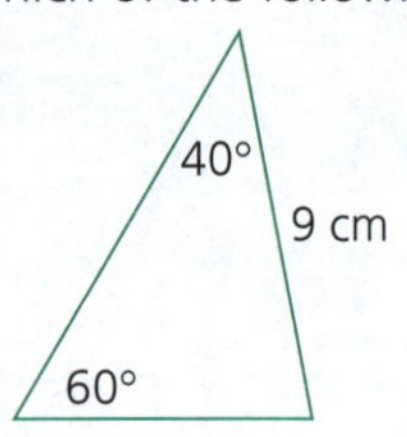

II

40° 80° 9 cm

III

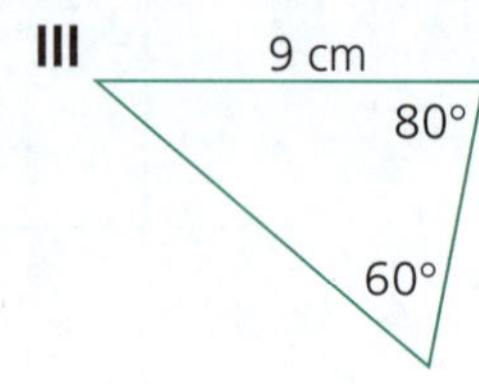

A I and II only **B** I and III only **C** I, II and III **D** II and III only (1 mark)

32 Simplify: $(4x^{-3})^2$

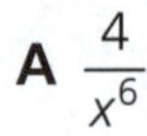

A $\frac{4}{x^6}$ **B** $\frac{16}{x^6}$ **C** $\frac{8}{x^5}$ **D** $\frac{8}{x^6}$ (1 mark)

33 The graph shows the goals scored by a soccer team in their games throughout the season.

How many games were played during the season?

A 5 **B** 14

C 12 **D** 16

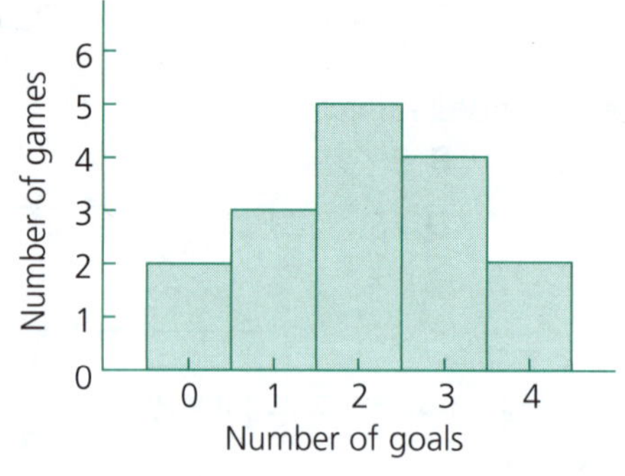

(1 mark)

34 In a box there are five blue, ten red, four green and four black balls. A ball is chosen at random. The probability that a green or black ball is chosen is:

A $\frac{1}{2}$ **B** $\frac{8}{23}$ **C** $\frac{4}{19}$ **D** $\frac{8}{15}$ (1 mark)

35 A cube has a volume of 27 cm^3. What is the area of the shaded face of this cube?

A 9 cm^2 **B** 81 cm^2

C 72 cm^2 **D** $\sqrt{27}$ cm^2

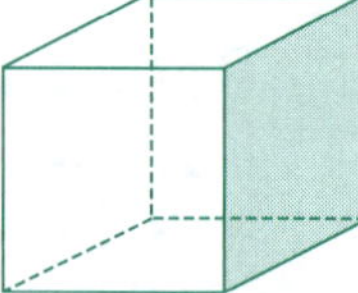

(1 mark)

36 $\frac{5bc}{3} \div \frac{25bc}{9} = \square$

A $\frac{5b}{3}$ **B** $\frac{125b^2c^2}{27}$ **C** $\frac{b^2c}{5}$ **D** $\frac{3}{5}$ (1 mark)

37 In the diagram, line l has the equation $y = 4 - 2x$. The area of △AOB is:

A 8 **B** 4

C 16 **D** 2

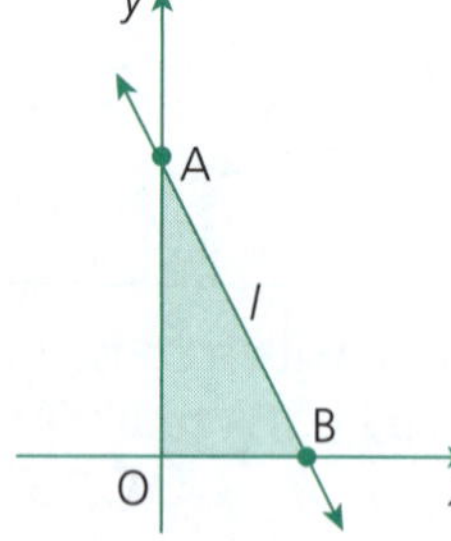

(1 mark)

38 In each of the scatterplots the link between x and y is shown.
Which scatterplot shows a strong negative relationship?

A

B

C

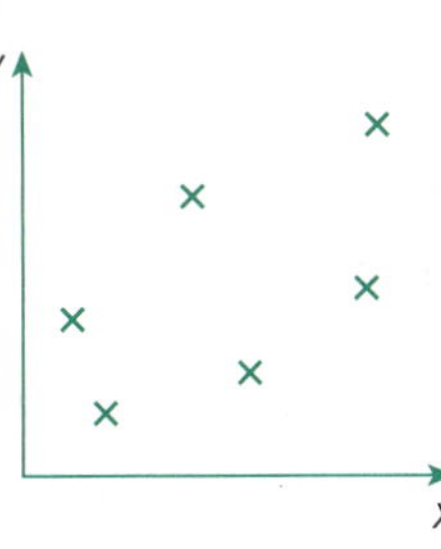

D

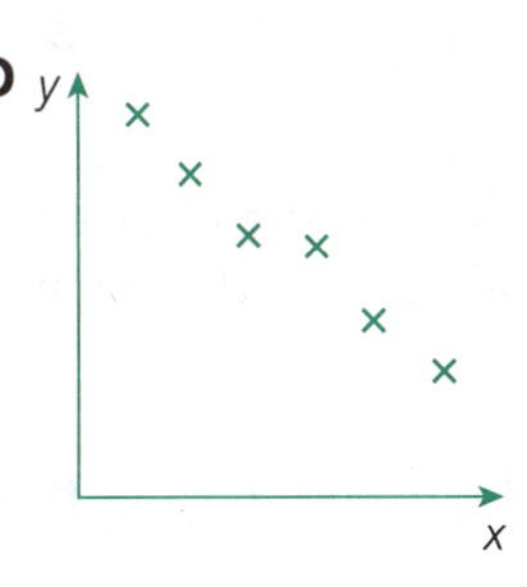

(1 mark)

39 The length of BC to the nearest cm is:

A 34 **B** 45
C 20 **D** 39

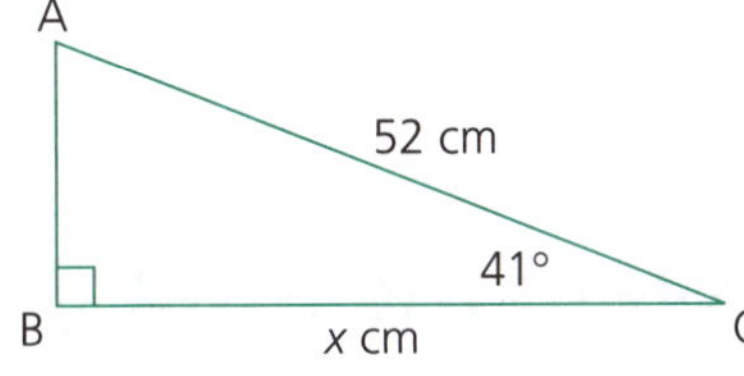

(1 mark)

40 The value of a motor vehicle depreciates by 17% p.a. If a vehicle is purchased for \$25 400, what is its value at the end of the first year?

A \$4318 **B** \$21 082 **C** \$22 420 **D** \$29 718

(1 mark)

41 The area of this figure is:

A 42 cm^2 **B** 54 cm^2
C 66 cm^2 **D** 84 cm^2

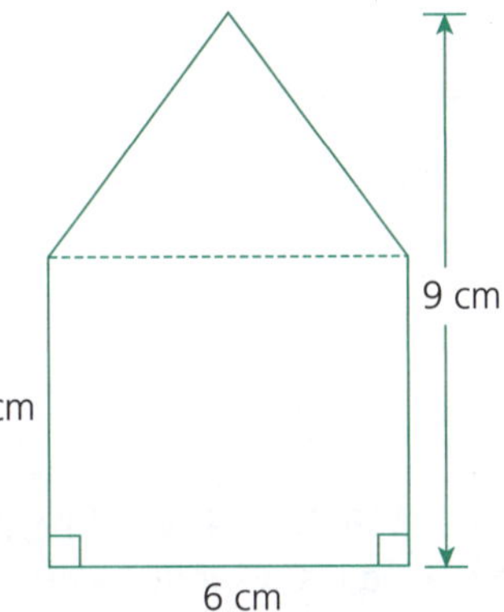

(1 mark)

42 For the frequency distribution table given, the median is:

x	f
3	1
4	2
5	0
6	4
7	5
8	1

A 4 **B** 5
C 6 **D** 7

(1 mark)

43 Which of the following graphs could best represent $y = 2x^2 - 1$?

A

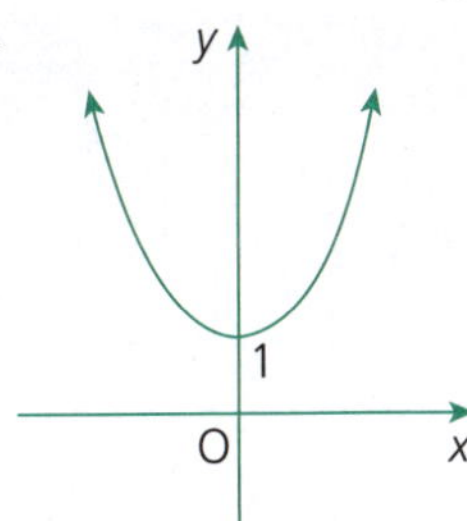

B

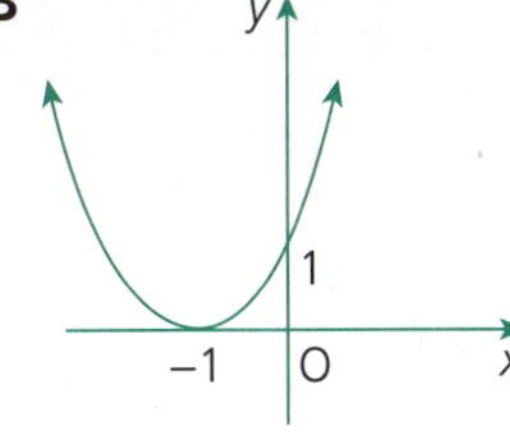

C

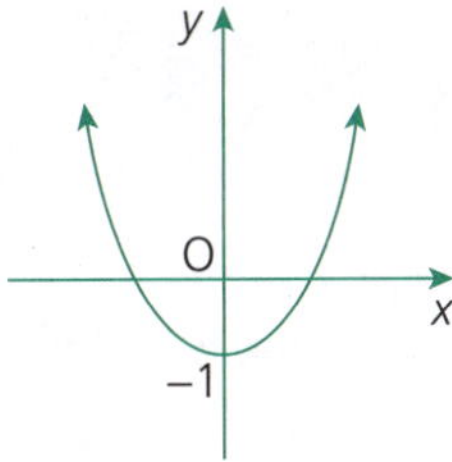

D

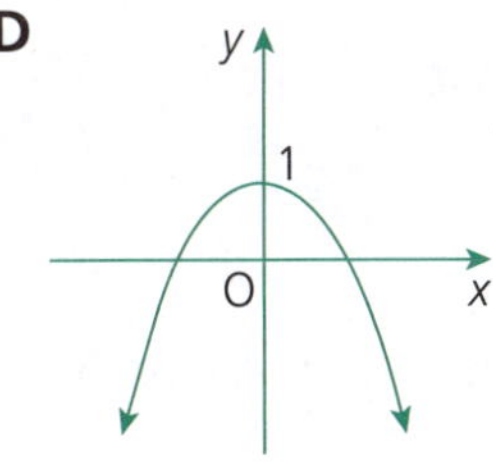

(1 mark)

44 The expression that represents 'half the product of x and y' is:

A $\frac{1}{2}(x + y)$ **B** $\frac{xy}{2}$ **C** $\frac{x}{2y}$ **D** $\frac{2x}{y}$ (1 mark)

45 PQR is an isosceles triangle with PQ = PR.
Angles RSQ and STR are right-angles.
TR = TS.
$x = ?$

A $22\frac{1}{2}$ **B** 45

C 60 **D** $67\frac{1}{2}$

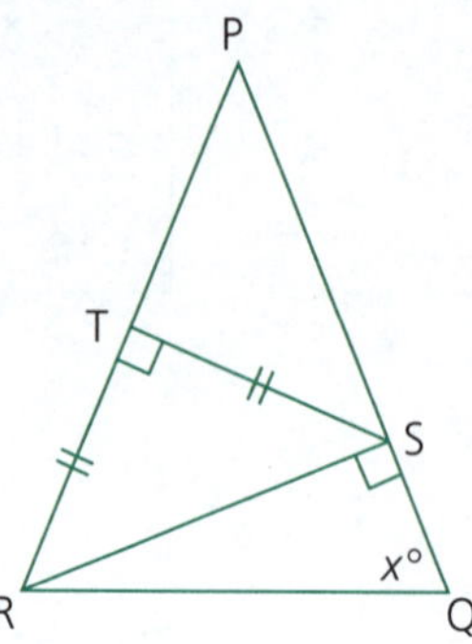

(1 mark)

Part B (Short-Answer Questions) (30 marks)

46 a A bag contains three red balls and two white balls. Two balls are chosen at random, the first not replaced before the second is drawn.

i Complete the tree diagram: (1 mark)

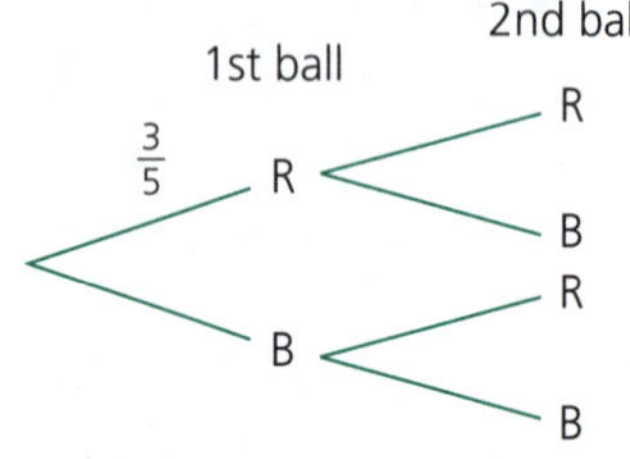

ii Find the probability of selecting:

A two red balls (1 mark)

B different colours. (1 mark)

b Paul receives a wage at the normal rate of $18.42 an hour. If he works under very wet conditions he receives an additional $3.45 per hour. Find his wage for a 40-hour week where 14 hours were worked under wet conditions. (1 mark)

c Find the area of the shaded part of the shape in the diagram. (1 mark)

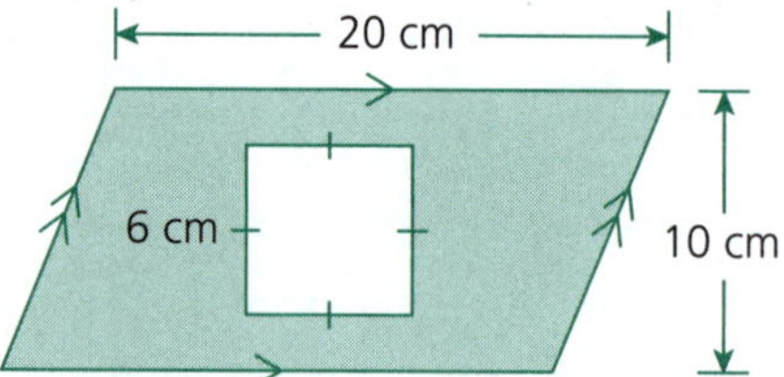

d In the triangle ABC, find the size of angle C to the nearest whole degree. (1 mark)

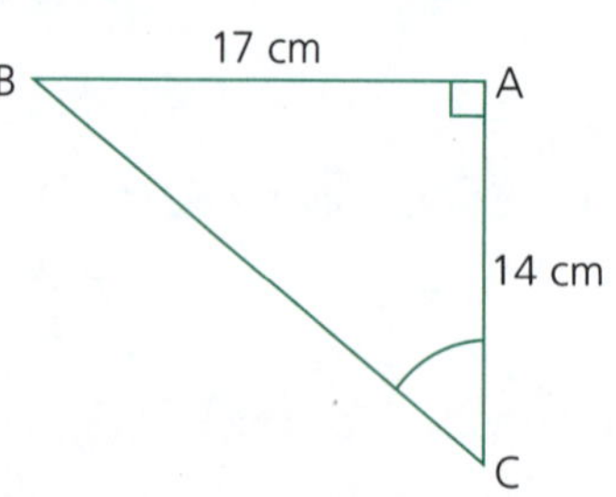

e The bearing of A from B is 080°. Find the bearing from B to A. (1 mark)

f Find the volume of a cylinder with a base radius of 6 cm and height of 40 cm. (Give your answer to the nearest cm^3.) (1 mark)

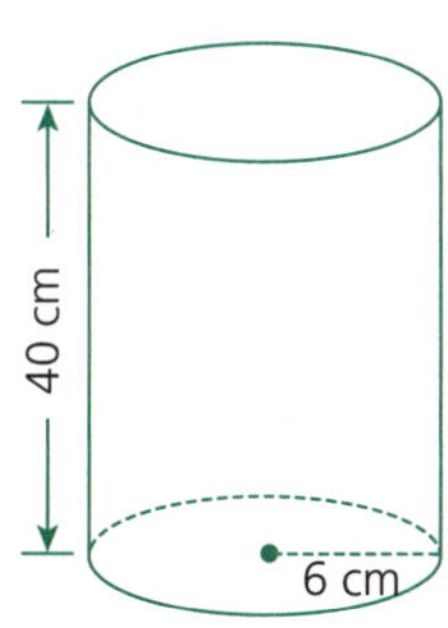

g A television set was priced at \$1489 and was bought on a deposit of \$250 with a repayment period of 2 years. If the instalments are to be \$70 per month, how much would be saved by paying the cash price instead of the instalments? (1 mark)

h Vance has a taxable income of \$188 400. Use the tax table to find the amount of tax he will pay. (1 mark)

Taxable income	Tax on this income
0–\$18 200	Nil
\$18 201–\$37 000	19c for each \$1 over \$18,200
\$37 001–\$80 000	\$3572 plus 32.5c for each \$1 over \$37 000
\$80 001–\$180 000	\$17 547 plus 37c for each \$1 over \$80 000
\$180 001 and over	\$54 547 plus 45c for each \$1 over \$180 000

i A 4-litre can of paint costs \$28.20 and has a normal coverage rate of 16 m^2 per litre. Find the cost of painting an area of 96 m^2 with two coats of paint. (1 mark)

j Calculate the simple interest on an investment of \$5400 at 8.5% p.a. for 4 years. (1 mark)

k Calculate the surface area of the rectangular prism in the diagram: (1 mark)

4 cm

2 cm

6 cm

l Convert 90 m/s to km/h. (1 mark)

m Peter earns \$1500 per week. He pays $\frac{1}{3}$ of this in tax, keeps $\frac{1}{2}$ of the remainder for bills and banks the rest. Find the amount of money Peter banks per week. (1 mark)

47 A man's normal wage is \$1004 for a 40-hour week. By working overtime he earned \$1229.90.

a Find his normal hourly rate. (1 mark)

b How many hours overtime did he work if he was paid time and a half for the overtime worked? (1 mark)

c The man's rate is increased by 5%. Find the new hourly rate for normal hours, to the nearest cent. (1 mark)

48 The cost c, of hiring an electrician is given by the formula $c = 85 + \frac{75n}{4}$, where c is the cost in dollars and n is the number of hours worked.

a How much does the electrician charge for 5 hours' work?

b If the cost was \$235, how long did the electrician work? (3 marks)

49 The graph is of the line which passes through the points A(3, 5) and B(–1, –3):

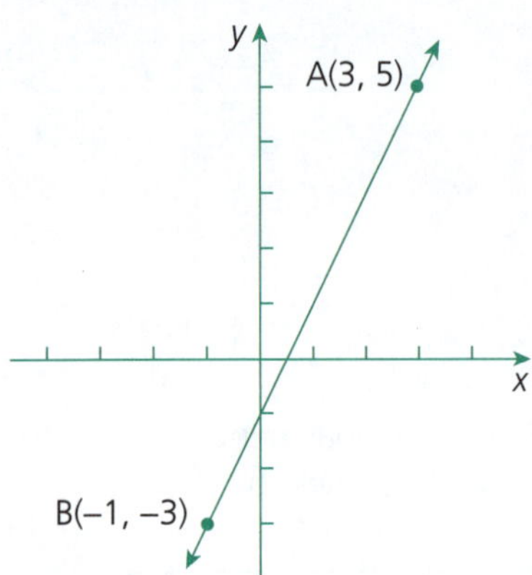

a Calculate the midpoint of the interval AB.
b Calculate the gradient of the line AB.
c Write the equation of the line AB. (3 marks)

50 The Venn Diagram shows the results of a survey of the activities of people last Boxing Day:

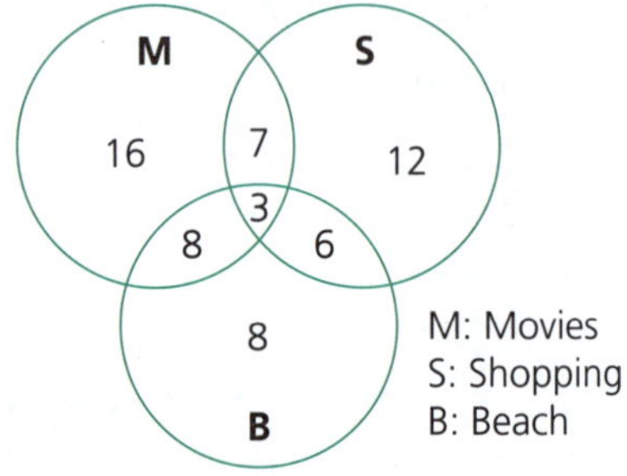

a If a person is chosen at random, what is the probability that they went to the movies and shopped? (1 mark)
b From the people who went to the beach, a person is chosen at random. What is the probability that this person saw a movie? (1 mark)
c From the people who went to the beach or to the movies, a person is chosen at random. What is the probability that this person also shopped? (1 mark)

51 In the diagram, AB II CD.
Angle BAC = 65°.
Angle BCD = 50°.
a Find the value of y, giving reasons.
b Find the value of z, giving reasons.
c Find the value of x, giving reasons. (3 marks)

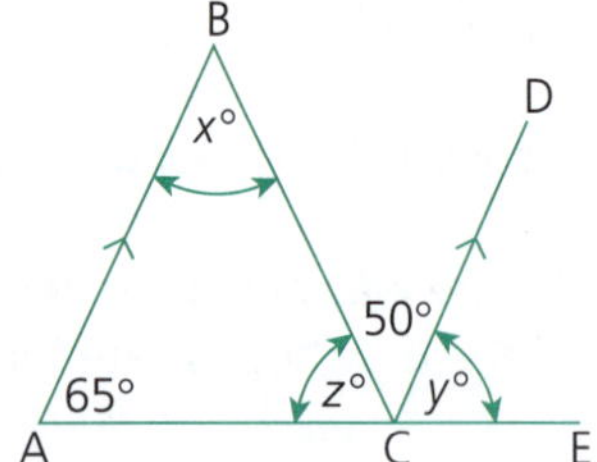

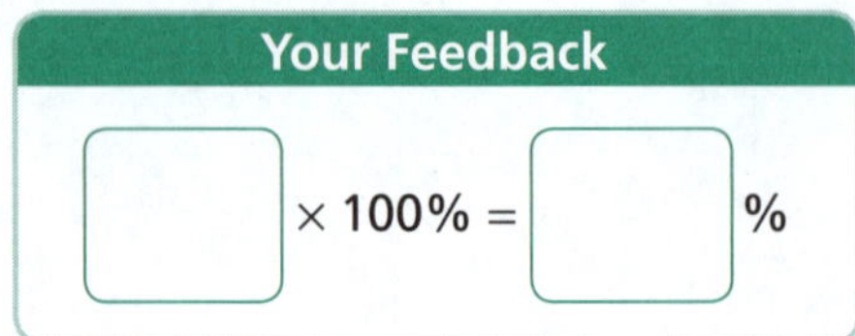

QA PAGE 225
WS PAGE 276

Quick Answers to Practise Practise

Chapter 1 Financial Mathematics pp. 9–11

1 \$1484.62 **2** \$526.50 **3** \$2015 **4** \$19.45 **5** \$196.50 **6** \$230.40 **7** \$878 **8** \$51 **9** a \$34 900 b \$1342.31 **10** a \$73.60 b \$73.60 c 15 h **11** \$168.75 **12** \$108 **13** \$548.23, \$504.75, \$585.25 **14** \$667.80 **15** \$17.20 **16** \$636 **17** a \$16 055 b \$507 160 **18** a \$728 b \$4888 **19** \$727.41 **20** a \$87 559.68 b \$63 174.68 c \$2429.80 **21** \$8124 **22** a Nil b \$18 003.95 c \$5061.48 d \$60 640 **23** a \$39 650 b \$4433.25 c \$506.75 **24** a See worked solution b \$52.50 c 9 weeks **25** \$2760.60 **26** \$388 800 **27** a \$630 b \$114.75 c \$162 **28** \$1248 **29** \$264 **30** \$1176 **31** a \$27 200 b \$566.67 **32** a \$352 800 b \$72 800 **33** a \$2542 b \$1047 **34** B **35** a \$21.60 b \$261.60 c \$16.80 **36** \$35.14 **37** \$1044 **38** A **39** \$68.28 **40** a \$1222.50 b \$2467.92 c \$330.62 d \$4435.33 **41** \$20 755.57 **42** 23 534 **43** \$982.60 **44** \$6043.06 **45** 204

Chapter 2 Algebraic Techniques pp. 25–26

1 a $x + y$ b $p - q$ c $7y$ d $\frac{a}{b}$ **2** a $100p$ cents b $7w$ days c $1000k$ mL d $10x$ mm e $\frac{m}{1000}$ kg f $\frac{y}{12}$ years **3** a $\frac{x+y+z}{3}$ b $\frac{2x+1}{2}$ c $\frac{a+b+c+d}{4}$ **4** $x + 1, x + 2$ **5** $x + 2, x + 4, x + 6$ **6** $x + 1, x + 3, x + 5$ **7** $4(x + 3)$ cm^2 **8** pq cents **9** $(m - n)$ mL **10** a -6 b -7 c -25 d -21 e -13 f -2 g -1 h 50 i 1 j 9 **11** a $6x$ b $6xy$ c $8a + 2b$ d $19a$ e $6x^2 + 4x$ f $19a$ g 0 h $4a$ i $6y$ j ab **12** a $3y$ b $2q$ c a d $7xy$ e $10y^2$ f $-2a$ g $-6y$ h p i $-8a^2$ j $-6m$ **13** a $7a + 9b$ b $4y + 2x$ c $13ab - 5a$ d $9 + 3x$ e $ab + a + b$ f $10cd - 3d$ g $-6p - q$ h $-2m + 5n$ i $-g - 8k$ j $3a - 6b$ **14** a $12a$ b $30y$ c $6sr$ d $4m^2$ e $-20a$ f $-18y$ g $-10p^2$ h k^3 i $abcd$ j $9b^2$ **15** a $4w$ b $7y$ c 2 d $-2y$ e $3g$ f $2ab$ g 6 h $2x$ i -5 j $8w$ **16** a $10y$ b $6ab - 12a$ c $4mn$ d 2 **17** a $\frac{5x}{6}$ b $\frac{13y}{15}$ c $\frac{17a}{12}$ d $\frac{7y}{10}$ e $\frac{ad+bc}{bd}$ f $\frac{5}{m}$ g $\frac{11}{2x}$ h $\frac{13y}{30}$ **18** a $1\frac{1}{2}$ b 2 c $1\frac{2}{3}$ d $\frac{9y}{4}$ e $6b$ **19** a 2 b $1\frac{1}{2}$ c $2\frac{2}{3}$ d 9 e $\frac{a}{x}$ **20** a $2a + 2b$ b $3y - 18$ c $10a - 5$ d $4y - 12$ e $6p - 10q$ f $28a + 49b$ g $-2m + 2$ h $-3p + 12$ i $-12r - 30$ j $-4g + 1$ k $-2x - 3$ l $-6 + 4y$ m $-x - 7$ n $x^2 + 2x$ o $6a^2 - 3a$ p $8y - 12y^2$ q $-3a^2 + 21a$ **21** a $6a + 10$ b $11y + 12$ c $24m - 5$ d $41a - 67$ e $3y - 9$ f $21 - 9y$ g $9a - 12$ h $7 - 9y$ i $7p - 4 - p^2$ j $3x^2 - 4x$ **22** a $1 + 2a$ b $12y - 38$ c $6 - 4m$ d $13 - 4r$ e $9b - 3$ **23** a $5a + 5$ b $8x + 4$ **24** a $m - 5$ b $2y - 3$ c $9q - 4$ d $9p - 4$ **25** a $x^2 + 6x + 8$ b $y^2 + 4y - 5$ c $c^2 + 4c + 3$ d $2y^2 + 5y - 3$ e $3n^2 + 3n - 6$ f $3q^2 - 12$ g $6t^2 + 7t - 5$ h $8x^2 - 34x + 21$ i $6w^2 - 11w - 10$ j $8y^2 - 26y + 21$ **26** a $x^2 + 6x + 9$ b $b^2 - 2b + 1$ c $4a^2 - 12a + 9$ d $16y^2 + 40y + 25$ e $16 - 8n + n^2$ **27** a $x^2 - 4$ b $y^2 - 9$ c $4a^2 - 1$ d $25 - 4p^2$ e $9a^2 - 25b^2$ **28** a $4(y - 2)$ b $3(a - 4b)$ c $2(3m - 5)$ d $4(3p - 2)$ e $2a(2b - 3)$ f $5x(y - 3)$ g $2q(3p - 2)$ h $a(a - 3 - b)$ i $y(y - 6)$ j $2a(a - 6b)$ **29** a $(x + y)(m + n)$ b $(m + n)(p + q)$ c $(b + c)(a + 2)$ d $(p + q)(3 + a)$ e $(a - b)(c - d)$ f $(x - y)(y - 7)$ g $(c + 6)(a - t)$ h $(g - 3)(g^2 + 2)$ **30** a $(p + 4)(p - 4)$ b $(y + 6)(y - 6)$ c $(m + 11)(m - 11)$ d $(2a + b)(2a - b)$ e $(3y + 4)(3y - 4)$ f $(5x + 8y)(5x - 8y)$ g $(9 + 11m)(9 - 11m)$ h $2(p + 1)(p - 1)$ i $3(z + 3)(z - 3)$ j $10(h + 7)(h - 7)$ **31** a $(x + 3)(x + 1)$ b $(y + 2)(y + 1)$ c $(a - 5)(a + 4)$ d $(y - 6)(y + 3)$ e $(m + 5)(m - 3)$ f $(k + 6)(k - 1)$ g $(x - 2)(x - 1)$ h $(y - 3)(y - 4)$ i $(t - 3)^2$ j $(y - 4)^2$

Chapter 3 Indices p. 34

1 a $2^4 \times 3^2$ b $2^4 \times 5^2$ c $13^2 \times 2^2$ **2** a p^8 b y^7 c a^9 d c^7 e x^{10} f y^{14} **3** a $8y^7$ b $14a^5$ c $6p^6$ d $-50m^7$ e $-12y^7$ f $20a^6b^4$ **4** a w^4 b x^2 c p^8 d y^3 e a^3 f m^2n^2 **5** a $2y$ b $3c^2$ c $7p^4$ d $2a^2$ e $-xy^2$ f cd^2 g $2y^2z$ h $6a^2b$ i $\frac{4p^3q^2}{3}$ **6** a p^5q^3 b $5x^8y^5$ **7** a $6a^2$ b 3 c $2m^6n$ **8** a y^6 b a^{12} c m^{30} d $9a^8$ e $8y^{15}$ f $49a^4b^6$ g $9m^8$ h $25p^{12}$ i $a^{12}b^9$ **9** a y^{-2} b a^{-5} c y^{-1} d 1 e b^8 f g^{-2} g k^2 h p^{-6} i m^{-6} j $4p^{-10}$ k $2ab^{-3}$ l $-5p^{-2}$ **10** a 1 b 1 c 1 **11** a 1 b 2 c 1 **12** a 4.28×10^3 b 3.6×10^5 c 5×10^2 **13** a 6.8×10^{-2} b 3.5×10^{-4} c 8.6×10^{-6} **14** a 2600 b 15 100 c 46 300 000 **15** a 0.058 b 0.007 06 c 0.000 084 **16** a 1.58×10^{-1} b 1.01×10^4 c 2.24×10^6 d 2.65×10^{-4} e 3.04×10^2 f 5.33×10^{-3} **17** 1.04×10^{20} **18** 5.62×10^{15}

Chapter 4 Equations .... pp. 47–48

1 a $x = 8$ b $a = -4$ c $y = 5\frac{1}{2}$ d $a = 6\frac{1}{2}$ e $x = 3$ f $y = 11\frac{1}{2}$ g $x = -4$ h $a = -\frac{4}{13}$ i $y = 4$ j $y = 4\frac{2}{5}$ k $x = -8$ l $a = -7\frac{1}{2}$ m $y = 8$ n $x = 1\frac{2}{13}$ o $x = \frac{1}{3}$ p $y = \frac{1}{2}$ **2** a $x < 3\frac{1}{2}$ b $a \geq 5$ c $y > 2\frac{2}{5}$ d $y > \frac{1}{2}$ e $y < -1\frac{3}{7}$ f $y \leq -7$ g $y \geq -6$ h $y < -2\frac{1}{2}$

3 a −3 −2 −1 b 2 3 4 c 0 $\frac{1}{3}$ $\frac{2}{3}$ 1 d −4 2

e 2 3 **4** a $-2 \leq x < 3$ b $x < -2, x \geq 1$ **5** a $x > 3$ 2 3 4 b $x \leq -4$ −5 −4 −3

6 a $D = 720$ b $A = 90$ c $S = 1062.35$ d $A = 281.71$ e $B = 0.03$ **7** $V = 9047.789$ **8** 45°C **9** a $y = 13 - 2x$ b $y = 3x - 11$ c $y = \frac{12 - 5x}{4}$ d $y = \frac{7x - 18}{3}$ e $y = \frac{19}{3x}$ f $y = \frac{x}{5}$ **10** a $t = 0.61$ b $S = 540$ c $I = 1.19$ d $T = 6.85$ e $h = 2.51$ f $k = 1.06$

11 a 30, 31, 32 b $x = 2$ c $x = 3$ d 8 e −7 f 30 years old g 8 years old **12** a $a = -2, b = 5$ b $m = 1, n = -2$ **13** a $p = 5, q = 1$ b $x = 1, y = 1$ **14** a 7 and 5 b i $x = 3, y = 1$ ii $x = 2, y = -1$ c Pen 25c, pencil 20c d Hamburger $4.10, can $1.80 **15** a $x = 1, -3$ b $p = -5, 2$ c $a = 0, 2$ **16** a $x = -2, -1$ b $m = 3, 1$ c $k = 4, 3$ d $b = -4, 1$ e $t = 6, -2$ f $w = 5, -3$ g $c = 0, -2$ h $y = 4, -4$ i $q = 0, 7$ j $h = 0, 10$ k $q = -5, 1$ l $x = 5, 4$

Chapter 5 Linear Relationships .... pp. 63–66

1 A(2, 2) B(−1, −1) C(−3, 1) D(2, −1) E(1, −3) F(3, 1) G(−2, 3) H(−2, 0) I(0, 1) J(2, 0) K(0, −2) L(−3, −2) **2** a 5 b $\sqrt{41}$ **3** a $\sqrt{26}$ b $\sqrt{50}$ c $\sqrt{50}$ d $\sqrt{26}$ e 5 f 5 **4** a $(1, \frac{1}{2})$ b (0, 1) **5** a $(\frac{-1}{2}, 3)$ b (0, 2) c (0, 0) d (2, −1) e (1, −2) f $(2, 1\frac{1}{2})$ **6** a $\frac{-3}{2}$ b $\frac{-1}{4}$ c $\frac{-3}{4}$ d $\frac{1}{2}$ e $\frac{2}{3}$ f $\frac{4}{3}$ **7** a $\frac{2}{5}$ b −1 c $\frac{1}{2}$ d $\frac{-1}{2}$ e 0 f Undefined **8** a $\sqrt{29}$ b $\sqrt{32}$ c $\sqrt{80}$ d $\sqrt{20}$ e 4 f 5

9 a (−1, 3) b (0, −1) c (0, 0) d (2, −1) e (1, −2) f $(2, 1\frac{1}{2})$ **10** a $\frac{2}{5}$ b −1 c $\frac{1}{2}$ d $\frac{-1}{2}$ e 0 f Undefined **11** a $\sqrt{41}, \sqrt{50}$ b $(1\frac{1}{2}, \frac{1}{2}), (-1, 2\frac{1}{2})$ c $\frac{-4}{5}, \frac{3}{4}$ **12** See solutions for diagrams and tables **13** See solutions for diagrams **14** a $y = 4x - 1$ b $y = -2x + 3$ c $y = \frac{1}{2}x + 2$ d $y = \frac{3}{4}x - 4$ **15** a $m = 3, b = 7$ b $m = -4, b = \frac{1}{2}$ c $m = -3, b = 4$ d $m = 2, b = \frac{2}{5}$ e $m = \frac{2}{3}, b = -2$ f $m = -2, b = 3$ g $m = 2, b = -3$ h $m = -\frac{3}{2}, b = 3$ i $m = \frac{2}{3}, b = -2$ j $m = 2, b = -\frac{7}{2}$

16 a $m = \frac{-3}{2}, y = \frac{-3}{2}x + 3$ b $m = -1, y = -x - 3$ c $m = \frac{2}{3}, y = \frac{2}{3}x + 2$ d $m = 2, y = 2x - 4$ e $y = 3x + 3$ f $y = \frac{-5}{2}x - 5$

17 See solutions for diagrams **18** (a), (b), (c), (f) are parallel with gradient $\frac{1}{2}$; (c) and (f) pass through origin.

19 a Yes b No c No d Yes e Yes f No **20** a No b Yes c Yes d Yes **21** a $\sqrt{40}, \sqrt{40}$, isosceles b $\frac{-1}{3}$ c (−1, 1)

22 a $y = -3x + 4$ b $y = 4x - 4$ **23** a Not perpendicular b Perpendicular c Perpendicular **24** $k = -2\frac{2}{3}$

25 A, yes B, no C, yes **26** Collinear, $m = -1$ **27** $d = 3\frac{1}{4}$ **28** a $y = 3x - 1$ b $y = -2x + 2$ c $y = \frac{1}{2}x$ d $y = \frac{-2}{3}x$ e $y = 3x - 6$ f $y = 2x - 2$ g $y = \frac{-3}{2}x - 4$ h $y = x + 1$ i $y = -\frac{1}{2}x + 4$ j $y = \frac{5}{3}x - 1$ **29** a $C = 50 + 70n$

b

n	0	2	4	6
C	50	190	330	470

c See solution d i $260 ii $610 e i 5 hours ii 10 hours

30 a i n ii P b

n	1000	1500	2000	2500	3000
P	200	1700	3200	4700	6200

(See solution for graph) c $6950 d 4320 pens

Chapter 6 Non-linear Relationships .... pp. 78–79

1 See solutions for diagrams and tables **2** a Concave up b Concave down c Concave down d Concave up **3** a Minimum b Maximum c Minimum d Maximum **4** See solutions **5** See solutions **6** See solutions **7** See solutions **8** a $x = 0$ b $x = -3$ c $x = 1$ d $x = -2$ e $x = 4$ **9** a Lies on parabola b Does not lie on parabola c Lies on parabola **10** a Does not pass through b Passes through (2, 3) c Does not pass through **11** $b = -1$ **12** $c = -4$ **13** a $y = 3x^2$ b $y = -2x^2$ c $y = x^2 + 2$ d $y = -3x^2 - 1$ e $y = 2(x - 1)(x - 3)$ f $y = -5x(x - 4)$ **14** See solution **15** See solution **16** a 7 units b 9 units c 4 units d $\sqrt{7}$ units e $\sqrt{29}$ units **17** See solutions **18** a $x^2 + y^2 = 16$ b $x^2 + y^2 = 3$ **19** See solution **20** See solutions **21** a (2, 4) and (−1, 1) b (4, 18) and (1, 3)

Chapter 7 Rates and Proportion .... pp. 88–90

1 a $27\frac{7}{9}$ m/s b 144 km/h c 5L/100 km d 10 km/L e 1.008 km/h f 0.125 kg/$ g 62.5 L/$100 h $3\frac{1}{3}$ h/L i 1.8 s/mL **2** a $C = 12d$ b Increases c $420 d 42 metres e See solution f 12 **3** a $W = 16.5n$ b Increases c $495 d 36 hours e See solution f 16.5 **4** a $D = 9n$ b 315 km c $\frac{1}{4}$ of petrol remains **5** a See solution b i 100 °F ii 15 °C **6** a i See solution b i 32 km ii 25 miles **7** a

Kilograms	0	5	10	15	20
Pounds	0	11	22	33	44

b See solution c i 17.6 pounds ii 330 pounds

8 a i €60 ii £80 b i €36 ii $70 iii £48 iv €25 c i €3000 ii $A4000 iii £600 iv €125 d i £48 ii $A125 **9** a i $y = kx$ ii $y = 3x$ iii $y = 60$ iv $x = 6$ **10** a $P = kt$ b $P = 5t$ c $P = 80$ d $t = 6$ **11** a $W = 18n$ b $486 c 36 hours **12** a $P = 5.2m$ b $18.72 c 2.8 kg **13** a $y = 3.6x$ b

x	1.3	2.9	4.5
y	4.68	10.44	16.2

14 a $F = 12m$ b 74.4 Newtons c 3.6 kg **15** a $y = k\sqrt{x}$ b $y = 8\sqrt{x}$ c $y = 32$ d $x = 900$ **16** a $y = kx^2$ b $y = 3x^2$ c $y = 108$ d $x = 14$ **17** See solution

Chapter 8 Pythagoras and Trigonometry .... pp. 112–118

1 a $x = 20$ b $y = 24$ c $a = 8$ d $b = 35$ e $c = 0.9$ f $d = 1.2$ g $n = 32$ h $q = 48$ i BD is 2.5 cm j PQ is 40 cm **2** a △ is right-angled b △ is not right-angled c △ is not right-angled d △ is right-angled e △ is right-angled f △ is right-angled g △ is not right-angled h △ is right-angled **3** a Triad is {7, 24, 25} b Triad is {9, 12, 15} c Triad is {18, 24, 30} d Triad is {9, 40, 41} e Triad is {13, 84, 85} f Triad is {21, 72, 75} **4** a $a = 5.7$ b $b = 4.1$ c $c = 2.6$ d $d = 8.5$ e $e = 7.9$ f $f = 11.2$ g $g = 14.4$ h $h = 8.7$ **5** a $n = \sqrt{50}$ b $m = \sqrt{301}$ c $= \sqrt{132}$ d $s = \sqrt{297}$ **6** Length of diagonal is 20 cm **7** Diagonal is 4.2 cm **8** Brace reaches 28 cm up the wall **9** Foot of ladder is 2.8 m from the wall **10** Wires are attached 10.96 m from the top **11** Length of side is 12.7 cm **12** Width is 15 cm **13** Length of each side is 13 cm **14** $y = 11.2$ **15** Third side is 24 m **16** Perimeter is 36 cm **17** a $\frac{3}{5}$ b $\frac{4}{5}$ c $\frac{3}{4}$ **18** a $\frac{4}{5}$ b $\frac{3}{5}$ c $\frac{3}{5}$ **19** $\frac{12}{5}$ **20** a 0.9573 b 0.4932 c 4.8077 **21** a 100°44′ b 51°46′ c 93°12′ d 103°45′ e 247°24′ **22** a 5.22 b 16.60 c 101.48 d 6.56 e 8.95 f 14.80 g 112.27 h 18.41 i 92.93 j 101.04 **23** a 21.2 b 23.3 c 2.3 d 47.5 e 125.9 f 39.2 g 100.9 h 5.5 i 40.7 **24** a 25 b 58 c 54 d 27 e 47 f 67 **25** a 54°22′ b 77°2′ c 60°57′ d 0°7′ e 29°45′ f 55°9′ g 34°51′ **26** a 28°4′ b 69°47′ c 41°25′ d 30°13′ e 27°54′ f 68°26′ **27** 69° **28** 31° **29** 10.6 cm **30** 29 m **31** 25 m **32** 8.58 m **33** 4.7 m **34** 9.7 m **35** 19°17′ **36** 3.1 m, 31°2′ **37** 35° **38** a Between 0° and 21°48′ b 0°–14°56′ **39** a 130° b 235° c 290° d 050° e 280° f 240° **40** 146° **41** 12.86 km **42** 157.57 km **43** 319 km **44** 153 km **45** 300°

Chapter 9 Measurement .... pp. 132–133

1 a 41.133 cm b 45.708 cm **2** a 58 m b 78 cm **3** 12 m **4** 606 cm **5** 30 cm^2 **6** a 108π cm^2 b 24 cm^2, 40 cm **7** 836 cm^2 **8** 2640 cm^2 **9** a 180 b 618 cm^2 **10** 3 cm **11** 220 cm^2 **12** 1379.084 cm^3 **13** a 16 889.20 cm^3 b 669.79 cm^3 c 1710.95 cm^3 d 923.63 cm^3 **14** 36π cm^3 **15** 23 411 L **16** Water is 3.125 cm deep **17** 1.571 L **18** 3054 kL **19** a $3515.40 b 217 m^3 c 217 kL **20** a 23.4 m^2 b 234 000 cm^2 c 40 **21** Mass is about 3 kg **22** a 7.6 m^2 b 2080 L c 1.56 kL

Chapter 10 Geometry pp. 154–158

1 a $x = 18$ b $x = 29$ c $y = 93, x = 87, z = 87$ d $x = 90$ e $x = 74$ f $x = 61$ g $x = 60$ h $x = 63$ i $x = 113$ j $x = 133$ k $x = 117$ l $x = 38$ m $x = 121$ n $x = 55$ o $x = 68$ p $x = 120$ q $x = 108$ r $x = 55, y = 125$ s $x = 68$ t $x = 82$ **2** $y = 67\frac{1}{2}$ **3** $m = 83$ **4** $x = 129$ **5** $x = 113$ **6** a i 108° ii $128\frac{4}{7}$° iii 150° iv 162° b i 45° ii 40° c i 15 ii 18 **7** a SAS b AAS c SSS d RHS e AAS **8** a $\triangle PRS \equiv \triangle PRQ$ (SSS) b $\triangle ABX \equiv \triangle DCX$ (SAS) c $\triangle BAE \equiv \triangle CDE$ (AAS) d $\triangle ABC \equiv \triangle ADC$ (RHS) **9** a $\triangle BAD \equiv \triangle DCB$ (AAS) b i $AD = BC$ ii $\angle BAD = \angle DCB$ **10** a $\triangle ABX \equiv \triangle DCX$ (SAS) b $\angle ABX = \angle DCX$, $AB \parallel CD$ **11** a $\triangle ABC \equiv \triangle DCB$ (AAS) b $AB = DC$ **12** a $\triangle ABD \equiv \triangle CBD$ (SAS) b $AB = CB$ c $\angle ABD = \angle CBD$ **13** a $x = 9$ b $x = 9, y = 15, z = 12$ c $x = 7, y = 6$ d $x = 9, y = 7\frac{1}{2}$ **14** a $x = 6$ b $x = 10$ c $x = 12$ d $x = 8$ e $x = 5\frac{1}{7}$ f $x = 4$ **15** a 20 m **16** 30 m

Chapter 11 Statistics pp. 178–181

1 a i 5 ii 13 iii 7 b Range = 5, Mode = 4, 5 c i 25% ii 65% d Mean = 3.9, Median = 4 e See solutions f $\frac{1}{4}$ **2** a Range = 7, Mode = 4, 6 b i 5 ii 13 iii 8 c i 20% ii 68% d i $\frac{1}{5}$ ii $\frac{8}{25}$ e Mean = 4.28, Median = 4 f See solution g See solution **3** a i 6 ii 5, 8 iii 6.5 b 1, 0.5 c 3 **4** a 5.3, 5 b 5.8, 5.5 **5** 95 **6** See solution **7** See solution **8** See solution **9** a See solution b Range = 38, Mean = 38, Mode = 39, 47, Median = 48.5 **10** a 6.5 b 7 c 16 d 10 **11** a 38 b 124 **12** a 4, 5.5, 6.5, 8, 9, see solutions b 5, 6, 7, 9, 9, see solutions c 12, 14, 16, 18, 19, see solutions d 7, 8, 10, 12, 14, see solutions **13** a i 16 ii 3 b i 63 ii 16 **14** a 38, 24, 47, 10, 54, see solutions b 124, 114.5, 135.5, 105, 147, see solutions **15** a See solutions b 167.5, 161, 172, 11 c 149, 182, see solutions **16** a Positively skewed b Normally distributed (symmetrical) c Positively skewed d Negatively skewed e Symmetrical—without knowing mean and mode we cannot say that it is normally distributed **17** a See solution b Mean = $4\frac{4}{15}$, Mode = 3, Median = 4 (8th score) c Positively skewed **18** a Negatively skewed b Positively skewed **19** a Normally distributed b Median = 8, Mode = 8, Mean = 8 (as the data is normally distributed, it will be the same as the median and mode) **20** The trigonometry test is positively skewed and the algebra test is negatively skewed **21** a See solution b See solution c See solution **22** a *r* b *n* c *s* **23** a Weak positive b Strong negative **24** a See solution b Strong positive relationship c About 166–167 cm d The length of the arm span increases as the height increases **25** a See solution b About 41 cm **26** a See solution b Negative c Strong

Chapter 12 Probability pp. 195–198

1 a $\frac{1}{13}$ b $\frac{1}{13}$ c $\frac{1}{4}$ d $\frac{1}{4}$ e $\frac{1}{2}$ f $\frac{3}{13}$ g $\frac{10}{13}$ h $\frac{5}{13}$ or $\frac{6}{13}$ i $\frac{1}{52}$ j $\frac{7}{13}$ k $\frac{1}{26}$ **2** $\frac{3}{10}$ **3** a $\frac{7}{15}$ b $\frac{8}{15}$ c $\frac{1}{3}$ d $\frac{2}{3}$ e 0 **4** a $\frac{1}{10}$ b $\frac{1}{2}$ c $\frac{3}{5}$ d $\frac{3}{10}$ **5** $\frac{2}{5}$ **6** a $\frac{9}{50}$ b $\frac{17}{100}$ c $\frac{11}{25}$ **7** a 33 b i $\frac{8}{33}$ ii $\frac{4}{33}$ iii $\frac{16}{33}$ **8** a $\frac{1}{3}$ b $\frac{1}{12}$ c $\frac{13}{36}$ **9** See solution **10** a 20 b i $\frac{4}{5}$ ii $\frac{1}{5}$ iii $\frac{3}{10}$ **11** a 30 b i $\frac{3}{5}$ ii $\frac{3}{10}$ iii $\frac{1}{6}$ iv $\frac{2}{5}$ **12** a See solution b i $\frac{1}{12}$ ii $\frac{1}{4}$ **13** a $\frac{1}{9}$ b $\frac{1}{9}$ c $\frac{1}{2}$ d $\frac{1}{6}$ e $\frac{5}{6}$ f $\frac{8}{9}$ g $\frac{11}{18}$ h $\frac{1}{18}$ **14** a $\frac{1}{18}$ b $\frac{1}{4}$ c $\frac{19}{36}$ d $\frac{23}{36}$ e $\frac{5}{9}$ f $\frac{5}{36}$ g $\frac{2}{3}$ **15** a $\frac{11}{36}$ b $\frac{1}{36}$ c $\frac{25}{36}$ d $\frac{2}{3}$ e $\frac{7}{12}$ f $\frac{17}{36}$ g $\frac{1}{12}$ h $\frac{35}{36}$ i $\frac{4}{9}$ j 0 **16** a See worked solutions b i $\frac{7}{25}$ ii $\frac{23}{100}$ c i $\frac{3}{10}$ ii $\frac{7}{10}$ **17** See solution **18** See solution **19** See solution **20** a $\frac{1}{8}$ b $\frac{3}{8}$ c $\frac{3}{8}$ d $\frac{1}{8}$ e $\frac{7}{8}$ **21** a $\frac{3}{10}$ b $\frac{1}{10}$ c $\frac{3}{5}$ d $\frac{2}{5}$ **22** a $\frac{5}{18}$ b $\frac{5}{9}$ c $\frac{5}{18}$ d $\frac{1}{6}$ e $\frac{5}{6}$ **23** a $\frac{1}{50}$ b $\frac{1}{4950}$ c $\frac{49}{2475}$ d $\frac{197}{4950}$ e $\frac{4753}{4950}$ f $\frac{197}{4950}$ **24** a $\frac{1}{3}$ b $\frac{2}{3}$ **25** a $\frac{5}{26}$ b $\frac{1}{5}$ c $\frac{1}{21}$ **26** a $\frac{6}{25}$ b $\frac{10}{27}$ c $\frac{8}{13}$ **27** a See worked solutions b i $\frac{14}{65}$ ii $\frac{12}{65}$ c i $\frac{11}{30}$ ii $\frac{3}{5}$ d $\frac{7}{20}$ **28** a $\frac{1}{2}$ b $\frac{1}{2}$ c $\frac{1}{2}$ **29** a $\frac{11}{23}$ b $\frac{4}{7}$ c $\frac{1}{2}$ **30** a See solution b $\frac{9}{17}$ c $\frac{4}{9}$

Quick Answers to Tests

Chapter 1 Financial Mathematics

Intermediate Test p. 13

1 \$206.80 **2** \$835.20 **3** \$1015 **4** \$600 **5** \$6300 **6** \$1450.20 **7** \$3471.30 **8** \$190.49 **9** \$189 **10** \$2400 **11** \$4830 **12** a \$3840 b \$330 **13** \$945.75 **14** \$8852.18 **15** \$7025.96

Advanced Test p. 14

1 \$15.63 **2** \$304 **3** \$4747 **4** 2.5% **5** \$470 **6** a \$680.05 b \$13 621.40 **7** a \$58 730 b \$10 634.25 **8** \$80 **9** a \$35 b \$228.38 **10** 6% **11** \$983.40 **12** $P = 36$ **13** \$152.90 **14** \$481.79

Chapter 2 Algebraic Techniques

Intermediate Test p. 28

1 a $8x$ b $6xy$ c $400x$ cents **2** a -48 b 3 c 20 d $\frac{1}{5}$ **3** a $10m$ b $36y^2$ c $-7ab$ d $16y^2$ e $3y$ f $\frac{b}{2a}$ **4** a $\frac{6x}{y}$ b $\frac{-5m}{6}$ c $\frac{19}{6x}$ d x^2 e $\frac{4}{25}$ **5** a $15x - 10y$ b $15 - 6x$ c 7 **6** a $x^2 + 2x - 15$ b $4x^2 - 25$ c $4x^2 + 20x + 25$ **7** a $7(y - 3)$ b $5m(m + 3n)$ c $(m + 3)(m - 3)$ d $(a + 8)(a - 5)$ e $(y - 9)(y - 2)$ f $(y + 9)(y + 2x)$ g $a(a - 4)(a + 1)$

Advanced Test p. 29

1 a $3y$ b $24a^2$ c $\$\frac{3x}{100}$ **2** a $-2t$ b $-42mn$ c $-17m^2$ d $3b$ e $27m^2n$ **3** a $5b^2$ b $\frac{19a}{10}$ c $\frac{16}{3m}$ d $\frac{9}{y}$ e $\frac{5y}{2}$ **4** a $35m - 21m^2$ b $-t + 40$ c $-3a + 40$ **5** a $3x^2 + 13x - 10$ b $25x^2 - 9$ c $25x^2 - 30x + 9$ d $-x$ e $-8x$ **6** a $9(6 - a)$ b $(m - 8)(m + 8)$ c $9y(3y - 1)$ d $(p + 7)(p - 6)$ e $(p - 9)(p + 8)$ f $10(x - 5)(x + 4)$ g $(m - 8)(m - 7t)$ h $(a^2 + b^2)(a + b)(a - b)$

Chapter 3 Indices

Intermediate Test p. 36

1 a $5^3 \times 2^2$ b $(3^2 \times 2^2)^2$ **2** a $12x^6$ b $40a^5$ c $15m^7n^7$ d $18p^5q^{10}$ **3** a $4b^2$ b $2a$ c $5k^3$ d $5g^3$ **4** a $4b^6$ b $9w^{10}$ c $64a^{12}$ d $9y^8$ e $-8m^9$ **5** a $2a$ b $2k^4$ c $2m$ d b^2c **6** a 1 b 1 c 5 d 1 **7** a 1 b 1 **8** a $\frac{1}{25}$ b $\frac{1}{36}$ c $\frac{1}{8}$ d 5 e 9 **9** a x^{-4} b a^{-2} c 1 d $6a^3$ e $35q^{-2}$ f p^5 g d^{-7} h y^4 i $2a^5$ j $3x^4$ **10** a 3.5×10^4 b 2.624×10^5 c 3.11×10^{-3} d 8.05×10^{-1} **11** a 257 000 b 19 100 c 0.025 31 d 0.0084

Advanced Test p. 37

1 a $6x^7y^9$ b $60a^5b^{13}$ c $490p^4q^6r^4$ d $63x^5y^4z^{11}$ **2** a $9a^2$ b $5pq$ c $8a$ d $3m^2$ e $1.5x^2y$ f $\frac{3}{5q^2}$ **3** a $4b$ b $5g^7$ c a^2b^4 d $4m$ e $-c$ **4** a $4y^6$ b $25c^8$ c $9m^{10}$ **5** a 1 b 1 c -3 d 5 e 1 f 81 g 1 h $25m^6$ **6** a $\frac{1}{x^2}$ b $\frac{6}{a^2}$ c $\frac{1}{8t^3}$ d p e $\frac{q^2}{9}$ f $4a^2$ g $\frac{3c^4}{b^2}$ h $\frac{2a^3}{3b^4}$ **7** a x^{-1} b 1 c $15p^{-4}$ d $4m^8$ e $5z^3$ f $\frac{m^{10}}{9}$ **8** a $\frac{4b^2}{cd}$ b $\frac{10p^2q}{3x^3y}$ **9** a 5.71×10^{-1} b 1.07×10^1 c 5.49×10^{13}

Chapter 4 Equations

Intermediate Test p. 50

1 a $x = 2$ b $a = 15$ c $a = 8$ d $x = 1\frac{3}{11}$ e $a = 9$ f $p = 1\frac{4}{23}$ g $y = 27$ **2** a $a > 4$ b $p \leq -2$ c $a \geq 8$ d $q \geq 2$

3 a −10 −9 −8 −7 −6 b −1 0 1 2 3 4 **4** a $P = 20$ b $l = 7$ **5** a $y = \frac{11 - 5x}{2}$ b $y = \frac{3x - 10}{7}$

c $y = 2x - 15$ **6** a $x + 1$ b 12 and 13 **7** a $x = 40$ b 130° **8** $x = 2, y = 1$ **9** a $x = 5, 1$ b $x = 4, -2$ c $x = 0, 5$

Advanced Test p. 51

1 a $x = -2$ b $x = -1$ c $x = 1\frac{3}{7}$ d $x = \frac{11}{23}$ e $y = 12\frac{1}{2}$ **2** a $p = -\frac{3}{4}$ b $a = 13$ c $x = \frac{12}{17}$ **3** a 2 3 4 5 6

b −5 −4 −3 −2 −1 **4** −2 −1 0 1 2 3 4 5 6 **5** a Area = $4(3x + 1)$ b $x = 5$ c 40 cm **6** a $y = \frac{3ax - z}{5b}$

b $y = \frac{mT^2}{a}$ c $y = \frac{4a + 6}{P}$ d $y = \frac{aK + 2}{3x}$ **7** 7.95 cm **8** $x = 1, y = 1$ **9** \$2.50 **10** a $x = 5, 1$ b $x = 5, -3$ **11** 22 cm

Chapter 5 Linear Relationships

Intermediate Test p. 68

1 See worked solutions a 5 b $(-1, \frac{-1}{2})$ **2**

x	0	1	2
y	−3	−1	1

See solution for diagram **3** See solution **4** x-intercept = 3; y-intercept = 6 **5** $m = \frac{2}{3}, b = -4$ **6** $y = -4x + 1$ **7** a $m = \frac{2}{3}$ b $b = -2$ c $y = \frac{2}{3}x - 2$ **8** See solution for sketch

9 a 5 b $(\frac{1}{2}, 3)$ c $\frac{4}{3}$ **10** Yes **11** Both gradients = $\frac{1}{2}$ **12** $y = -2x - 7$ **13** $y = 4x + 2$ **14** a $C = 15 + 3n$

b

n	0	4	8	12
C	15	27	39	51

c See solution d \$80

Advanced Test p. 69

1 See worked solutions a $\sqrt{73}$ b 2 **2**

x	0	1	2
y	3	1	−1

See solution for diagrams **3** See solution for diagram

4 x-intercept = $\frac{4}{3}$; y-intercept = −2 **5** $m = \frac{-2}{3}, b = 2$ **6** a $m = \frac{-3}{4}$ b $b = -3$ c $y = \frac{-3}{4}x - 3$ **7** $y = 2x - 2$ **8** a $\sqrt{74}$ b $(-2, \frac{-1}{2})$ c $m = \frac{-2}{5}$ **9** $k = 2\frac{1}{2}$ **10** $y = 4x - 1$ **11** a See solution b $x = 1, y = 3$ **12** Midpoint AB = (−1, 5); Gradient AB = 1, $y = -x + 4$ **13** $k = -2$ **14** a $C = 120 + 12n$ b See solution c \$516

Chapter 6 Non-linear Relationships

Intermediate Test p. 81

1 See solution **2** a C b B c H d F e G f E g A h D **3** See solution **4** See solution **5** $a = \frac{1}{2}$ **6** C **7** B

Advanced Test p. 82

1 See solution **2** a $x^2 + y^2 = 25$ b $y = 3^x$ c $y = -3x(x - 6)$ d $y = -\frac{2}{x}$ **3** a (6, 7) and (1, −3) b (−5, 39) and (1, 9)

Chapter 7 Rates and Proportion

Intermediate Test p. 92

1 14.6 cm/year **2** a 200 kg/ha b $800/m^3$ **3** Cheetah is faster by 23 km/h **4** a

h	0	1	2	3	5
D	0	90	180	270	360

b $D = 90h$ c 90 km/h d 112.5 km e See solution f First car travelled 105 km more g Second car takes 1 hour longer
5 a $C = 1175A$ b $364 250 **6** a $d = 3.2\sqrt{h}$ b 6.25 metres c As 13.58 > 12, then the person can see the island

Advanced Test p. 93

1 160 g/L **2** If x is doubled, then y is doubled **3** If x is doubled, then y is multiplied by 4 **4** a $k = 2$
b

x	4	16	49	144
y	4	8	14	24

5 a $D = 80t$; $D = 100t$ b 80 km more c After 3 hours **6** a $C = 7.2n^2$ b 2.5 cm c $133.13
7 $160

Chapter 8 Pythagoras and Trigonometry

Intermediate Test pp. 120–121

1 $PR^2 = PQ^2 + QR^2$ **2** a 20 b 1.7 **3** a $\sqrt{104}$ b $\sqrt{55}$ **4** a No b Yes **5** 8.5 cm **6** a 0.122 b 1.342 c 19.700
d 48.801 **7** a 18° b 37° **8** a 8.2 cm b 23.7 cm c 19.2 cm d 21.0 cm **9** a 65° b 59° c 63° **10** 38°
11 39 metres **12** See solutions **13** 1.6 km South

Advanced Test pp. 122–123

1 a 9.4 b 5.4 **2** 30 cm **3** 30.4 cm **4** 120.1 metres **5** Hypotenuse = 17 a $\frac{8}{15}$ b $\frac{8}{17}$ c $\frac{15}{17}$ **6** a 60°57′ b 77°42′
c 63°28′ d 54°52′ **7** Other side = 5, $\tan y° = \frac{5}{12}$ **8** a 19.4 cm b 3.1 cm c 14.7 cm d 17.1 cm **9** a 49°4′ b 56°46′
10 6.5 metres **11** 196.5 metres **12** 10.4 km North **13** 175.4 km

Chapter 9 Measurement

Intermediate Test pp. 135–136

1 a 30.85 cm b 60.41 cm **2** 54.8 cm **3** a 120 cm^2 b 168 cm^2 **4** 4π cm^2 **5** 1187.52 cm^2 **6** 9 m^2 **7** 132 cm^2
8 a 85.73 cm^3 b 904.78 cm^3 **9** 160π cm^3 **10** 4264 kL **11** 360 mL **12** 4 cm

Advanced Test pp. 137–138

1 27.42 cm **2** a $(8\pi + 32)$ cm b $(5\pi + 14)$ cm **3** 146 cm^2 **4** $\frac{2\pi}{3}$ $unit^2$ **5** 54 cm^2 **6** 753.982 cm^2 **7** 960 cm^3 **8** 121.8 kL
9 13 cm **10** 985.75 cm^3 **11** $28a^2(a + 1)$ cm^3 **12** 400 litres **13** Volume of 84.82 cm^3; Surface area of 122.96 cm^2

Chapter 10 Geometry

Intermediate Test pp. 160–161

1 a $x = 44$ b $x = 38$ c $x = 80$ d $x = 129$ e $x = 107$ f $x = 45$ g $x = 120$ h $x = 60$ i $x = 55$ **2** D **3** a Congruent b Similar **4** C **5** True **6** $x = 50, y = 4$ **7** $m = 47, n = 3$ **8** a 2340° b 156° c 24° **9** 8 **10** a SSS b AAS c SAS **11** $y = 59, x = 4, z = 21$ **12** a $y = 65$ (corresponding) b $x = 54$ (alternate) c $z = 61$ (angle sum of triangle) **13** a △ABX ||| △ZYK (equiangular) b $x = 5$ **14** a △ADC ≡ △CBA (SSS) b ∠ADC = ∠CBA (corresponding angles)

Advanced Test pp. 162–163

1 $x = 30$ (angle sum of △) **2** $x = 140$ (co-interior) **3** $x = 72$ **4** ∠CAB = 30° **5** $x = 50$ **6** B **7** B **8** I and II **9** $x = 24$ **10** a Congruent b True c NP d ∠DFE e True **11** a True b False c False d True **12** a $147\frac{3}{11}$° b 30° c 40 **13** a △ABC ≡ △QPC (AAS) b $a = b = \sqrt{136}, c = 6$ **14** a △ABC ||| △ADE (equiangular) b DB = 6 cm

Chapter 11 Statistics

Intermediate Test p. 183

1 C **2** a Mean number of points is 31 b Team needs to score 31 points c Mean after 6 games is 32 d Team lost by 10 points **3** a Mean = 4, Mode = 5, Range = 5 b Median = 4, Lower quartile = 3, Upper quartile = 5, IQR = 2 c See solution **4** a Highest was 24 out of 25; lowest was 3 out of 25. b i 50% ii 75% iii 25% c 7 students d Bottom half e Negatively skewed **5** a i Wildfires has a bigger range ii Diamonds has a bigger mode iii Diamonds has bigger median b Wildfires is positively skewed; Diamonds is negatively skewed **6** a See solution b y c Positive d Strong

Advanced Test pp. 184–185

1 a Normally distributed b Bi-modal c Negatively skewed d Positively skewed **2** a Mode = 10; Range = 35 b 18 c Interquartile range = 21 d See solution **3** a 30 b Negatively skewed c The class mean is $9\frac{1}{6}$. This means this class' mean is higher than the overall mean **4** a $10 b 75% c 100 books d Shop A had IQR of 25, which is higher than Shop B's IQR of 15 **5** a See solution b Positive c Strong relationship. The higher the temperature, the more people at the beach **6** Interquartile range = 2.7 **7** a Weak b 50 c $a = 45$; $b = 40$

Chapter 12 Probability

Intermediate Test pp. 200–201

1 a $\frac{1}{4}$ b $\frac{3}{4}$ **2** a See solution b i $\frac{25}{64}$ ii $\frac{15}{32}$ **3** a $\frac{1}{12}$ b $\frac{1}{3}$ **4** a $\frac{27}{100}$ b $\frac{16}{25}$ c $\frac{137}{200}$ **5** a $\frac{37}{80}$ b $\frac{43}{80}$ **6** a $\frac{1}{12}$ b $\frac{1}{4}$ c $\frac{1}{6}$ **7** a See solution b i $\frac{1}{9}$ ii $\frac{1}{9}$ iii $\frac{5}{9}$ iv $\frac{1}{3}$ v $\frac{2}{3}$ **8** a See solution b i $\frac{2}{7}$ ii $\frac{2}{7}$ iii $\frac{1}{7}$ **9** a $\frac{1}{4}$ b $\frac{1}{2}$ c $\frac{3}{5}$

Advanced Test pp. 202–203

1 a See solution b i $\frac{8}{15}$ ii $\frac{13}{15}$ **2** a 0.45 b 0.425 **3** a i 0.61 ii 0.31 b i $\frac{5}{9}$ ii $\frac{8}{27}$ **4** a See solution b $\frac{1}{5}$ c $\frac{3}{11}$ **5** a $\frac{1}{5}$ b $\frac{2}{5}$ **6** a See solution b i $\frac{1}{8}$ ii $\frac{3}{8}$ iii $\frac{7}{8}$ **7** a See solution b i $\frac{15}{77}$ ii $\frac{39}{77}$ iii $\frac{38}{77}$ iv $\frac{62}{77}$ **8** a See solution b i 0.504 ii 0.054 iii 0.398

Quick Answers to Sample Examinations

Examination Paper 1 pp. 204–208

1 a \$656.60 b \$3944 c \$11 833.50 d i \$64 ii \$22.40 e \$51.68 f \$457.23 g \$6762.23 **2** a i $-1 - x$ ii $-4x - 14$ b i $4a(3b - 1)$ ii $x^6(x^6 + 1)$ c i $\frac{11x}{12}$ ii $\frac{4x}{by}$ d $2x$ e $6x^2 - 5x - 4$ **3** a $x \leq 1$ 

b $y = 5$ c -84 d $y = \pm 5$ e $x = 2, 1$ f \$3.80 **4** a i $\sqrt{50}$ units ii $(\frac{1}{2}, 1\frac{1}{2})$ iii -1 b See worked solution c $p = 13$ d $y = -2x + 4$ e D **5** a 108 km/h b i 20 L/min ii $V = 20t$ iii See solution iv 1200 L v 300 minutes, or 5 hours c i $y = kx$ ii $y = 4x$ iii $y = 60$ d $W = 18.5n$ **6** a $\sqrt{44}$ b i 22.91 ii 30.05 c 42° d 16 m e 23.0 km **7** a 51.42 cm b 96.37 cm² c 5183.63 cm² d 2714.34 cm³ e 3016 L **8** a i 72 ii 135 b 36° c ASA d Equiangular (2 ∠s equal) e i $6\frac{2}{3}$ ii 6 f 30° **9** a i α 77 β 76 ii See worked solution b $\frac{2}{5}$ c i $\frac{1}{36}$ ii 0 iii $\frac{1}{6}$ d 1

Examination Paper 2 pp. 209–216

Part A

1 C **2** D **3** C **4** B **5** A **6** B **7** A **8** B **9** D **10** C **11** A **12** B **13** A **14** B **15** D **16** C **17** B **18** C **19** A **20** D **21** D **22** C **23** C **24** B **25** A **26** B **27** B **28** A **29** B **30** D **31** C **32** B **33** D **34** B **35** A **36** D **37** B **38** D **39** D **40** B **41** A **42** C **43** C **44** B **45** D

Part B

46 a i See solution ii A $\frac{3}{10}$ B $\frac{3}{5}$ b \$785.10 c 164 cm² d 51° e 260° f 4524 cm³ g \$441 h \$58 327 i \$84.60 j \$1836 k 88 cm² l 324 km/h m \$500 **47** a \$25.10 b 6 c \$26.36 **48** a \$178.75 b 8 h **49** a (1, 1) b 2 c $y = 2x - 1$ **50** a $\frac{1}{6}$ b $\frac{11}{25}$ c $\frac{1}{3}$ **51** See worked solutions

Worked Solutions to Practise Practise

Chapter 1 . . . . pp. 9–11

Financial Mathematics

1 Pay = 77 200 ÷ 52
= 1484.62 (2 dec. pl.)
∴ Her weekly pay is $1484.62.

2 Wage = 16.20 × 5 × 6.5
= 526.50
∴ Susan paid $526.50.

3 Earnings = 52 390 ÷ 26
= 2015
∴ Julie is paid $2015 per fortnight.

4 Rate = 739.10 ÷ 38
= 19.45
∴ Hourly rate of $19.45.

5 Pay = 26.20 × 5 × 1.5
= 196.5
∴ Peter would earn $196.50.

6 Earnings = $14.40 × 10 + $14.40 × 1.5 × 4
= $230.40
∴ Joanne earned $230.40.

7 Earnings = $17.56 × 35 + $17.56 × 1.5 × 6 + $17.56 × 2 × 3
= $878
∴ John earned $878.

8 Earnings = 36 + 15
= 51
∴ Jerry earns $51.

9 a Net salary = 43 500 − 8600
= 34 900
∴ Paul's annual net salary is $34 900.

b Fortnightly amount
= 34 900 ÷ 26
= 1342.31 (2 dec. pl.)
∴ Paul's fortnightly pay is $1342.31.

10 a Pay = 9.20 × 8
= 73.6
∴ The boy is paid $73.60.

b Pay = 9.20 × 5 + 9.20 × 1.5 × 2
= 73.6
∴ He would earn $73.60.

c Hours = 138 ÷ 9.2
= 15
∴ He would need to work 15 hours.

11 Pay = 18.75 × 3 + 18.75 × 1.5 × 4
= 168.75
∴ Amanda earns $168.75.

12 Earnings = 18 × 2 + 18 × 2 × 2
= 108
∴ Loukia earns $108.

13 Net pay
= 18.93 × 35 − 114.32
= 548.23
Net pay
= 16.15 × 30 + 16.15 × 1.5 × 6 − 125.10
= 504.75
Net pay
= 17.10 × 32 + 17.10 × 1.5 × 7 − 141.50
= 585.25
∴ $548.23, $504.75, $585.25.

14 Wage = 6 × 5 × 18.55 + 4 ×1.5 × 18.55
= 667.8
∴ Eleri would be paid $667.80.

15 Hours = 36 + 6 × 1.5
= 45
∴ Rate = 774 ÷ 45
= 17.2
∴ John's hourly rate is $17.20.

16 Wage = 345 + 0.06 × 4850
= 636
∴ The sales person earns $636.

17 a Commission
= 2900 + 0.03 × (528 500 − 90 000)
= 16 055
∴ Commission of $16 055.

b Amount
= 528 500 − 16 055 − (0.01 × 528 500)
= 507 160
∴ The owner receives $507 160.

18 a Holiday loading = 1040 × 4 × 0.175
= 728
∴ Holiday loading of $728.

b Holiday pay = 1040 × 4 + 728
= 4888
∴ Don's holiday pay is $4888.

19 Gross pay = 17.62 × 35 + 17.62 × 1.5 × 6 + 17.62 × 2 × 5
= 951.48
Dedn = 175.64 + 21.08 + 15 + 12.35
= 224.07
Net pay = 951.48 − 224.07
= 727.41
∴ Mr Brown's net pay is $727.41 per week.

20 a Salary = 1683.84 × 52
= 87 559.68
∴ His salary is $87 559.68.

b Net income = 87 559.68 − 24 385
= 63 174.68
∴ His net income is $63 174.68.

c Fortnightly pay = 63 174.68 ÷ 26
= 2429.80 (2 dec. pl.)
∴ His fortnightly pay is $2429.80.

21 Tax = 2850 + 0.30 × (42 580 − 25 000)
= 8124
∴ Tax of $8124.

22 a Nil tax

b 17 547 + 0.37 × 1235
= 18 003.95
∴ Tax of $18 003.95.

c 3572 + 0.325 × 4583
= 5061.475
= 5061.48 (2 dec. pl.)
∴ Tax of $5061.48.

d 54 547 + 0.45 × 13 540
= 60 640
∴ Tax of $60 640.

23 a 41 520 − 1870 = 39 650
∴ Taxable income of $39 650.

b 3572 + 0.325 × 2650
= 4433.25
∴ Tax of $4433.25.

c 190 × 26 = 4940
Refund = 4940 − 4433.25
= 506.75
∴ Tax refund of $506.75.

24 a

Income		Expenses	
S'ship	$152.00	Rent	$45.00
Part-time job	$76.00	Food	$58.00
		Travelling expenses	$12.00
		Clothing	$17.00
		Entertainment	$43.50
		Balance	$52.50
Total	$228.00	Total	$228.00

b Since the student saves the balance of his weekly budget, he saves $52.50 per week.

c No. of weeks = 472.5 ÷ 52.5
= 9
∴ Student will save for 9 weeks.

25 Rates = 430 000 × 0.642 cents
= 276 060 cents
= $2760.60

26 Total amount paid = $2700 × 144
= $388 800

27 a I = 1800 × 0.07 × 5
= 630 ∴ $630

b I = 450 × 0.085 × 3
= 114.75 ∴ $114.75

c I = 1200 × 0.09 × 1.5
= 162 ∴ $162

28 Interest earned = 5200 × 0.12 × 2
= 1248 ∴ $1248

29 Interest = 1200 × 0.11 × 2
= 264 ∴ $264

30 Amount paid for the sofa
= 49 × 24
= 1176 ∴ $1176

31 a Interest charged = \$7200
Amount to be repaid
= Amount borrowed + Interest
= 20 000 + 7200
= 27 200 ∴ \$27 200

b Monthly repayment
= 27 200 ÷ 48
= 566.67 (to 2 dec. pl.) ∴ \$566.67

32 a Total amount he will pay
= 1400 × 12 × 21
= 352 800 ∴ \$352 800

b Interest he will pay over 21 years
= 352 800 − 280 000
= 72 800 ∴ \$72 800

33 a Cost on terms
= 150 + 11.50 × 208
= 2542 ∴ \$2542

b Amount of interest charged
= 2542 − 1495
= 1047 ∴ \$1047

34 A Total cost
= 250 × 12
= 3000 ∴ \$3000

B Total cost
= 1500 + 100 × 12
= 2700 ∴ \$2700

C Total cost
= 1000 + 150 × 12
= 2800 ∴ \$2800

D Total cost
= 0.2 × 2400 + 210 × 12
= 3000 ∴ \$3000

∴ **B** is the cheapest buy.

35 Cash price = \$240
Deposit = \$60
Balance = Cash price − Deposit
= \$240 − \$60
= \$180

a $I = 0.12 \times 180$
= 21.60 ∴ \$21.60

b Total amount to be repaid
= Cash price + Interest
= 240 + 21.60
= 261.60 ∴ \$261.60

c Deposit = \$60
Amount owed = \$261.60 − \$60
= \$201.60
Instalment = 201.60 ÷ 12
= 16.80 ∴ \$16.80

36 Cash price = \$670
Deposit = 20% of \$670
= \$134
Balance = \$670 − \$134
= \$536
Interest charged = \$96.48
Total amount to be repaid
= Cash price + Interest
= \$670 + \$96.48
= \$766.48
Amount still owed = \$766.48 − \$134
= \$632.48
Monthly repayment
= 632.48 ÷ 18
= 35.14 ∴ \$35.14

37 Price = 0.8 × 0.9 × 1450
= 1044
∴ The price paid is \$1044.

38 Let product cost \$100
∴ A: 0.85 × 0.75 × 100 = 63.75
B: 0.65 × 100 = 65
∴ A is a better deal as the product will cost \$63.75 compared to \$65 for deal B.

39 Cost = 0.85 × 0.9 × 11.90 × 7.5
= 68.27625
= 68.28 (to 2 dec. pl.)
Steak costs \$68.28.

40 a Interest = $6400 \times 1.06^3 - 6400$
= 1222.5024
= 1222.50 (to 2 dec. pl.)
∴ \$1222.50

b Interest = $4900 \times 1.085^5 - 4900$
= 2467.917 782
= 2467.92 (to 2 dec. pl.)
∴ \$2467.92

c Monthly:
6% p.a. = 0.5% per month
2 years = 24 months
∴ Interest = $2600 \times 1.005^{24} - 2600$
= 330.6154181
= 330.62 (to 2 dec. pl.)
∴ \$330.62

d Six monthly:
7% p.a. = 3.5% per 6 months
4 years = 48 months
∴ Interest
= $14\,000 \times 1.035^8 - 14\,000$
= 4435.326 517
= 4435.33 (to 2 dec. pl.)
∴ \$4435.33

41 Quarterly:
8% p.a. = 2% per quarter
10 years = 40 quarters
∴ Amount = 9400×1.02^{40}
= 20 755.572 84
= 20 755.57 (to 2 dec. pl.)
∴ \$20 755.57 after 10 years.

42 Pop. = $11\,240 \times 1.03^{25}$
= 23 534.063 93
= 23 534 (to nearest whole number)
∴ The population is 23 534.

43 Value = 1600×0.85^3
= 982.6
∴ The computer is valued at \$982.60.

44 Time = 5 years
∴ Value = $16\,300 \times 0.82^5$
= 6043.059 444
= 6043.06 (to 2 dec. pl.)
∴ The motorbike is valued at \$6043.06.

45 Number = 340×0.95^{10}
= 203.570 5593
= 204 (to nearest whole number)
∴ The projected number is 204.

Chapter 2 . . . . pp. 25–26

Algebraic Techniques

1 a $x + y$

b $p - q$

c $7y$

d $\frac{a}{b}$

2 a $100p$ cents

b $7w$ days

c $1000k$ mL

d $10x$ mm

e $\frac{m}{1000}$ kg

f $\frac{y}{12}$ years

3 a $\frac{x + y + z}{3}$

b $\frac{x + x + 1}{2} = \frac{2x + 1}{2}$

c $\frac{a + b + c + d}{4}$

4 $x + 1, x + 2$

5 $x + 2, x + 4, x + 6$

6 $x + 1, x + 3, x + 5$

7 $4(x + 3)$ cm^2

8 pq cents

9 $(m - n)$ mL

10 a $ab = -2 \times 3$
$= -6$

b $a + c = -2 + (-5)$
$= -7$

c $bc - ac = 3 \times (-5) - (-2) \times (-5)$
$= -15 - 10$
$= -25$

d $a^2 - c^2 = (-2)^2 - (-5)^2$
$= 4 - 25$
$= -21$

e $2c - b = 2 \times (-5) - 3$
$= -10 - 3$
$= -13$

f $a^2 - 2b = (-2)^2 - 2 \times 3$
$= 4 - 6$
$= -2$

g $(b - a) \div c = (3 - (-2)) \div (-5)$
$= 5 \div (-5)$
$= -1$

h $2c^2 = 2 \times (-5)^2$
$= 2 \times 25$
$= 50$

i $\frac{ab+1}{c} = \frac{(-2)\times 3 + 1}{(-5)}$
$= \frac{-5}{-5}$
$= 1$

j $(c-a)^2 = (-5-(-2))^2$
$= (-3^2)$
$= 9$

11 a $6x$
b $6xy$
c $8a + 2b$
d $19a$
e $6x^2 + 4x$
f $19a$
g 0
h $4a$
i $6y$
j ab

12 a $3y$
b $2q$
c a
d $7xy$
e $10y^2$
f $-2a$
g $-6y$
h p
i $-8a^2$
j $-6m$

13 a $7a + 9b$
b $4y + 2z$
c $13ab - 5a$
d $9 + 3x$
e $ab + a + b$
f $10cd - 3d$
g $-6p - q$
h $-2m + 5n$
i $-g - 8k$
j $3a - 6b$

14 a $12a$
b $30y$
c $6sr$
d $4m^2$
e $-20a$
f $-18y$
g $-10p^2$
h k^3
i $abcd$
j $9b^2$

15 a $4w$
b $7y$
c 2
d $-2y$
e $3g$
f $2ab$
g 6
h $2x$
i -5
j $8w$

16 a $4y + 6y = 10y$
b $6ab - 12a$
c $12mn - 8mn = 4mn$
d $\frac{10t - 4t}{3t} = \frac{6t}{3t}$
$= 2$

17 a $\frac{x}{2} + \frac{x}{3} = \frac{3x + 2x}{6}$
$= \frac{5x}{6}$

b $\frac{2y}{3} + \frac{y}{5} = \frac{10y + 3y}{15}$
$= \frac{13y}{15}$

c $\frac{5a}{3} - \frac{a}{4} = \frac{20a - 3a}{12}$
$= \frac{17a}{12}$

d $\frac{6y}{5} - \frac{y}{2} = \frac{12y - 5y}{10}$
$= \frac{7y}{10}$

e $\frac{a}{b} + \frac{c}{d} = \frac{ad + bc}{bd}$

f $\frac{2}{m} + \frac{3}{m} = \frac{2+3}{m}$
$= \frac{5}{m}$

g $\frac{3}{x} + \frac{5}{2x} = \frac{6+5}{2x}$
$= \frac{11}{2x}$

h $\frac{4y}{3} - \frac{2y}{5} - \frac{y}{2} = \frac{40y - 12y - 15y}{30}$
$= \frac{13y}{30}$

18 a $\frac{x}{2} \times \frac{3}{x} = \frac{3}{2}$
$= 1\frac{1}{2}$

b $\frac{4y}{3} \times \frac{3}{2y} = 2$

c $\frac{5a}{2} \times \frac{2}{3a} = \frac{5}{3}$
$= 1\frac{2}{3}$

d $\frac{6xy}{5} \times \frac{15}{8x} = \frac{9y}{4}$

e $\frac{10ab}{7b} \times \frac{21b}{5a} = 6b$

19 a $\frac{x}{2} \div \frac{x}{4} = \frac{x}{2} \times \frac{4}{x}$
$= 2$

b $\frac{2a}{3} \div \frac{4a}{9} = \frac{2a}{3} \times \frac{9}{4a}$
$= \frac{3}{2}$
$= 1\frac{1}{2}$

c $\frac{4y}{5} \div \frac{3y}{10} = \frac{4y}{5} \times \frac{10}{3y}$
$= \frac{8}{3}$
$= 2\frac{2}{3}$

d $\frac{6p}{5} \div \frac{2p}{15} = \frac{6p}{5} \times \frac{15}{2p}$
$= 9$

e $\frac{7ab}{5xy} \div \frac{21b}{15y} = \frac{7ab}{5xy} \times \frac{15y}{21b}$
$= \frac{a}{x}$

20 a $2(a + b) = 2a + 2b$
b $3(y - 6) = 3y - 18$
c $5(2a - 1) = 10a - 5$
d $4(y - 3) = 4y - 12$
e $2(3p - 5q) = 6p - 10q$
f $7(4a + 7b) = 28a + 49b$
g $-2(m - 1) = -2m + 2$
h $-3(p - 4) = -3p + 12$
i $-6(2r + 5) = -12r - 30$
j $-(4g - 1) = -4g + 1$
k $-(2x + 3) = -2x - 3$
l $-(6 - 4y) = -6 + 4y$

m $-(x + 7) = -x - 7$

n $x(x + 2) = x^2 + 2x$

o $3a(2a - 1) = 6a^2 - 3a$

p $4y(2 - 3y) = 8y - 12y^2$

q $-3a(a - 7) = -3a^2 + 21a$

21 a $2(a + 3) + 4(a + 1)$
$= 2a + 6 + 4a + 4$
$= 6a + 10$

b $5(y + 2) + 2(3y + 1)$
$= 5y + 10 + 6y + 2$
$= 11y + 12$

c $3(4m - 1) + 2(6m - 1)$
$= 12m - 3 + 12m - 2$
$= 24m - 5$

d $7(5a - 7) + 3(2a - 6)$
$= 35a - 49 + 6a - 18$
$= 41a - 67$

e $4(2y - 1) - 5(y + 1)$
$= 8y - 4 - 5y - 5$
$= 3y - 9$

f $6(4 - 2y) - 3(1 - y)$
$= 24 - 12y - 3 + 3y$
$= 21 - 9y$

g $5(2a - 1) - (a + 7)$
$= 10a - 5 - a - 7$
$= 9a - 12$

h $6(1 - y) - (3y - 1)$
$= 6 - 6y - 3y + 1$
$= 7 - 9y$

i $4(2p - 1) - p(p + 1)$
$= 8p - 4 - p^2 - p$
$= 7p - 4 - p^2$

j $x(2x - 1) - x(3 - x)$
$= 2x^2 - x - 3x + x^2$
$= 3x^2 - 4x$

22 a $3 + 2(a - 1)$
$= 3 + 2a - 2$
$= 1 + 2a$

b $4 + 6(2y - 7)$
$= 4 + 12y - 42$
$= 12y - 38$

c $2 - 4(m - 1)$
$= 2 - 4m + 4$
$= 6 - 4m$

d $6 - (4r - 7)$
$= 6 - 4r + 7$
$= 13 - 4r$

e $4b - (3 - 5b)$
$= 4b - 3 + 5b$
$= 9b - 3$

23 a $2a + 1 + 3a + 4 = 5a + 5$

b $5x - 7 + 3x + 11 = 8x + 4$

24 a $2m + 1 - (m + 6) = 2m + 1 - m - 6$
$= m - 5$

b $5y - 7 - (3y - 4) = 5y - 7 - 3y + 4$
$= 2y - 3$

c $7q - (4 - 2q) = 7q - 4 + 2q$
$= 9q - 4$

d $5p - 1 - (3 - 4p) = 5p - 1 - 3 + 4p$
$= 9p - 4$

25 a $(x + 4)(x + 2) = x^2 + 2x + 4x + 8$
$= x^2 + 6x + 8$

b $(y + 5)(y - 1) = y^2 - y + 5y - 5$
$= y^2 + 4y - 5$

c $(c + 3)(c + 1) = c^2 + c + 3c + 3$
$= c^2 + 4c + 3$

d $(2y - 1)(y + 3) = 2y^2 + 6y - y - 3$
$= 2y^2 + 5y - 3$

e $(3n + 6)(n - 1) = 3n^2 - 3n + 6n - 6$
$= 3n^2 + 3n - 6$

f $(q - 2)(3q + 6) = 3q^2 + 6q - 6q - 12$
$= 3q^2 - 12$

g $(2t - 1)(3t + 5) = 6t^2 + 10t - 3t - 5$
$= 6t^2 + 7t - 5$

h $(4x - 3)(2x - 7) = 8x^2 - 28x - 6x + 21$
$= 8x^2 - 34x + 21$

i $(2w - 5)(3w + 2)$
$= 6w^2 + 4w - 15w - 10$
$= 6w^2 - 11w - 10$

j $(4y - 7)(2y - 3) = 8y^2 - 12y - 14y + 21$
$= 8y^2 - 26y + 21$

26 a $(x + 3)^2 = x^2 + 6x + 9$

b $(b - 1)^2 = b^2 - 2b + 1$

c $(2a - 3)^2 = 4a^2 - 12a + 9$

d $(4y + 5)^2 = 16y^2 + 40y + 25$

e $(4 - n)^2 = 16 - 8n + n^2$

27 a $(x + 2)(x - 2) = x^2 - 4$

b $(y + 3)(y - 3) = y^2 - 9$

c $(2a - 1)(2a + 1) = 4a^2 - 1$

d $(5 - 2p)(5 + 2p) = 25 - 4p^2$

e $(3a + 5b)(3a - 5b) = 9a^2 - 25b^2$

28 a $4y - 8 = 4(y - 2)$

b $3a - 12b = 3(a - 4b)$

c $6m - 10 = 2(3m - 5)$

d $12p - 8 = 4(3p - 2)$

e $4ab - 6a = 2a(2b - 3)$

f $5xy - 15x = 5x(y - 3)$

g $6pq - 4q = 2q(3p - 2)$

h $a^2 - 3a - ab = a(a - 3 - b)$

i $y^2 - 6y = y(y - 6)$

j $2a^2 - 12ab = 2a(a - 6b)$

29 a $m(x + y) + n(x + y) = (x + y)(m + n)$

b $p(m + n) + q(m + n) = (m + n)(p + q)$

c $ab + ac + 2b + 2c = a(b + c) + 2(b + c)$
$= (b + c)(a + 2)$

d $3p + 3q + ap + aq = 3(p + q) + a(p + q)$
$= (p + q)(3 + a)$

e $ca - cb - da + db = c(a - b) - d(a - b)$
$= (a - b)(c - d)$

f $xy - y^2 - 7x + 7y = y(x - y) - 7(x - y)$
$= (x - y)(y - 7)$

g $6a - ct + ac - 6t = 6a + ac - ct - 6t$
$= a(6 + c) - t(c + 6)$
$= (c + 6)(a - t)$

h $g^3 - 3g^2 + 2g - 6$
$= g^2(g - 3) + 2(g - 3)$
$= (g - 3)(g^2 + 2)$

30 a $p^2 - 16 = (p + 4)(p - 4)$

b $y^2 - 36 = (y + 6)(y - 6)$

c $m^2 - 121 = (m + 11)(m - 11)$

d $4a^2 - b^2 = (2a + b)(2a - b)$

e $9y^2 - 16 = (3y + 4)(3y - 4)$

f $25x^2 - 64y^2 = (5x + 8y)(5x - 8y)$

g $81 - 121m^2 = (9 + 11m)(9 - 11m)$

h $2p^2 - 2 = 2(p^2 - 1)$
$= 2(p + 1)(p - 1)$

i $3z^2 - 27 = 3(z^2 - 9)$
$= 3(z + 3)(z - 3)$

j $10h^2 - 490 = 10(h^2 - 49)$
$= 10(h + 7)(h - 7)$

31 a $x^2 + 4x + 3 = (x + 3)(x + 1)$

b $y^2 + 3y + 2 = (y + 2)(y + 1)$

c $a^2 - a - 20 = (a - 5)(a + 4)$

d $y^2 - 3y - 18 = (y - 6)(y + 3)$

e $m^2 + 2m - 15 = (m + 5)(m - 3)$

f $k^2 + 5k - 6 = (k + 6)(k - 1)$

g $x^2 - 3x + 2 = (x - 2)(x - 1)$

h $y^2 - 7y + 12 = (y - 3)(y - 4)$

i $t^2 - 6t + 9 = (t - 3)(t - 3)$
$= (t - 3)^2$

j $y^2 - 8y + 16 = (y - 4)(y - 4)$
$= (y - 4)^2$

Chapter 3 ...... p. 34

Indices

1 a

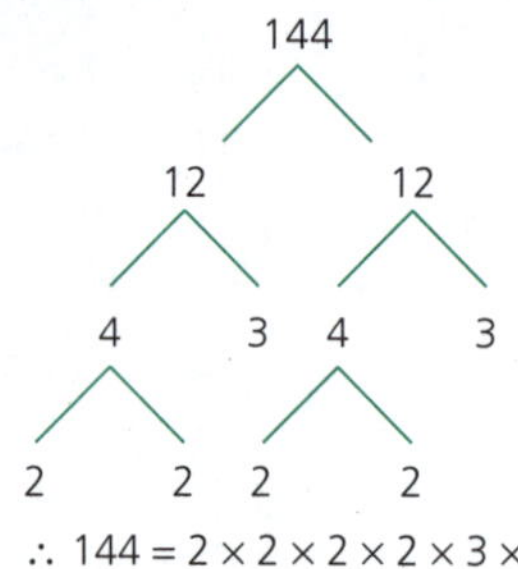

$\therefore 144 = 2 \times 2 \times 2 \times 2 \times 3 \times 3$
$= 2^4 \times 3^2$

b

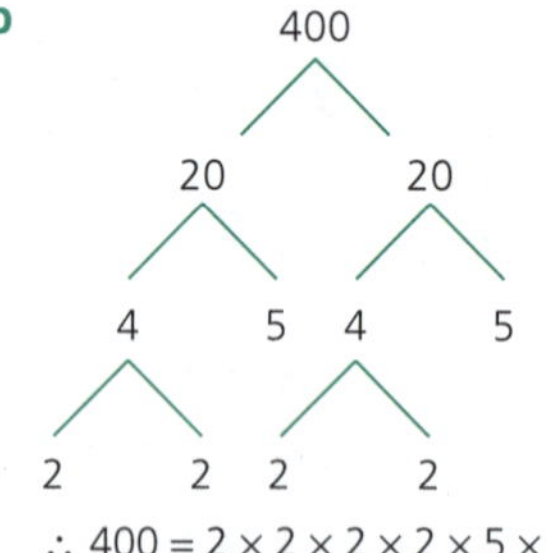

$\therefore 400 = 2 \times 2 \times 2 \times 2 \times 5 \times 5$
$= 2^4 \times 5^2$

c

676
4 169
2 2 13 13

$\therefore 676 = 13^2 \times 2^2$

2 a $p^5 \times p^3 = p^8$

b $y^6 \times y = y^7$

c $a^4 \times a^3 \times a^2 = a^9$

d $c^4 \times c^3 = c^7$

e $x^6 \times x \times x^3 = x^{10}$

f $y^9 \times y^5 = y^{14}$

3 a $4y^3 \times 2y^4 = 8y^7$

b $7a^4 \times 2a = 14a^5$

c $-2p^4 \times -3p^2 = 6p^6$

d $-5m^3 \times 10m^4 = -50m^7$

e $6y^3 \times (-2y^4) = -12y^7$

f $10a^2b^3 \times 2a^4b = 20a^6b^4$

4 a $w^7 \div w^3 = w^4$

b $x^7 \div x^5 = x^2$

c $p^9 \div p = p^8$

d $\dfrac{y^6}{y^3} = y^3$

e $\dfrac{a^8}{a^5} = a^3$

f $\dfrac{m^4n^3}{m^2n} = m^2n^2$

5 a $4y^3 \div 2y^2 = 2y$

b $15c^4 \div 5c^2 = 3c^2$

c $42p^9 \div 6p^5 = 7p^4$

d $10a^4b \div 5a^2b = 2a^2$

e $-5xy^3 \div 5y = -xy^2$

f $-10c^2d^3 \div -10cd = cd^2$

g $\dfrac{10y^4z}{5y^2} = 2y^2z$

h $\dfrac{48a^3b^2}{8ab} = 6a^2b$

i $\dfrac{24p^7q^5}{18p^4q^3} = \dfrac{4p^3q^2}{3}$

6 a $p^4q^3 \times p^2q \div pq = p^5q^3$

b $10x^7y^5 \div 2x^3y \times x^4y = 5x^8y^5$

7 a $\dfrac{4a \times 3a^2}{2a} = \dfrac{12a^3}{2a}$
$= 6a^2$

b $\dfrac{12p^3 \times 2p^4}{8p^7} = \dfrac{24p^7}{8p^7}$
$= 3$

c $\dfrac{10m^7n^3 \times 4m^2}{20m^3n^2} = \dfrac{40m^9n^3}{20m^3n^2}$
$= 2m^6n$

8 a $(y^3)^2 = y^6$

b $(a^4)^3 = a^{12}$

c $(m^6)^5 = m^{30}$

d $(3a^4)^2 = 9a^8$

e $(2y^5)^3 = 8y^{15}$

f $(7a^2b^3)^2 = 49a^4b^6$

g $(-3m^4)^2 = 9m^8$

h $(-5p^6)^2 = 25p^{12}$

i $(-a^4b^3)^3 = -a^{12}b^9$

9 a $y^3 \div y^5 = y^{-2}$

b $a^6 \div a^{11} = a^{-5}$

c $y^5 \div y^6 = y^{-1}$

d $x^{-3} \div x^{-3} = x^0 = 1$

e $b^5 \div b^{-3} = b^8$

f $g^{-4} \div g^{-2} = g^{-2}$

g $k^2 \div k^{-3} \div k^3 = k^2$

h $(p^3)^{-2} = p^{-6}$

i $(m^{-2})^3 = m^{-6}$

j $(2p^{-5})^2 = 4p^{-10}$

k $10a^4b \div 5a^3b^4 = 2ab^{-3}$

l $15p^4 \div -3p^6 = -5p^{-2}$

10 a $3^0 = 1$

b $(3x)^0 = 1$

c $(10a^3 \times 2a^3)^0 = 1$

11 a $p^0 = 2^0$
$= 1$

b $2p^0 = 2 \times 1$
$= 2$

c $(2p)^0 = (2 \times 2)^0$
$= 1$

12 a 4.28×10^3

b 3.6×10^5

c 5×10^2

13 a 6.8×10^{-2}

b 3.5×10^{-4}

c 8.6×10^{-6}

14 a 2600

b 15 100

c 46 300 000

15 a 0.058

b 0.007 06

c 0.000 084

16 a $\sqrt{0.025} = 0.158\,113\,883\ldots$
$= 1.58 \times 10^{-1}$ (3 sig. figs)

b $6.32^5 = 10\,082.898\,41\ldots$
$= 1.01 \times 10^4$ (3 sig. figs)

c $6985 \times 321 = 2\,242\,185$
$= 2.24 \times 10^6$ (3 sig. figs)

d $\dfrac{0.256}{965} = 0.000\,265\,284\ldots$
$= 2.65 \times 10^{-4}$ (3 sig. figs)

e $\dfrac{\sqrt{521}}{0.075} = 304.338\,992\,3\ldots$
$= 3.04 \times 10^2$ (3 sig. figs)

f $\dfrac{\sqrt{3.016 \times 99.3}}{56.98^2}$
$= 0.005\,330\,223\ldots$
$= 5.33 \times 10^{-3}$ (3 sig. figs)

17 Distance $= 9.4605 \times 10^{15} \times 11\,000$
$= 1.040\,655 \times 10^{20}$
$= 1.04 \times 10^{20}$ (3 sig. figs)

18 Number $= 65\,000\,000\,000 \times 60 \times 60 \times 24$
$= 6.5 \times 10^{10} \times 60 \times 60 \times 24$
$= 5.62 \times 10^{15}$

Chapter 4 . . . . pp. 47–48

Equations

1 a
$$4x - 2 = 3x + 6$$
$$4x - 3x = 6 + 2$$
$$x = 8$$

b
$$5a - 4 = 3a - 12$$
$$5a - 3a = -12 + 4$$
$$2a = -8$$
$$\frac{\cancel{2}a}{\cancel{2}} = \frac{-8}{2}$$
$$a = -4$$

c
$$4y = 2y + 11$$
$$4y - 2y = 11$$
$$2y = 11$$
$$\frac{\cancel{2}y}{\cancel{2}} = \frac{11}{2}$$
$$y = 5\frac{1}{2}$$

d
$$5(a + 1) = 3(a + 6)$$
$$5a + 5 = 3a + 18$$
$$5a - 3a = 18 - 5$$
$$2a = 13$$
$$\frac{\cancel{2}a}{\cancel{2}} = \frac{13}{2}$$
$$a = 6\frac{1}{2}$$

e
$$4 + 2(x + 1) = 12$$
$$4 + 2x + 2 = 12$$
$$2x + 6 = 12$$
$$2x = 12 - 6$$
$$2x = 6$$
$$\frac{\cancel{2}x}{\cancel{2}} = \frac{6}{2}$$
$$x = 3$$

f
$$7(2y - 3) = 4(2y + 12)$$
$$14y - 21 = 8y + 48$$
$$14y - 8y = 48 + 21$$
$$6y = 69$$
$$\frac{\cancel{6}y}{\cancel{6}} = \frac{69}{6}$$
$$y = 11\frac{1}{2}$$

g
$$4x - 2 = 7x + 10$$
$$4x - 7x = 10 + 2$$
$$-3x = 12$$
$$\frac{-\cancel{3}x}{-\cancel{3}} = \frac{12}{-3}$$
$$x = -4$$

h
$$3a - 4 = 16a$$
$$3a - 16a = 4$$
$$-13a = 4$$
$$\frac{-\cancel{13}a}{-\cancel{13}} = \frac{4}{-13}$$
$$a = -\frac{4}{13}$$

i
$$\frac{5y - 2}{3} = 6$$
$$\cancel{3}\left[\frac{5y - 2}{\cancel{3}}\right] = 3 \times 6$$
$$5y - 2 = 18$$
$$5y = 18 + 2$$
$$5y = 20$$
$$\frac{\cancel{5}y}{\cancel{5}} = \frac{20}{5}$$
$$y = 4$$

j
$$\frac{4y - 2}{3} = \frac{y + 6}{2}$$
$$\therefore 2(4y - 2) = 3(y + 6)$$
[by cross-multiplying]
$$\therefore 8y - 4 = 3y + 18$$
$$8y - 3y = 18 + 4$$
$$5y = 22$$
$$\frac{\cancel{5}y}{\cancel{5}} = \frac{22}{5}$$
$$y = 4\frac{2}{5}$$

k
$$\frac{5x}{2} - 3x = 4$$
$$\cancel{2}\left[\frac{5x}{\cancel{2}}\right] - 2(3x) = 2(4)$$
$$5x - 6x = 8$$
$$-x = 8$$
$$\frac{\cancel{-}x}{\cancel{-}1} = \frac{8}{-1}$$
$$x = -8$$

l
$$\frac{2a}{3} - 4 = \frac{6a}{5}$$
$${}^{5}\cancel{15}\left(\frac{2a}{\cancel{3}}\right) - 15(4) = {}^{3}\cancel{15}\left(\frac{6a}{\cancel{5}_1}\right)$$
$$10a - 60 = 18a$$
$$10a - 18a = 60$$
$$-8a = 60$$
$$\frac{-\cancel{8}a}{\cancel{8}} = \frac{60}{-8}$$
$$a = -7\frac{1}{2}$$

m
$$\frac{4 - 3y}{-2} = 10$$
$$-\cancel{2}\left[\frac{4 - 3y}{-\cancel{2}}\right] = -2(10)$$
$$4 - 3y = -20$$
$$-3y = -20 - 4$$
$$-3y = -24$$
$$\frac{-3y}{-3} = \frac{-24}{-3}$$
$$y = 8$$

n
$$5 - \frac{4x}{3} = 3x$$
$$3[5] - \cancel{3}\left[\frac{4x}{\cancel{3}}\right] = 3[3x]$$
$$15 - 4x = 9x$$
$$-4x - 9x = -15$$
$$-13x = -15$$
$$\frac{-\cancel{13}x}{-\cancel{13}} = \frac{-15}{-13}$$
$$x = 1\frac{2}{13}$$

o
$$\frac{4}{x} = 12$$
$$\therefore 4 = 12x \text{ [by cross-mult.]}$$
$$\therefore 12x = 4$$
$$\frac{\cancel{12}x}{\cancel{12}} = \frac{4}{12}$$
$$x = \frac{1}{3}$$

p
$$\frac{3}{y} - 2 = 4$$
$$\frac{3}{y} = 4 + 2$$
$$\therefore 3 = 6y \text{ [cross-mult.]}$$
$$\therefore 6y = 3$$
$$\frac{\cancel{6}y}{\cancel{6}} = \frac{3}{6}$$
$$y = \frac{1}{2}$$

2 a
$$4x - 2 < 12$$
$$4x < 12 + 2$$
$$4x < 14$$
$$\frac{\cancel{4}x}{\cancel{4}} < \frac{14}{4}$$
$$x < 3\frac{1}{2}$$

b
$$3a - 4 \geq 2a + 1$$
$$3a - 2a \geq 1 + 4$$
$$a \geq 5$$

c
$$\frac{5y - 4}{2} > 4$$
$$\cancel{2}\left[\frac{5y - 4}{\cancel{2}}\right] > 2(4)$$
$$5y - 4 > 8$$
$$5y > 8 + 4$$
$$5y > 12$$
$$\frac{\cancel{5}y}{\cancel{5}} > \frac{12}{5}$$
$$y > 2\frac{2}{5}$$

d
$$5 - 4y < 3$$
$$-4y < 3 - 5$$
$$-4y < -2$$
$$\frac{-\cancel{4}y}{-\cancel{4}} > \frac{-2}{-4}$$
$$y > \frac{1}{2}$$

Note: the symbol change—division by a negative.

e $\frac{7y-2}{4} < -3$

$4\left[\frac{7y-2}{4}\right] < 4(-3)$

$7y - 2 < -12$

$7y < -12 + 2$

$7y < -10$

$\frac{7y}{7} < \frac{-10}{7}$

$y < -1\frac{3}{7}$

f $5y - 2 \geq 7y + 12$

$5y - 7y \geq 12 + 2$

$-2y \geq 14$

$\frac{-2y}{-2} \leq \frac{14}{-2}$ (Symbol change)

$y \leq -7$

g $6y - 4 \leq 7y + 2$

$6y - 7y \leq 2 + 4$

$-y \leq 6$

$\frac{-y}{-1} \geq \frac{6}{-1}$ (Symbol change)

$y \geq -6$

h $\frac{4y-2}{-3} > 4$

$-3\left[\frac{4y-2}{-3}\right] < -3(4)$ (Symbol change)

$4y - 2 < -12$

$4y < -12 + 2$

$4y < -10$

$\frac{4y}{4} < \frac{-10}{4}$

$y < -2\frac{1}{2}$

3 a $x \leq -2$

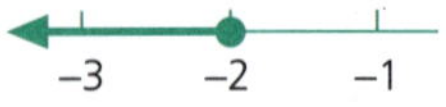

[number line: −3, −2, −1; closed dot at −2, arrow left]

b $x > 3$

[number line: 2, 3, 4; open circle at 3, arrow right]

c $x < \frac{2}{3}$

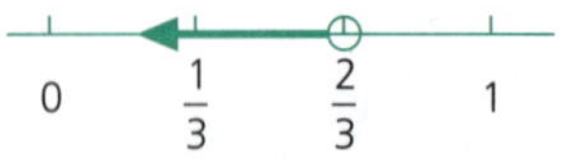

[number line: 0, $\frac{1}{3}$, $\frac{2}{3}$, 1; open circle at $\frac{2}{3}$, arrow left]

d $-4 < x \leq 2$

[number line: open circle at −4, closed dot at 2]

e $x < 2, x \geq 3$

[number line: open circle at 2 arrow left; closed dot at 3 arrow right]

4 a $-2 \leq x < 3$

b $x < -2, x \geq 1$

5 a $3x - 2 > x + 4$

$3x - x > 4 + 2$

$2x > 6$

$\frac{2x}{2} > \frac{6}{2}$

$x > 3$

[number line: 2, 3, 4; open circle at 3, arrow right]

b $4 - 2x \geq 12$

$-2x \geq 12 - 4$

$-2x \geq 8$

$\frac{-2x}{-2} \leq \frac{8}{-2}$ (Symbol change)

$x \leq -4$

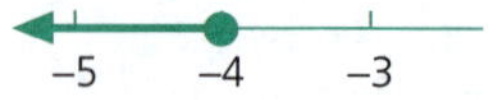

[number line: −5, −4, −3; closed dot at −4, arrow left]

6 a $D = 240 \times 3$

$= 720$

b $A = \frac{h}{2}(a + b)$

$= \frac{12}{2}(6 + 9)$

$= 90$

c $S = 1200(1 - 0.03)^4$

$= 1062.351\,372\ldots$

$= 1062.35$ (2 dec. pl.)

d $A = \pi(11.6^2 - 6.7^2)$

$= 281.706\,613\,2\ldots$

$= 281.71$ (2 dec. pl.)

e $B = \frac{10 \times 3 - 7.5 \times 2}{6.8 \times 82}$

$= 0.026\,901\,004\ldots$

$= 0.03$ (2 dec. pl.)

7 $V = \frac{4\pi r^2 h}{3}$

$= \frac{4 \times \pi \times 12^2 \times 15}{3}$

$= 9047.786\,842\ldots$

$= 9047.789$ (3 dec. pl.)

8 $C = \frac{5}{9}(F - 32)$

$= \frac{5}{9}(113 - 32)$

$= 45$ $\quad\therefore 45\,°C$

9 a $2x + y = 13$

$y = 13 - 2x$

b $3x - y = 11$

$3x - 11 = y$

$y = 3x - 11$

c $5x + 4y = 12$

$4y = 12 - 5x$

$y = \frac{12 - 5x}{4}$

d $7x - 3y = 18$

$7x - 18 = 3y$

$3y = 7x - 18$

$y = \frac{7x - 18}{3}$

e $3xy = 19$

$y = \frac{19}{3x}$

f $\frac{x}{y} = 5$

$x = 5y$

$5y = x$

$y = \frac{x}{5}$

10 a $v = u + at$

$12 = 6 + 9.8t$

$\therefore 6 + 9.8t = 12$

$9.8t = 12 - 6$

$9.8t = 6$

$\frac{9.8t}{9.8} = \frac{6}{9.8}$

$t = 0.612\,244\,898$ [from calculator]

$\therefore t = 0.61$ (to 2 dec. pl.)

b $D = \frac{S}{T}$

$\therefore 120 = \frac{S}{4.5}$

$\frac{S}{4.5} = 120$

$S = 120 \times 4.5$ [cross-multiply]

$\therefore S = 540$

c $V = \frac{1}{3}lbh$

$\therefore 320 = \frac{1}{3} \times l \times 60 \times 13\frac{1}{2}$

$\therefore \frac{1}{3} \times l \times 60 \times 13\frac{1}{2} = 320$

$270l = 320$

$\therefore \frac{270l}{270} = \frac{320}{270}$

$\therefore l = 1.185\,185\,185$ [from calculator]

$\therefore l = 1.19$ (to 2 dec. pl.)

d $K = \frac{4V}{T}$

$\therefore 7.26 = \frac{4 \times 12.43}{T}$

$\therefore 7.26 = \frac{49.72}{T}$

$\therefore 7.26 \times T = 49.72$ [cross-multiply]

$\therefore 7.26T = 49.72$

$\frac{7.26T}{7.26} = \frac{49.72}{7.26}$

$T = 6.848\,484\,8$ [from calculator]

$\therefore T = 6.85$ (to 2 dec. pl.)

e
$$V = \pi r^2 h$$
$$\therefore 126 = 3.14 \times 4^2 \times h$$
$$\therefore 3.14 \times 4^2 \times h = 126$$
$$50.24h = 126$$
$$\therefore \frac{50.24h}{50.24} = \frac{126}{50.24}$$
$$h = 2.507\,961\,783$$
[from calculator]
$$\therefore h = 2.51 \text{ (to 2 dec. pl.)}$$

f
$$w = \frac{3}{4}gk$$
$$7.4 = \frac{3}{4} \times 9.31 \times k$$
$$\therefore \frac{3}{4} \times 9.31 \times k = 7.4$$
$$6.9825k = 7.4$$
$$\frac{6.9825K}{6.9825} = \frac{7.4}{6.9825}$$
$$k = 1.059\,792\,338$$
[from calculator]
$$\therefore k = 1.06 \text{ (to 2 dec. pl.)}$$

11 a Let the numbers be x, $x + 1$, $x + 2$.
$$\therefore x + x + 1 + x + 2 = 93$$
$$3x + 3 = 93$$
$$3x = 93 - 3$$
$$3x = 90$$
$$\frac{\not{3}x}{\not{3}} = \frac{90}{3}$$
$$x = 30$$
$\therefore$ The numbers are 30, 31, 32.

b
$$P = x + 8 + 3x + 12 + 5x$$
$$= 9x + 20$$
$\therefore$ Perimeter $= (9x + 20)$ cm
$$\therefore 9x + 20 = 38$$
$$9x = 38 - 20$$
$$9x = 18$$
$$\frac{\not{9}x}{\not{9}} = \frac{18}{9}$$
$$x = 2$$

c The lengths are equal,
$$\therefore 4x + 8 = 2x + 14$$
$$4x - 2x = 14 - 8$$
$$2x = 6$$
$$\frac{\not{2}x}{\not{2}} = \frac{6}{2}$$
$$x = 3$$

d Let the number be x.
$$\therefore 2x = x + 8$$
$$2x - x = 8$$
$$x = 8$$
$\therefore$ The number is 8.

e Let the number be x.
$$\therefore 3(4 + x) = x - 2$$
$$12 + 3x = x - 2$$
$$3x - x = -2 - 12$$
$$2x = -14$$
$$\frac{\not{2}x}{\not{2}} = \frac{-14}{2}$$
$$x = -7$$
The number is −7.

f Let x be Fiona's age 15 years ago.
Now Fiona $= 2x$
$$\therefore 2x = x + 15$$
$$2x - x = 15$$
$$x = 15$$
$\therefore$ Fiona is now $2x = 2 \times 15$
$= 30$ years old.

g Let Laura's age now be x.
$\therefore$ Daniel's age is now $2x$
$\therefore$ Daniel's age 4 years ago
$= 2x - 4$
Also, Laura's age 4 years ago
$= x - 4$
$$\therefore 2x - 4 = 3(x - 4)$$
$$\therefore 2x - 4 = 3x - 12$$
$$\therefore 2x - 3x = -12 + 4$$
$$-x = -8$$
$$\therefore \frac{\not{-}x}{\not{-}1} = \frac{-8}{-1}$$
$$x = 8$$
$\therefore$ Laura is now 8 years old.

12 a
$$2a + b = 1 \quad (1)$$
$$a + 2b = 8 \quad (2)$$
From (1) $\quad b = 1 - 2a \quad (3)$
Subs. in (2)
$$a + 2(1 - 2a) = 8$$
$$a + 2 - 4a = 8$$
$$2 - 3a = 8$$
$$-3a = 8 - 2$$
$$\frac{\not{-3}a}{\not{-3}} = \frac{6}{-3}$$
$$a = -2$$
Subs. in (3)
$$b = 1 - 2(-2)$$
$$= 1 + 4$$
$$= 5$$
$$\therefore a = -2, b = 5$$

b
$$3m - 2n = 7 \quad (1)$$
$$2m + n = 0 \quad (2)$$
From (2) $\quad n = -2m \quad (3)$
Subs. in (1)
$$3m - 2(-2m) = 7$$
$$3m + 4m = 7$$
$$\frac{7m}{7} = \frac{7}{7}$$
$$m = 1$$
Subs. in (3) $\; n = -2\,(1)$
$$= -2$$
$$\therefore m = 1, n = -2$$

13 a
$$p - 3q = 2 \quad (1)$$
$$2p + 3q = 13 \quad (2)$$
(1) + (2) $\quad \frac{3p}{3} = \frac{15}{3}$
$$p = 5$$
Subs. in (1)
$$5 - 3q = 2$$
$$-3q = 2 - 5$$
$$\frac{-3q}{-3} = \frac{-3}{-3}$$
$$q = 1$$
$$\therefore p = 5, q = 1$$

b
$$x + 2y = 3 \quad (1)$$
$$3x + 4y = 7 \quad (2)$$
$3 \times (1) \quad 3x + 6y = 9 \quad (3)$
(3) − (2) $\quad \frac{2y}{2} = \frac{2}{2}$
$$y = 1$$
Subs. in (1) $\; x + 2 \times 1 = 3$
$$x + 2 = 3$$
$$x = 3 - 2$$
$$x = 1$$
$$\therefore x = 1, y = 1$$

14 a Let the numbers be x and y.
$$\therefore x + y = 12 \quad (1)$$
$$x - y = 2 \quad (2)$$
(1) + (2) $\quad \frac{2x}{2} = \frac{14}{2}$
$$x = 7$$
Subs. in (1) $\; 7 + y = 12$
$$y = 12 - 7$$
$$= 5$$
$$\therefore x = 7, y = 5$$
$\therefore$ The numbers are 7 and 5.

b i Equilateral △ means all sides equal.
$$\therefore 2x + y = 7 \quad (1)$$
$$3x - 2y = 7 \quad (2)$$
From (1) $\quad y = 7 - 2x \quad (3)$
Subs. in (2)
$$3x - 2(7 - 2x) = 7$$
$$3x - 14 + 4x = 7$$
$$7x - 14 = 7$$
$$7x = 7 + 14$$
$$\frac{7x}{7} = \frac{21}{7}$$
$$x = 3$$
Subs. in (3) $\; y = 7 - 2 \times 3$
$$= 7 - 6$$
$$= 1$$
$$\therefore x = 3, y = 1$$

ii Opposite sides of rectangle equal.
$$3x - 1 = 3 - 2y$$
$$\therefore 3x + 2y = 4 \quad (1)$$
and, $\quad 5x - 7 = y + 4$
$$5x - y = 11 \quad (2)$$
From (2) $\; 5x - 11 = y$
$$\therefore y = 5x - 11 \quad (3)$$
Subs. in (1)
$$3x + 2(5x - 11) = 4$$
$$3x + 10x - 22 = 4$$
$$13x = 4 + 22$$
$$\frac{\not{13}x}{\not{13}} = \frac{26}{13}$$
$$x = 2$$
Subs. in (3) $\quad y = 5(2) - 11$
$$= 10 - 11$$
$$= -1$$
$$\therefore x = 2, y = -1$$

c Let x = cost of 1 pen
y = cost of 1 pencil
$\therefore 6x + 5y = 250$ (1)
$3x + 2y = 115$ (2)
$2 \times (2)$ $6x + 4y = 230$ (3)
$(1) - (3)$ $y = 20$
Subs. in (2) $3x + 40 = 115$
$3x = 115 - 40$
$\frac{3x}{3} = \frac{75}{3}$
$x = 25$
∴ Pen costs 25c, pencil costs 20c.

d Let h = cost of a hamburger
d = cost of a drink
$\therefore 2h + d = 1000$ (1)
$3h + 2d = 1590$ (2)
$2 \times (1)$ $4h + 2d = 2000$ (3)
$(3) - (2)$ $h = 410$
Subs. in (1)
$2(410) + d = 1000$
$820 + d = 1000$
$d = 1000 - 820$
$= 180$
∴ Hamburger \$4.10, can \$1.80.

15 a $(x - 1)(x + 3) = 0$
$x = 1, -3$

b $(p + 5)(p - 2) = 0$
$p = -5, 2$

c $a(a - 2) = 0$
$a = 0, 2$

16 a $x^2 + 3x + 2 = 0$
$(x + 2)(x + 1) = 0$
$x = -2, -1$

b $m^2 - 4m + 3 = 0$
$(m - 3)(m - 1) = 0$
$m = 3, 1$

c $k^2 - 7k + 12 = 0$
$(k - 4)(k - 3) = 0$
$k = 4, 3$

d $b^2 + 3b - 4 = 0$
$(b + 4)(b - 1) = 0$
$b = -4, 1$

e $t^2 - 4t - 12 = 0$
$(t - 6)(t + 2) = 0$
$t = 6, -2$

f $w^2 - 2w - 15 = 0$
$(w - 5)(w + 3) = 0$
$w = 5, -3$

g $c^2 + 2c = 0$
$c(c + 2) = 0$
$c = 0, -2$

h $y^2 - 16 = 0$
$(y - 4)(y + 4) = 0$
$y = 4, -4$

i $q^2 - 7q = 0$
$q(q - 7) = 0$
$q = 0, 7$

j $h^2 = 10h$
$h^2 - 10h = 0$
$h(h - 10) = 0$
$h = 0, 10$

k $5 - 4q - q^2 = 0$
$q^2 + 4q - 5 = 0$
$(q + 5)(q - 1) = 0$
$q = -5, 1$

l $x^2 - 9x + 25 = 5$
$x^2 - 9x + 20 = 0$
$(x - 5)(x - 4) = 0$
$x = 5, 4$

Chapter 5 . . . . pp. 63–66

Linear Relationships

1 A (2, 2)
B (−1, −1)
C (−3, 1)
D (2, −1)
E (1, −3)
F (3, 1)
G (−2, 3)
H (−2, 0)
I (0, 1)
J (2, 0)
K (0, −2)
L (−3, −2)

2 a

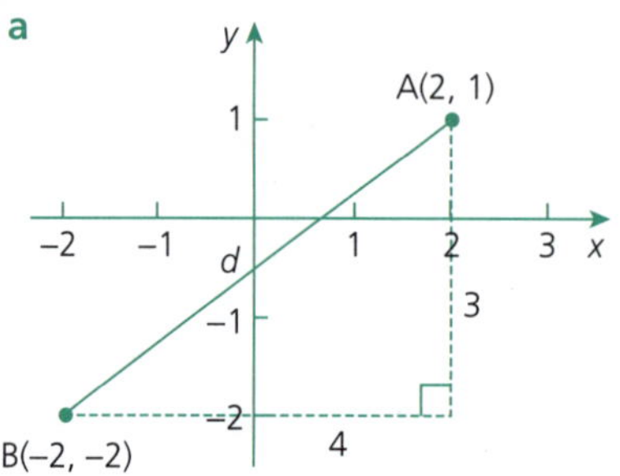

Using Pythagoras' Theorem
$d^2 = 3^2 + 4^2 = 25$
$\therefore d = \sqrt{25}$
$= 5$
Distance AB is 5 units.

b $\sqrt{41}$ units

3 a $\sqrt{26}$ units

b $\sqrt{50}$ units

c $\sqrt{50}$ units

d $\sqrt{26}$ units

e 5 units

f 5 units

4 a $\left(1, \frac{1}{2}\right)$

b (0, 1)

5 a $\left(-\frac{1}{2}, 3\right)$

b (0, 2)

c (0, 0)

d (2, −1)

e (1, −2)

f $\left(2, 1\frac{1}{2}\right)$

6 a $m = \frac{\text{Rise}}{\text{Run}} = -\frac{3}{2}$ (negative slope)

b $m = -\frac{1}{4}$

c $m = -\frac{3}{4}$

d $m = \frac{1}{2}$

e $m = \frac{2}{3}$

f $m = \frac{4}{3}$

7 a $m = \frac{2}{5}$

b $m = -1$

c $m = \frac{1}{2}$

d $m = -\frac{1}{2}$

e $m = 0$ (horizontal line)

f m = undefined (vertical line)

8 a $d = \sqrt{(x_2 - x_1)^2 + (y_2 - y_1)^2}$
$= \sqrt{(2 + 3)^2 + (4 - 2)^2}$
$= \sqrt{25 + 4}$
$= \sqrt{29}$
Length AB = $\sqrt{29}$ units.

b $d = \sqrt{(x_2 - x_1)^2 + (y_2 - y_1)^2}$
$= \sqrt{(-2 - 2)^2 + (1 + 3)^2}$
$= \sqrt{16 + 16}$
$= \sqrt{32}$
Length AB = $\sqrt{32}$ units.

c $\sqrt{80}$ units

d $\sqrt{20}$ units

e 4 units

f 5 units

9 a Midpoint $= \left(\frac{x_1 + x_2}{2}, \frac{y_1 + y_2}{2}\right)$
$= \left(\frac{-3 + 1}{2}, \frac{2 + 4}{2}\right)$
$= \left(\frac{-2}{2}, \frac{6}{2}\right)$
$= (-1, 3)$

b Midpoint $= \left(\frac{2+-2}{2}, \frac{-3+1}{2}\right)$

$= \left(\frac{0}{2}, \frac{-2}{2}\right)$

$= (0, -1)$

c (0, 0)

d (2, −1)

e (1, −2)

f $\left(2, 1\frac{1}{2}\right)$

10 a $m = \frac{y_2 - y_1}{x_2 - x_1}$

$= \frac{4-2}{2--3}$

$= \frac{2}{5}$

b $m = \frac{y_2 - y_1}{x_2 - x_1}$

$= \frac{1--3}{-2-2}$

$= \frac{4}{-4}$

$= -1$

c $m = \frac{y_2 - y_1}{x_2 - x_1}$

$= \frac{-2-2}{-4-4}$

$= \frac{-4}{-8}$

$= \frac{1}{2}$

d $m = -\frac{1}{2}$

e $m = 0$

f m is undefined $\left(\frac{5}{0}\right)$

11 a $AB = \sqrt{41}$ units

$AC = \sqrt{50}$ units

b Midpoint AC $= \left(1\frac{1}{2}, \frac{1}{2}\right)$

Midpoint BC $= \left(-1, 2\frac{1}{2}\right)$

c $m_{BA} = \frac{-4}{5}$

$m_{CB} = \frac{3}{4}$

12 a

x	0	1	2
y	0	2	4

b

x	0	1	2
y	0	−2	−4

c

x	0	1	2
y	1	−1	−3

d

x	0	1	2
y	−1	−3	−5

e

x	0	1	2
y	2	$2\frac{1}{2}$	3

f

x	0	1	2
y	−2	$-1\frac{1}{2}$	−1

g

x	0	1	2
y	0	$\frac{1}{2}$	1

h

x	0	1	2
y	0	$-\frac{1}{2}$	−1

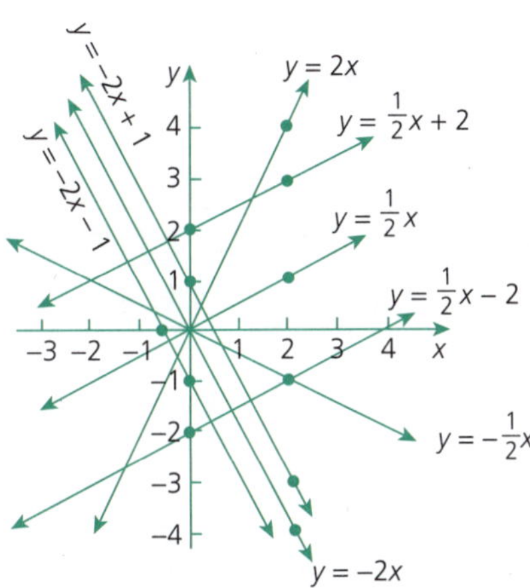

13

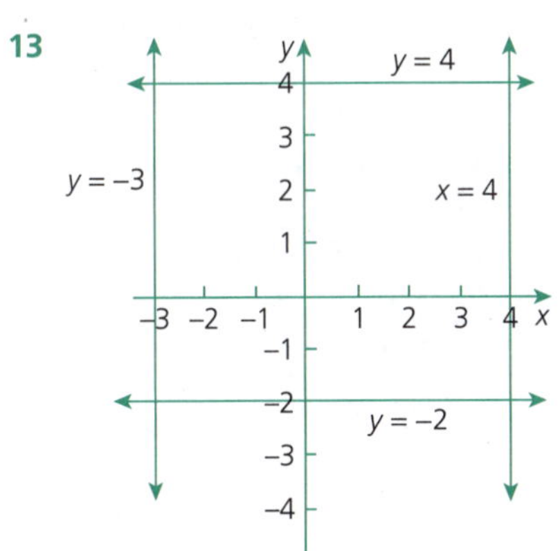

14 a $y = 4x - 1$

b $y = -2x + 3$

c $y = \frac{1}{2}x + 2$

d $y = \frac{3}{4}x - 4$

15 a $m = 3, b = 7$

b $m = -4, b = \frac{1}{2}$

c $m = -3, b = 4$

d $y = 2x + \frac{2}{5}$

$m = 2, b = \frac{2}{5}$

e $y = \frac{2}{3}x - 2$

$m = \frac{2}{3}, b = -2$

f $y = -2x + 3$

$m = -2, b = 3$

g $y = 2x - 3$

$m = 2, b = -3$

h $3x + 2y = 6$

$2y = -3x + 6$

$y = -\frac{3}{2}x + 3$

$m = -\frac{3}{2}, b = 3$

i $2x - 3y = 6$

$3y = 2x - 6$

$y = \frac{2}{3}x - 2$

$m = \frac{2}{3}, b = -2$

j $4x = 2y + 7$

$\therefore 4x - 7 = 2y$

$\therefore y = 2x - \frac{7}{2}$

$m = 2, b = -\frac{7}{2}$

16 a $m = \frac{\text{Rise}}{\text{Run}}$ [negative slope]

$= -\frac{3}{2}$

Now using $y = mx + b$

$y = -\frac{3}{2}x + 3$

b $m = \frac{\text{Rise}}{\text{Run}}$ [negative slope]

$= -\frac{3}{3}$

$= -1$

Now $y = mx + b$

$\therefore y = -x - 3$

c $m = \frac{2}{3}$ [positive slope]

Using $y = mx + b$

$\therefore y = \frac{2}{3}x + 2$

d $m = \frac{4}{2}$ [positive slope]

$= 2$

Using $y = mx + b$

$\therefore y = 2x - 4$

e $y = 3x + 3$

f $y = -\frac{5}{2}x - 5$

17 a

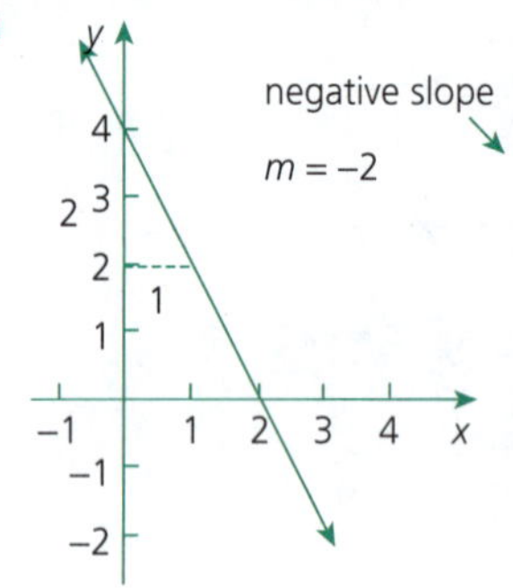

b

positive slope

$m = \frac{1}{3}$

c

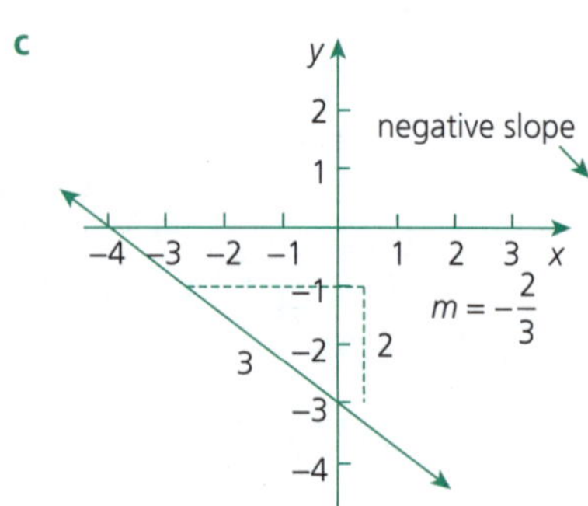

d

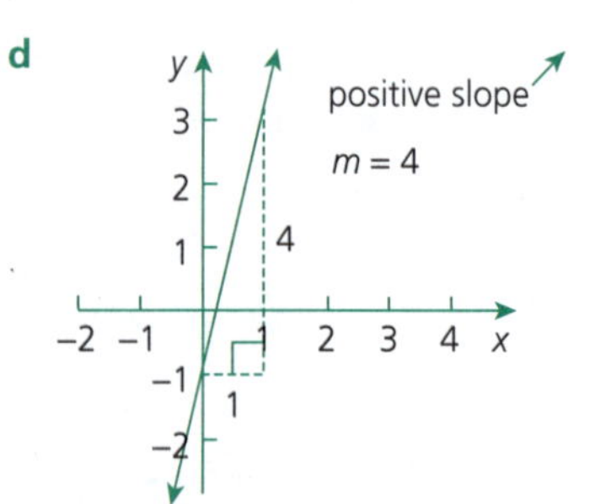

18 a $m = \frac{1}{2}$

b $y = \frac{1}{2}x + \frac{5}{2}$

$m = \frac{1}{2}$

c $y = \frac{1}{2}x$

$m = \frac{1}{2}$

d $m = 2$

e $2y = -x + 1$

$y = -\frac{1}{2}x + \frac{1}{2}$

$m = -\frac{1}{2}$

f $2y = x$

$\therefore y = \frac{1}{2}x$

$m = \frac{1}{2}$

$\therefore$ Parallel lines are (a), (b), (c) and (f) as they all have a gradient of $\frac{1}{2}$. Lines (c) and (f) pass through the origin. (They do not have a constant term.)

19 Substitute coordinates of points into $y = 3x - 1$.

a $(-1, -4)$ $-4 = -3 - 1$ True
$(-1, -4)$ lies on line.

b $(4, -1)$ $-1 \neq 12 - 1$
$(4, -1)$ does not lie on line.

c $(1, -4)$ $-4 \neq 3 - 1$
$(1, -4)$ does not lie on line.

d $(2, 5)$ $5 = 6 - 1$ True
$(2, 5)$ lies on line.

e $\left(\frac{1}{3}, 0\right)$ $0 = 1 - 1$ True
$\left(\frac{1}{3}, 0\right)$ lies on line.

f $(0, 1)$ $1 \neq 0 - 1$
$(0, 1)$ does not lie on line.

20 a $(-2, 1)$ in $y = 2x + 3$
$1 \neq -4 + 3$
$(-2, 1)$ does not lie on line.

b $y = 2x + 5$ $(-2, 1)$
$1 = -4 + 5$
$(-2, 1)$ lies on line.

c $x + 2y = 0$ $(-2, 1)$
$-2 + 2 = 0$
$(-2, 1)$ lies on line.

d $x - 2y + 4 = 0$ $(-2, 1)$
$-2 - 2 + 4 = 0$
$(-2, 1)$ lies on line.
$\therefore$ $(-2, 1)$ lies on $y = 2x + 5$, $x + 2y = 0$ and $x - 2y + 4 = 0$.

21

A(−2, 4), B(4, 2), C(0, −2)

a Distance AB

$= \sqrt{(x_2 - x_1)^2 + (y_2 - y_1)^2}$

$= \sqrt{(4 + 2)^2 + (2 - 4)^2}$

$= \sqrt{36 + 4}$

$= \sqrt{40}$

Distance AC

$= \sqrt{(x_2 - x_1)^2 + (y_2 - y_1)^2}$

$= \sqrt{(0 + 2)^2 + (-2 - 4)^2}$

$= \sqrt{4 + 36}$

$= \sqrt{40}$

$\therefore$ AB = AC
That is, $\triangle$ABC is isosceles.

b $m_{AB} = \frac{y_2 - y_1}{x_2 - x_1}$

$= \frac{4 - 2}{-2 - 4}$

$= \frac{2}{-6}$

$= -\frac{1}{3}$

c $MP_{AC} = \left(\frac{x_1 + x_2}{2}, \frac{y_1 + y_2}{2}\right)$

$= \left(\frac{-2 + 0}{2}, \frac{4 + -2}{2}\right)$

$= (-1, 1)$ [Call point P]

22 a $y = mx + b$,
$\therefore y = -3x + 4$

$m = -3$, $b = 4$ (0, 4)

b $y = 4x - 7$, $m = 4$
The line parallel has gradient = 4.
The required line has equation:
$y = 4x - 4$ as line cuts y-axis at $(0, -4)$
i.e. y-intercept is -4.

23 a $2x - 3y = 7$

$3y = 2x - 7$

$y = \frac{2}{3}x - \frac{7}{3}$

Gradient $= \frac{2}{3}$

$6x - 12y - 5 = 0$

$12y = 6x - 5$

$y = \frac{6}{12}x - \frac{5}{12}$

Gradient $= \frac{1}{2}$

The lines are not perpendicular.

b $x + 2y - 3 = 0$

$2y = -x + 3$

$y = -\frac{1}{2}x + \frac{3}{2}$

Gradient $= -\frac{1}{2}$

$y = 2x + 1$

Gradient = 2

As $-\frac{1}{2} \times 2 = -1$, the lines are perpendicular.

c $2x + 3y = 6$

$3y = -2x + 6$

$y = -\frac{2}{3}x + 2$

Gradient $= -\frac{2}{3}$

$4y = 6x - 5$

$y = \frac{3}{2}x - \frac{5}{4}$

Gradient $= \frac{3}{2}$

As $-\frac{2}{3} \times \frac{3}{2} = -1$, the lines are perpendicular.

24 Consider $2x - ky = 5$ (1)

$ky = 2x - 5$

$y = \frac{2}{k}x - \frac{5}{k}$

The gradient of line (1) $= \frac{2}{k}$

Consider $3x + 4y - 7 = 0$ (2)

$4y = -3x + 7$

$y = -\frac{3}{4}x + \frac{7}{4}$

The gradient of line (2) $= -\frac{3}{4}$

As the lines are parallel, their gradients are equal.

$\therefore \frac{2}{k} = -\frac{3}{4}$,

that is, $-3k = 8$

$k = -\frac{8}{3}$

$= -2\frac{2}{3}$

25 $2x + y = 1$

Substitute each point.

A(−3, 7) LHS = −6 + 7 = 1 = RHS

A(−3, 7) lies on $2x + y = 1$

B(3, −7) LHS = 6 − 7 = −1

B(3, −7) does not lie on the line.

C(−11, 23) LHS = −22 + 23 = 1 = RHS

C(−11, 23) lies on $2x + y = 1$

∴ A and C lie on $2x + y = 1$.

26 The gradient of the line joining (−1, 3) and (1, 1) is:

$m = \frac{3-1}{-1-1} = \frac{2}{-2} = -1$

The gradient of the line joining (1, 1) and (4, −2) is:

$m = \frac{1+2}{1-4} = \frac{3}{-3} = -1$

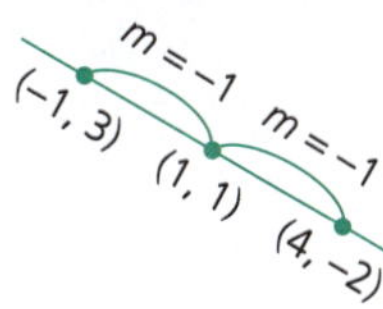

∴ The line must contain all three points.

27 (2*d*, 5) lies on $2x - y - 8 = 0$, so the coordinates must satisfy the equation.

Thus $2x - y - 8 = 0$

$2(2d) - 5 - 8 = 0$

$4d = 13$

$d = \frac{13}{4} = 3\frac{1}{4}$

28 a Gradient $= m = 3$

∴ Subs. (2, 5) in $y = 3x + c$

$5 = 3 \times 2 + c$

$5 = c + 6$

$c = 5 - 6$

$c = -1$

$\therefore y = 3x - 1$

b Gradient $= m = -2$

∴ Subs. (0, 2) in $y = -2x + c$

$2 = -2 \times 0 + c$

$2 = c$

$c = 2$

$\therefore y = -2x + 2$

c Gradient $= m = \frac{1}{2}$

∴ Subs. (2, 1) in $y = \frac{1}{2}x + c$

$1 = \frac{1}{2} \times 2 + c$

$1 = c + 1$

$c = 1 - 1$

$c = 0$

$\therefore y = \frac{1}{2}x$

d Gradient $= m = \frac{-2}{3}$

∴ Subs. (3, −2) in $y = \frac{-2}{3}x + c$

$-2 = \frac{-2}{3} \times 3 + c$

$-2 = c - 2$

$c = 0$

$\therefore y = \frac{-2}{3}x$

e Gradient $= m = 3$

∴ Subs. (4, 6) in $y = 3x + c$

$6 = 3 \times 4 + c$

$6 = c + 12$

$c = 6 - 12$

$c = -6$

$\therefore y = 3x - 6$

f $2x - y + 9 = 0$

$y = 2x + 9$

Gradient $= m = 2$

∴ Subs. (3, 4) in $y = 2x + c$

$4 = 2 \times 3 + c$

$4 = c + 6$

$c = 4 - 6$

$c = -2$

$\therefore y = 2x - 2$

g $3x + 2y - 5 = 0$

$2y = -3x + 5$

$y = \frac{-3x + 5}{2}$

Gradient $= m = \frac{-3}{2}$

∴ Subs. (−4, 2) in $y = \frac{-3}{2}x + c$

$2 = \frac{-3}{2} \times -4 + c$

$2 = c + 6$

$c = 2 - 6$

$c = -4$

$\therefore y = \frac{-3}{2}x - 4$

h $x + y - 2 = 0$

$y = -x + 2$, has gradient of −1

∴ Gradient of perp.: $m = 1$

∴ Subs. (−3, −2) in $y = x + c$

$-2 = -3 + c$

$-2 = c - 3$

$c = -2 + 3$

$c = 1$

$\therefore y = x + 1$

i $4x - 2y = 1$

$2y = 4x - 1$

$y = \frac{4x - 1}{2}$, has gradient of 2

∴ Gradient of perp.: $m = -\frac{1}{2}$

∴ Subs. (6, 1) in $y = -\frac{1}{2}x + c$

$1 = -\frac{1}{2} \times 6 + c$

$1 = c - 3$

$c = 1 + 3$

$c = 4$

$\therefore y = -\frac{1}{2}x + 4$

j $3x + 5y + 1 = 0$

$5y = -3x - 1$

$y = \frac{-3x - 1}{5}$,

has gradient of $-\frac{3}{5}$

∴ Gradient of perp.: $m = \frac{5}{3}$

∴ Subs. (3, 4) in $y = \frac{5}{3}x + c$

$4 = \frac{5}{3} \times 3 + c$

$4 = c + 5$

$c = 4 - 5$

$c = -1$

$\therefore y = \frac{5}{3}x - 1$

29 a $C = 50 + 70n$

b

n	0	2	4	6
C	50	190	330	470

c

d **i** Subs. $n = 3$ in $C = 50 + 70n$:
$C = 50 + 70 \times 3$
$= 260$ $\quad \therefore$ \$260

ii Subs. $n = 8$ in $C = 50 + 70n$:
$C = 50 + 70 \times 8$
$= 610$ $\quad \therefore$ \$610

e **i** Subs. $C = 400$ in $C = 50 + 70n$:
$400 = 50 + 70n$
$70n = 350$
$n = 5$ $\quad \therefore$ 5 hours

ii Subs. $C = 750$ in $C = 50 + 70n$:
$750 = 50 + 70n$
$70n = 700$
$n = 10$ $\quad \therefore$ 10 hours

30 a **i** n

ii P

b

n	1000	1500	2000	2500	3000
P	200	1700	3200	4700	6200

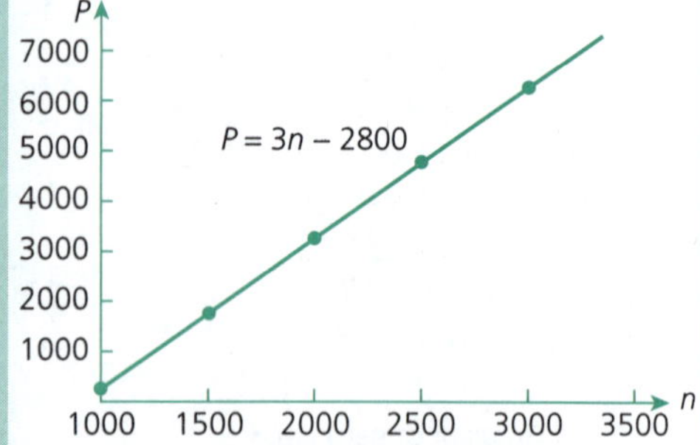

c Subs. $n = 3250$ in $P = 3n - 2800$:
$P = 3 \times 3250 - 2800$
$= 6950$ $\quad \therefore$ \$6950

d Subs. $P = 10\,160$ in $P = 3n - 2800$:
$10\,160 = 3n - 2800$
$3n = 10\,160 + 2800$
$3n = 12\,960$
$n = 4320$ $\quad \therefore$ 4320 pens

Chapter 6 . . . . pp. 78–79

Non-linear Relationships

1 a $y = x^2 + 1$

x	−2	−1	0	1	2
y	5	2	1	2	5

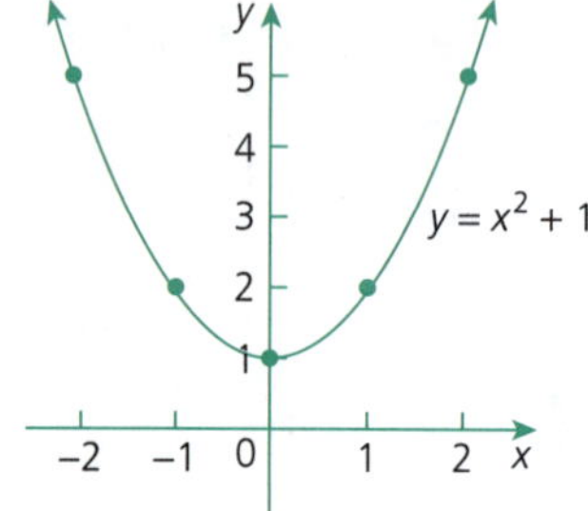

b $y = 2x^2$

x	−2	−1	0	1	2
y	8	2	0	2	8

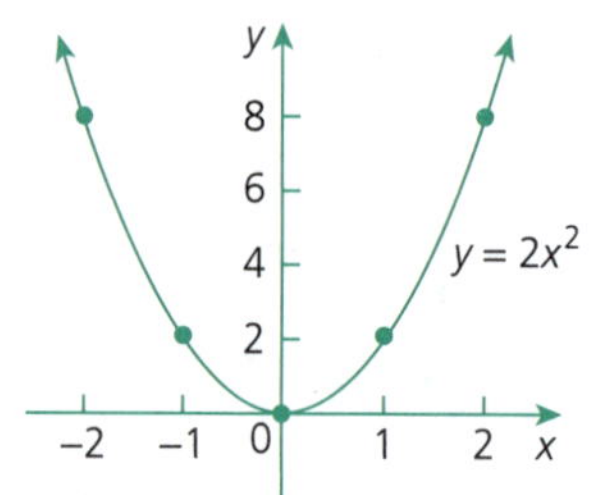

c $y = 2 - x^2$

x	−2	−1	0	1	2
y	−2	1	2	1	−2

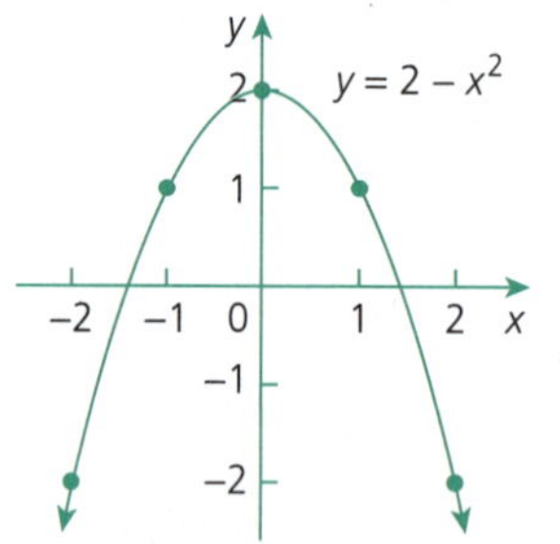

2 a In $y = x^2 + 8$, the x^2 is positive ($a > 0$).
$\therefore$ The parabola is concave up.

b In $y = 3 - x^2$, the x^2 is negative ($a < 0$).
$\therefore$ The parabola is concave down.

c In $y = -5x^2$, the x^2 is negative ($a < 0$).
$\therefore$ The parabola is concave down.

d In $y = 2x^2 + x - 9$, the x^2 is positive ($a > 0$).
$\therefore$ The parabola is concave up.

3 a In $y = 3x^2$, the x^2 is positive ($a > 0$).
$\therefore$ The parabola is concave up.
$\therefore$ The parabola has a minimum.

b In $y = 5 - x^2$, the x^2 is negative ($a < 0$).
$\therefore$ The parabola is concave down.
$\therefore$ The parabola has a maximum.

c In $y = x^2 - 4x + 5$, the x^2 is positive ($a < 0$).
$\therefore$ The parabola is concave up.
$\therefore$ The parabola has a minimum.

d In $y = 3 + x - x^2$, the x^2 is negative ($a < 0$).
$\therefore$ The parabola is concave down.
$\therefore$ The parabola has a maximum.

4 a

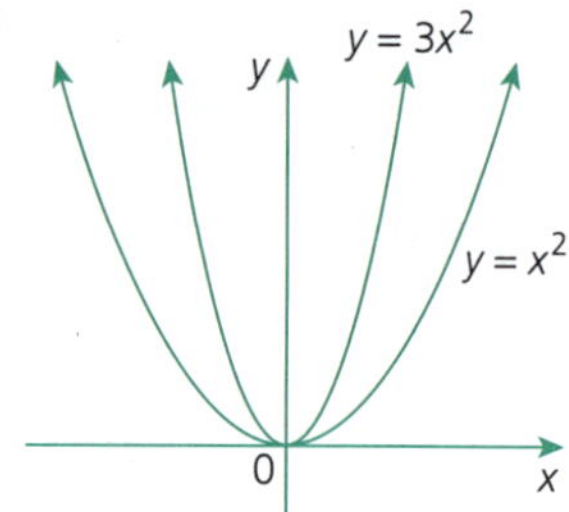

b

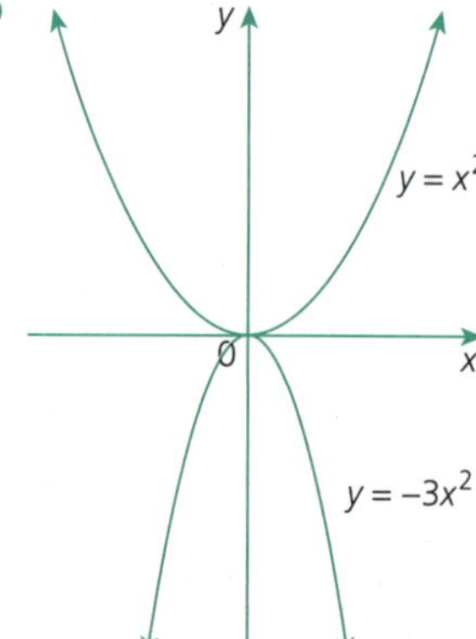

c

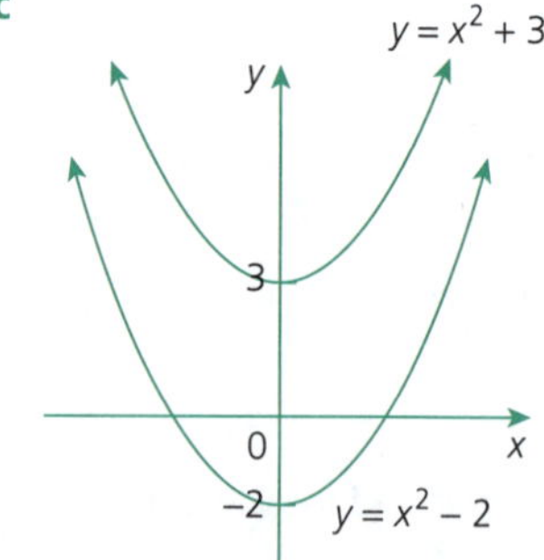

d

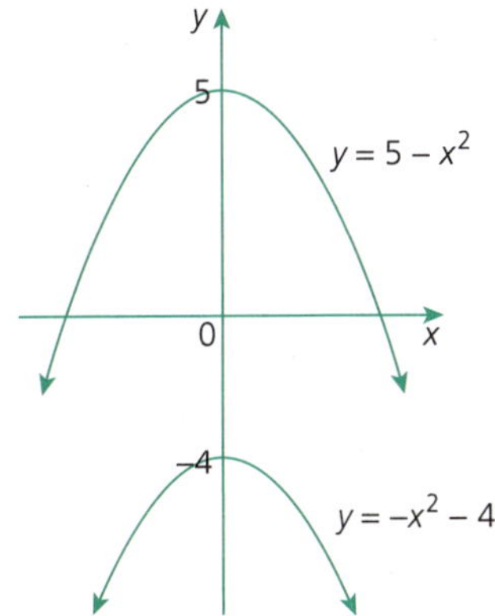

e

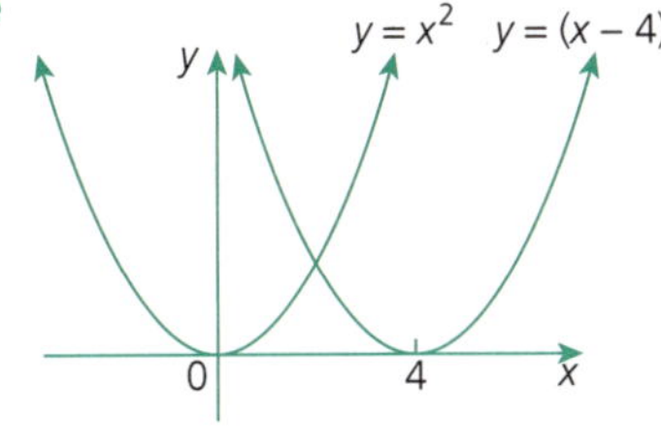

f

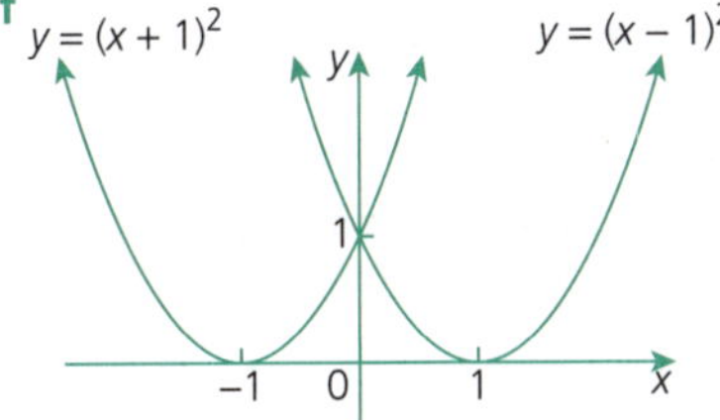

5

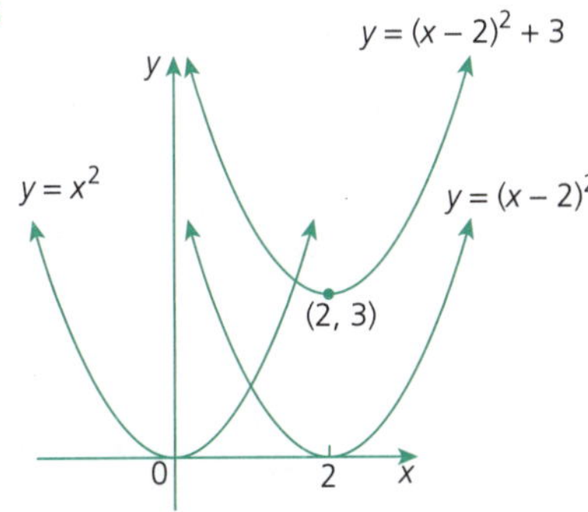

6 a As $y = (x + 1)(x - 4)$

$\therefore$ cuts the x-axis (x-intercepts) at -1 and 4.

Also, $y = x^2 - 3x - 4$

$\therefore$ cuts the y-axis (y-intercept) at -4.

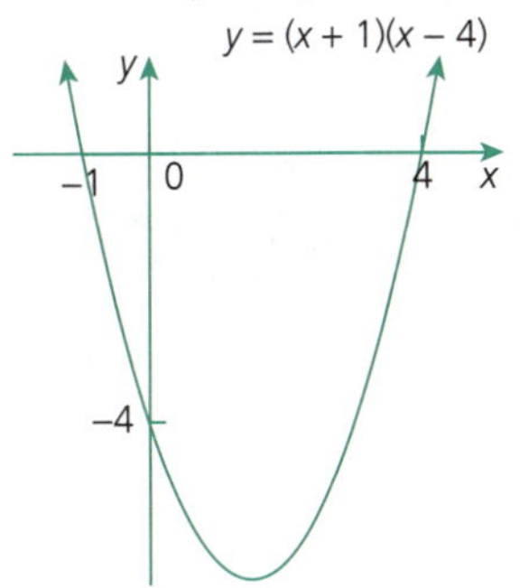

b As $y = (x - 2)(x - 6)$

$\therefore$ cuts the x-axis (x-intercepts) at 2 and 6.

Also, $y = x^2 - 8x + 12$

$\therefore$ cuts the y-axis (y-intercept) at 12.

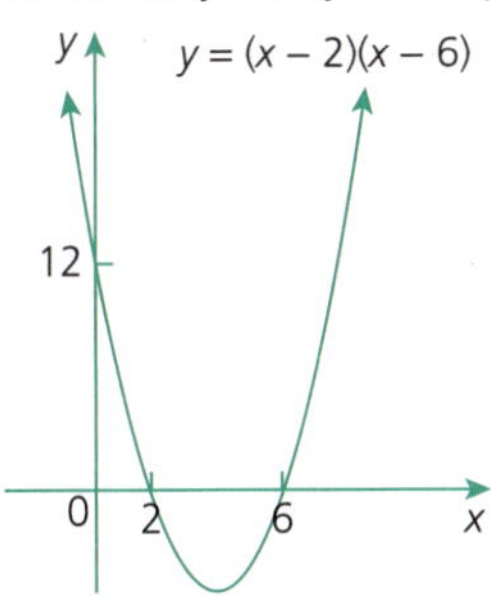

c As $y = x(x - 2)$

$\therefore$ cuts the x-axis (x-intercepts) at 0 and 2.

Also, $y = x^2 - 2x$

$\therefore$ cuts the y-axis (y-intercept) at 0.

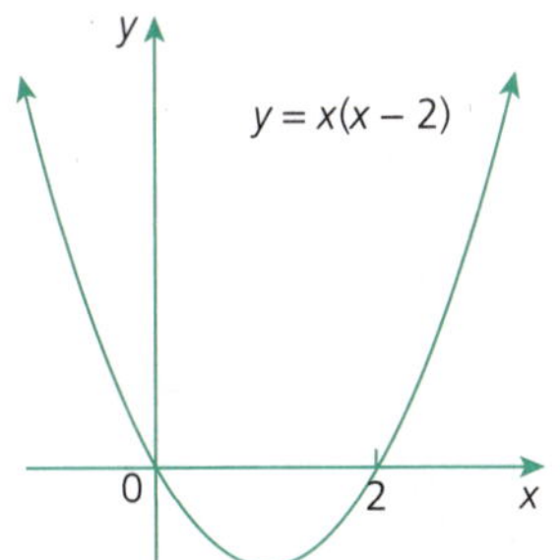

d As $y = x(x + 4)$

$\therefore$ cuts the x-axis (x-intercepts) at 0 and -4.

Also, $y = x^2 + 4x$

$\therefore$ cuts the y-axis (y-intercept) at 0.

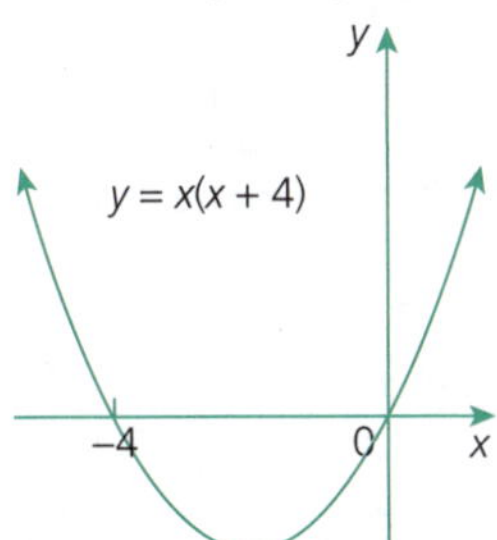

e As $y = (x + 4)^2$

$\therefore$ touches the x-axis (x-intercept) at -4.

Also, $y = x^2 + 8x + 16$

$\therefore$ cuts the y-axis (y-intercept) at 16.

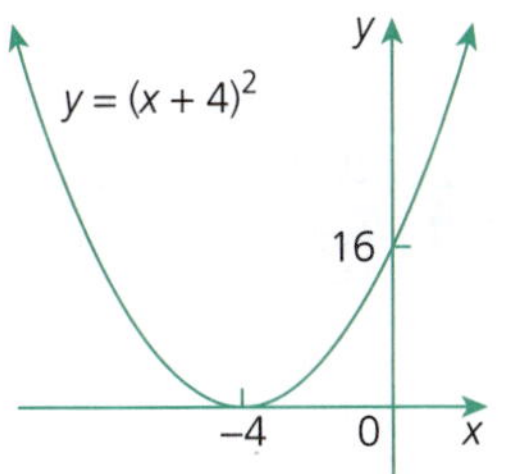

7 a $y = x^2 - 9$

$\therefore$ cuts the y-axis (y-intercept) at -9.

As $y = (x - 3)(x + 3)$

$\therefore$ cuts the x-axis (x-intercepts) at 3 and -3.

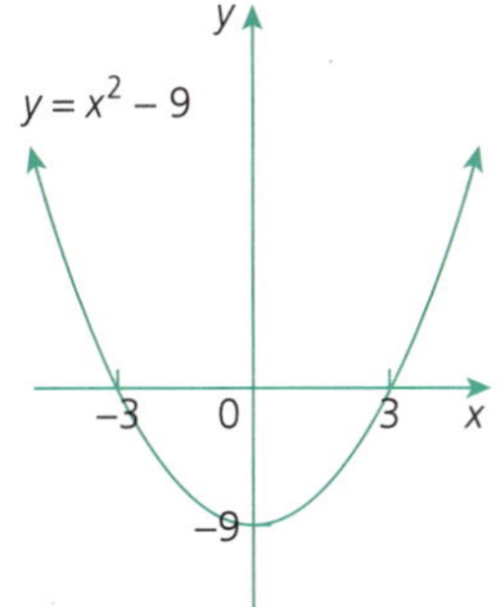

b $y = x^2 - 3x - 4$

$\therefore$ cuts the y-axis (y-intercept) at -4.

As $y = (x - 4)(x + 1)$

$\therefore$ cuts the x-axis (x-intercepts) at 4 and -1.

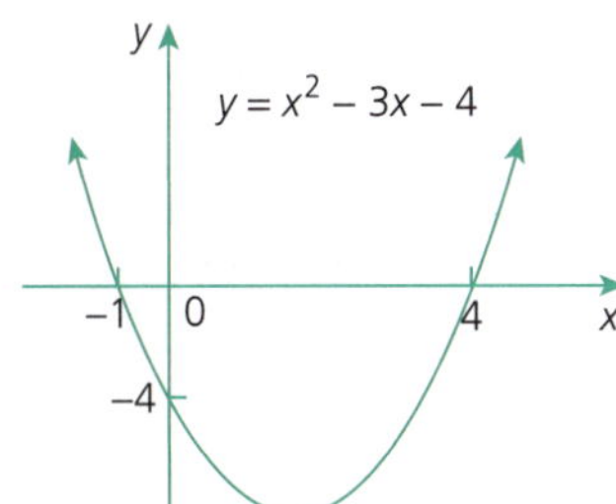

c $y = x^2 - x - 6$

$\therefore$ cuts the y-axis (y-intercept) at -6.

As $y = (x - 3)(x + 2)$

$\therefore$ cuts the x-axis (x-intercepts) at 3 and -2.

d $y = x^2 + 2x + 1$

$\therefore$ cuts the y-axis (y-intercept) at 1.

As $y = (x + 1)(x + 1)$

$\therefore y = (x + 1)^2$

$\therefore$ touches the x-axis (x-intercept) at -1.

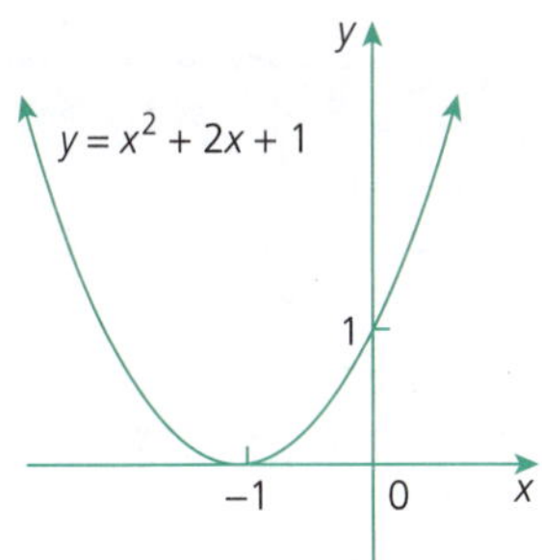

e $y = -x^2 + 4x - 4$
$\therefore$ cuts the y-axis (y-intercept) at -4.
As $y = -(x^2 - 4x + 4)$
$\therefore y = -(x-2)^2$
$\therefore$ touches the x-axis (x-intercept) at 2.

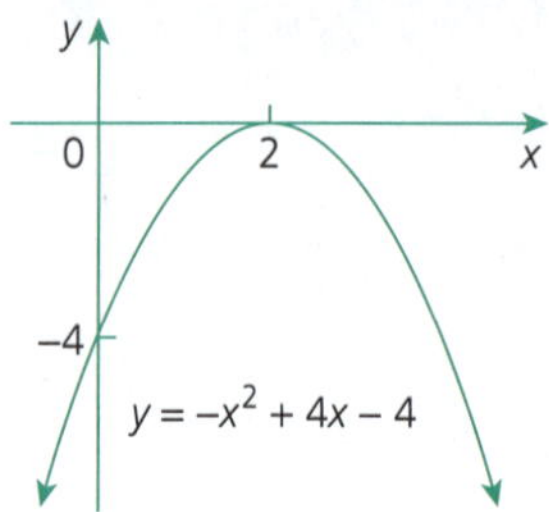

8 a $(x+2)(x-2) = 0$
$\therefore x = -2, 2$
$\therefore$ cuts the x-axis (x-intercepts) at -2 and 2.
Axis of symmetry is middle of -2 and 2.
$\therefore$ axis of symmetry of $x = 0$

b $(x+4)(x+2) = 0$
$\therefore x = -4, -2$
$\therefore$ cuts the x-axis (x-intercepts) at -4 and -2.
Axis of symmetry is middle of -4 and -2.
$\therefore$ axis of symmetry of $x = -3$

c $(x-3)(x+1) = 0$
$\therefore x = 3, -1$
$\therefore$ cuts the x-axis (x-intercepts) at 3 and -1.
Axis of symmetry is middle of 3 and -1.
$\therefore$ axis of symmetry of $x = 1$

d $x(x+4) = 0$
$\therefore x = 0, -4$
$\therefore$ cuts the x-axis (x-intercepts) at 0 and -4.
Axis of symmetry is middle of 0 and -4.
$\therefore$ axis of symmetry of $x = -2$

e $-x(x-8) = 0$
$\therefore x = 0, 8$
$\therefore$ cuts the x-axis (x-intercepts) at 0 and 8.
Axis of symmetry is middle of 0 and 8.
$\therefore$ axis of symmetry of $x = 4$

9 a $y = x^2 - 4$
$0 = 2^2 - 4?$
$= 0?$ $\therefore$ lies on parabola

b $y = x^2 - 4$
$3 = (-1)^2 - 4?$
$= -3?$ $\therefore$ does not lie on parabola

c $y = x^2 - 4$
$5 = 3^2 - 4?$
$= 5?$ $\therefore$ lies on parabola

10 a $y = x^2 - 3x + 1$
$3 = 2^2 - 3(2) + 1$
$= -1?$ $\therefore$ does not pass through

b $y = x^2 + 4x - 9$
$3 = 2^2 + 4(2) - 9?$
$= 3?$ $\therefore$ passes through (2, 3)

c $y = x^2 - x - 2$
$3 = 2^2 - 2 - 2?$
$= 0?$ $\therefore$ does not pass through

11 Subs. $(1, b)$ in $y = 2x^2 - 3$:
$b = 2(1)^2 - 3$
$= -1$

12 Subs. (1, 3) in $y = x^2 - cx - 2$:
$3 = 1^2 - c(1) - 2$
$3 = 1 - c - 2$
$3 = -1 - c$
$c = -1 - 3$
$= -4$

13 a Subs. (2, 12) in $y = ax^2$:
$12 = a(2)^2$
$12 = 4a$
$a = 3$ $\therefore y = 3x^2$

b Subs. $(-2, -8)$ in $y = ax^2$:
$-8 = a(-2)^2$
$-8 = 4a$
$a = -2$ $\therefore y = -2x^2$

c Subs. (3, 11) in $y = ax^2 + 2$
$11 = a(3)^2 + 2$
$11 = 9a + 2$
$9a = 11 - 2$
$9a = 9$
$a = 1$ $\therefore y = x^2 + 2$

d Subs. $(2, -13)$ in $y = ax^2 - 1$
$-13 = a(2)^2 - 1$
$-13 = 4a - 1$
$4a = -13 + 1$
$4a = -12$
$a = -3$ $\therefore y = -3x^2 - 1$

e Subs. (0, 6) in $y = a(x-1)(x-3)$
$6 = a(0-1)(0-3)$
$6 = 3a$
$a = 2$ $\therefore y = 2(x-1)(x-3)$

f Subs. (3, 15) in $y = ax(x-4)$
$15 = a(3)(3-4)$
$15 = -3a$
$a = -5$ $\therefore y = -5x(x-4)$

14

x	-2	-1	0	1	2
y	$\frac{1}{16}$	$\frac{1}{4}$	1	4	16

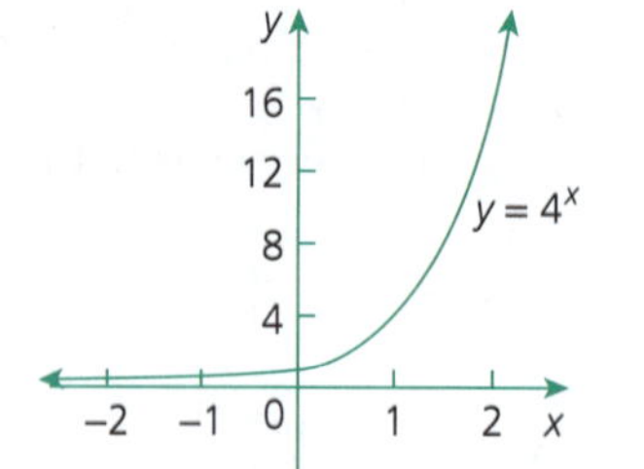

15

$y = 2^x$
$y = 2^{-x}$

16 a As $\sqrt{49} = 7$, then radius is 7 units.

b As $\sqrt{81} = 9$, then radius is 9 units.

c $x^2 + y^2 = 16$
As $\sqrt{16} = 4$, then radius is 4 units.

d Radius is $\sqrt{7}$ units.

e Radius is $\sqrt{29}$ units.

17 a centre (0, 0), radius = $\sqrt{9} = 3$

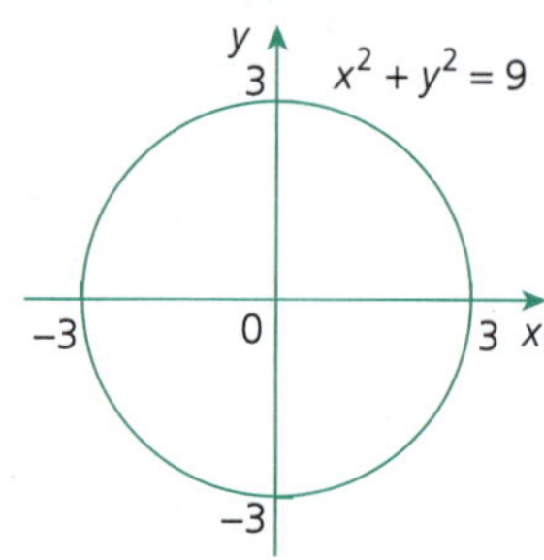

b centre (0, 0), radius = $\sqrt{5}$

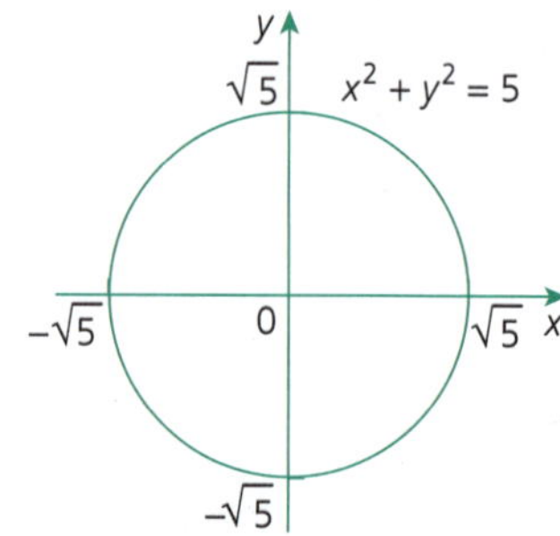

18 a centre (0, 0) and radius = 4 units:
$\therefore x^2 + y^2 = 4^2$
$\therefore x^2 + y^2 = 16$

b centre (0, 0) and radius = $\sqrt{3}$ units:
$\therefore x^2 + y^2 = (\sqrt{3})^2$
$\therefore x^2 + y^2 = 3$

19

x	-2	-1	$-\frac{1}{2}$	0	$\frac{1}{2}$	1	2
y	-1	-2	-4	undef.	4	2	1

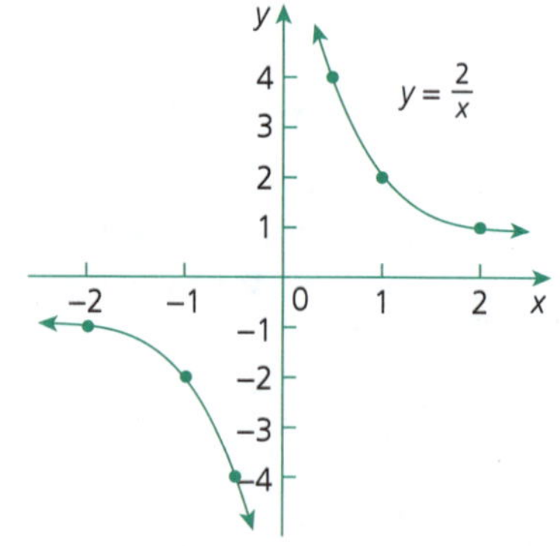

20 $xy = 3 \therefore y = \frac{3}{x}$

x	-3	-1	$-\frac{1}{3}$	0	$\frac{1}{3}$	1	3
y	-1	-3	-9	undef.	9	3	1

$xy = -3 \therefore y = -\frac{3}{x}$

x	−3	−1	$-\frac{1}{3}$	0	$\frac{1}{3}$	1	3
y	1	3	9	undef.	−9	−3	−1

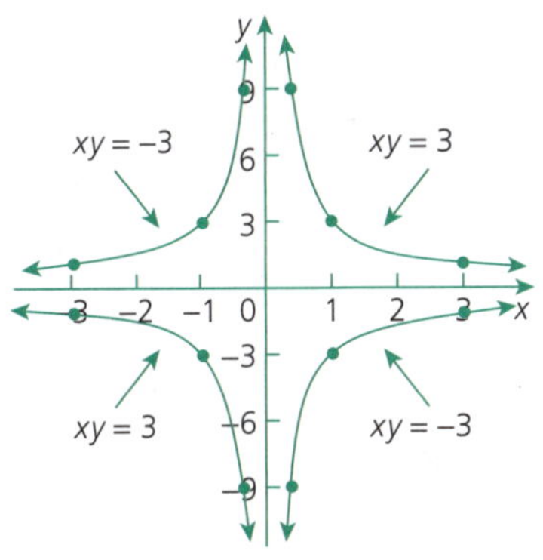

21 a $x^2 = x + 2$

$x^2 - x - 2 = 0$

$(x-2)(x+1) = 0$

$x = 2, -1$

Subs. $x = 2$ in $y = x^2 \therefore y = 4$

Subs. $x = -1$ in $y = x^2 \therefore y = 1$

$\therefore$ Points of intersection are (2, 4) and (−1, 1).

b $x^2 + 2 = 5x - 2$

$x^2 - 5x + 4 = 0$

$(x-4)(x-1) = 0$

$x = 4, 1$

Subs. $x = 4$ in $y = x^2 + 2 \therefore y = 18$

Subs. $x = 1$ in $y = x^2 + 2 \therefore y = 3$

$\therefore$ Points of intersection are (4, 18) and (1, 3).

Chapter 7 . . . . pp. 88–90

Rates and Proportion

1 a 100 km = 100 000 m

$100000 \div 60 \div 60 = 27\frac{7}{9}$

$\therefore 27\frac{7}{9}$ m/s

b $40 \times 60 \times 60 = 144000$

144 000 m = 144 km

$\therefore$ 144 km/h

c 20 km/L = 1 L/ 20 km

Multiply by 5: 5 L/100 km

d 10 L/100 km = 100 km/10 L

Divide by 10: 10 km/L

e $28 \times 60 \times 60 = 100800$

100 800 cm = 1008 m = 1.008 km

$\therefore$ 1.008 km/h

f 80c/100 g = 100 g/80c

Divide by 80, then multiply by 100:

$100 \div 80 \times 100 = 125$

$\therefore$ 125 g/\$ = 0.125 kg/\$

g 160c/L = 1 L/160c

Divide by 160 and multiply by 100:

$1 \div 160 \times 100 = 0.625$

$\therefore$ 0.625 L/\$ = 62.5 L/\$100

h 5 mL/min = 1 min/5 mL

Multiply by 200:

200 min/1000 mL = 200 min/L

$\therefore 3\frac{1}{3}$ h/L

i 2 L/h = 1 h/2 L

As $60 \times 60 = 3600$, then

3600 s/2000 mL

$\therefore$ 1.8 s/mL

2 a $C = 12d$

b Increases

c Subs. $d = 35$ in $C = 12d$:

$C = 12 \times 35$

$= 420$ $\therefore$ \$420

d Subs. $C = 504$ in $C = 12d$:

$504 = 12d$

$d = \frac{504}{12}$

$= 42$ $\therefore$ 42 metres

e

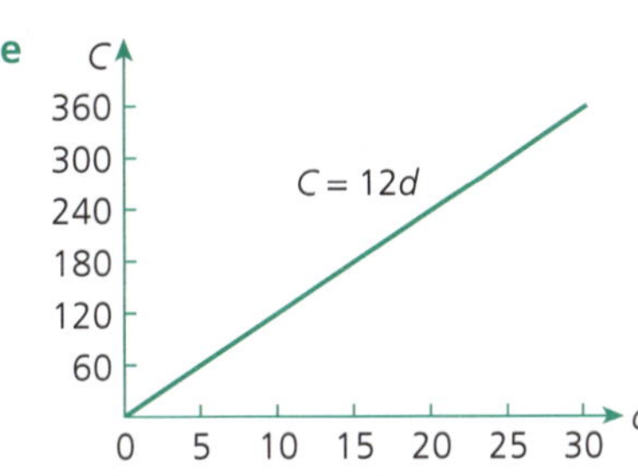

f Gradient $= \frac{\text{Rise}}{\text{Run}}$

$= \frac{360}{30}$

$= 12$

OR: From $C = 12d$, gradient = 12

3 a $W = 16.5n$

b Increases

c Subs. $n = 30$ in $W = 16.5n$:

$W = 16.5 \times 30$

$= 495$ $\therefore$ \$495

d Subs. $W = 594$ in $W = 16.5n$:

$594 = 16.5n$

$n = \frac{594}{16.5}$

$= 36$ $\therefore$ 36 hours

e

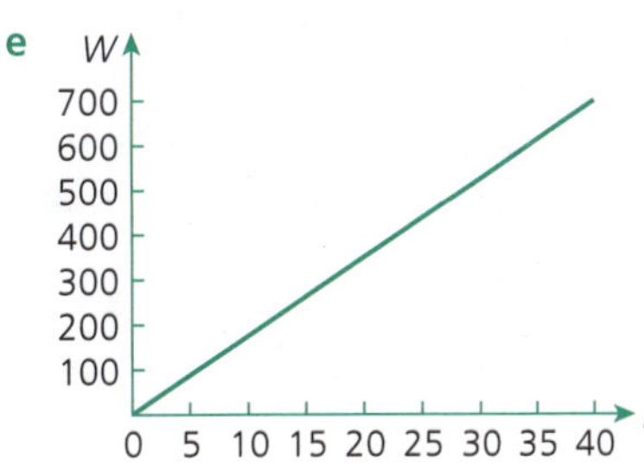

f From $W = 16.5n$, gradient = 16.5

OR: Gradient $= \frac{\text{Rise}}{\text{Run}}$

$= \frac{660}{40}$

$= 16.5$

4 a $D = 9n$

b Subs. $n = 35$ in $D = 9n$:

$D = 9 \times 35$

$= 315$ $\therefore$ 315 km

c Subs. $D = 405$ in $D = 9n$:

$405 = 9n$

$n = \frac{405}{9}$

$= 45$ $\therefore$ used 45 L

Fraction used $= \frac{45}{60}$

$= \frac{3}{4}$

$\therefore \frac{1}{4}$ of the petrol remains

5 a

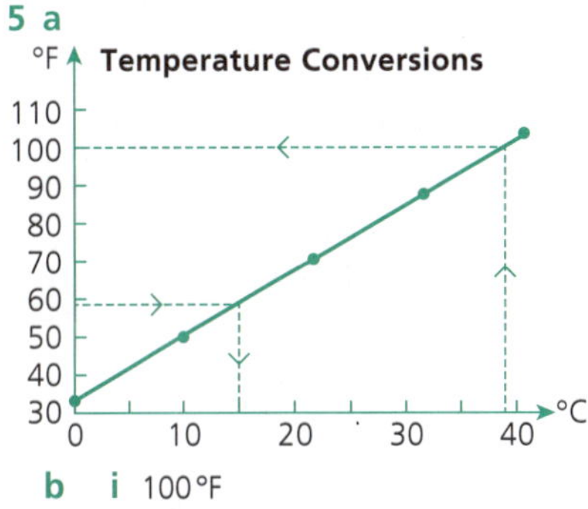

b **i** 100 °F

ii 15 °C

6 a **i**

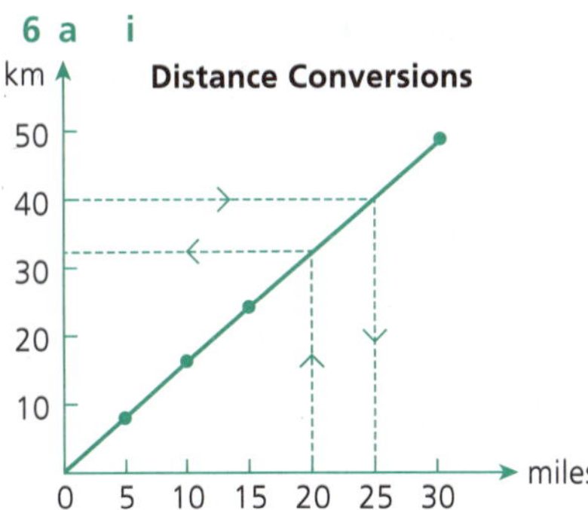

b **i** 32 km

ii 25 miles

7 a

Kilograms	0	5	10	15	20
Pounds	0	11	22	33	44

b

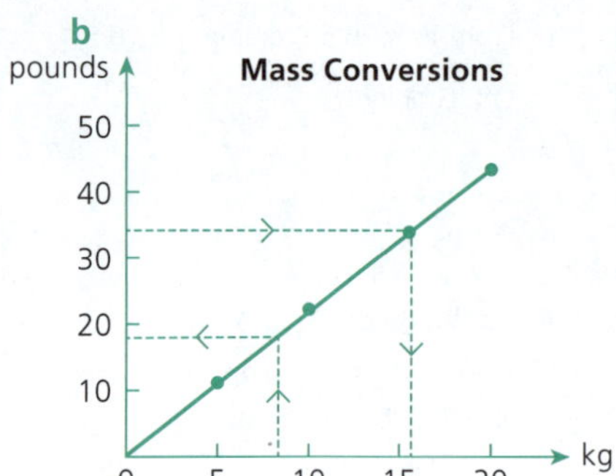

c **i** 17.6 pounds

ii As 15 kg = 33 pounds, then 150 kg = 330 pounds

8 a **i** \$A100 = €60

ii €100 = £80

b **i** \$A60 = €36

ii €42 = \$70

iii €60 = £48

iv £20 = €25

c **i** Use \$A100 = €60
Multiply by 50:
\$A5000 = €3000

ii Use €60 = \$A100
Multiply by 40:
€2400 = \$A4000

iii Use €100 = £80
Multiply by 7.5:
€750 = £600

iv Use £80 = €100
Multiply by 1.25:
£100 = €125

d **i** \$A100 = €60 = £48

ii £60 = €75 = \$A125

9 a **i** $y = kx$

ii Subs. $y = 12$ and $x = 4$ in $y = kx$:
$12 = 4k$
$k = 3$ $\therefore y = 3x$

iii Subs. $x = 20$ in $y = 3x$:
$y = 3 \times 20$
$= 60$

iv Subs. $y = 18$ in $y = 3x$:
$18 = 3x$
$x = \dfrac{18}{3}$
$= 6$

10 a $P = kt$

b Subs. $P = 10$ and $t = 2$ in $P = kt$:
$10 = 2k$
$k = 5$ $\therefore P = 5t$

c Subs. $t = 16$ in $P = 5t$:
$P = 5 \times 16$
$= 80$

d Subs. $P = 30$ in $P = 5t$:
$30 = 5t$
$t = \dfrac{30}{5}$
$= 6$

11 a $W = kn$
Subs. $W = 468$ and $n = 26$ in $W = kn$:
$468 = 26k$
$k = \dfrac{468}{26}$
$= 18$ $\therefore W = 18n$

b Subs. $n = 27$ in $W = 18n$:
$W = 18 \times 27$
$= 486$ $\therefore$ \$486

c Subs. $W = 648$ in $W = 18n$:
$648 = 18n$
$n = \dfrac{648}{18}$
$= 36$ $\therefore$ 36 hours

12 a $P = km$
Subs. $P = 7.8$ and $m = 1.5$ in $P = km$:
$7.8 = 1.5k$
$k = \dfrac{7.8}{1.5}$
$= 5.2$ $\therefore P = 5.2m$

b Subs. $m = 3.6$ in $P = 5.2m$:
$P = 5.2 \times 3.6$
$= 18.72$ $\therefore$ \$18.72

c Subs. $P = 14.56$ in $P = 5.2m$:
$14.56 = 5.2m$
$m = \dfrac{14.56}{5.2}$
$= 2.8$ $\therefore$ 2.8 kg

13 a $y = kx$
Subs. $y = 16.2$ and $x = 4.5$ in $y = kx$:
$16.2 = 4.5k$
$k = \dfrac{16.2}{4.5}$
$= 3.6$ $\therefore y = 3.6x$

b Subs. $x = 1.3$ in $y = 3.6x$:
$y = 3.6 \times 1.3$
$= 4.68$ $\therefore y = 4.68$
Subs. $y = 10.44$ in $y = 3.6x$:
$10.44 = 3.6x$
$x = \dfrac{10.44}{3.6}$
$= 2.9$ $\therefore x = 2.9$

x	1.3	2.9	4.5
y	4.68	10.44	16.2

14 a $F = km$
Subs. $F = 36$ and $m = 3$ in $F = km$:
$36 = 3k$
$k = 12$ $\therefore F = 12m$

b Subs. $m = 6.2$ in $F = 12m$:
$F = 12 \times 6.2$
$= 74.4$ $\therefore$ 74.4 Newtons

c Subs. $F = 43.2$ in $F = 12m$:
$43.2 = 12m$
$m = \dfrac{43.2}{12}$
$= 3.6$ $\therefore$ 3.6 kg

15 a $y = k\sqrt{x}$

b Subs. $y = 24$ and $x = 9$ in $y = k\sqrt{x}$:
$24 = k\sqrt{9}$
$3k = 24$
$k = 8$ $\therefore y = 8\sqrt{x}$

c Subs. $x = 16$ in $y = 8\sqrt{x}$:
$y = 8 \times \sqrt{16}$
$= 32$

d Subs. $y = 240$ in $y = 8\sqrt{x}$:
$240 = 8\sqrt{x}$
$\sqrt{x} = \dfrac{240}{8}$
$= 30$
$x = 30^2$
$= 900$

16 a $y = kx^2$

b Subs. $y = 48$ and $x = 4$ in $y = kx^2$:
$48 = k \times 4^2$
$16k = 48$
$k = 3$ $\therefore y = 3x^2$

c Subs. $x = 6$ in $y = 3x^2$:
$y = 3 \times 6^2$
$= 108$

d Subs. $y = 588$ in $y = 3x^2$:
$588 = 3x^2$
$x^2 = \dfrac{588}{3}$
$= 196$
$x = 14$ (as $x > 0$)

17

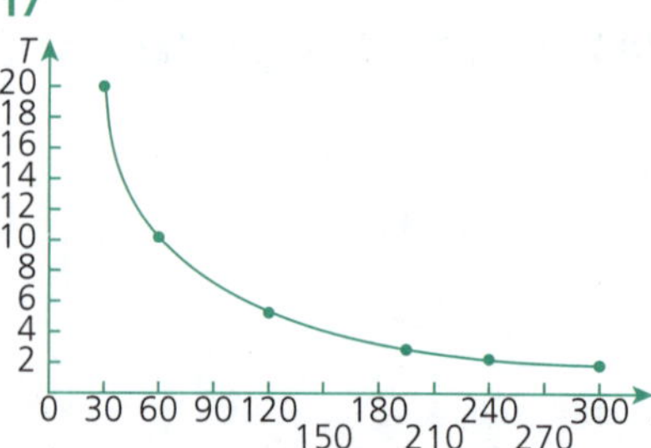

T is inversely proportional to s because the gradient of the curve is negative and, as s increases, T decreases.

Chapter 8 . . . pp. 112–118

Pythagoras and Trigonometry

1 a $x^2 = 16^2 + 12^2$
$= 256 + 144$
$= 400$
$\therefore x = \sqrt{400}$
$= 20$

b $y^2 + 10^2 = 26^2$
$\therefore y^2 = 26^2 - 10^2$
$= 676 - 100$
$= 576$
$\therefore y = \sqrt{576}$
$= 24$

c $a^2 + 15^2 = 17^2$
$\therefore a^2 = 17^2 - 15^2$
$= 289 - 225$
$= 64$
$\therefore a = \sqrt{64}$
$= 8$

d $b^2 = 21^2 + 28^2$
$= 441 + 784$
$= 1225$
$\therefore b = \sqrt{1225}$
$= 35$

e $c^2 + 1.2^2 = 1.5^2$
$\therefore c^2 = 1.5^2 - 1.2^2$
$= 2.25 - 1.44$
$= 0.81$
$\therefore c = \sqrt{0.81}$
$= 0.9$

f $d^2 + 1.6^2 = 2^2$
$\therefore d^2 = 2^2 - 1.6^2$
$= 4 - 2.56$
$= 1.44$
$\therefore d = \sqrt{1.44}$
$= 1.2$

g $n^2 + 24^2 = 40^2$
$\therefore n^2 = 40^2 - 24^2$
$= 1600 - 576$
$= 1024$
$\therefore n = \sqrt{1024}$
$= 32$

h $q^2 + 14^2 = 50^2$
$\therefore q^2 = 50^2 - 14^2$
$= 2500 - 196$
$= 2304$
$\therefore q = \sqrt{2304}$
$= 48$

I

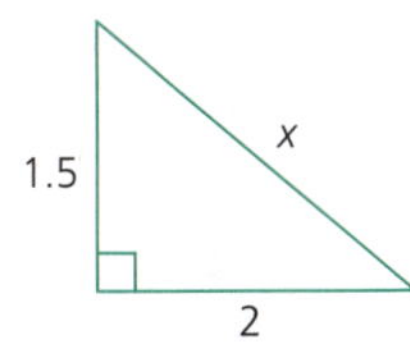

Let diagonal = x cm
$\therefore x^2 = 1.5^2 + 2^2$
$= 2.25 + 4$
$= 6.25$
$\therefore x = \sqrt{6.25}$
$= 2.5$
The diagonal BD is 2.5 cm.

j

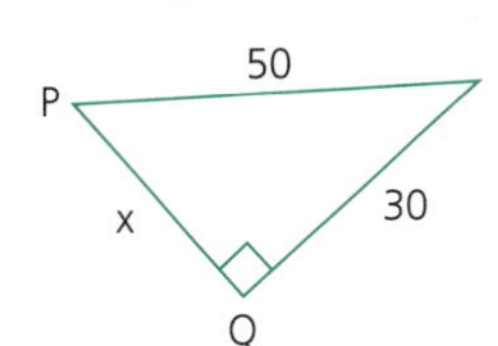

As OP = 25
then PR = 50
(diameter = 2 × radius)
Let PQ = x cm
$\therefore x^2 + 30^2 = 50^2$
$x^2 = 50^2 - 30^2$
$= 2500 - 900$
$= 1600$
$\therefore x = \sqrt{1600}$
$= 40$
PQ is 40 cm.

2 a To prove that:
$10^2 = 6^2 + 8^2$
LHS = 100
RHS = 36 + 64 = 100
$\therefore 10^2 = 6^2 + 8^2$
△ is right-angled.

b To prove that:
$14^2 = 12^2 + 9^2$
LHS = 14^2 = 196
RHS = $12^2 + 9^2$
$= 144 + 81$
$= 225$
$\therefore 14^2 \neq 12^2 + 9^2$
△ is not right-angled.

c To prove that:
$18^2 = 15^2 + 8^2$
LHS = 18^2
$= 324$
RHS = $15^2 + 8^2$
$= 225 + 64$
$= 289$
$\therefore 18^2 \neq 15^2 + 8^2$
△ is not right-angled.

d To prove that:
$25^2 = 24^2 + 7^2$
LHS = 625
RHS = $24^2 + 7^2$
$= 576 + 49$
$= 625$
$\therefore 25^2 = 24^2 + 7^2$
△ is right-angled.

e To prove that:
$6.1^2 = 6^2 + 1.1^2$
LHS = 6.1^2
$= 37.21$
RHS = $6^2 + 1.1^2$
$= 36 + 1.21$
$= 37.21$
$\therefore 6.1^2 = 6^2 + 1.1^2$
△ is right-angled.

f To prove that:
$41^2 = 40^2 + 9^2$
LHS = 41^2
$= 1681$
RHS = $40^2 + 9^2$
$= 1600 + 81$
$= 1681$
$\therefore 41^2 = 40^2 + 9^2$
△ is right-angled.

g To prove that:
$34^2 = 28^2 + 21^2$
LHS = 34^2
$= 1156$
RHS = $28^2 + 21^2$
$= 784 + 441$
$= 1225$
$\therefore 34^2 \neq 28^2 + 21^2$
△ is not right-angled.

h To prove that:
$85^2 = 84^2 + 13^2$
LHS = 85^2
$= 7225$
RHS = $84^2 + 13^2$
$= 7056 + 169$
$= 7225$
$\therefore 85^2 = 84^2 + 13^2$
△ is right-angled.

3 a $a^2 = 7^2 + 24^2$
$= 49 + 576$
$= 625$
$\therefore a = \sqrt{625}$
$= 25$
Triad is {7, 24, 25}.

b $y^2 = 9^2 + 12^2$
$= 81 + 144$
$= 225$
$\therefore y = \sqrt{225}$
$= 15$
Triad is {9, 12, 15}.

c $18^2 + n^2 = 30^2$
$\therefore n^2 = 30^2 - 18^2$
$= 900 - 324$
$= 576$
$n = \sqrt{576}$
$= 24$
Triad is {18, 24, 30}.

d $9^2 + t^2 = 41^2$
$\therefore t^2 = 41^2 - 9^2$
$= 1681 - 81$
$= 1600$
$\therefore t = \sqrt{1600}$
$= 40$
Triad is {9, 40, 41}.

e $13^2 + N^2 = 85^2$
$N^2 = 85^2 - 13^2$
$= 7225 - 169$
$= 7056$
$\therefore N = \sqrt{7056}$
$= 84$
Triad is {13, 84, 85}.

f $21^2 + 72^2 = p^2$
$\therefore p^2 = 441 + 5184$
$= 5625$
$\therefore p = \sqrt{5625}$
$= 75$
Triad is {21, 72, 75}.

4 a $a^2 + 4^2 = 7^2$
$\therefore a^2 = 7^2 - 4^2$
$= 49 - 16$
$= 33$
$\therefore a = \sqrt{33}$
$= 5.7$ (to 1 dec. pl.)

b $b^2 + 8^2 = 9^2$
$\therefore b^2 = 9^2 - 8^2$
$= 81 - 64$
$= 17$
$\therefore b = \sqrt{17}$
$= 4.1$ (to 1 dec. pl.)

c $c^2 + 3^2 = 4^2$
$\therefore c^2 = 4^2 - 3^2$
$= 16 - 9$
$= 7$
$\therefore c = \sqrt{7}$
$= 2.6$ (to 1 dec. pl.)

d $d^2 + 7^2 = 11^2$
$\therefore d^2 = 11^2 - 7^2$
$= 121 - 49$
$= 72$
$\therefore d = \sqrt{72}$
$= 8.5$ (to 1 dec. pl.)

e $e^2 + 9^2 = 12^2$
$\therefore e^2 = 12^2 - 9^2$
$= 144 - 81$
$= 63$
$\therefore e = \sqrt{63}$
$= 7.9$ (to 1 dec. pl.)

f $f^2 + 10^2 = 15^2$
$\therefore f^2 = 15^2 - 10^2$
$= 225 - 100$
$= 125$
$\therefore f = \sqrt{125}$
$= 11.2$ (to 1 dec. pl.)

g $g^2 + 9^2 = 17^2$
$\therefore g^2 = 17^2 - 9^2$
$= 289 - 81$
$= 208$
$\therefore g = \sqrt{208}$
$= 14.4$ (to 1 dec. pl.)

h $h^2 + 11^2 = 14^2$
$\therefore h^2 = 14^2 - 11^2$
$= 196 - 121$
$= 75$
$\therefore h = \sqrt{75}$
$= 8.7$ (to 1 dec. pl.)

5 a $n^2 = 5^2 + 5^2$
$= 25 + 25$
$= 50$
$\therefore n = \sqrt{50}$

b $m^2 + 18^2 = 25^2$
$m^2 = 25^2 - 18^2$
$= 625 - 324$
$= 301$
$\therefore m = \sqrt{301}$

c $r^2 + 8^2 = 14^2$
$\therefore r^2 = 14^2 - 8^2$
$= 196 - 64$
$= 132$
$\therefore r = \sqrt{132}$

d $s^2 + 8^2 = 19^2$
$\therefore s^2 = 19^2 - 8^2$
$= 361 - 64$
$= 297$
$\therefore s = \sqrt{297}$

6

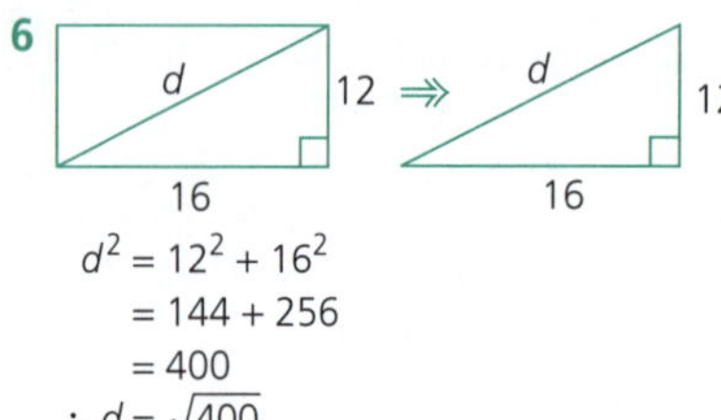

$d^2 = 12^2 + 16^2$
$= 144 + 256$
$= 400$
$\therefore d = \sqrt{400}$
$= 20$
The length of diagonal is 20 cm.

7

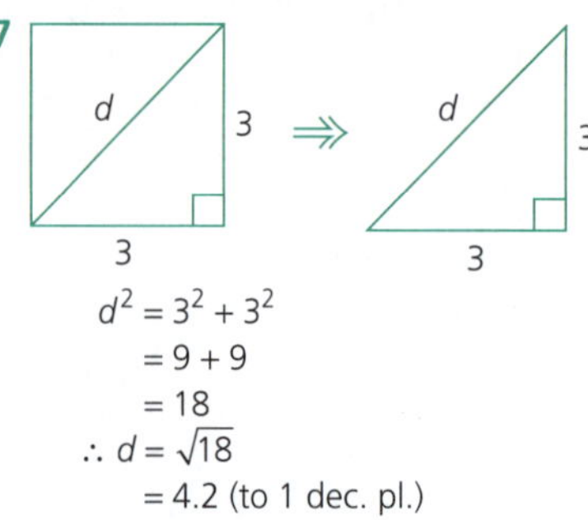

$d^2 = 3^2 + 3^2$
$= 9 + 9$
$= 18$
$\therefore d = \sqrt{18}$
$= 4.2$ (to 1 dec. pl.)
The diagonal is 4.2 cm.

8

h
35
21

$h^2 + 21^2 = 35^2$
$\therefore h^2 = 35^2 - 21^2$
$= 1225 - 441$
$= 784$
$\therefore h = \sqrt{784}$
$= 28$
The brace reaches 28 cm up wall.

9

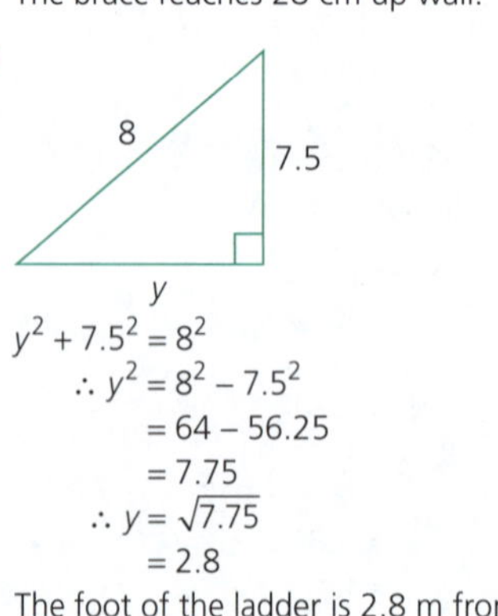

$y^2 + 7.5^2 = 8^2$
$\therefore y^2 = 8^2 - 7.5^2$
$= 64 - 56.25$
$= 7.75$
$\therefore y = \sqrt{7.75}$
$= 2.8$
The foot of the ladder is 2.8 m from the wall.

10

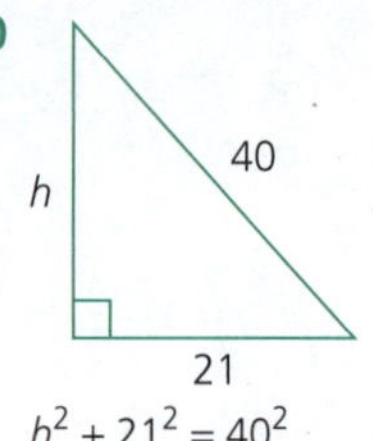

$h^2 + 21^2 = 40^2$
$h^2 = 40^2 - 21^2$
$= 1600 - 441$
$= 1159$
$\therefore h = \sqrt{11.59}$
$= 34.044089$
Distance from top $= 45 - 34.044089$
$= 10.955911$
$= 10.96$ (to 4 sig. fig.)
Wires are attached 10.96 m from the top.

11

18
x
x
18
x
x

Let length of square be x cm.
$\therefore x^2 + x^2 = 18^2$
$2x^2 = 324$
$\therefore x^2 = 162$
$x = \sqrt{162}$
$= 12.7$ (to 1 dec. pl.)
The length of the side is 12.7 cm.

12 $w^2 + 20^2 = 25^2$
$\therefore w^2 = 25^2 - 20^2$
$= 625 - 400$
$= 225$
$\therefore w = \sqrt{225}$
$= 15$
The width is 15 cm.

13

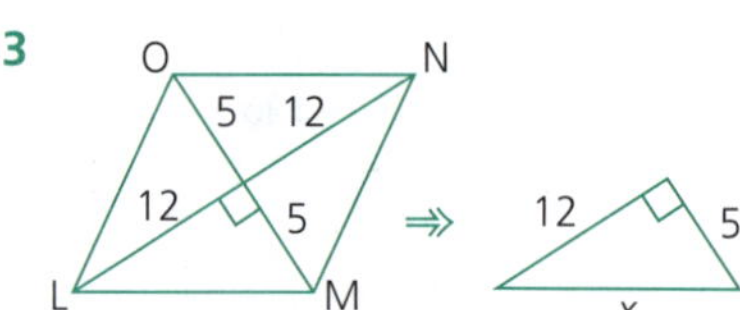

Let length of side be x cm.
$x^2 = 5^2 + 12^2$
$= 169$
$\therefore x = \sqrt{169}$
$= 13$
The length of each side is 13 cm.

14

15
y
x
8
6
x
8
6
15
y
10

$x^2 = 6^2 + 8^2$
$= 36 + 64$
$= 100$
$\therefore x = \sqrt{100}$
$= 10$

From second △,
$y^2 + 10^2 = 15^2$
$\therefore y^2 = 15^2 - 10^2$
$= 225 - 100$
$= 125$
$\therefore y = \sqrt{125}$
$= 11.2$ (to 3 sig. fig.)

15

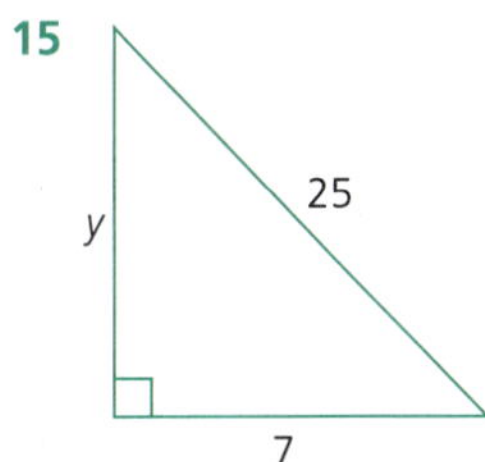

Let third side by y m.
$y^2 + 7^2 = 25^2$
$\therefore y^2 = 25^2 - 7^2$
$= 625 - 49$
$= 576$
$\therefore y = \sqrt{576}$
$= 24$
The third side is 24 m.

16

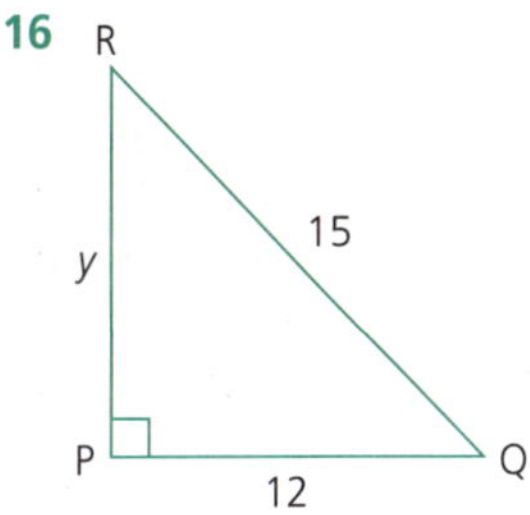

Let PR be y m.
$\therefore y^2 + 12^2 = 15^2$
$\therefore y^2 = 15^2 - 12^2$
$= 225 - 144$
$= 81$
$\therefore y = \sqrt{81}$
$= 9$
$\therefore$ PR = 9 m
Perimeter = 9 + 12 + 15 = 36
The perimeter is 36 m.

17

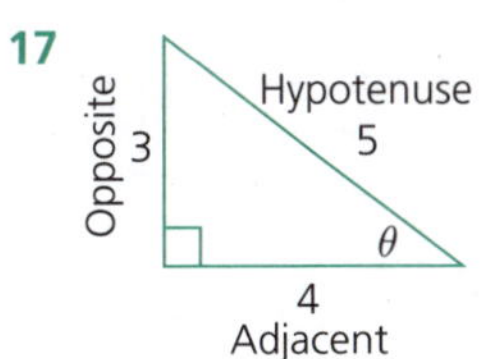

a $\sin\theta = \frac{3}{5}$

b $\cos\theta = \frac{4}{5}$

c $\tan\theta = \frac{3}{4}$

18

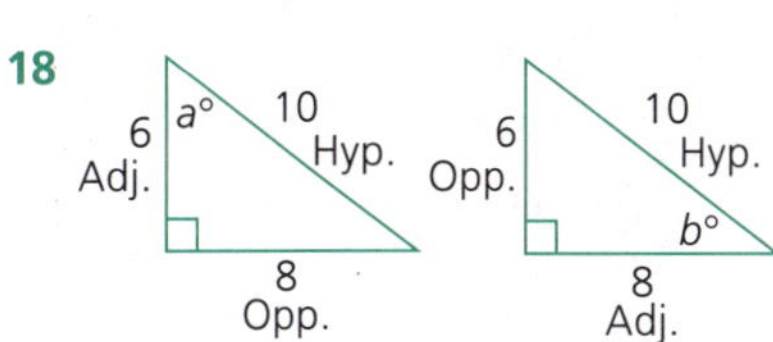

a $\sin a° = \frac{8}{10} = \frac{4}{5}$

b $\sin b° = \frac{6}{10} = \frac{3}{5}$

c $\cos a° = \frac{6}{10} = \frac{3}{5}$

19

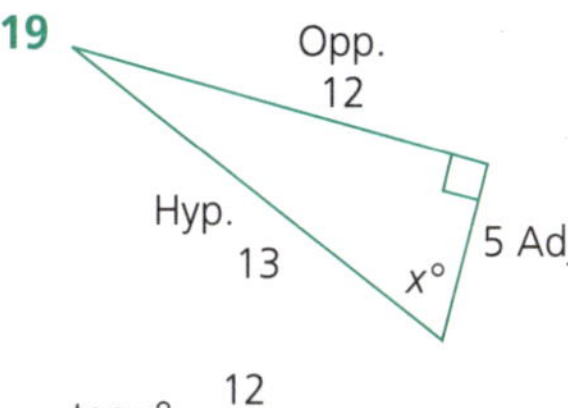

$\tan x° = \frac{12}{5}$

20 a 0.9573

b 0.4932

c 4.8077

21 a 100°44′

b 51°46′

c 93°12′

d 103°45′

e 247°24′

22 a

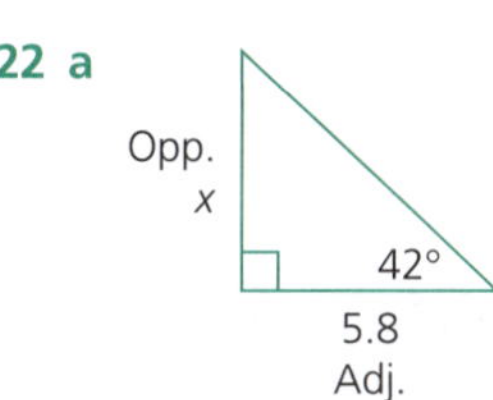

$\frac{x}{5.8} = \tan 42°$
$x = 5.8 \times \tan 42°$
≈ 5.22

b

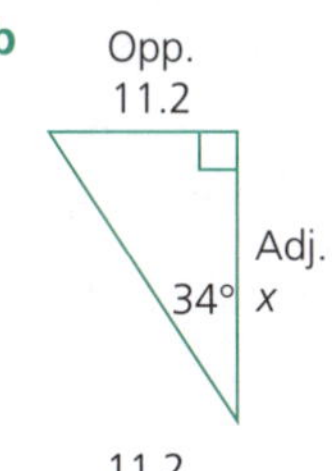

$\frac{11.2}{x} = \tan 34°$
$\frac{11.2}{\tan 34°} = x$
$\therefore x = 16.60$

c

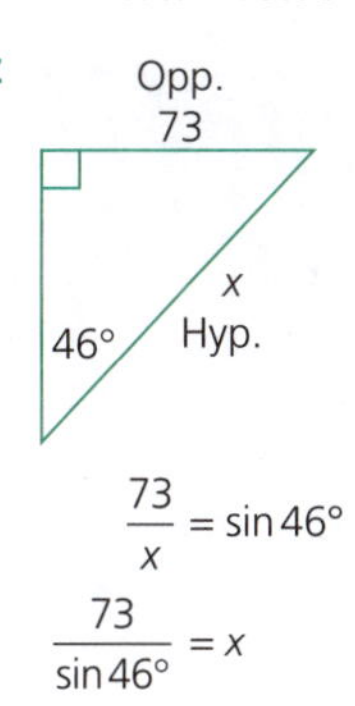

$\frac{73}{x} = \sin 46°$
$\frac{73}{\sin 46°} = x$
$\therefore x = 101.48$

d

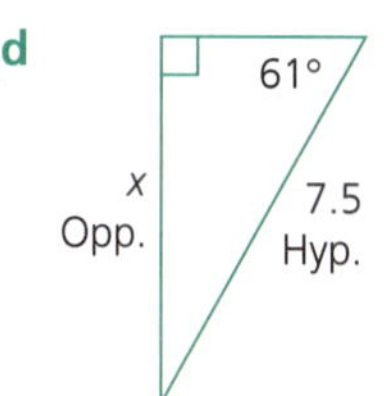

$\frac{x}{7.5} = \sin 61°$
$x = 7.5 \times \sin 61°$
$x \approx 6.56$

e

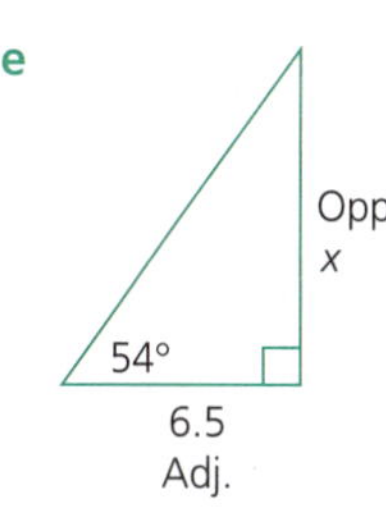

$\frac{x}{6.5} = \tan 54°$
$x = 6.5 \times \tan 54°$
≈ 8.95

f

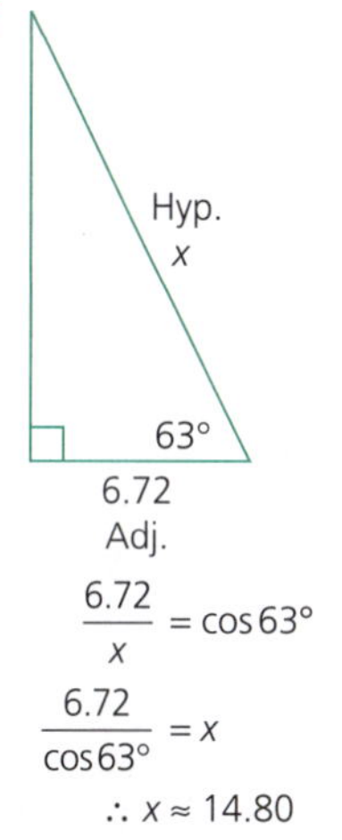

$\frac{6.72}{x} = \cos 63°$
$\frac{6.72}{\cos 63°} = x$
$\therefore x \approx 14.80$

g

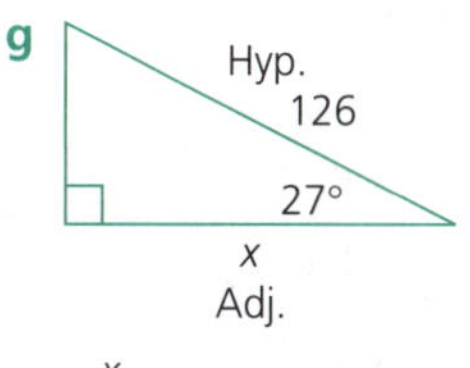

$\frac{x}{126} = \cos 27°$
$x = 126 \times \cos 27°$
≈ 112.27

h

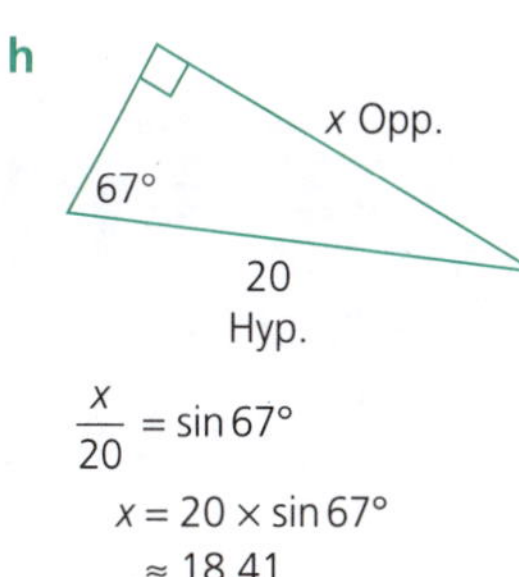

$\frac{x}{20} = \sin 67°$
$x = 20 \times \sin 67°$
≈ 18.41

i

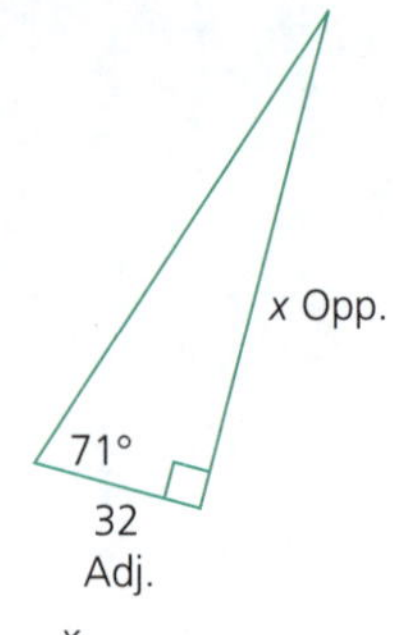

$\frac{x}{32} = \tan 71°$

$x = 32 \times \tan 71°$

≈ 92.93

j

Opp. 12

102.3
Hyp.

81°

$\frac{x}{102.3} = \sin 81°$

$x = 102.3 \times \sin 81°$

≈ 101.04

23 a $\frac{y}{48} = \sin 26°12'$

$\therefore y = 48 \times \sin 26°12'$

$= 21.192\,281$

$= 21.2$ (to 1 dec. pl.)

b $\frac{y}{24} = \tan 44°9'$

$\therefore y = 24 \times \tan 44°9'$

$= 23.298\,265$

$= 23.3$ (to 1 dec. pl.)

c $\frac{y}{4.9} = \cos 62°18'$

$\therefore y = 4.9 \times \cos 62°18'$

$= 2.277\,726$

$= 2.3$ (to 1 dec. pl.)

d $\frac{18}{y} = \sin 22°17'$

$\therefore \frac{y}{18} = \frac{1}{\sin 22°17'}$

Reciprocals of both sides.

$\therefore y = \frac{18}{\sin 22°17'}$

$= 47.469\,98$

$= 47.5$ (to 1 dec. pl.)

e $\frac{y}{42} = \tan 71°33'$

$90° - 18°27' = 71°33'$
Note use of opposite angle.

$\therefore y = 42 \times \tan 71°33'$

$= 125.889\,76$

$= 125.9$ (to 1 dec. pl.)

f $\frac{27}{y} = \cos 46°29'$

$\therefore \frac{y}{27} = \frac{1}{\cos 46°29'}$

$\therefore y = \frac{27}{\cos 46°29'}$

$= 39.211\,953$

$= 39.2$ (to 1 dec. pl.)

g $\frac{29}{y} = \sin 16°42'$

$\therefore \frac{y}{29} = \frac{1}{\sin 16°42'}$

$\therefore y = \frac{29}{\sin 16°42'}$

$= 100.918\,53$

$= 100.9$ (to 1 dec. pl.)

h $\frac{y}{6.5} = \tan 40°8'$

$90° - 49°52' = 40°8'$
Note use of opposite angle.

$\therefore y = 6.5 \times \tan 40°8'$

$= 5.479\,9744$

$= 5.5$ (to 1 dec. pl.)

i $\frac{15.8}{y} = \cos 67°8'$

$\therefore \frac{y}{15.8} = \frac{1}{\cos 67°8'}$

$\therefore y = \frac{15.8}{\cos 67°8'}$

$= 40.660\,035$

$= 40.7$ (to 1 dec. pl.)

24 a

6
x°
12.9

$\tan x = \frac{6}{12.9}$

$\therefore x = 25$

b

4.8
x°
9

$\cos x = \frac{4.8}{9}$

$\therefore x = 58$

c

4.3
x°
5.9

$\tan x = \frac{5.9}{4.3}$

$\therefore x = 54$

d

14
31
x°

$\sin x = \frac{14}{31}$

$\therefore x = 27$

e

42
x°
39

$\tan x = \frac{42}{39}$

$\therefore x = 47$

f

3.7
x°
9.6

$\cos x = \frac{3.7}{9.6}$

$\therefore x = 67$

25 a 54°22′

b 77°2′

c 60°57′

d 0°7′

e 29°45′

f 55°9′

g 34°51′

26 a $\sin \theta = \frac{8}{17}$

$\therefore \theta = 28°4'$

b $\tan x = \frac{19}{7}$

$\therefore x = 69°47'$

c $\cos \phi = \frac{18}{24}$

$\therefore \phi = 41°25'$

d $\sin y = \frac{7.5}{14.9}$

$\therefore y = 30°13'$

e $\cos \omega = \frac{19}{21.5}$

$\therefore \omega = 27°54'$

f $\tan \alpha = 43$

$\therefore \alpha = 68°26'$

27

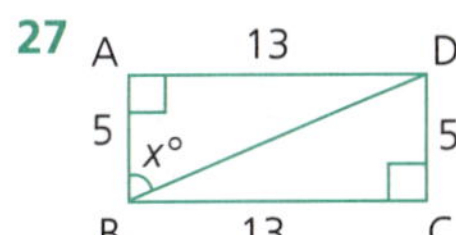

In $\triangle ABD$ let $\angle ABD = x°$

$\therefore \tan x = \dfrac{13}{5}$

$\therefore x = 69$

$\therefore \angle ABD = 69°$

28 $\sin\theta = \dfrac{1.2}{1.3}$

$\therefore \theta = 31$

$\therefore$ The angle between the ladder and the wall is 31°.

29

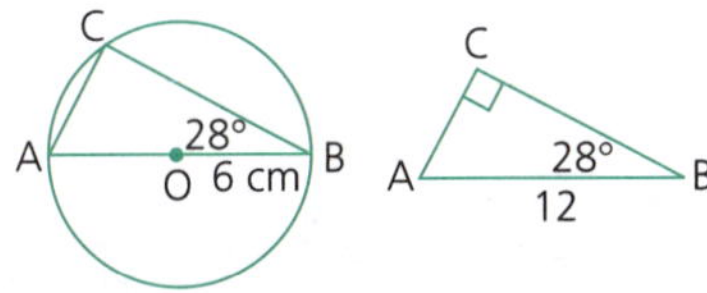

Since AB is a diameter, $\angle BCA = 90°$. (Angle in a semicircle.)

$\dfrac{BC}{12} = \cos 28°$

$\therefore$ BC $= 12 \times \cos 28°$

≈ 10.6

$\therefore$ The length of BC = 10.6 cm.

30

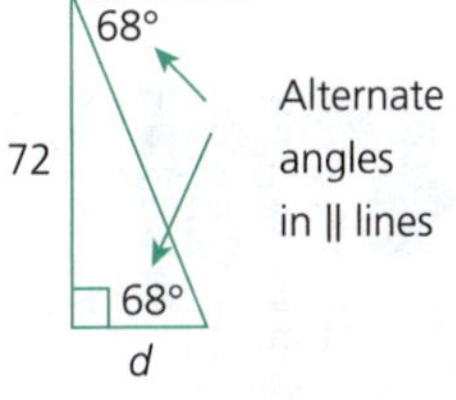

Let d = distance from woman to base.

$\dfrac{72}{d} = \tan 68°$

$\dfrac{72}{\tan 68°} = d$

$\therefore d \approx 29$

$\therefore$ The distance from the woman to the base of the tower is 29 m.

31

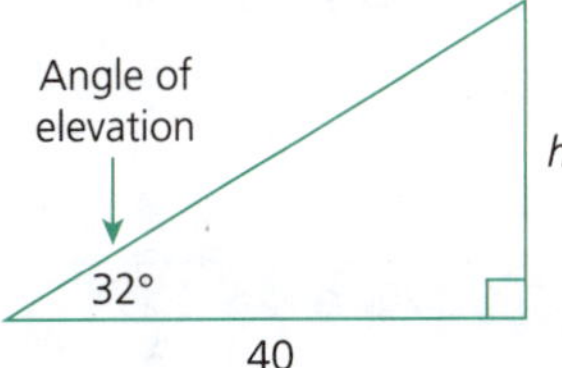

$\dfrac{h}{40} = \tan 32°$

$h = 40 \times \tan 32°$

≈ 25

$\therefore$ Height of the building is 25 metres.

32

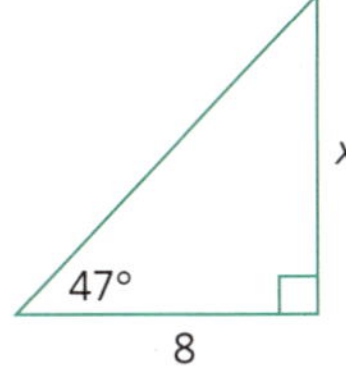

$\dfrac{h}{8} = \tan 47°$

$h = 8 \times \tan 47°$

$\therefore h \approx 8.58$

$\therefore$ Height of the pole is 8.58 metres.

33

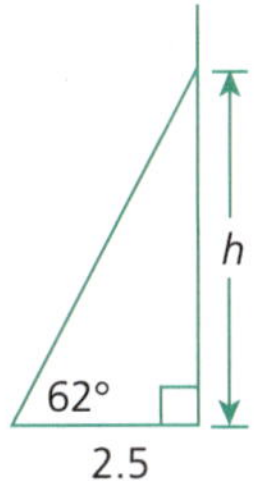

$\dfrac{h}{2.5} = \tan 62°$

$h = 2.5 \times \tan 62°$

$\therefore h \approx 4.7$

$\therefore$ The ladder reaches 4.7 metres up the wall.

34

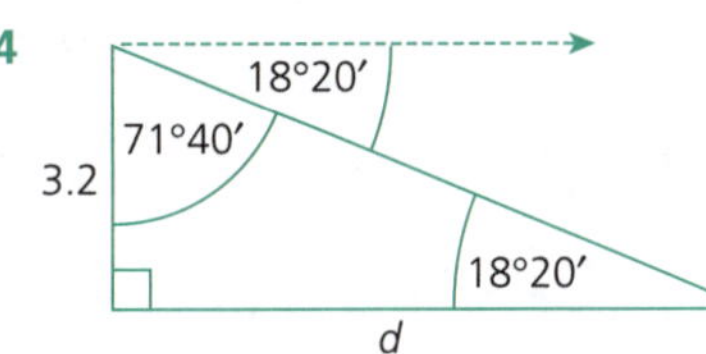

Let distance be d metres.

$\angle$depression = $\angle$elevation

$\dfrac{d}{3.2} = \tan 71°40'$

$\therefore d = 3.2 \times \tan 71°40'$

$= 9.657\,0564$

$= 9.7$ (to 1 dec. pl.)

(90° – 18°20′ = 71°40′)

The crocodile is 9.7 m from the ship.

35

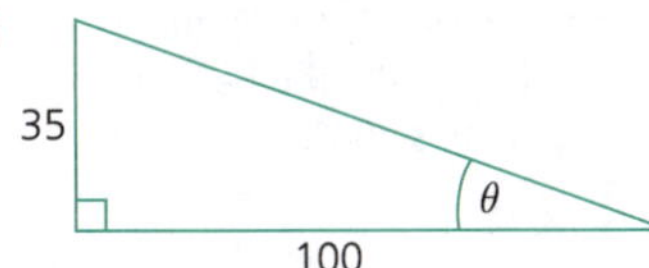

Let angle be $\theta°$.

$\tan\theta = \dfrac{35}{100}$

$\therefore \theta = 19°17'$

Angle of elevation is 19°17′.

36

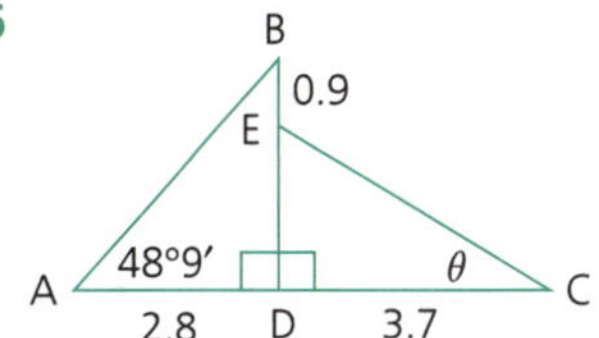

From $\triangle ADB$, $\dfrac{BD}{2.8} = \tan 48°9'$

$\therefore BD = 2.8 \times \tan 48°9'$

$= 3.126\,1349$

$DE = 3.126\,1349 - 0.9 = 2.226\,1349$

Let $\angle DCE = \theta°$

From $\triangle DCE$, $\tan\theta = \dfrac{DE}{3.7}$

$= \dfrac{2.2261349}{3.7}$

$\therefore \theta = 31°2'$

Then BD = 3.1 m (to 1 dec. pl.) and $\angle DCE = 31°2'$.

37

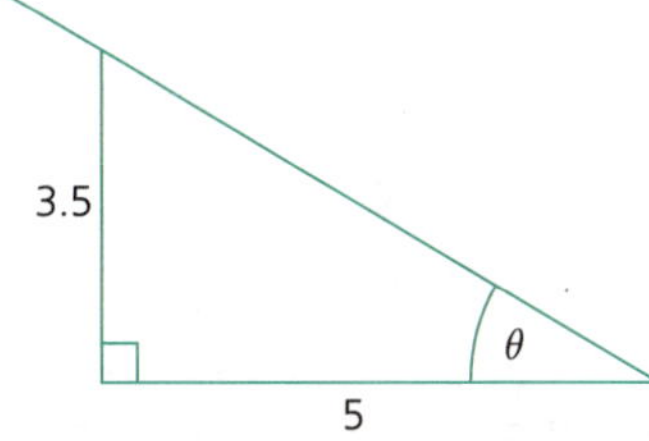

Let the angle of elevation of the sun be $\theta°$.

$\tan\theta = \dfrac{3.5}{5}$

$\therefore \theta = 35°$ (to the nearest degree)

38 a

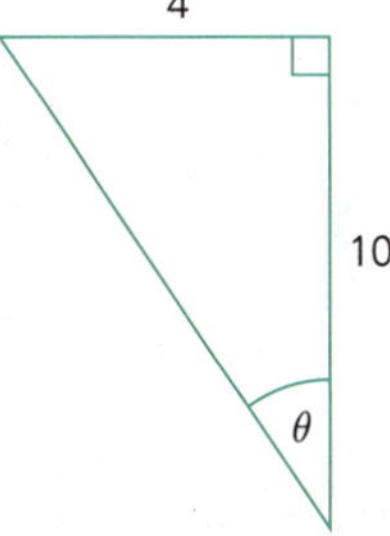

Let θ be the angle that the ball must be kicked to strike the outer can.

$\tan\theta = \dfrac{4}{10}$

$\therefore \theta = 21°48'$

The ball can be kicked at any angle between 0° and 21°48′ relative to the nearest can.

b As for (a), but $\tan\theta = \dfrac{4}{15}$

$\theta = 14°56'$

The range of angles is now within 0° and 14°56′.

39 a 130°

b $180 + 55 = 235$ $\therefore$ 235°

c $360 - 70 = 290$ $\therefore$ 290°

d $90 - 40 = 50$ $\therefore$ 050°

e

$360 - 80 = 280$ $\quad\therefore 280°$

f

$180 + 60 = 240$ $\quad\therefore 240°$

40

$\tan\theta = \dfrac{12}{8}$

$\therefore \theta = 56°$ (nearest degree)

$90 + 56 = 146$ $\quad\therefore 146°$

41

$\dfrac{x}{20} = \sin 40°$

$x = 20 \sin 40°$

$= 12.86$ (2 dec. pl.)

$\therefore$ It is 12.86 km north.

42 Distance $= 80 \times 2$

$= 160$ $\quad\therefore 160$ km

$\dfrac{x}{160} = \cos 10°$

$x = 160 \cos 10°$

$= 157.57$ (2 dec. pl.)

$\therefore$ The car is 157.57 km west.

43

$\dfrac{300}{x} = \cos 20°$

$x = \dfrac{300}{\cos 20°}$

$= 319$ (nearest whole)

$\therefore$ The plane has travelled 319 km.

44

$\dfrac{x}{200} = \cos 40°$

$x = 200 \cos 40°$

$= 153$ (nearest whole)

$\therefore$ The plane is 153 km south.

45

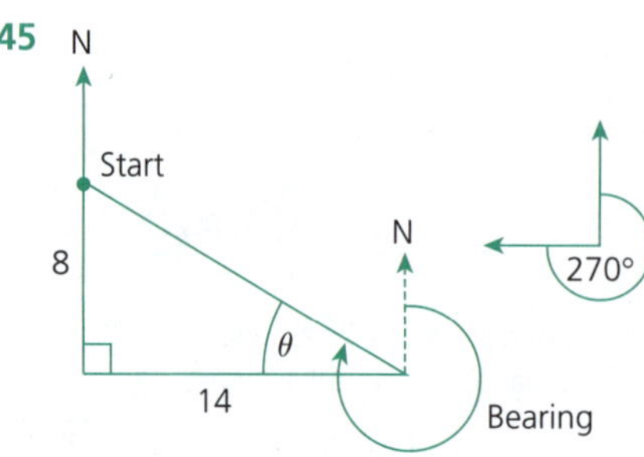

From $\triangle$, $\tan\theta = \dfrac{8}{14}$

$\theta = 30°$ (to nearest degree)

Bearing is measured clockwise from N to the direction of the starting point.

This angle is $270° + \theta°$, that is, $270° + 30° = 300°$. The bearing of the starting point is 300°.

Chapter 9 . . . pp. 132–133

Measurement

1 a 16 cm, $r = 8$

Perimeter $= 16 + \frac{1}{2} \times 2 \times 8 \times \pi$

$= 16 + 25.13274123$

$= 41.13274123$

$= 41.133$ (to 3 dec. pl.)

$\therefore$ The perimeter is 41.133 cm.

b

Perimeter $= 15 + 15 + \frac{1}{6} \times 2 \times \pi \times 15$

$= 30 + 15.707963$

$= 45.707963$

$= 45.708$ (to 3 dec. pl.)

$\therefore$ The perimeter is 45.708 cm.

2 a

Using Pythagoras' Theorem to find x:

$x^2 = 5^2 + 12^2 = 169$

$\therefore x = \sqrt{169} = 13$

Perimeter $= 14 + 12 + 19 + 13$

$= 58$

Perimeter is 58 m.

b

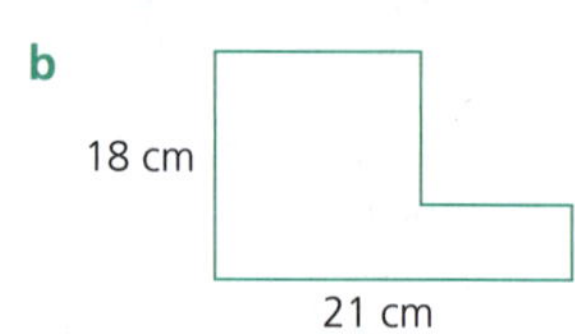

Perimeter $= 2(18 + 21)$

$= 2(39)$

$= 78$

$\therefore$ The perimeter is 78 cm.

3

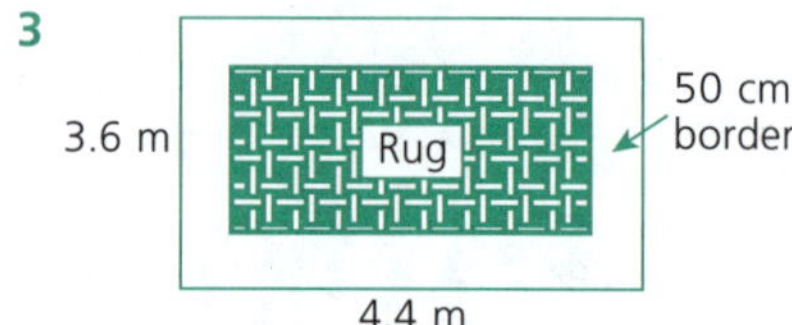

Rug length $= 4.4 - 2(0.5)$

$= 4.4 - 1$

$= 3.4$ m

Rug width $= 3.6 - 2(0.5)$

$= 3.6 - 1$

$= 2.6$ m

$\therefore$ Rug perimeter $= 2(3.4 + 2.6)$

$= 2(6)$

$= 12$

$\therefore$ The rug's perimeter is 12 m.

4

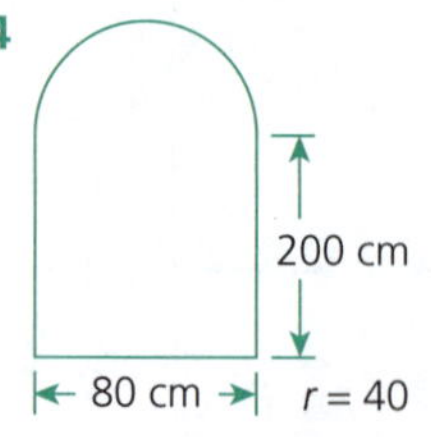

$P = 200 + 80 + 200 + \frac{1}{2} \times 2 \times \pi \times 40$

$= 480 + 125.663706$

$= 605.663706$

$= 606$ (to the nearest whole number)

$\therefore$ The perimeter is 606 cm.

5

AC = 6 cm
BD = 10 cm

$A = \frac{1}{2}$(product of diagonals)
$= \frac{1}{2}(6 \times 10)$
$= \frac{1}{2}(60)$
$= 30$
$\therefore$ The area is 30 cm^2.

6 a

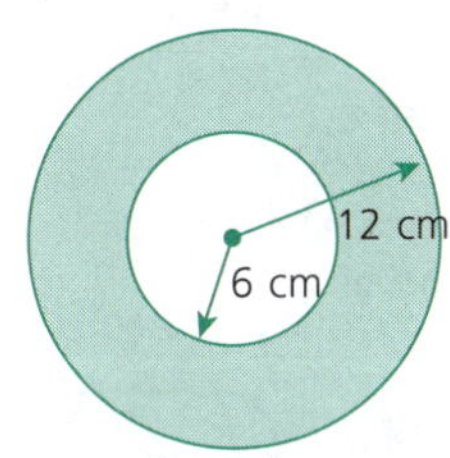

$A = \pi R^2 - \pi r^2$
$= \pi \times 12^2 - \pi \times 6^2$
$= \pi[12^2 - 6^2]$
$= \pi[144 - 36]$
$= \pi[108]$
$= 108\pi$
$\therefore$ The area is 108π cm^2.

b

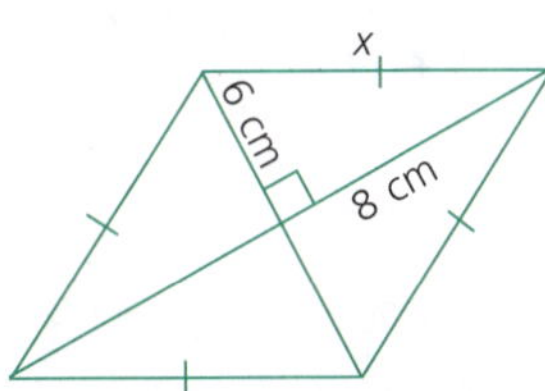

Diagonals are 12 cm and 16 cm.
$\therefore A = \frac{1}{2}$[product of diagonals]
$= \frac{1}{2}(6 \times 8)$
$= \frac{1}{2}(48)$
$= 24$
$\therefore$ The area is 24 cm^2.
Now, diagonals meet at right angles
$\therefore$ find x, using Pythagoras' Theorem.
$x^2 = 6^2 + 8^2$
$= 36 + 64$
$= 100$
$x = \sqrt{100}$
$= 10$
$\therefore$ Length of one side = 10 cm
$\therefore$ Perimeter = 4 × 10 = 40
$\therefore$ The perimeter is 40 cm.

7

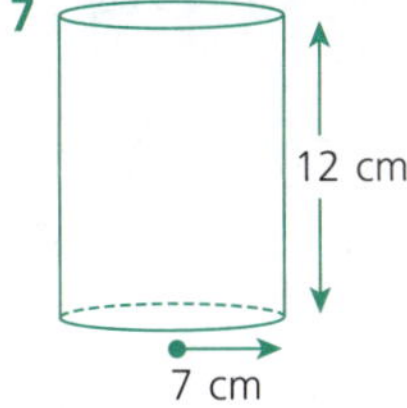

Total surface area
$= 2\pi r^2 + 2\pi rh$
$= 2 \times \pi \times 7^2 + 2 \times \pi \times 7 \times 12$
$= 835.663\,6459$
$= 836$ (correct to 3 sig. fig.)
$\therefore$ Surface area is 836 cm^2.

8

24 cm, x, 10, 20 cm, 30 cm; 24, x, 10

First find x.
$\therefore x^2 = 24^2 + 10^2$
$= 576 + 100$
$= 676$
$x = \sqrt{676}$
$= 26$
$\therefore$ Surface area
$= 2 \times \frac{1}{2} \times 20 \times 24 + 2 \times 26 \times 30 + 20 \times 30$
$= 480 + 1560 + 600$
$= 2640$
$\therefore$ The surface area is 2640 cm^2.

9

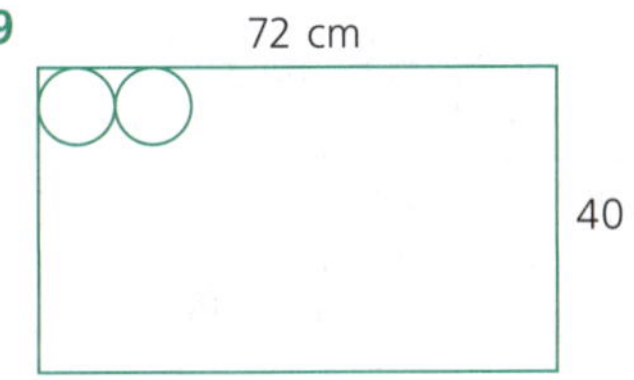

a Circles:
Lengthways = 72 ÷ 4
= 18
Sideways = 40 ÷ 4
= 10,
$\therefore$ Number of circles = 18 × 10
= 180
$\therefore$ The maximum number is 180.

b Amount of scrap metal
= (Area of sheet) – (Area of all circles)
$= 72 \times 40 - 180(\pi \times 2^2)$
$= 2880 - 2261.946\,711$
$= 618.053\,2895$
= 618 (to the nearest whole number)
$\therefore$ The amount of scrap metal is 618 cm^2.

10

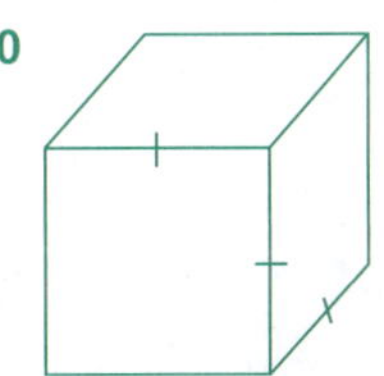

Surface area $= 6s^2 = 54$
$\therefore s^2 = 9$
$s = 3$ [cannot be $s = -3$]
$\therefore$ Each side is 3 cm.

11

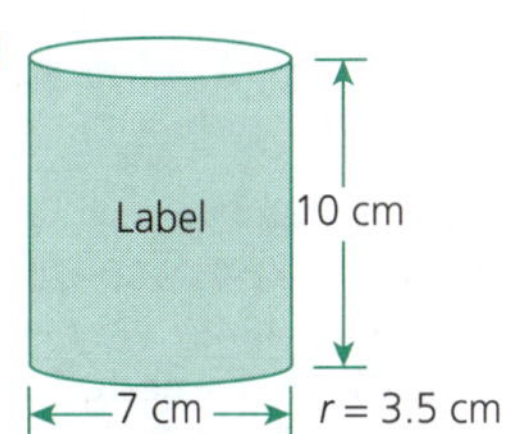

Area of label
= Curved surface area of cylinder
$\therefore$ Curved surface area
$= 2\pi rh$
$= 2 \times \pi \times 3.5 \times 10$
$= 219.911\,4858$
= 220 (to the nearest whole number)
$\therefore$ The area of label is 220 cm^2.

12

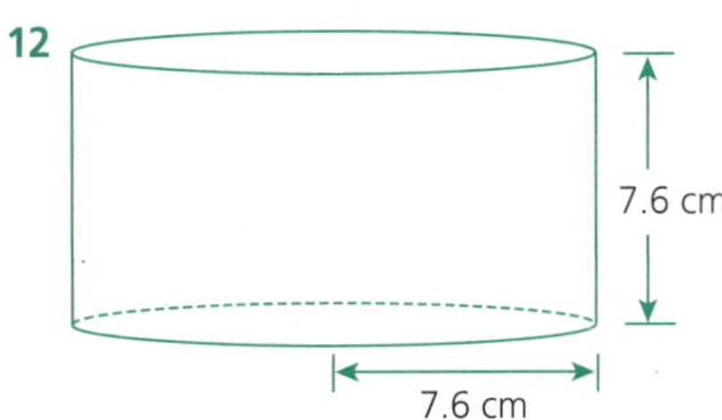

$V = \pi r^2 h$
$= \pi \times 7.6^2 \times 7.6$
$= 1379.083\,777$
$= 1379.084$ (to 3 dec. pl.)
$\therefore$ The volume is 1379.084 cm^3.

13 a $V = \pi r^2 h$
$= \pi \times 16^2 \times 21$
$= 16\,889.202\,11\ldots$
$= 16\,889.20$ (2 dec. pl.)
$\therefore 16\,889.20$ cm^3

b $V = \pi r^2 h$
$= \pi \times 4.5^2 \times 11$
$= 699.789\,7636\ldots$
$= 669.79$ (2 dec. pl.)
$\therefore 669.79$ cm^3

c $V = 23 \times 10 \times 10 - 3 \times \pi \times 2.5^2 \times 10$
$= 1710.951\,377\ldots$
$= 1710.95$ (2 dec. pl.)
$\therefore 1710.95$ cm^3

d $V = \frac{1}{2} \times \pi r^2 h$
$= \frac{1}{2} \times \pi \times 7^2 \times 12$
$= 923.628\,2402\ldots$
$= 923.63$ (2 dec. pl.)
$\therefore 923.63$ cm^3

14 $V = \pi r^2 h$
$= \pi \times 3^2 \times 4$
$= 36\pi$
$\therefore 36\pi$ cm^3

15

2.3 m
1.8 m

$V = \pi r^2 h$
$= \pi \times 1.8^2 \times 2.3$
$= 23.411\,148\,45$
$\therefore$ Volume is $23.411\,148\,45\ m^3$.
$\therefore$ Capacity $= 23.411\,148\,45 \times 1000$
$= 23\,411.148\,45$
$= 23\,411$ (to the nearest litre)
$\therefore$ The capacity is 23 411 litres.

16 $1\ L = 1000\ mL = 1000\ cm^3$
$V = lbh$
$1000 = 16 \times 20 \times h$
$1000 = 320h$
$h = 3.125$
$\therefore$ The water is 3.125 cm deep.

17 $V = \pi r^2 h$
$= \pi \times 20^2 \times 30$
$= 37\,699.111\,84\ cm^3$

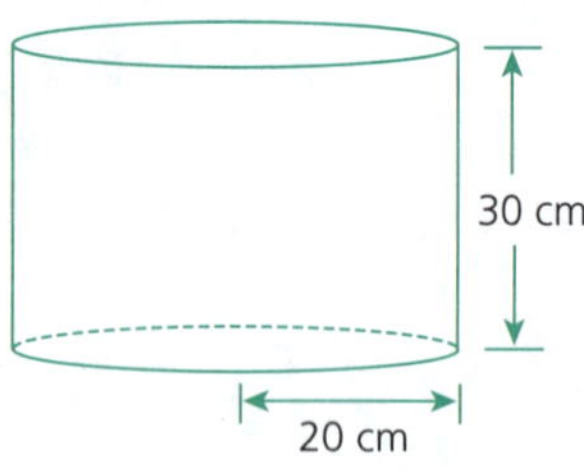

$\therefore$ The capacity $= 2356.194\,49 \div 1000$
$= 2.356\,194\,49$ litres
But if only $\frac{2}{3}$ full,
$\therefore \frac{2}{3} \times 2.356\,194\,49$
$= 1.570\,796\,327$
$= 1.571$ (3 dec. pl.)
$\therefore$ It requires 1.571 litres.

18

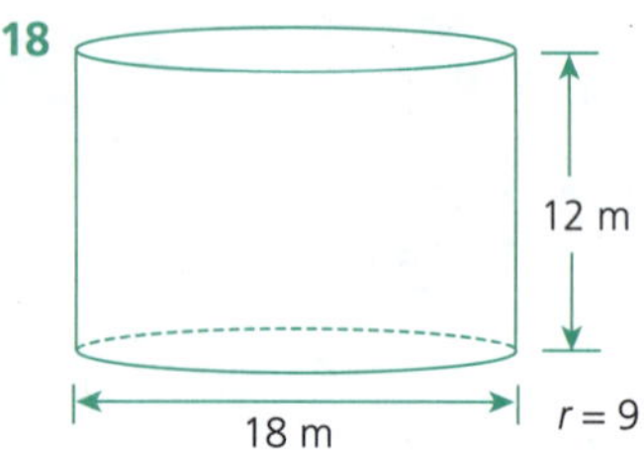

Volume $= \pi r^2 h$
$= \pi \times 9^2 \times 12$
$= 3053.628\,059$
Capacity $= 3053.628\,059$ kilolitres
$= 3054$ (to the nearest kilolitre)
$\therefore$ The capacity is 3054 kL.

19

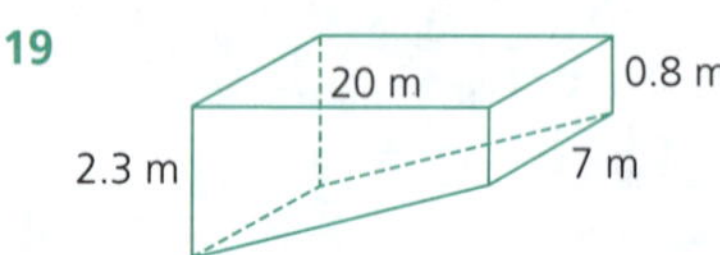

a Area of walls
= 2 rectangles + 2 trapezia
$\therefore$ Area $= 2.3 \times 7 + 0.8 \times 7$ ← Ends
$+ 2 \times \frac{1}{2} \times 20(0.8 + 2.3)$
$= 16.1 + 5.6 + 62$ ↑
$= 83.7\ m^2$, sides Sides
$\therefore$ Cost $= \$42 \times 83.7$
$= \$3515.40$,
$\therefore$ The cost is \$3515.40.

b

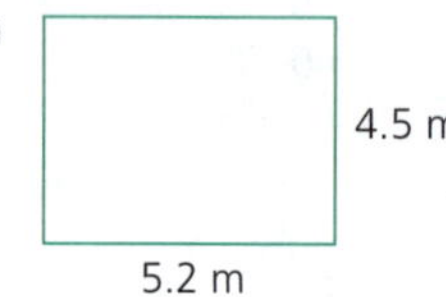

$V = Ah$
= Area of trapezium × Height
$= [\frac{1}{2} \times 20(2.3 + 0.8)] \times 7$
$= 31 \times 7$
$= 217$
$\therefore$ The volume is $217\ m^3$.

c Volume $= 217\ m^3$
$\therefore$ Capacity $= 217 \times 1000$
$= 217\,000$ litres
$= 217$ kilolitres
$\therefore$ The capacity is 217 kilolitres.

20 Pergola area

4.5 m
5.2 m

a Area $= 5.2 \times 4.5$
$= 23.4$
$\therefore$ The area is $23.4\ m^2$.

b $1\ m^2 = 100\ cm \times 100\ cm$
$= 10\,000\ cm^2$
Area $= 23.4 \times 10\,000$
$= 234\,000\ cm^2$
$\therefore$ The area is $234\,000\ cm^2$.

c Area of paver $= 22.5 \times 10$
$= 225\ cm^2$
Number of pavers
$= 234\,000 \div 225$
$= 1040$
Pavers still needed $= 1040 - 1000$
$= 40$
Still needs 40 pavers.

21 Outer radius = 1.5 cm
Inner radius = 1.4 cm
Also, 2 metres = 200 cm
$V = \pi \times 1.5^2 \times 200 - \pi \times 1.4^2 \times 200$
$= 182.212\,3739\ldots$
$= 182.21$ (2 dec. pl.)
$\therefore$ Volume of $182.21\ cm^3$.
Mass $= 16.5 \times 182.21$
$= 3006.465$
$\therefore$ Mass of 3006.465 g.
$\therefore$ Mass is about 3 kg.

22

1 m
1 m
1 m
40 cm

1 m
0.4 m

a Surface area of vinyl sheet
= Base + 6 rectangles
$= 5.2 + 6 \times 1 \times 0.4$
$= 5.2 + 2.4$
$= 7.6$
$\therefore$ The total surface area is $7.6\ m^2$.

b Volume = Area of base × Height
$= 5.2 \times 0.4$
$= 2.08$
$\therefore$ The volume is $2.08\ m^3$.
$= 2.08 \times 1000$ litres
$= 2080$ litres.
$\therefore$ The capacity is 2080 litres.

c Three quarters capacity
$= \frac{3}{4} \times 2080$
$= 1560$ litres
$= 1.56$ kilolitres
$\therefore$ The volume when $\frac{3}{4}$ full is 1.56 kilolitres.

Chapter 10 . . pp. 154–158

Geometry

1 a $x = 18$ (complementary angles)

b $x = 29$ (straight angle)

c $y = 93$ (∠BOD is vertically opposite ∠AOC)
$x = 87$ (supplementary to ∠BOD)
$z = 87$ (∠DOA is vertically opposite ∠COB)

d $x = 90$ (angles about point O)

e ∠ABC = 53° (base angles of isosceles △ABC)
$x = 74$ (angle sum of △ABC)

f ∠BCA = x° (base angles of isosceles △ABC)
$\therefore x + x + 58 = 180$ (angle sum of △ABC)
$\therefore x = 61$

g $x = 60$ (angle in an equilateral triangle)

h $x = 63$ (alternate to ∠FGA and AB ∥ CD)

i $x = 113$ (corresponding to ∠JKH and EF ∥ GH)

j $x = 133$ (co-interior to ∠TSM and MN ∥ PQ)

k ∠FGA = 63° (vertically opposite ∠HGB)
$x = 117$ (co-interior to ∠FGA and AB ∥ CD)

l

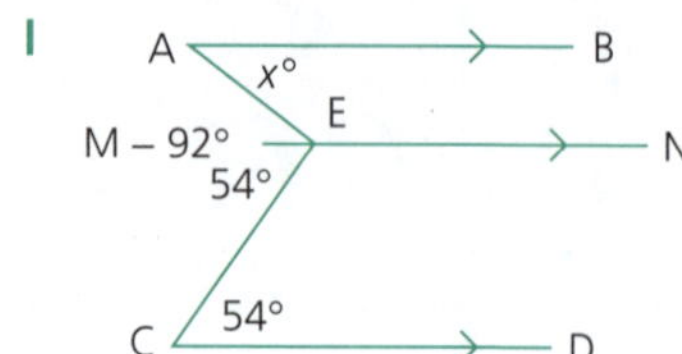

Construct a line MN through E and parallel to AB. AB || CD and AB || MN,
$\therefore$ CD is also || MN.
$\angle CEM = 54°$ (alternate to $\angle ECD$ and MN || CD),
$\therefore \angle MEA = (92–54)°$
$= 38°$
$x = 38$ (alternate to $\angle MEA$ and AB || MN)

m $x = (49 + 72)$ (exterior angle of $\triangle ABC$)
$\therefore x = 121$

n $x + 68 = 123$ (exterior angle of $\triangle ABC$)
$\therefore x = 55$

o $x + 90 + 120 + 82 = 360$
(angle sum of quadrilateral ABCD)
$\therefore x = 68$

p $\angle BCE = 60°$ (angle sum of quadrilateral CEAB)
$x = 120$ (supplementary to $\angle BCE$)

q Angle sum of pentagon
$= (n – 2) \times 180$ (pentagon $n = 5$)
$= (5 – 2) \times 180$
$= 3 \times 180$
$= 540$
$x = \dfrac{540}{5}$
$= 108$ (interior angle of regular pentagon)

r $x = 55$ (opposite angles of parallelogram)
$y = 125$ ($\angle CDA$ and $\angle DAB$ are co-interior and CD || BA)

s $\angle BCA = 56°$ (alternate to $\angle DAC$ and AD || BC, since ABCD is a rhombus)
$\angle CAB = 56°$ (base angles of isosceles $\triangle ABC$, since ABCD is a rhombus, then AB = BC)
$x + 56 + 56 = 180$ (angle sum of $\triangle ABC$)
$\therefore x = 68$

t $\angle CDB = 49°$ (complementary to $\angle BDA$)
$\angle AXD = (49 + 49)°$ (exterior angle of $\triangle DXC$)
$= 98°$
$x = 82$ (supplementary to $\angle AXD$)

2 $\angle DCA = 45°$ (diagonal of a square bisects the angle that it passes through)
$\angle EDC = y°$ (base angles of isosceles $\triangle CED$)
$y + y + 45 = 180$ (angle sum of $\triangle CED$)
$y = 67\frac{1}{2}$

3 $\angle ACB + \angle CBE = 180°$
(co-interior angles and AC || ED)
$\therefore m + 44 + 53 = 180$
$m = 83$

4 $\angle HGA = 129°$ (supplementary to $\angle AGE$)
$x = 129$ (corresponding to $\angle HGA$ and AB || CD)

5 $\angle CAB = \angle BCA$ (base angles of isosceles $\triangle ABC$)
$\therefore 2\angle BCA + 46 = 180$ (angle sum of $\triangle ABC$)
$2\angle BCA = 134$
$\therefore \angle BCA = 67°$
$x = 180 – 67$
(supplementary to $\angle BCA$)
$\therefore x = 113$

6 a i Pentagon $\Rightarrow n = 5$
Angle sum $= (n – 2) \times 180°$
$= 3 \times 180°$
$= 540°$
Each angle $= 540° \div 5$
$= 108°$

ii Heptagon $\Rightarrow n = 7$
Angle sum $= (n – 2) \times 180°$
$= 5 \times 180°$
$= 900°$
Each angle $= 900° \div 7$
$= 128\frac{4}{7}°$

iii Dodecagon $\Rightarrow n = 12$
Angle sum $= (n – 2) \times 180°$
$= 10 \times 180°$
$= 1800°$
Each angle $= 1800° \div 12$
$= 150°$

iv Angle sum $= (n – 2) \times 180°$
$= 18 \times 180°$
$= 3240°$
Each angle $= \dfrac{3240°}{20}$
$= 162°$

b i Exterior angle $= \dfrac{360°}{n}$
(octagon $\Rightarrow n = 8$) $= \dfrac{360°}{8}$
$= 45°$

ii Exterior angle $= \dfrac{360°}{n}$
(nonagon $\Rightarrow n = 9$) $= \dfrac{360}{9}$
$= 40°$

c i Interior angle $= \dfrac{(n – 2) \times 180}{n}$
$\therefore 156 = \dfrac{(n – 2) \times 180}{n}$
$156n = 180n – 360$
$24n = 360$
$n = 15$
The polygon has 15 sides.

ii Exterior angle $= \dfrac{360}{n}$
$\therefore 20 = \dfrac{360}{n}$
$20n = 360$
$n = \dfrac{360}{20}$
$n = 18$
The polygon has 18 sides.

7 a (SAS)
b (AAS)
c (SSS)
d (RHS)
e (AAS)

8 a In $\triangle$s PSR and PQR

PS = PQ	(S)	(given)
SR = QR	(S)	(given)
PR = PR	(S)	(common)
$\therefore \triangle PRS \equiv \triangle PRQ$	(SSS)	

b

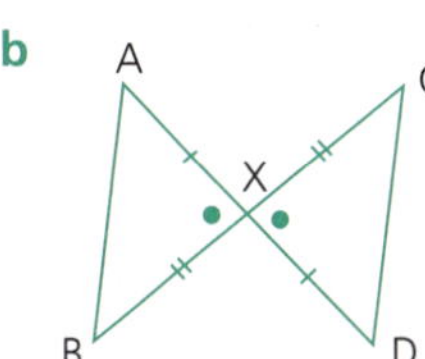

In $\triangle$s ABX and DCX

AX = DX	(S)	(given)
$\angle BXA = \angle CXD$	(A)	(vertically opposite)
BX = CX	(S)	(given)
$\therefore \triangle ABX \equiv \triangle DCX$	(SAS)	

c

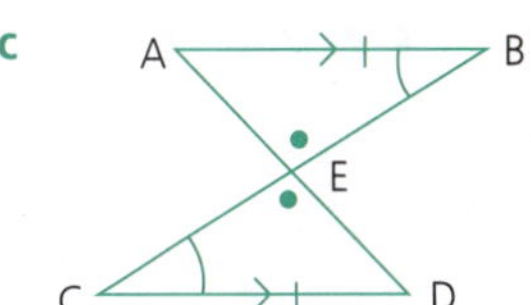

In $\triangle$s BAE and CDE

$\angle AEB = \angle DEC$	(A)	(vertically opposite angles)
$\angle EBA = \angle ECD$	(A)	(alternate $\angle$s and AB || CD)
AB = DC	(S)	(given)
$\therefore \triangle BAE \equiv \triangle CDE$	(AAS)	

d

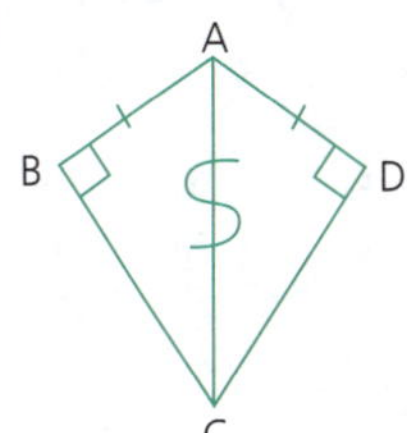

In $\triangle$s ABC and ADC

$\angle ABC = \angle ADC$	(R)	(= 90°, given)
AC = AC	(H)	(common)
AB = AD	(S)	(given)
$\therefore \triangle ABC \equiv \triangle ADC$	(RHS)	

9

a In $\triangle$s BAD and DCB

$\angle ADB = \angle CBD$	(A)	(alternate $\angle$s, BC || AD)
$\angle DBA = \angle BDC$	(A)	(alternate $\angle$s, AB || DC)
BD = DB	(S)	(common)
$\therefore \triangle BAD \equiv \triangle DCB$	(AAS)	

b **i** AD = BC (corresponding sides in congruent triangles)

ii ∠BAD = ∠DCB (corresponding angles in congruent triangles)

10

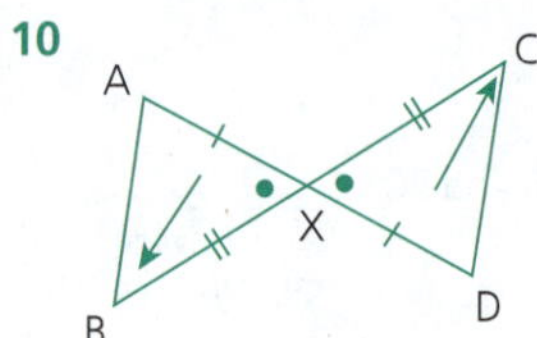

a In △s ABX and DCX

AX = DX	(S)	(given)
∠BXA = ∠CXD	(A)	(vertically opposite)
BX = CX	(S)	(given)
∴ △ABX ≡ △DCX	(SAS)	

b ∠ABX = ∠DCX (corresponding angles in congruent triangles)
∴ AB ∥ CD (a pair of alternate angles ABX and DCX are equal)

11

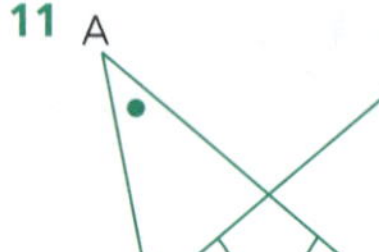
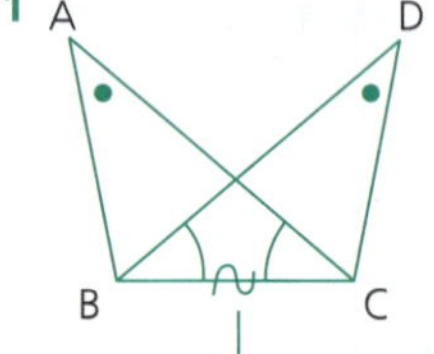

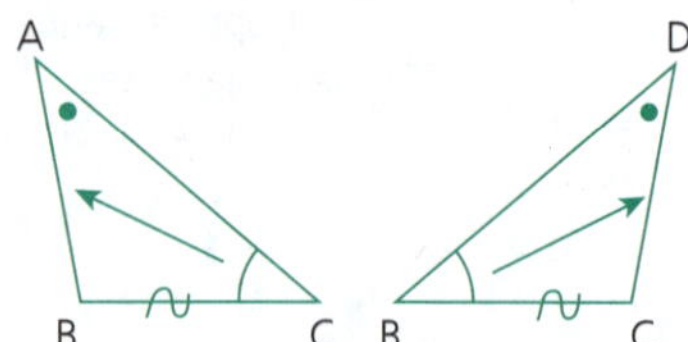

a In △s ABC and DCB

∠CAB = ∠BDC	(A)	(given)
∠BCA = ∠CBD	(A)	(given)
BC = CB	(S)	(common)
∴ △ABC ≡ △DCB	(AAS)	

b AB = DC (corresponding sides in congruent triangles)

12

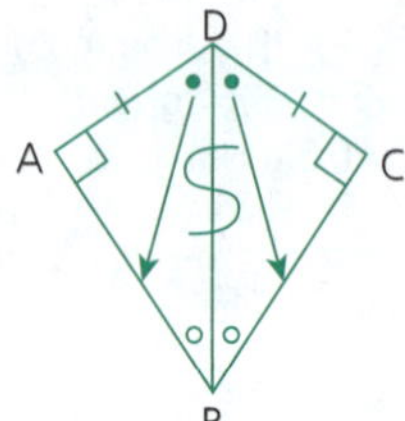

a In △s DAB and DCB

AD = CD	(S)	(given)
∠BDA = ∠BDC	(A)	(given)
BD = BD	(S)	(common)
∴ △ABD ≡ △CBD	(SAS)	

b AB = CB (corresponding sides in congruent triangles)

c ∠ABD = ∠CBD
(corresponding angles in congruent triangles)
Now, since ∠ABD = ∠CBD, then DB bisects ∠ABC.

13 a

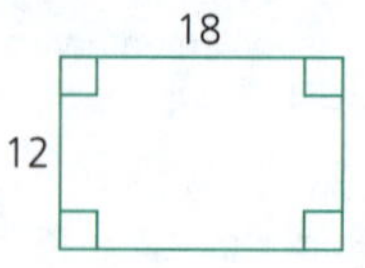

$\frac{x}{18} = \frac{6}{12}$

$12x = 6 \times 18$ (cross-multiply)

$\therefore x = \frac{6 \times 18}{12}$

$\therefore x = 9$

b

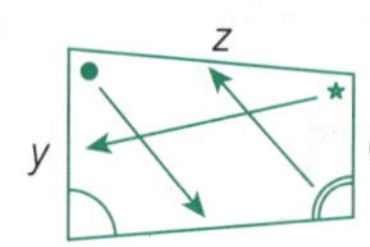

$\frac{x}{3} = \frac{y}{5} = \frac{z}{4} = \frac{6}{2} = \frac{3}{1}$

That is, $\frac{x}{3} = 3 \quad \therefore x = 9$

$\frac{y}{5} = 3 \quad \therefore y = 15$

$\frac{z}{4} = 3 \quad \therefore z = 12$

c

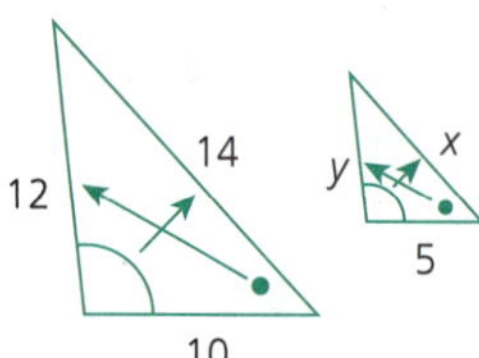

$\frac{x}{14} = \frac{y}{14} = \frac{5}{10}$

$\therefore \frac{x}{14} = \frac{5}{10}$ and $\frac{y}{12} = \frac{5}{10}$

$x = \frac{5}{10} \times 14 \qquad y = \frac{5}{10} \times 12$

$x = 7 \qquad y = 6$

d

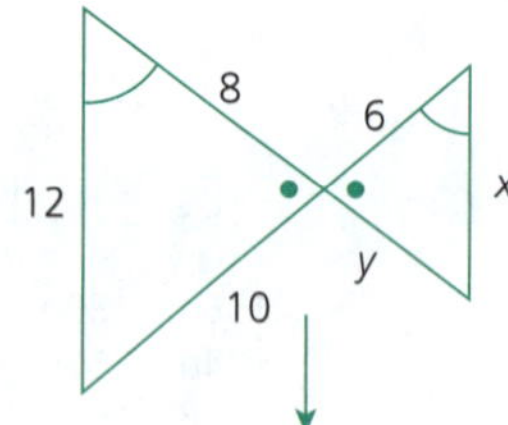

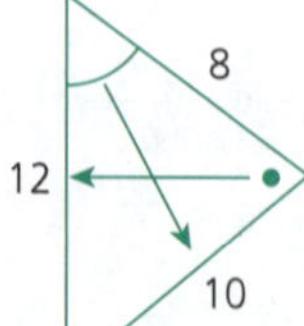

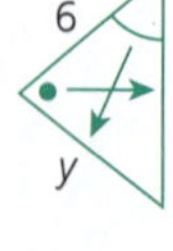

$\frac{x}{12} = \frac{y}{10} = \frac{6}{8}$

$\therefore \frac{x}{12} = \frac{6}{8}$ and $\frac{y}{10} = \frac{6}{8}$

$x = \frac{6}{8} \times 12 \qquad y = \frac{6}{8} \times 10$

$x = 9 \qquad y = 7\frac{1}{2}$

14 a

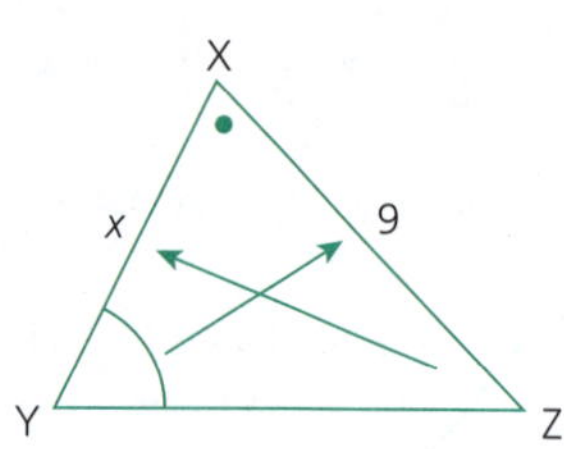

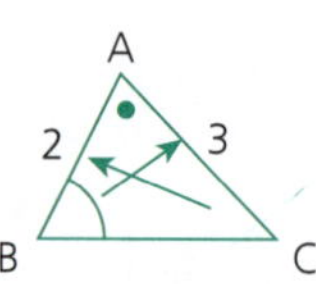

In △s ABC and XYZ

∠ABC = ∠XYZ	(given)
∠CAB = ∠ZXY	(given)
∴ △ABC ||| △XYZ	(equiangular)

$\frac{x}{2} = \frac{9}{3}$

(corresponding sides of similar triangles are in the same ratio)

$x = \frac{9}{3} \times 2$

$x = 6$

b

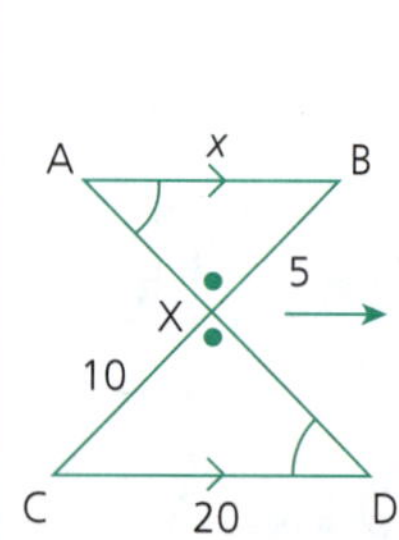

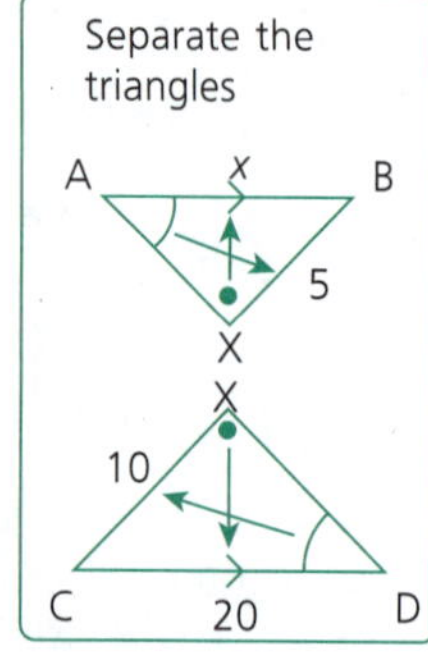

In △s AXB and DXC
∠AXB = ∠DXC
(vertically opposite angles)
∠BAX = ∠CDX
(alternate ∠s and AB ∥ DE)
∴ △AXB ||| △DXC (equiangular)

Then $\frac{x}{20} = \frac{5}{10}$

(corresponding sides of similar triangles are in the same ratio)

$x = \frac{5}{10} \times 20$

$x = 10$

c

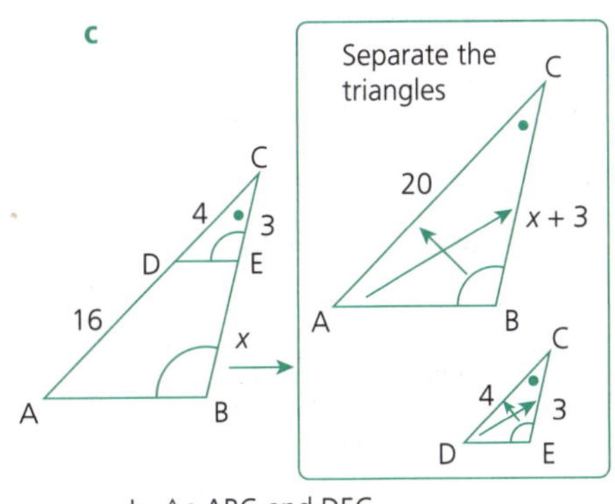

In $\triangle$s ABC and DEC

$\angle ABC = \angle DEC$

(corresponding $\angle$s and AB || DE)

$\angle BCA = \angle ECD$ (common)

$\therefore \triangle ABC \ ||| \ \triangle DEC$ (equiangular)

$\frac{x+3}{3} = \frac{20}{4}$ (corresponding sides of similar triangles are in same ratio)

$4(x + 3) = 60$

$4x + 12 = 60$

$4x = 48$

$x = 12$

d

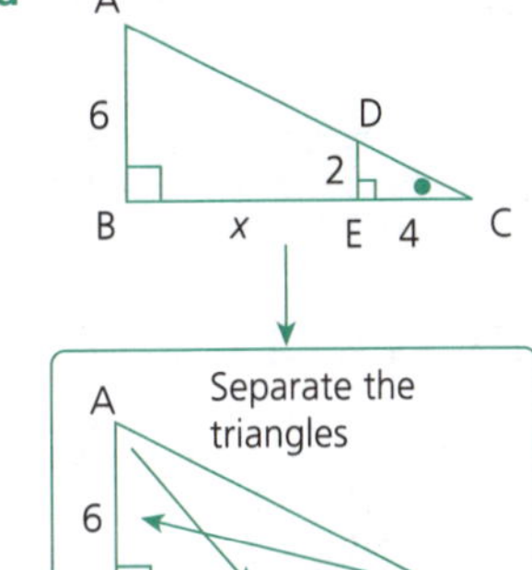

In $\triangle$s ABC and DEC

$\angle ABC = \angle DEC = 90°$ (given)

$\angle BCA = \angle ECD$ (common)

$\therefore \triangle ABC \ ||| \ \triangle DEC$ (equiangular)

Then $\frac{x+4}{4} = \frac{6}{2}$

(corresponding sides in similar triangles are in the same ratio)

$2(x + 4) = 24$ (solve equation)

$2x + 8 = 24$

$2x = 16$

$x = 8$

e

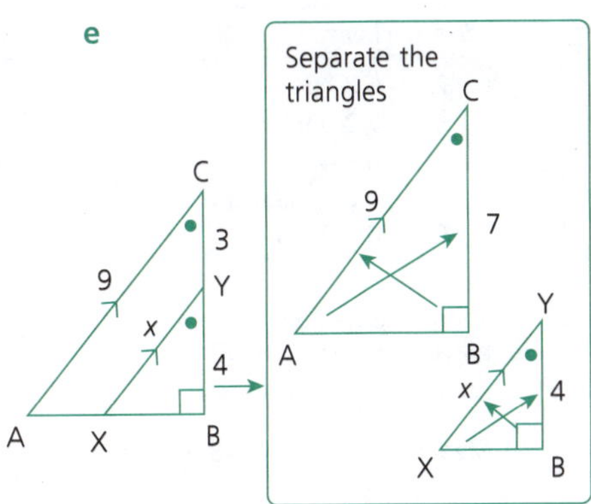

In $\triangle$s ABC and XBY

$\angle ABC = \angle XBY = 90°$ (given)

$\angle BCA = \angle BYX$

(corresponding $\angle$s, AC || XY)

$\therefore \triangle ABC \ ||| \ \triangle XBY$ (equiangular)

Then $\frac{x}{9} = \frac{4}{7}$

(corresponding sides in similar triangles are in proportion)

$x = \frac{4}{7} \times 9$

$x = 5\frac{1}{7}$

Note: 'in proportion' means 'in the same ratio'.

f

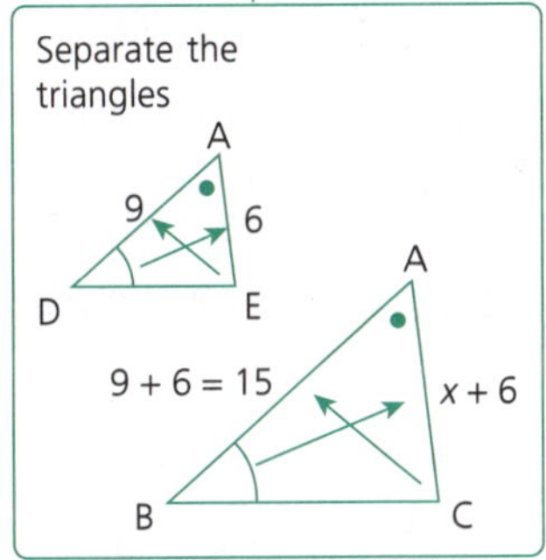

In $\triangle$s ABC and ADE

$\angle ABC = \angle ADE$

(corresponding $\angle$s, DE || BC)

$\angle CAB = \angle EAD$ (common)

$\therefore \triangle ABC \ ||| \ \triangle ADE$ (equiangular)

Then $\frac{x+6}{6} = \frac{15}{9}$

(corresponding sides in similar triangles are in proportion)

$9(x + 6) = 90$

$9x + 54 = 90$

$9x = 36$

$x = 4$

15 a

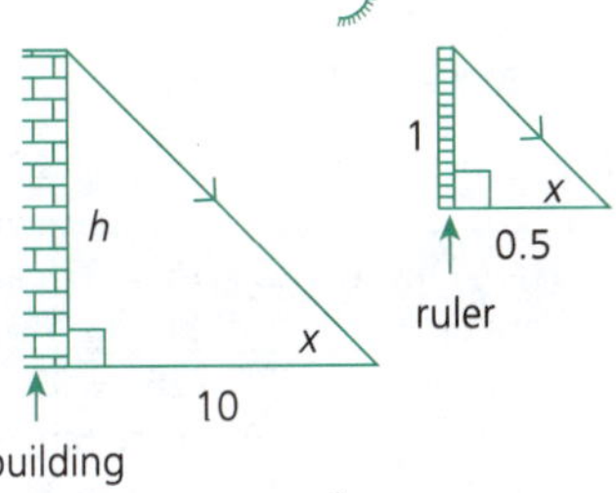

b Let the height of the building be h metres. The two triangles are similar, therefore, $\frac{h}{1} = \frac{10}{0.5}$

(corresponding sides in similar triangles are in proportion)

$\therefore h = \frac{10}{0.5}$, that is, $h = 20$

Therefore, the height of the building is 20 metres.

16

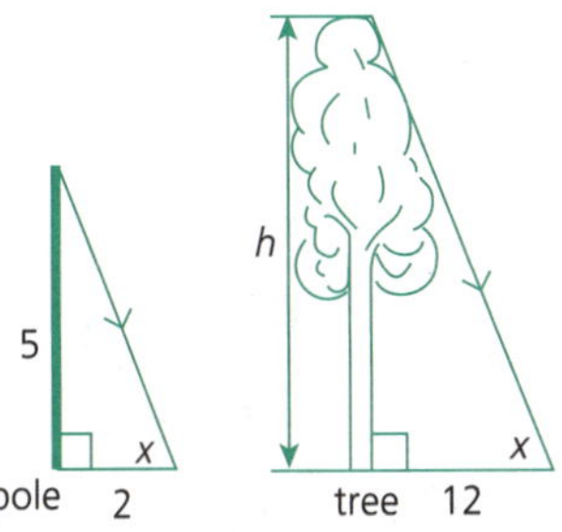

Note: draw a neat diagram displaying all given information.
Angle of elevation of the sun is the same for the pole as for the tree.

$\frac{h}{5} = \frac{12}{2}$ (corresponding sides in similar triangles are in proportion)

$h = \frac{12}{2} \times 5 = 30$

$\therefore$ The tree is 30 m high.

Chapter 11 . . pp. 178–181

Statistics

1

Number of notes (x)	Tally	Frequency (f)	cf	xf
2	IIII	4	4	8
3	IIII	4	8	12
4	𝍸	5	13	20
5	𝍸	5	18	25
6	I	1	19	6
7	I	1	20	7
	Σ	20	Σ	78

a **i** 5

ii 13 (2, 3, 4 notes)

iii 7 (5, 6, 7 notes)

b Range = 7 − 2 = 5, Mode = 4, 5

c **i** $\frac{5}{20} \times 100 = 25\%$

ii $\frac{13}{20} \times 100 = 65\%$

d Mean = $\frac{78}{20} = 3.9$

Median = 4

(Between 10th and 11th scores)

e Frequency polygon

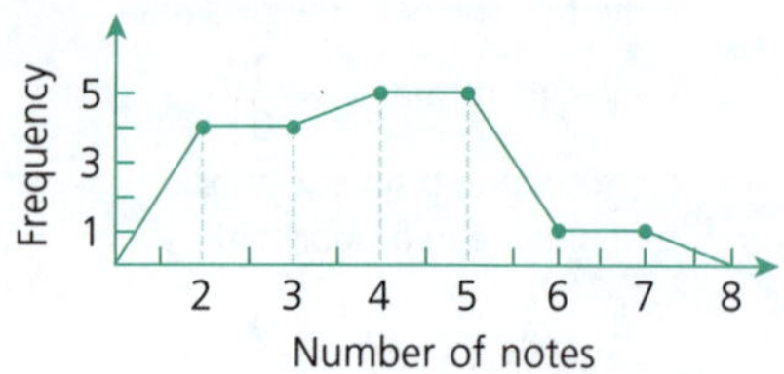

Cumulative frequency histogram

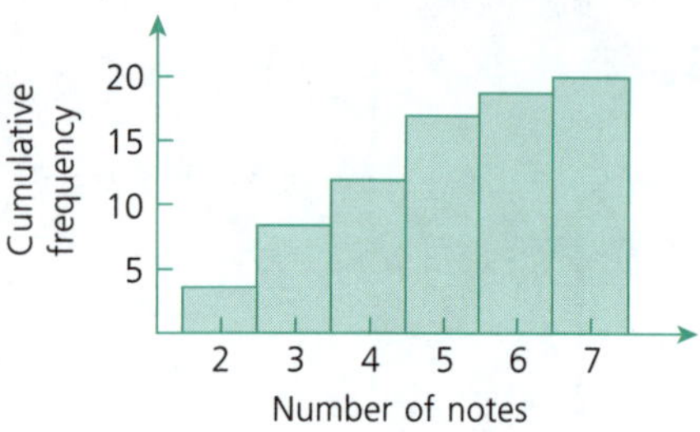

f Probability

$= \frac{\text{No. of favourable outcomes}}{\text{Total no. of outcomes}}$

$= \frac{5}{20} = \frac{1}{4}$

2

Number of lumps (x)	Tally	Frequency (f)	cf	xf
0	I	1	1	0
1	I	1	2	1
2	III	3	5	6
3	III	3	8	9
4	卌	5	13	20
5	IIII	4	17	20
6	卌	5	22	30
7	III	3	25	21
	Σ	25	Σ	107

a Range = 7 – 0 = 7, Mode = 4 and 6

b **i** 5

ii 13

iii 8 [3 or less]

c **i** $\frac{5}{25} \times 100 = 20\%$

ii $\frac{17}{25} \times 100 = 68\%$

d **i** Probability $= \frac{5}{25} = \frac{1}{5}$

ii $\frac{8}{25}$

e Mean $= \frac{\Sigma xf}{\Sigma f} = \frac{107}{25} = 4.28$

Median = 4 [13th score from c.f.]

f

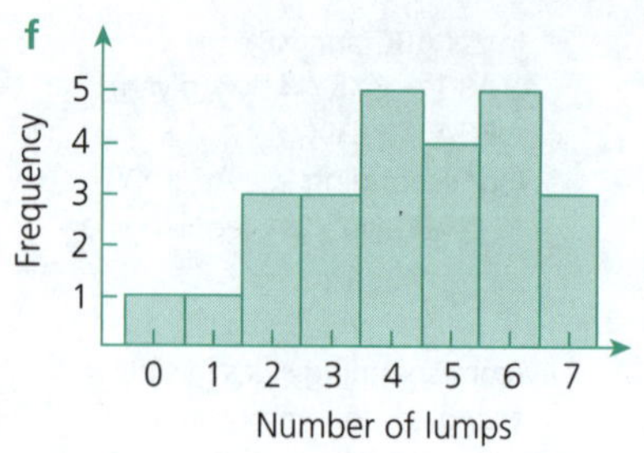

g

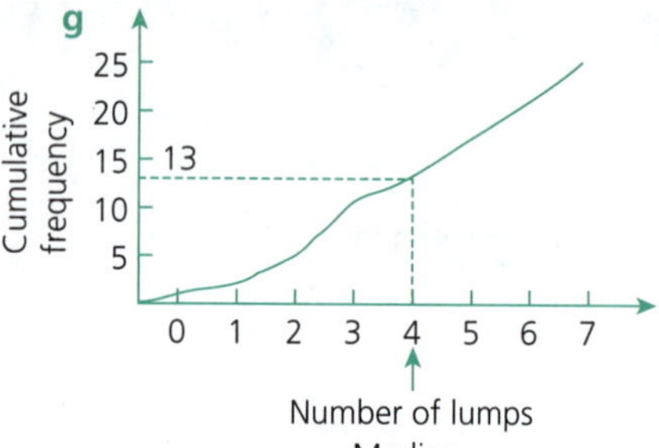

3 **a** **i** Mean $= \frac{8+7+4+8+6+3}{6} = 6$

ii Range = 8 – 3 = 5

Mode = 8

iii 3, 4, (6, 7), 8, 8

Median $= \frac{6+7}{2} = 6.5$

b Scores are now 3, 4, 6, 7, 8, 8, 13

Mean $= \frac{49}{7} = 7$

Mean has increased by one.

3, 4, 6, (7), 8, 8, 13

Median = 7

Median has increased by 0.5.

c Total no. of goals in 8 seasons $= 6.5 \times 8 = 52$.

Gai scored (52 – 49) goals in the 8th season, that is, 3 goals.

4 **a** Mean $= \frac{48}{9} = 5.3$ (one dec. pl.)

3, 4, 4, 4, (5), 6, 7, 7, 8

Median = 5

b Mean $= \frac{58}{10} = 5.8$

2, 3, 4, 4, (5, 6), 7, 7, 9, 11

Median = 5.5

5 Total over 4 exams $= 4 \times 70 = 280$.

Needs to average 75 over 5 exams, that is, total must be $(5 \times 75) = 375$ after 5 exams.

Mark needed = 375 – 280 = 95

Zoltan must score 95 in the 5th exam.

6

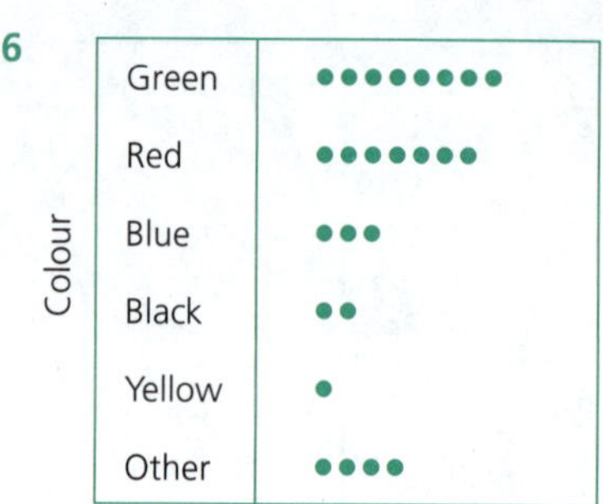

7

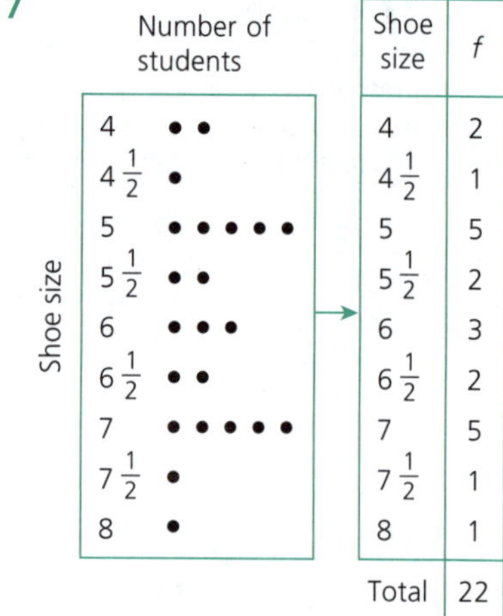

Shoe size	f
4	2
$4\frac{1}{2}$	1
5	5
$5\frac{1}{2}$	2
6	3
$6\frac{1}{2}$	2
7	5
$7\frac{1}{2}$	1
8	1
Total	22

8 Initial plot:

Stem	Leaf
14	8 9 7 9
15	0 1 5 4 9 2 8 4 0 2
16	3 2 5 7 8 0

Final plot:

Stem	Leaf
14	7 8 9 9
15	0 0 1 2 2 4 4 5 8 9
16	0 2 3 5 7 8

9 Initial plot:

Stem	Leaf
2	6 8
3	9 3 9 9 3
4	7 9 7 6 4 8 3 0 7
5	3 7 2 0 2 4 5 0
6	2 4 0 4 1 2

a Final plot:

Stem	Leaf
2	6 8
3	3 3 9 9 9
4	0 3 4 6 7 7 7 (8 9)
5	0 0 2 2 3 4 5 7
6	0 1 2 2 4 4

b Range = 64 – 26 = 38

Mode = 39, 47

Median between 15th and 16th scores

Median $= \frac{48+49}{2} = 48.5$

10 a Rearranged numbers
4, 5, 6, 6, 7, 8, 8, 9
↑

Median = $\frac{6+7}{2}$ = 6.5

(Even number of scores)

b 5, 6, 6, 6, 7, 7, 8, 8, 9, 9, 9
↑
Median = 7
(Odd number of scores)

c 12, 14, 14, 15, 16, 16, 17, 18, 18, 19,
↑
Median = 16

d 7, 7, 7, 8, 8, 9, 9, 10, 11, 11, 12, 12, 12, 13, 14
↑
Median = 10

11 a 23 scores—middle is 12th score.

Stem	Leaf
1	0 7 8
2	1 2 4 6 9
3	7 7 7 8 8 9
4	0 0 6 7 8 8 9
5	1 4

◯ Median = 38

b 16 scores—middle is between 8th and 9th scores.

Stem	Leaf
10	5 8
11	3 3 6 6 9
12	4 4 7
13	5 5 6 8 8
17	7

◯ Median = 124

12 a Rearranged numbers
4, 5, 6, 6, 7, 8, 8, 9
↑

Median = $\frac{6+7}{2}$ = 6.5

(Even number of scores)
Consider 4, 5, 6, 6
↑

Lower quartile = $\frac{5+6}{2}$ = 5.5

Consider 7, 8, 8, 9
↑

Upper quartile = 8
Interquartile range = 8 − 5.5
= 2.5
Lowest = 4
Highest = 9
Lower quartile = 5.5
Median = 6.5
Upper quartile = 8

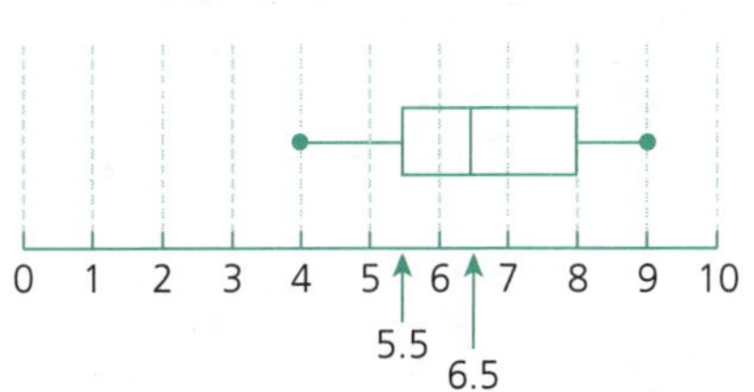

b 1 5, 6, 6, 6, 7, 7, 8, 8, 9, 9, 9
↑

Median = 7
Consider 5, 6, 6, 6, 7
↑

Lower quartile = 6
Consider 8, 8, 9, 9, 9
↑

Upper quartile = 9
Interquartile range = 9 − 6 = 3
Highest = 9
Lowest = 5
Lower quartile = 6
Median = 7
Upper quartile = 9

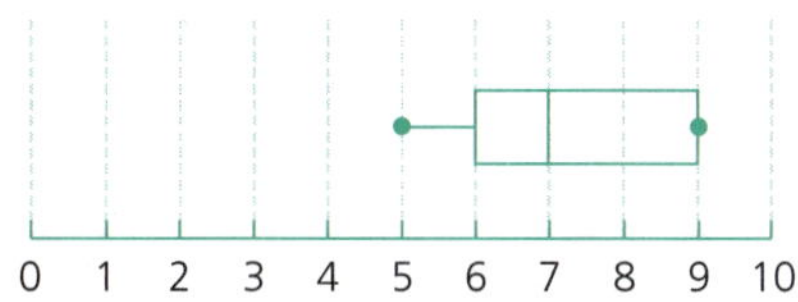

c 12, 14, 14, 15, 16, 16, 17, 18, 18, 19,
↑
Median = 16
Consider 12, 14, 14, 15, 16
↑

Lower quartile = 14
Consider 16, 17, 18, 18, 19
↑

Upper quartile = 18
Interquartile range = 18 − 14
= 4
Lowest score = 12
Highest score = 19
Median = 16
Lower quartile = 14
Upper quartile = 18

d 7, 7, 7, 8, 8, 9, 9, 10, 11, 11, 12, 12, 12, 13, 14
↑
Median = 10
Consider 7, 7, 7, 8, 8, 9, 9
↑

Lower quartile = 8
Consider 11, 11, 12, 12, 12, 13, 14
↑

Upper quartile = 12
Interquartile range = 12 − 8
= 4
Lowest score = 7
Highest score = 14
Median = 10
Lower quartile = 8
Upper quartile = 12

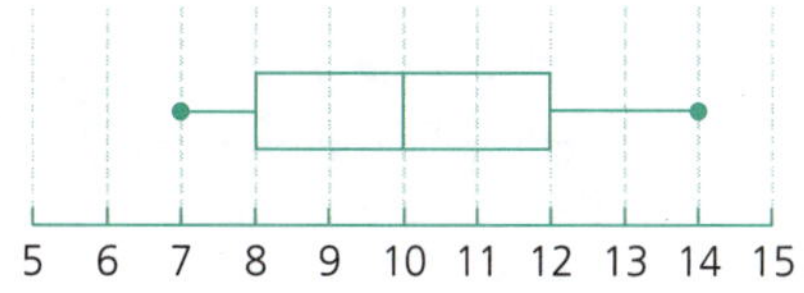

13 i a Median = 16

b Interquartile range = 18 − 15
= 3

ii a Median = 63

b Interquartile range = 73 − 57
= 16

14 a 23 scores—middle is 12th score.

Stem	Leaf
1	0 7 8
2	1 2 4 6 9
3	7 7 7 8 8 9
4	0 0 6 7 8 8 9
5	1 4

◯ Median = 38
□ Lower quartile = 24
□ Upper quartile = 47
Lowest score = 10
Highest score = 54

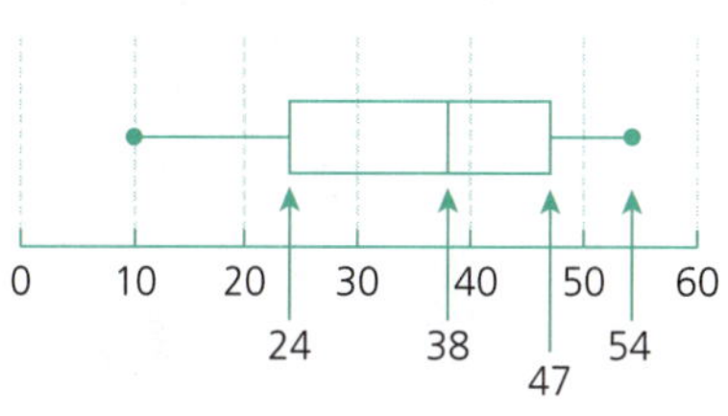

b 16 scores—middle is between 8th and 9th scores.

Stem	Leaf
10	5 8
11	3 3 6 6 9
12	4 4 7
13	5 5 6 8 8
14	7

◯ Median = 124

□ Lower quartile = $\frac{113+116}{2}$
= 114.5

□ Upper quartile = $\frac{135+136}{2}$
= 135.5

Lowest score = 105
Highest score = 147

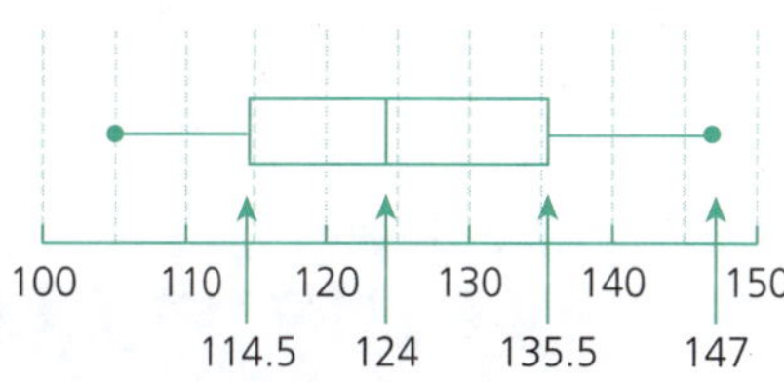

15 a Initial plot:

Stem	Leaf
14	9
15	0 0 4 9 8
16	1 1 7 9 7 8 8 6 2 0 9 6 4
17	2 4 0 5 4 2 0
18	0 2 2 1

Final plot:

Stem	Leaf
14	9
15	0 0 4 8 9
16	0 1 1 2 4 6 6 7 7 8 8 9 9
17	0 0 2 2 4 4 5
18	0 1 2 2

b 30 scores—middle is between 15th and 16th scores.

$$\text{Median} = \frac{167 + 168}{2} = 167.5$$

Lower quartile = 161
Upper quartile = 172
Interquartile range = 172 − 161
= 11

c Lowest score = 149
Highest score = 182

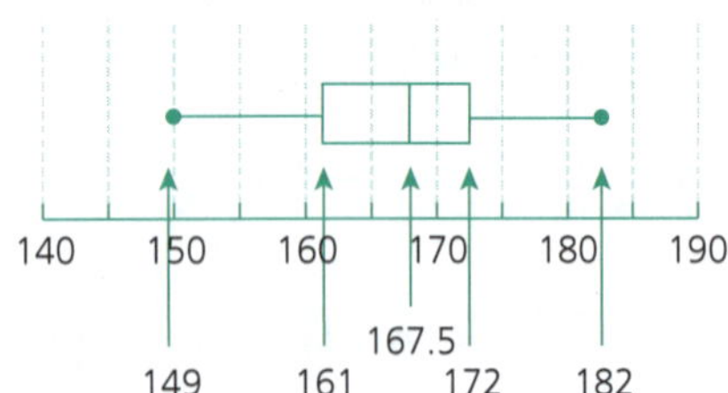

16 a Positively skewed

b Normally distributed (symmetrical)

c Positively skewed

d Negatively skewed

e Symmetrical—without knowing mean and mode we cannot say that it is normally distributed

17 a

Score	Frequency
3	6
4	4
5	3
6	0
7	1
8	1

b Mean = $4\frac{4}{15}$
Mode = 3
Median = 4 (8th score)

c Positively skewed

18 a Negatively skewed

b Positively skewed

19 a Normally distributed

b Median = 8
Mode = 8
Mean = 8 (as the data is normally distributed, it will be the same as the median and mode)

20 The trigonometry test is positively skewed and the algebra test is negatively skewed.

21 a 9, 13, 15, 16, 16, 17, 17, 17, 18
Lowest mark: 9
Lower quartile: 14
Median: 16
Upper quartile: 17
Highest mark: 18

Negatively skewed

b 8, 9, 10, 10, 10, 11, 18
Lowest mark: 8
Lower quartile: 9
Median: 10
Upper quartile: 11
Highest mark: 18

Positively skewed

c 10, 12, 12, 14, 14, 14, 16, 16, 18
Lowest mark: 10
Lower quartile: 12
Median: 14
Upper quartile: 16
Highest mark: 18

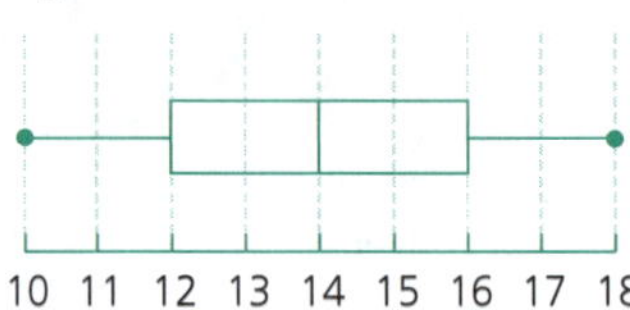

Symmetrical

22 a *r*

b *n*

c *s*

23 a Weak positive

b Strong negative

24 a

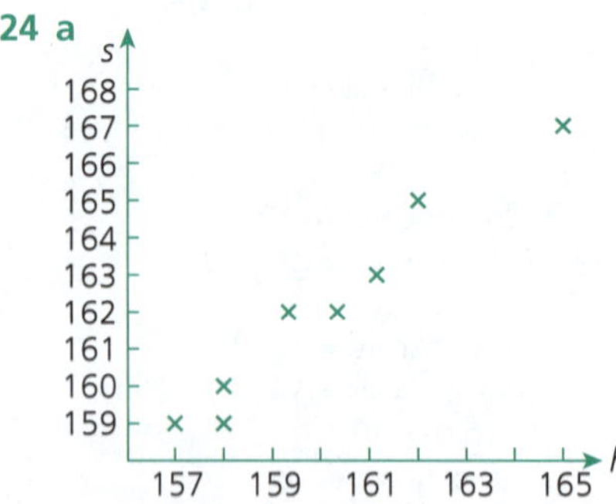

b Strong positive relationship

c About 166–167 cm

d The length of the arm span increases as the height increases.

25 a

b About 41 cm

26 a

b Negative

c Strong

Chapter 12 . . pp. 195–198

Probability

1 a $\frac{4}{52} = \frac{1}{13}$

b $\frac{4}{52} = \frac{1}{13}$

c $\frac{13}{52} = \frac{1}{4}$

d $\frac{13}{52} = \frac{1}{4}$

e $\frac{26}{52} = \frac{1}{2}$

f $\frac{12}{52} = \frac{3}{13}$

g $1 - \frac{3}{13} = \frac{10}{13}$

h Includes 2, 3, 4, 5, 6
[ace may be considered as 1]
$P = \frac{5}{13}$ (No Ace)
or $P = \frac{6}{13}$ (Ace is 1)

i $\frac{1}{52}$

j 26 red cards + 2 black 6s,
$P = \frac{28}{52} = \frac{7}{13}$

k $\frac{2}{52} = \frac{1}{26}$

2 $P(\text{no rain}) = 1 - \frac{7}{10} = \frac{3}{10}$

3 7R, 5B, 3W. Total no. = 15

a $\frac{7}{15}$

b $1 - \frac{7}{15} = \frac{8}{15}$

c $\frac{5}{15} = \frac{1}{3}$

d $\frac{10}{15} = \frac{2}{3}$

e 0 [There are no greens.]

4 a $\frac{1}{10}$

b $\frac{5}{10} = \frac{1}{2}$

c $\frac{6}{10} = \frac{3}{5}$

d $\frac{3}{10}$

5 Robyn needs a 50c coin.
$P(50c) = \frac{4}{10} = \frac{2}{5}$

6 Total = 200

a $\frac{36}{200} = \frac{9}{50}$

b As 20 + 14 = 34
then $\frac{34}{200} = \frac{17}{100}$

c As 200 − 112 = 88
then $\frac{88}{200} = \frac{11}{25}$

7 a Number = 1 + 3 + 5 + 8 + 10 + 4 + 2
= 33

b **i** $\frac{8}{33}$

ii As 1 + 3 = 4
then $\frac{4}{33}$

iii As 10 + 4 + 2 = 16
then $\frac{16}{33}$

8 a $\frac{120}{360} = \frac{1}{3}$

b Missing ∠ is 30°
then $\frac{30}{360} = \frac{1}{12}$

c 40 + 90 = 130
$\frac{130}{360} = \frac{13}{36}$

9

Colour	Frequency	Relative frequency
Red	48	0.48
Green	19	0.19
Blue	33	0.33
Total	100	1

10 a 10 + 6 + 4 = 20
∴ 20 students

b **i** $\frac{16}{20} = \frac{4}{5}$

ii $\frac{4}{20} = \frac{1}{5}$

iii $\frac{6}{20} = \frac{3}{10}$

11 a Sum = 30

b **i** As 4 + 5 + 2 + 7 = 18
then $\frac{18}{30} = \frac{3}{5}$

ii As 4 + 5 = 9
then $\frac{9}{30} = \frac{3}{10}$

iii $\frac{5}{30} = \frac{1}{6}$

iv As 5 + 3 + 4 = 12
then $\frac{12}{30} = \frac{2}{5}$

12 a

		Die					
		1	2	3	4	5	6
Coin	Head	H1	H2	H3	H4	H5	H6
	Tail	T1	T2	T3	T4	T5	T6

b **i** $P(H4) = \frac{1}{12}$

ii $P(T1, T3, T5) = \frac{3}{12} = \frac{1}{4}$

13 See table on page 189.

a $P(9) = \frac{4}{36} = \frac{1}{9}$

b $P(5) = \frac{4}{36} = \frac{1}{9}$

c $P(\text{odd}) = \frac{18}{36} = \frac{1}{2}$

d $P(> 9) = \frac{6}{36} = \frac{1}{6}$

e $P(\leq 9) = 1 - \frac{1}{6} = \frac{5}{6}$

f $P(\neq 9) = 1 - \frac{1}{9} = \frac{8}{9}$

g 6 less than 5 + 16 odd,
$P = \frac{22}{36} = \frac{11}{18}$

h $P = \frac{2}{36} = \frac{1}{18}$ [only the threes]

14 See table on page 190.

a $\frac{2}{36} = \frac{1}{18}$

b $\frac{9}{36} = \frac{1}{4}$

c $\frac{19}{36}$

d $\frac{23}{36}$

e $\frac{20}{36} = \frac{5}{9}$

f $\frac{5}{36}$

g $\frac{2}{3}$

15

	1	2	3	4	5	6
1	1	2	3	4	5	6
2	2	2	3	4	5	6
3	3	3	3	4	5	6
4	4	4	4	4	5	6
5	5	5	5	5	5	6
6	6	6	6	6	6	6

a $\frac{11}{36}$

b $\frac{1}{36}$

c $1 - \frac{11}{36} = \frac{25}{36}$

d $\frac{24}{36} = \frac{2}{3}$

e $\frac{21}{36} = \frac{7}{12}$

f $\frac{17}{36}$

g $\frac{3}{36} = \frac{1}{12}$

h $\frac{35}{36}$

i $\frac{16}{36} = \frac{4}{9}$

j 0

16 a

	Born in Australia	Born overseas	Total
Female	74	46	120
Male	56	24	80
Totals	130	70	200

b **i** $\frac{56}{200} = \frac{7}{25}$

ii $\frac{46}{200} = \frac{23}{100}$

c **i** $\frac{24}{80} = \frac{3}{10}$

ii $\frac{56}{80} = \frac{7}{10}$

17

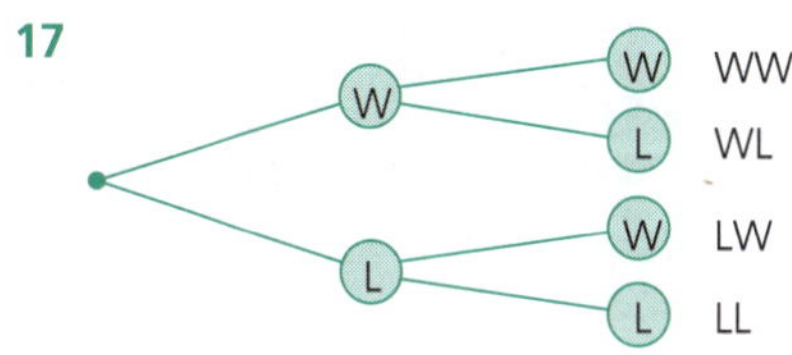

18

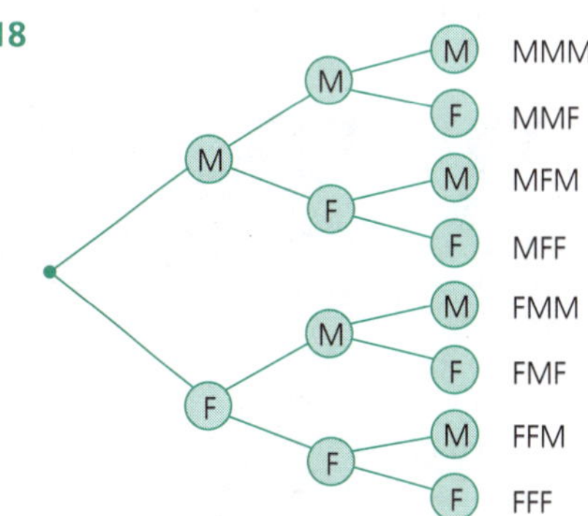

19

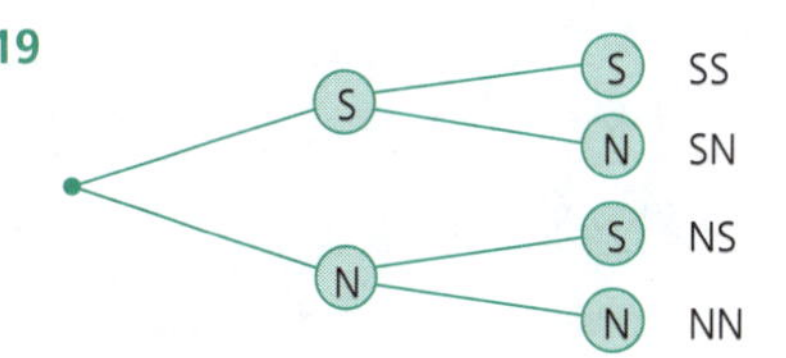

20

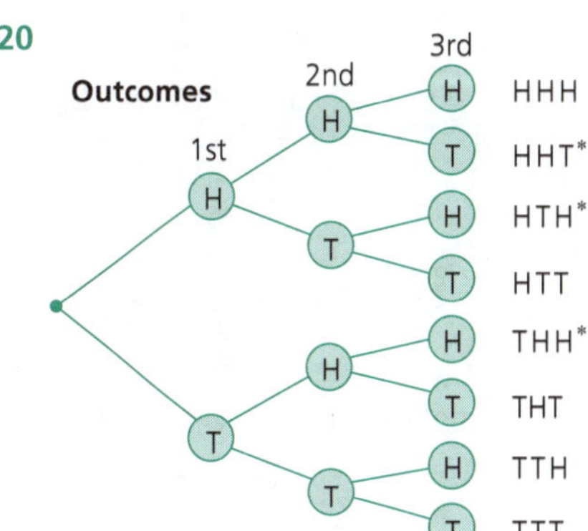

No. of outcomes = 8

a $P(\text{HHH}) = \frac{1}{8}$

b $P(2\text{H}) = \frac{3}{8}$ (marked*)

c $P(1\text{H}) = \frac{3}{8}$

d $P(\text{TTT}) = \frac{1}{8}$

e $P(\text{at least 1H}) = 1 - P(\text{TTT}) = \frac{7}{8}$

21

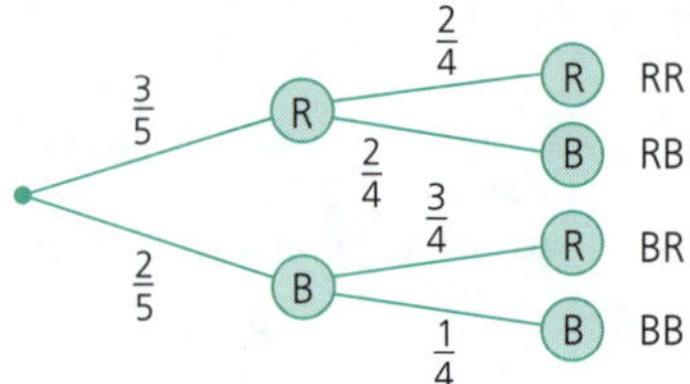

a $P(\text{RR}) = \frac{3}{5} \times \frac{2}{4} = \frac{3}{10}$

b $P(\text{BB}) = \frac{2}{5} \times \frac{1}{4} = \frac{1}{10}$

c $P(\text{RB or BR}) = \frac{3}{5} \times \frac{2}{4} + \frac{2}{5} \times \frac{3}{4} = \frac{3}{5}$

d $P(\text{RR or BB}) = \frac{3}{10} + \frac{1}{10} = \frac{2}{5}$

22

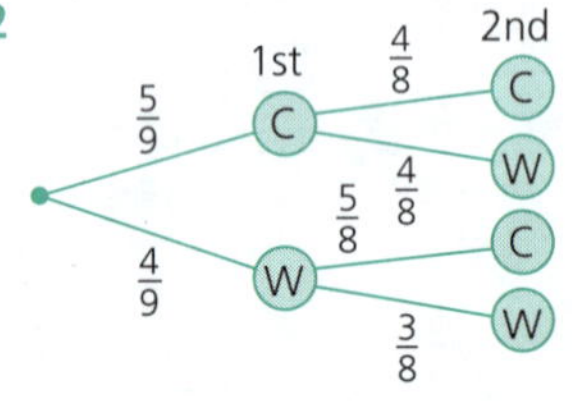

Outcome	Probability
CC	$P(\text{CC}) = \frac{5}{9} \times \frac{4}{8} = \frac{5}{18}$
CW	$P(\text{CW}) = \frac{5}{9} \times \frac{4}{8} = \frac{5}{18}$
WC	$P(\text{WC}) = \frac{4}{9} \times \frac{5}{8} = \frac{5}{18}$
WW	$P(\text{WW}) = \frac{4}{9} \times \frac{3}{18} = \frac{1}{6}$

From the table:

a $P(\text{CC}) = \frac{5}{18}$

b $P(\text{one C}) = P(\text{WC}) + P(\text{CW})$
$= \frac{5}{18} + \frac{5}{18}$
$= \frac{5}{9}$

c $P(\text{CW}) = \frac{5}{18}$

d $P(\text{WW}) = \frac{1}{6}$

e $P(\text{at least one C}) = 1 - P(\text{WW})$
$= 1 - \frac{1}{6} = \frac{5}{6}$

23

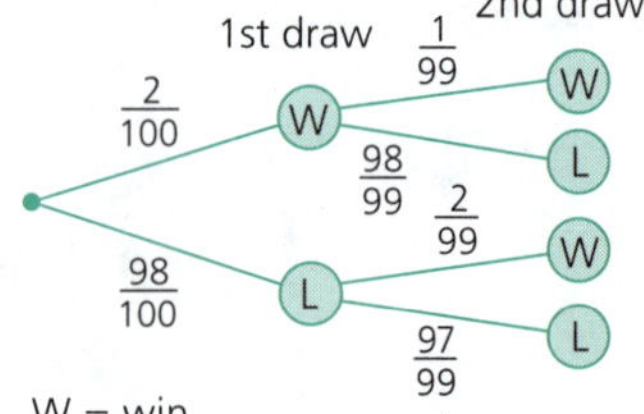

W = win
L = not win

Outcome	Probability
WW	$P(\text{WW}) = \frac{2}{100} \times \frac{1}{99} = \frac{1}{4950}$
WL	$P(\text{WL}) = \frac{2}{100} \times \frac{98}{99} = \frac{98}{4950}$
LW	$P(\text{LW}) = \frac{98}{100} \times \frac{2}{99} = \frac{98}{4950}$
LL	$P(\text{LL}) = \frac{98}{100} \times \frac{97}{99} = \frac{4753}{4950}$

From the diagram:

a $P(\text{first prize}) = \frac{2}{100} = \frac{1}{50}$

b $P(\text{WW}) = \frac{1}{4950}$

c $P(\text{LW}) = \frac{98}{4950} = \frac{49}{2475}$

d $P(\text{a prize}) = P(\text{WW}) + P(\text{WL}) + P(\text{LW})$
or
$1 - P(\text{LL})$
$= 1 - \frac{4753}{4950} = \frac{197}{4950}$

e $P(\text{LL}) = \frac{4753}{4950}$

f $\frac{197}{4950}$ [same as (d)]

24 a $P(\text{4, given even}) = \frac{1}{3}$

b $P(\text{less than 4, given even}) = \frac{2}{3}$

25 a $P(\text{vowel}) = \frac{5}{26}$

b $P(\text{E, given vowel}) = \frac{1}{5}$

c $P(\text{P, given consonant}) = \frac{1}{21}$

26 a $P(\text{13- or 17-year-old female}) = \frac{12}{50}$
$= \frac{6}{25}$

b $P(\text{18 and over, given female}) = \frac{10}{27}$

c $P(\text{male, given 12 and under}) = \frac{8}{13}$

27 a

	Low	Normal	High	Totals
Men	16	28	26	70
Women	24	22	14	60
Totals	40	50	40	130

b i $\frac{28}{130} = \frac{14}{65}$

ii $\frac{24}{130} = \frac{12}{65}$

c i $\frac{22}{60} = \frac{11}{30}$

ii As $22 + 14 = 36$,
then $\frac{36}{60} = \frac{3}{5}$

d $\frac{14}{40} = \frac{7}{20}$

28 a $P(\text{diamond, given a red}) = \frac{1}{2}$

b $P(\text{black, given a 7}) = \frac{1}{2}$

c $P(\text{queen, of hearts, given a red queen}) = \frac{1}{2}$

29 a $P(\text{coffee}) = \frac{11}{23}$

b $P(\text{pastry, given a tea drinker}) = \frac{4}{7}$

c $P(\text{coffee drinker, given a pastry eater})$
$= \frac{6}{12} = \frac{1}{2}$

30 a

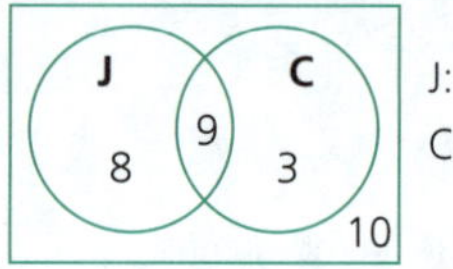

b $\frac{9}{17}$

c $\frac{8}{18} = \frac{4}{9}$

Worked Solutions to Tests

Chapter 1

Financial Mathematics

Intermediate Test

p.13

1 Pay $= 9.4 \times 22$
$= 206.8$
$\therefore$ Pay is \$206.80. ✓

2 Pay $= 17.4 \times 24 + 17.4 \times 8 \times 1.5 + 17.4 \times 6 \times 2$ ✓
$= 835.2$
$\therefore$ Ryan's pay is \$835.20. ✓

3 Loading $= 1450 \times 4 \times 0.175$ ✓
$= 1015$
$\therefore$ Holiday loading of \$1015. ✓

4 Income $= 420 + 0.025 \times 7200$ ✓
$= 600$
$\therefore$ Sheri's income is \$600. ✓

5 Commission $= 0.03 \times 210\,000$ ✓
$= 6300$
$\therefore$ Commission of \$6300. ✓

6 Deductions $= 421.4 + 86.7 + 75.4 + 11.3$
$= 594.8$ ✓
Net pay $= 2045 - 594.8$
$= 1450.2$
$\therefore$ Net pay is \$1450.20. ✓

7 Tax $= 0.19 \times (36\,470 - 18\,200)$ ✓
$= 3471.3$
$\therefore$ Tax of \$3471.30. ✓

8 New price $= 249 \times 0.85 \times 0.90$ ✓
$= 190.485$
$= 190.49$ (2 dec. pl.)
$\therefore$ New price of \$190.49. ✓

9 $I = Prn$
$= 700 \times 0.09 \times 3$
$= 189$
$\therefore$ Interest of \$189. ✓

10 $I = Prn$
$= 6400 \times 0.075 \times 5$ ✓
$= 2400$
$\therefore$ Interest of \$2400. ✓

11 $I = Prn$
$= 4200 \times 0.05 \times 3$
$= 630$ ✓
$\therefore$ Total $= 4200 + 630$
$= 4830$
$\therefore$ The total repaid is \$4830. ✓

12 a $I = Prn$
$= 12\,000 \times 0.08 \times 4$ ✓
$= 3840$
$\therefore$ Interest of \$3840. ✓

b Total repayment $= 12\,000 + 3840$
$= 15\,840$ ✓
Monthly repayment $= 15\,840 \div 48$
$= 330$
$\therefore$ Repayment of \$330. ✓

13 $A = P(1 + r)^n$
$= 6000 \times 1.05^3$
$= 6945.75$ ✓
$\therefore$ Interest $= 6945.75 - 6000$
$= 945.75$
$\therefore$ Compound interest of \$945.75. ✓

14 $A = P(1 + r)^n$
$= 4500 \times 1.07^{10}$ ✓
$= 8852.181\,108$
$= 8852.18$ (to 2 dec. pl.)
$\therefore$ The amount is \$8852.18. ✓

15 Value $= P(1 - r)^n$
$= 20\,000 \times 0.84^6$ ✓
$= 7025.960\,633$
$= 7025.96$ (to 2 dec. pl.)
$\therefore$ The value is \$7025.96. ✓

Advanced Test

p. 14

1 Rate $= 250.08 \div 16$
$= 15.63$
$\therefore$ Christie's hourly rate is \$15.63. ✓

2 Pay $= 15.20 \times 8 + 15.20 \times 4 \times 1.5 + 15.20 \times 3 \times 2$ ✓
$= 304$
$\therefore$ Ryan's pay is \$304. ✓

3 Holiday pay $= 2020 \times 2 + 2020 \times 2 \times 0.175$ ✓
$= 4747$
$\therefore$ Sung's holiday pay is \$4747. ✓

4 Rate $= \dfrac{15\,500}{620\,000} \times 100\%$ ✓
$= 2.5$
$\therefore$ The rate of commission is 2.5%. ✓

5 Commission $= 0.02 \times 10\,000 + 0.015 \times 18\,000$ ✓
$= 470$
$\therefore$ A commission of \$470. ✓

6 a Deductions $= 186.5 + 69.2 + 6.25$
$= 261.95$
$\therefore$ Net earnings $= 942 - 261.95$
$= 680.05$
$\therefore$ Sam's weekly earnings are \$680.05. ✓

b Total $= 261.95 \times 52$
$= 13\,621.4$
$\therefore$ His total deductions are \$13 621.40. ✓

7 a Taxable income $= 63\,490 - 4760$
$= 58\,730$
$\therefore$ Taxable income is \$58 730. ✓

b Tax $= 3572 + 0.325(58\,730 - 37\,000)$
$= 10\,634.25$ ✓
$\therefore$ Chris will pay tax of \$10 634.25. ✓

8 New price $= 200 \times 0.75 \times 0.80$ ✓
$= 120$
Discount $= 200 - 120$
$= 80$
$\therefore$ Discount of \$80. ✓

9 a $I = Prn$
$= 750 \times 0.08 \times \frac{7}{12}$ ✓
$= 35$
$\therefore$ Interest of \$35. ✓

b $I = Prn$
$= 1050 \times 0.0725 \times 3$ ✓
$= 228.375$
$= 228.38$ (to 2 dec. pl.)
$\therefore$ Interest of \$228.38. ✓

10 $I = Prn$
$1536 = 6400 \times r \times 4$ ✓
$1536 = 25\,600r$
$\therefore 25\,600r = 1536$
$r = \dfrac{1536}{25\,600}$
$= 0.06$
$\therefore$ Interest rate of 6%. ✓

11 6% p.a. = 0.005/month for 36 months
$A = P(1 + r)^n$
$= 5000 \times 1.005^{36}$ ✓
$= 5983.402\,624$
$= 5983.40$ (2 dec. pl.)
$\therefore$ Interest $= 5983.40 - 5000$
$= 983.40$
$\therefore$ Compound interest of \$983.40. ✓

12 Let product cost \$100.
$\therefore$ New cost $= 0.8 \times 0.8 \times 100$
$= 64$ ✓
$\therefore$ Reduction of \$36
i.e. discount of 36%
$\therefore P = 36$. ✓

13 $I = Prn$
$= 4000 \times 0.06 \times 5$
$= 1200$
$A = P(1 + r)^n$
$= 4000 \times 1.06^5$
$= 5352.902\,31$
$= 5352.90$ (to 2 dec. pl.) ✓
∴ Compound interest
$= 5352.90 - 4000$
$= 1352.90$
∴ Difference $= 1352.90 - 1200$
$= 152.90$
∴ \$152.90 more compound interest than simple interest. ✓

14 Value $= P(1 - r)^n$
$= 2490 \times 0.72^5$ ✓
$= 481.794\,4904$
$= 481.79$ (to 2 dec. pl.)
∴ The computer's value is \$481.79. ✓

Chapter 2

Algebraic Techniques

Intermediate Test

p. 28

1 a $3x + 5x = 8x$ ✓
b $6x \times y = 6xy$ ✓
c $\$4x = 400x$ cents ✓

2 a $2 \times -6 \times 4 = -48$ ✓
b $5 + (-6) + 4 = 3$ ✓
c $(-6)^2 - 4^2 = 36 - 16 = 20$ ✓
d $\dfrac{-6+4}{-6-4} = \dfrac{-2}{-10} = \dfrac{1}{5}$ ✓

3 a $10m$ ✓
b $36y^2$ ✓
c $5ab - 12ab = -7ab$ ✓
d $16y^2$ ✓
e $\dfrac{{}^{3}\cancel{21}\cancel{x}y}{{}_{1}\cancel{7}\cancel{x}} = 3y$ ✓
f $\dfrac{{}^{1}\cancel{9}\cancel{a}b}{{}_{2}\cancel{18}\cancel{a}a} = \dfrac{b}{2a}$ ✓

4 a $\dfrac{{}^{6}\cancel{42}\cancel{x}x\cancel{y}}{{}_{1}\cancel{7}\cancel{x}\cancel{y}y} = \dfrac{6x}{y}$ ✓
b $\dfrac{2(2m) - 3(3m)}{6} = \dfrac{4m - 9m}{6}$
$= \dfrac{-5m}{6}$ ✓
c $\dfrac{3(11) - 2(7)}{6x} = \dfrac{33 - 14}{6x}$
$= \dfrac{19}{6x}$ ✓
d $\dfrac{2x}{5} \times \dfrac{10x}{4} = \dfrac{20x^2}{20} = x^2$ ✓
e $\dfrac{2x}{5} \times \dfrac{4}{10x} = \dfrac{8x}{50x} = \dfrac{4}{25}$ ✓

5 a $15x - 10y$ ✓
b $5 - 6x + 10 = 15 - 6x$ ✓✓
c $12x - 9 - 12x + 16 = 7$ ✓✓

6 a $x^2 - 3x + 5x - 15 = x^2 + 2x - 15$ ✓
b $4x^2 - 25$ ✓
c $4x^2 + 2(2x)(5) + 25$
$= 4x^2 + 20x + 25$ ✓

7 a $7(y - 3)$ ✓
b $5m(m + 3n)$ ✓
c $(m + 3)(m - 3)$ ✓
d $(a + 8)(a - 5)$ ✓
e $(y - 9)(y - 2)$ ✓
f $y(y + 2x) + 9(y + 2x)$ ✓
$= (y + 9)(y + 2x)$ ✓
g $a^3 - 3a^2 - 4a$
$= a(a^2 - 3a - 4)$ ✓
$= a(a - 4)(a + 1)$ ✓

Advanced Test

p. 29

1 a $8y - 5y = 3y$ ✓
b $8a \times 3a = 24a^2$ ✓
c $\$\dfrac{3x}{100}$ ✓

2 a $-2t$ ✓
b $-42mn$ ✓
c $18m^2 - 35m^2 = -17m^2$ ✓
d $\dfrac{{}^{3}\cancel{42}\cancel{a}b}{{}_{1}\cancel{14}\cancel{a}} = 3b$ ✓
e $27m^2n$ ✓

3 a $\dfrac{{}^{5}\cancel{35}\cancel{ab}bb}{{}_{1}\cancel{7}\cancel{ab}} = 5b^2$ ✓
b $\dfrac{25a - 6a}{10} = \dfrac{19a}{10}$ ✓
c $\dfrac{21 - 5}{3m} = \dfrac{16}{3m}$ ✓
d $\dfrac{{}^{1}\cancel{15}}{\cancel{x}y} \times \dfrac{3\cancel{x}}{\cancel{5}_{1}} = \dfrac{9}{y}$ ✓
e $\dfrac{{}^{5}\cancel{25}\cancel{xy}}{{}_{2}\cancel{14}\cancel{x}} \times \dfrac{{}^{1}\cancel{7}\cancel{x}x}{{}_{1}\cancel{5}\cancel{xy}} = \dfrac{5y}{2}$ ✓

4 a $35m - 21m^2$ ✓
b $15t - 16t + 40$ ✓
$= 40 - t$ ✓
c $12a + 30 - 15a + 10$ ✓
$= 40 - 3a$ ✓

5 a $3x^2 + 15x - 2x - 10$
$= 3x^2 + 13x - 10$ ✓
b $25x^2 - 9$ ✓
c $25x^2 - 2(5x)(3) + 9$
$= 25x^2 - 30x + 9$ ✓
d $2x^2 - x - 6 - 2x^2 + 6$ ✓
$= -x$ ✓
e $4x^2 - 4x + 1 - (4x^2 + 4x + 1)$ ✓
$= 4x^2 - 4x + 1 - 4x^2 - 4x - 1$
$= -8x$ ✓

6 a $9(6 - a)$ ✓
b $(m - 8)(m + 8)$ ✓
c $9y(3y - 1)$ ✓
d $(p + 7)(p - 6)$ ✓
e $(p - 9)(p + 8)$ ✓
f $10x^2 - 10x - 200$
$10(x^2 - x - 20)$
$= 10(x - 5)(x + 4)$ ✓
g $m(m - 8) - 7t(m - 8)$ ✓
$= (m - 8)(m - 7t)$ ✓
h $(a^2 + b^2)(a^2 - b^2)$ ✓
$= (a^2 + b^2)(a + b)(a - b)$ ✓

Chapter 3

Indices

Intermediate Test

p. 36

1 a

500
25 20
5 5 4 5
2 2 ✓

∴ $500 = 5 \times 5 \times 5 \times 2 \times 2$
$= 5^3 \times 2^2$ ✓

b 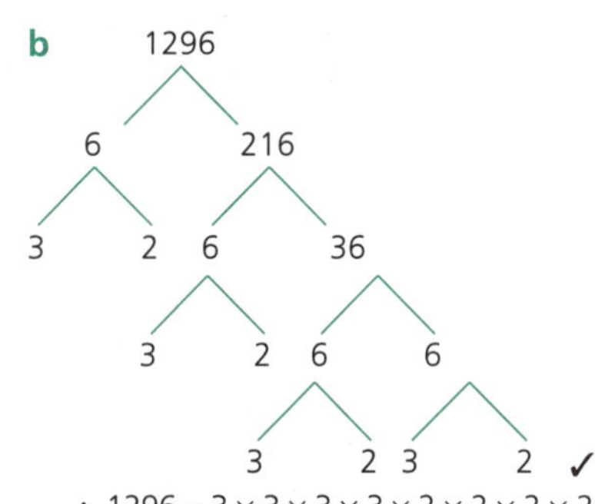

$\therefore 1296 = 3 \times 3 \times 3 \times 3 \times 2 \times 2 \times 2 \times 2$
$= 3^4 \times 2^4$ ✓

2 a $4x^2 \times 3x^4 = 12x^6$ ✓

b $10a^3 \times 4a^2 = 40a^5$ ✓

c $5m^3n^2 \times 3m^4n^5 = 15m^7n^7$ ✓

d $3p^3q^4 \times 6p^2q^6 = 18p^5q^{10}$ ✓

3 a $8b^5 \div 2b^3 = 4b^2$ ✓

b $14a^3 \div 7a^2 = 2a$ ✓

c $25k^5 \div 5k^2 = 5k^3$ ✓

d $30g^4 \div 6g = 5g^3$ ✓

4 a $(2b^3)^2 = 4b^6$ ✓

b $(3w^5)^2 = 9w^{10}$ ✓

c $(8a^6)^2 = 64a^{12}$ ✓

d $(-3y^4)^2 = 9y^8$ ✓

e $(-2m^3)^3 = -8m^9$ ✓

5 a $\frac{8a^2}{4a} = 2a$ ✓

b $\frac{20k^5}{10k} = 2k^4$ ✓

c $\frac{14m^3}{7m^2} = 2m$ ✓

d $\frac{b^7c^2}{b^5c} = b^2c$ ✓

6 a $4^0 = 1$ ✓

b $a^0 = 1$ ✓

c $5p^0 = 5$ ✓

d $(3 + 8)^0 = 1$ ✓

7 a $(a^6)^0 = a^0$
$= 1$ ✓

b $(m^0)^6 = m^0$
$= 1$ ✓

8 a $5^{-2} = \frac{1}{25}$ ✓

b $6^{-2} = \frac{1}{36}$ ✓

c $2^{-3} = \frac{1}{8}$ ✓

d $\left(\frac{1}{5}\right)^{-1} = 5$ ✓

e $\left(\frac{1}{3}\right)^{-2} = 3^2$
$= 9$ ✓

9 a $x^{-2} \times x^{-2} = x^{-4}$ ✓

b $a^{-4} \times a^2 = a^{-2}$ ✓

c $g^2 \times g^{-2} = g^0$
$= 1$ ✓

d $3a^5 \times 2a^{-2} = 6a^3$ ✓

e $7q^3 \times 5q^{-5} = 35q^{-2}$ ✓

f $p^3 \div p^{-2} = p^5$ ✓

g $d^{-4} \div d^3 = d^{-7}$ ✓

h $y^2 \div y^{-2} = y^4$ ✓

i $8a^2 \div 4a^{-3} = 2a^5$ ✓

j $15x^3 \div 5x^{-1} = 3x^4$ ✓

10 a $35\,000 = 3.5 \times 10^4$ ✓

b $262\,400 = 2.624 \times 10^5$ ✓

c $0.003\,11 = 3.11 \times 10^{-3}$ ✓

d $0.805 = 8.05 \times 10^{-1}$ ✓

11 a $2.57 \times 10^5 = 257\,000$ ✓

b $1.91 \times 10^4 = 19\,100$ ✓

c $2.531 \times 10^{-2} = 0.025\,31$ ✓

d $8.4 \times 10^{-3} = 0.0084$ ✓

Advanced Test

p. 37

1 a $3x^2y^5 \times 2x^5y^4 = 6x^7y^9$ ✓

b $12a^2b^8 \times 5a^3b^5 = 60a^5b^{13}$ ✓

c $70p^3q^2r \times 7pq^4r^3 = 490p^4q^6r^4$ ✓

d $7x^4yz^2 \times 9xy^3z^9 = 63x^5y^4z^{11}$ ✓

2 a $18a^7 \div 2a^5 = 9a^2$ ✓

b $25p^3q^2 \div 5p^2q = 5pq$ ✓

c $32a^8b^3 \div 4a^7b^3 = 8a$ ✓

d $24m^3n^4 \div 8mn^4 = 3m^2$ ✓

e $9x^3y^2 \div 6xy = \frac{3x^2y}{2} = 1.5x^2y$ ✓

f $15pq^4 \div 25pq^6 = \frac{3}{5q^2}$ ✓

3 a $\frac{12b^2}{3b} = 4b$ ✓

b $\frac{50g^9}{10g^2} = 5g^7$ ✓

c $\frac{a^7b^6}{a^5b^2} = a^2b^4$ ✓

d $\frac{12m^3n^2}{3m^2n^2} = 4m$ ✓

e $\frac{-5a^2bc}{5a^2b} = -c$ ✓

4 a $(2y^3)^2 = 4y^6$ ✓

b $(5c^4)^2 = 25c^8$ ✓

c $(-3m^5)^2 = 9m^{10}$ ✓

5 a $3^0 = 1$ ✓

b $(a + b)^0 = 1$ ✓

c $(4a)^0 - 4a^0 = 1 - 4$
$= -3$ ✓

d $10p^3q^2 \div 2p^3q^2 = 5$ ✓

e $(a^0)^6 = 1$ ✓

f $(3h^0)^4 = 81h^0$
$= 81$ ✓

g $(2s^7)^0 = 1$ ✓

h $(5m^3)^2 \div (5m^3)^0 = 25m^6 \div 1$
$= 25m^6$ ✓

6 a $x^{-2} = \frac{1}{x^2}$ ✓

b $6a^{-2} = \frac{6}{a^2}$ ✓

c $(2t)^{-3} = \frac{1}{8t^3}$ ✓

d $\left(\frac{1}{p}\right)^{-1} = p$ ✓

e $\left(\frac{3}{q}\right)^{-2} = \frac{q^2}{9}$ ✓

f $\frac{4}{a^{-2}} = 4a^2$ ✓

g $3b^{-2}c^4 = \frac{3c^4}{b^2}$ ✓

h $\frac{2}{3}a^3b^{-4} = \frac{2a^3}{3b^4}$ ✓

7 a $x^{-4} \times x^3 = x^{-1}$ ✓

b $a^7 \times a^{-7} = a^0 = 1$ ✓

c $5p^{-2} \times 3p^{-2} = 15p^{-4}$ ✓

d $12m^3 \div 3m^{-5} = 4m^8$ ✓

e $10z^{-3} \div 2z^{-6} = 5z^3$ ✓

f $(3m^{-5})^{-2} = \frac{m^{10}}{9}$ ✓

8 a $\frac{10a^2b^4}{3c^4d^3} \times \frac{6c^3d^2}{5a^2b^2} = \frac{4b^2}{cd}$ ✓✓

b $\frac{8x^2y^3}{9p^5q} \div \frac{4x^5y^4}{15p^7q^2}$
$= \frac{8x^2y^3}{9p^5q} \times \frac{15p^7q^2}{4x^5y^4}$
$= \frac{10p^2q}{3x^3y}$ ✓✓

9 a $\dfrac{\sqrt{11.7-3.84}}{12.91-8.004}$

$= 0.571\,457\,226\ldots$ ✓

$= 5.71 \times 10^{-1}$ (3 sig. figs.) ✓

b $\sqrt{\dfrac{19.3105-11.02}{3.1 \div 42.87}}$

$= 10.707\,454\,97\ldots$ ✓

$= 1.07 \times 10^{1}$ (3 sig. figs.) ✓

c $\dfrac{(3.24 \times 10^5)^3}{6.2 \times 10^2}$

$= 5.485\,842\,581\ldots \times 10^{13}$ ✓

$= 5.49 \times 10^{13}$ (sig. figs.) ✓

Chapter 4

Equations

Intermediate Test

p. 50

1 a $3x + 1 = 7$

$3x = 7 - 1$

$3x = 6$

$x = 2$ ✓

b $2a - 3 = a + 12$

$2a - a = 12 + 3$ ✓

$a = 15$ ✓

c $\dfrac{2a-1}{5} = 3$

$2a - 1 = 15$ ✓

$2a = 15 + 1$

$2a = 16$

$a = 8$ ✓

d $\dfrac{5x}{2} + 3x = 7$

$2 \times \dfrac{5x}{2} + 2 \times 3x = 2 \times 7$

$5x + 6x = 14$ ✓

$11x = 14$

$\dfrac{11x}{11} = \dfrac{14}{11}$

$x = 1\frac{3}{11}$ ✓

e $2(a - 3) = 12$

$2a - 6 = 12$ ✓

$2a = 12 + 6$

$2a = 18$

$a = 9$ ✓

f $4(5p - 2) + 3p = 19$

$20p - 8 + 3p = 19$ ✓

$23p = 19 + 8$

$23p = 27$

$\dfrac{23p}{23} = \dfrac{27}{23}$

$p = 1\frac{4}{23}$ ✓

g $2(2y - 1) - 3(y + 3) = 16$

$4y - 2 - 3y - 9 = 16$ ✓

$y - 11 = 16$ ✓

$y = 16 + 11$

$y = 27$ ✓

2 a $3a + 2 > a + 10$

$3a - a > 10 - 2$ ✓

$2a > 8$

$a > 4$ ✓

b $2p - 3 \le p - 5$

$2p - p \le -5 + 3$ ✓

$p \le -2$ ✓

c $\dfrac{2a-1}{5} \ge 3$

$2a - 1 \ge 15$ ✓

$2a \ge 15 + 1$

$2a \ge 16$

$a \ge 8$ ✓

d $q - 5 \le 3q - 9$

$q - 3q \le -9 + 5$ ✓

$-2q \le -4$

$\dfrac{-2q}{-2} \ge \dfrac{-4}{-2}$

$q \ge 2$ ✓

3 a $2x + 6 \le x - 2$

$2x - x \le -2 - 6$

$x \le -8$ ✓

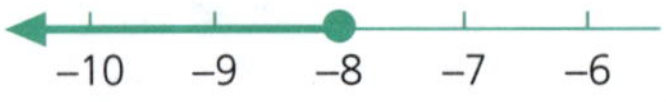

✓

b $x - 3 \le 4x - 6$

$x - 4x \le -6 + 3$

$-3x \le -3$

$\dfrac{-3x}{-3} \ge \dfrac{-3}{-3}$

$x \ge 1$ ✓

✓

4 a $P = 2(l + b)$

$= 2(6 + 4)$ ✓

$= 20$ ✓

b $P = 2(l + b)$

$20 = 2(l + 3)$ ✓

$20 = 2l + 6$

$2l + 6 = 20$

$2l = 20 - 6$

$2l = 14$

$l = 7$ ✓

5 a $5x + 2y = 11$

$2y = 11 - 5x$ ✓

$y = \dfrac{11-5x}{2}$ ✓

b $3x - 7y = 10$

$3x - 10 = 7y$ ✓

$7y = 3x - 10$

$y = \dfrac{3x-10}{7}$ ✓

c $\dfrac{2x-y}{3} = 5$

$2x - y = 15$ ✓

$2x - 15 = y$

$y = 2x - 15$ ✓

6 a $x + 1$ ✓

b $x + (x + 1) = 25$

$2x + 1 = 25$ ✓

$2x = 25 - 1$

$2x = 24$

$x = 12$

∴ The numbers are 12 and 13. ✓

7 a Opposite ∠s equal

∴ $3x + 10 = 2x + 50$ ✓

$3x - 2x = 50 - 10$

$x = 40$ ✓

b $\angle DAB = (3 \times 40 + 10)°$

$= 130°$ ✓

8 $2x + y = 5$ (1)

$2x + 3y = 7$ (2)

(2) − (1) $2y = 2$

$y = 1$ ✓

Subs. in (1)

$2x + 1 = 5$

$2x = 5 - 1$

$2x = 4$

$x = 2$

∴ $x = 2$ and $y = 1$ ✓

9 a $x^2 - 6x + 5 = 0$

$(x - 5)(x - 1) = 0$ ✓

$x = 5, 1$ ✓

b $x^2 - 2x - 8 = 0$

$(x - 4)(x + 2) = 0$ ✓

$x = 4, -2$ ✓

c $x^2 - 5x = 0$

$x(x - 5) = 0$ ✓

$x = 0, 5$ ✓

Advanced Test

p. 51

1 a $3x - 4 = 4x - 2$

$3x - 4x = -2 + 4$ ✓

$-x = 2$

$\dfrac{-x}{-1} = \dfrac{2}{-1}$

$x = -2$ ✓

b $1 - x = 3x + 5$

$-x - 3x = 5 - 1$ ✓

$-4x = 4$

$\dfrac{-4x}{-4} = \dfrac{4}{-4}$

$x = -1$ ✓

c $3(4x - 3) + 2x = 11$
$12x - 9 + 2x = 11$
$14x - 9 = 11$ ✓
$14x = 11 + 9$
$14x = 20$
$\frac{14x}{14} = \frac{20}{14}$
$x = 1\frac{3}{7}$ ✓

d $5(3a - 1) + 4(2a + 1) = 10$
$15a - 5 + 8a + 4 = 10$ ✓
$23a - 1 = 10$
$23a = 10 + 1$
$23a = 11$
$\frac{23a}{23} = \frac{11}{23}$
$a = \frac{11}{23}$ ✓

e $2(3y - 2) - (4y + 3) = 18$
$6y - 4 - 4y - 3 = 18$ ✓
$2y - 7 = 18$
$2y = 18 + 7$
$2y = 25$
$\frac{2y}{2} = \frac{25}{2}$
$y = 12\frac{1}{2}$ ✓

2 a $\frac{2p - 3}{2} = 3p$
$2p - 3 = 6p$ ✓
$2p - 6p = 3$
$-4p = 3$
$p = -\frac{3}{4}$ ✓

b $\frac{2a - 1}{5} = \frac{a - 3}{2}$
$10 \times \frac{2a - 1}{5} = 10 \times \frac{a - 3}{2}$
$2(2a - 1) = 5(a - 3)$ ✓
$4a - 2 = 5a - 15$
$4a - 5a = -15 + 2$
$-a = -13$
$a = 13$ ✓

c $\frac{5x}{2} + \frac{x}{3} = 2$
$6 \times \frac{5x}{2} + 6 \times \frac{x}{3} = 6 \times 2$
$15x + 2x = 12$ ✓
$17x = 12$
$x = \frac{12}{17}$ ✓

3 a $2x + 2 \leq 4x - 6$
$2x - 4x \leq -6 - 2$ ✓
$-2x \leq -8$
$\frac{-2x}{-2} \geq \frac{-8}{-2}$
$x \geq 4$ ✓

2 3 4 5 6 ✓

b $\frac{1 - 3x}{5} > 2$
$1 - 3x > 10$ ✓
$-3x > 10 - 1$
$-3x > 9$
$\frac{-3x}{-3} < \frac{9}{-3}$
$x < -3$ ✓

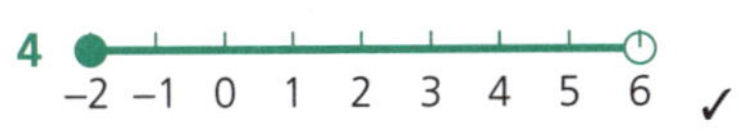

✓

4 −2 −1 0 1 2 3 4 5 6 ✓

5 a Area $= 4(3x + 1)$ ✓

b $4(3x + 1) = 64$
$12x + 4 = 64$ ✓
$12x = 64 - 4$
$12x = 60$
$x = 5$ ✓

c Length $= 3 \times 5 + 1$
$= 16$ ✓
∴ Dimensions are 16 cm × 4 cm
∴ Per. $= 2(16 + 4)$
$= 2(20)$
$= 40$
∴ 40 cm ✓

6 a $3ax - 5by = z$
$3ax - z = 5by$ ✓
$5by = 3ax - z$
$y = \frac{3ax - z}{5b}$ ✓

b $T = \sqrt{\frac{ay}{m}}$
$T^2 = \frac{ay}{m}$ ✓
$mT^2 = ay$
$ay = mT^2$
$y = \frac{mT^2}{a}$ ✓

c $P = \frac{4a + b}{y}$
$Py = 4a + b$ ✓
$y = \frac{4a + b}{P}$ ✓

d $k = \frac{3xy - 2}{a}$
$ak = 3xy - 2$ ✓
$3xy = ak + 2$
$y = \frac{ak + 2}{3x}$ ✓

7 $A = \pi(R^2 - r^2)$
$148.3 = \pi(R^2 - 4^2)$ ✓
$148.3 = \pi R^2 - 16\pi$
$\pi R^2 = 148.3 + 16\pi$
$R = \sqrt{\frac{148.3 + 16\pi}{\pi}}$ ✓
$= 7.950\,179\,628\ldots$
$= 7.95$ (2 dec. pl.)
∴ Radius of larger circle is 7.95 cm. ✓

8 $4x - y = 3$ (1)
$2x + y = 3$ (2)
(1) + (2) $6x = 6$
$x = 1$ ✓
Subs. in (2) $2(1) + y = 3$
$2 + y = 3$
$y = 3 - 2$
$y = 1$
∴ $x = 1, y = 1$ ✓

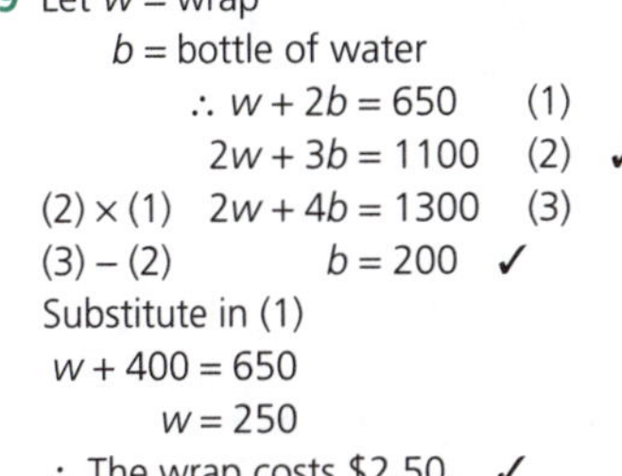

9 Let w = wrap
b = bottle of water
∴ $w + 2b = 650$ (1)
$2w + 3b = 1100$ (2) ✓
(2) × (1) $2w + 4b = 1300$ (3)
(3) − (2) $b = 200$ ✓
Substitute in (1)
$w + 400 = 650$
$w = 250$
∴ The wrap costs $2.50. ✓

10 a $x^2 - 6x + 5 = 0$
$(x - 5)(x - 1) = 0$
$x = 5, 1$ ✓

b $x^2 = 2x + 15$
$x^2 - 2x - 15 = 0$ ✓
$(x - 5)(x + 3) = 0$
$x = 5, -3$ ✓

11 $(x + 5)(x - 2) = 18$
$x^2 + 3x - 10 = 18$ ✓
$x^2 + 3x - 28 = 0$
$(x + 7)(x - 4) = 0$
$x = -7, 4$ ✓
But here $x > 2$, therefore $x = 4$.
Dimensions are 9 cm and 2 cm.
Per. $= 2(9 + 2)$
$= 22$
∴ Perimeter is 22 cm. ✓

Chapter 5
Linear Relationships

Intermediate Test

p. 68

1

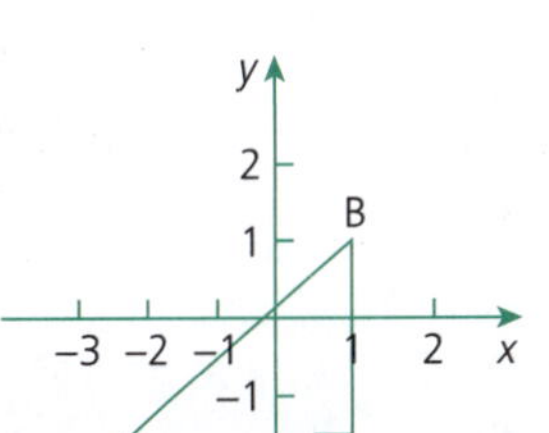

✓

a $AB^2 = 3^2 + 4^2$
$= 25$
$AB = \sqrt{25} = 5$ ✓

b For midpoint, average coordinates

$X = \frac{-3+1}{2} = \frac{-2}{2} = -1$

$Y = \frac{-2+1}{2} = \frac{-1}{2}$

midpoint $= \left(-1, -\frac{1}{2}\right)$ ✓

2 $y = 2x - 3$

x	0	1	2
y	−3	−1	1

✓✓

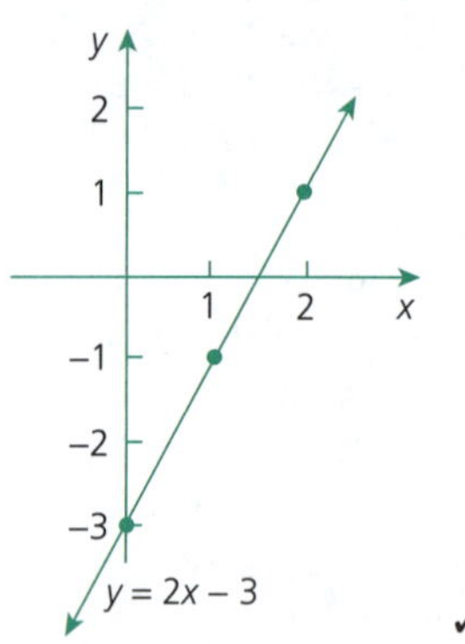

✓✓

3

y

$x = 4$

1 2 3 4 x

✓

4 $x = 0, y = 6$

y-intercept $= 6$ ✓

$y = 0,\ 2x = 6$

$x = 3$

x-intercept $= 3$ ✓

5 $y = \frac{2}{3}x - 4$

$m = \frac{2}{3}$ ✓

$b = -4$ ✓

6 $m = -4,\ b = 1$ ($y = mx + b$)

$y = -4x + 1$ ✓✓

7 a $m = \frac{\text{Rise}}{\text{Run}}$ (↗ positive)

$= \frac{2}{3}$ ✓

b $b = -2$ ✓

c Equation is

$y = \frac{2}{3}x - 2$ ✓ ($y = mx + b$)

8

y
3
2
1
1
2
1 2 3 4 x
y-intercept

Rise = 1

Run = 2

↗ positive ✓✓

9 a $d = \sqrt{(x_2 - x_1)^2 (y_2 - y_1)^2}$

$d = \sqrt{(1-2)^2 + (-1-5)^2}$ ✓

$= \sqrt{9 + 16}$

$= \sqrt{25} = 5$ ✓

b $\text{MP} = \left(\frac{x_1 + x_2}{2}, \frac{y_1 + y_2}{2}\right)$

$= \left(\frac{2 + -1}{2}, \frac{5+1}{2}\right)$ ✓

$= \left(\frac{1}{2}, 3\right)$ ✓

c $m = \frac{y_2 - y_1}{x_2 - x_1}$

$= \frac{1-5}{-1-2} = \frac{-4}{-3}$ ✓

$= \frac{4}{3}$ ✓

10 (3, 1) $\quad y = 2x - 5$

$x = 3, y = 1$ in $\quad 1 = (2 \times 3) - 5$ ✓

$= 6 - 5$

$= 1$

(3, 1) lies on line. ✓

11 $y = \frac{1}{2}x - 5 \qquad m_1 = \frac{1}{2}$ ✓

$x = 2y - 3$

$\therefore 2y = x + 3$

$y = \frac{1}{2}x + \frac{3}{2} \qquad m_2 = \frac{1}{2}$ ✓

Lines are parallel as they have equal gradients. ✓

12 Gradient $= m = -2$

$\therefore$ Subs. (−3, −1) in $y = -2x + c$ ✓

$-1 = -2 \times -3 + c$

$-1 = c + 6$

$c = -1 - 6$

$c = -7$

$\therefore y = -2x - 7$ ✓

13 Gradient $= m = 4$

$\therefore$ Subs. (1, 6) in $y = 4x + c$ ✓

$6 = 4 \times 1 + c$

$6 = c + 4$

$c = 6 - 4$

$c = 2$

$\therefore y = 4x + 2$ ✓

14 a $C = 15 + 3n$ ✓

b

n	0	4	8	12
C	15	27	39	51

✓

c

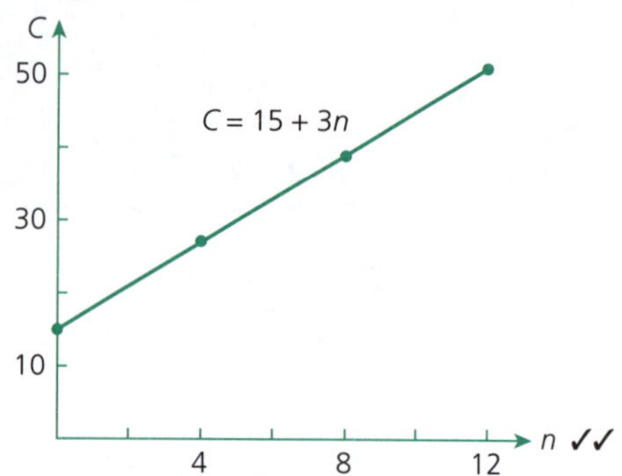

✓✓

d Substitute $n = 25$:

$C = 15 + 3 \times 25$ ✓

$= 15 + 75$

$= 80 \qquad \therefore \$80$ ✓

Advanced Test

p. 69

1

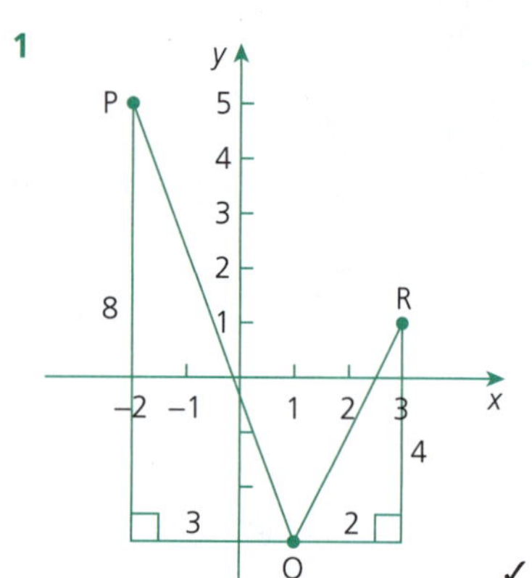

✓

a $PQ^2 = 8^2 + 3^2$

$= 64 + 9$

$= 73$

$\therefore PQ = \sqrt{73}$ ✓✓

b Gradient $QR = \frac{\text{Rise}}{\text{Run}}$ (↗ +)

$= \frac{4}{2}$

$= 2$ ✓

2

x	0	1	2
y	3	1	−1

✓✓

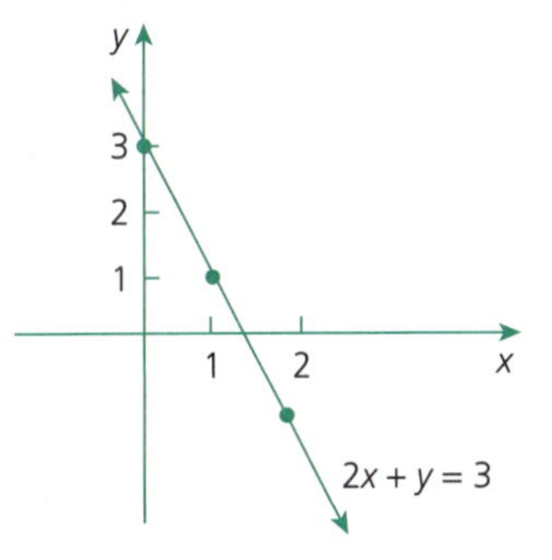

✓

3

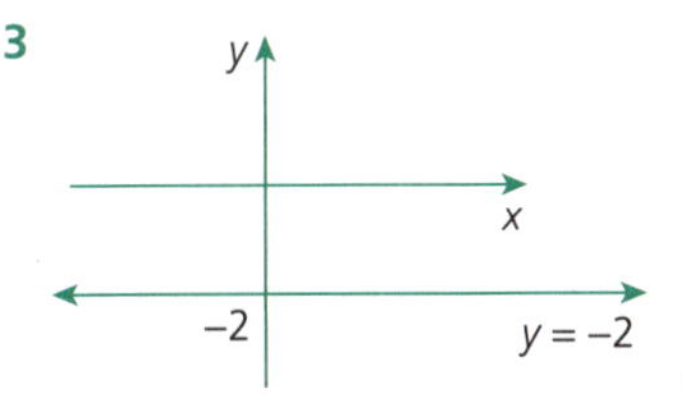

✓

4 $3x - 2y = 4$

$x = 0,\quad -2y = 4$

$y = -2$

y-intercept $= -2$ ✓

$y = 0,\quad 3x = 4$

$x = \frac{4}{3}$

x-intercept $= \frac{4}{3}$ ✓

5 $2x + 3y = 6$

$3y = -2x + 6$

$y = \frac{-2}{3}x + 2$ ✓

$y = mx + b$

$m = \frac{-2}{3},\ b = 2$ ✓✓

6 a Gradient $= \frac{\text{Rise}}{\text{Run}}$

$= -\frac{3}{4}$ ✓

b y-intercept $= -3$ ✓

c Equation is

$y = -\frac{3}{4}x - 3$ ✓

7 y-intercept $= -2$

$m = 2$

Using $y = mx + b$

$y = 2x - 2$ ✓✓

8 a $d = \sqrt{(-3-4)^2 + (-4-1)^2}$ ✓

$= \sqrt{49 + 25}$

$= \sqrt{74}$ ✓

b $\text{MP} = \left(\frac{-1+-3}{2}, \frac{3+-4}{2}\right)$ ✓

$= \left(\frac{-4}{2}, \frac{-1}{2}\right)$

$= \left(-2, \frac{-1}{2}\right)$ ✓

c $m = \frac{3-1}{-1-4} = \frac{2}{-5} = \frac{-2}{5}$ ✓✓

9 $(-2, 3k)$ $\quad 3x + 2y = 9$

$3(-2) + 2(3k) = 9$ ✓

$-6 + 6k = 9$ ✓

$6k = 15$

$k = \frac{15}{6} = \frac{5}{2}$ ✓

10 $3y = 12x - 4$

$y = 4x - \frac{4}{3}$

$m = 4$ ✓

Gradient of parallel line = 4

y-intercept $= -1$ ✓

Equation is $y = 4x - 1$ ✓

11 a $y = 2x + 1$:

x	0	1	2
y	1	3	5

✓

$y = 4 - x$:

x	0	1	2
y	4	3	2

✓

✓

b From Part (a), solution is $x = 1$, $y = 3$. ✓

12 Midpoint AB $= \left(\frac{2-4}{2}, \frac{8+2}{2}\right)$

$= (-1, 5)$ ✓

Gradient AB $= \frac{2-8}{-4-2}$

$= \frac{-6}{-6}$

$= 1$

If perpendicular, then gradient $= m = -1$ ✓

The line is of the form $y = -x + c$

Subs. $(-1, 5)$ in $y = -x + c$

$5 = -(-1) + c$ ✓

$5 = 1 + c$

$c = 4$

$\therefore y = -x + 4$ ✓

13 $2x - y - 3 = 0$

$y = 2x - 3$

$\therefore m = 2$ ✓

$4x + ky + 11 = 0$

$ky = -4x - 11$

$y = \frac{-4x - 11}{k}$

$\therefore m = \frac{-4}{k}$ ✓

If lines parallel, $\frac{-4}{k} = 2$

$2k = -4$

$k = -2$ ✓

14 a $C = 120 + 12n$ ✓

b

n	0	5	10	15	20	25
C	120	180	240	300	360	420

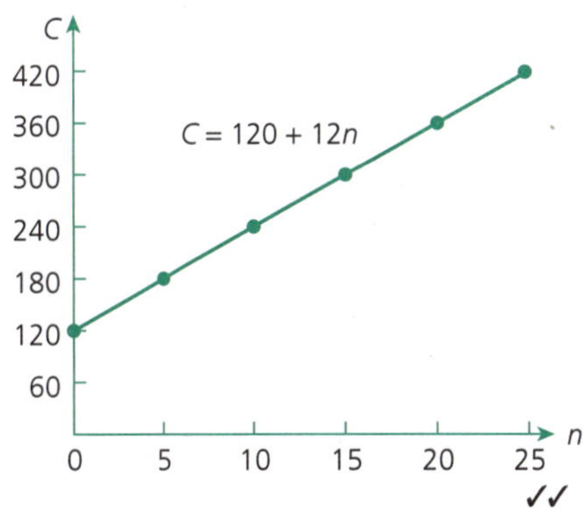

✓✓

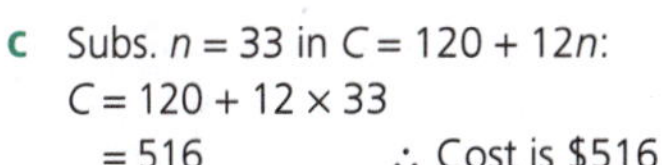

c Subs. $n = 33$ in $C = 120 + 12n$:

$C = 120 + 12 \times 33$

$= 516$ $\quad\therefore$ Cost is \$516. ✓

Chapter 6

Non-linear Relationships

Intermediate Test

p. 81

1 $y = 3x^2 - 4$

x	–2	–1	0	1	2
y	8	–1	–4	–1	8

✓✓

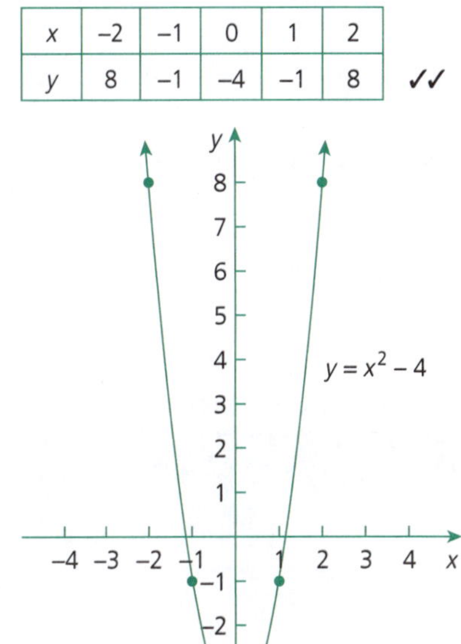

✓

2 a C ✓

b B ✓

c H ✓

d F ✓

e G ✓

f E ✓

g A ✓

h D ✓

3

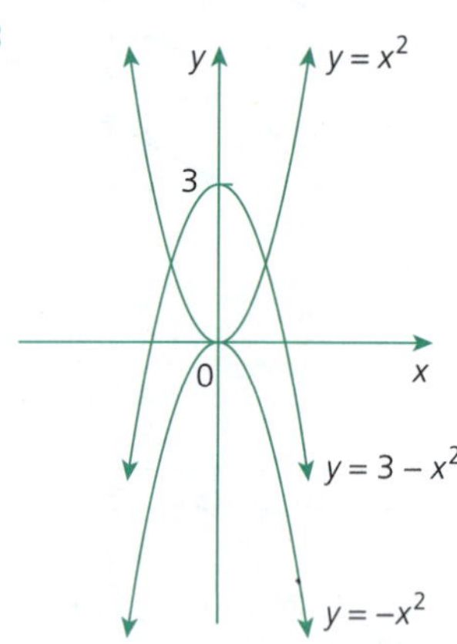

✓✓✓

4

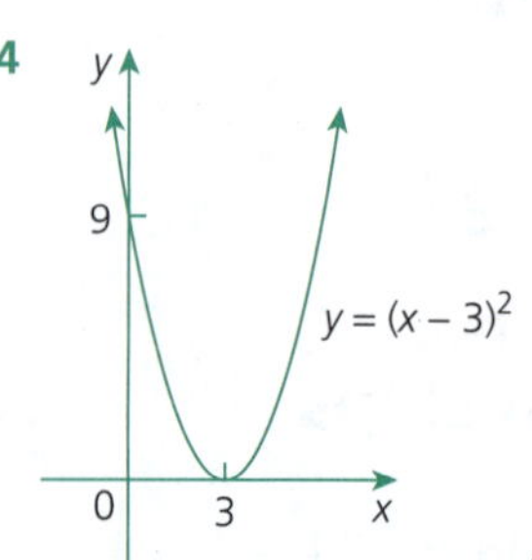

✓

Any parabola of form $y = a(x - 3)^2$, where $a > 0$. For example, $a = 1$:
$y = (x - 3)^2$ ✓

5 Subs. $(8, a)$ in $xy = 4$:

$8a = 4$ ✓

$a = \frac{1}{2}$ ✓

6 For $y = \frac{2}{x}$, subs. in (2, 1)

$1 = \frac{2}{2} = 1$

$\therefore y = \frac{2}{x}$

$\therefore$ C ✓

7 $y = 4 - x$ is straight line

$\therefore$ B ✓

Advanced Test

p. 82

1 a $y = (x - 1)(x + 2)$

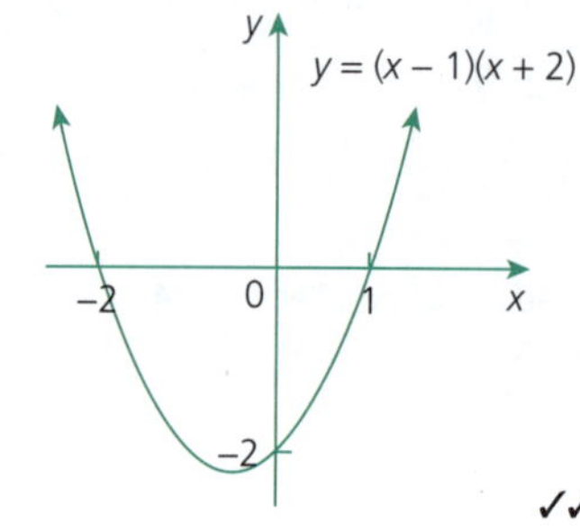

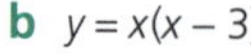

✓✓

b $y = x(x - 3)$

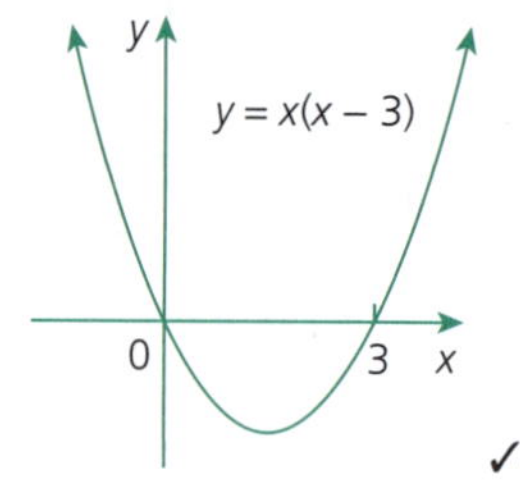

✓✓

c $y = (2 - x)(x + 3)$

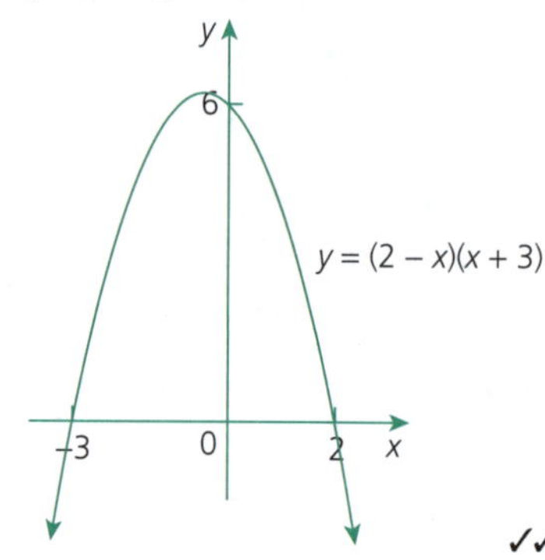

✓✓

d $y = x^2 - 3x - 4$

$\therefore y = (x - 4)(x + 1)$

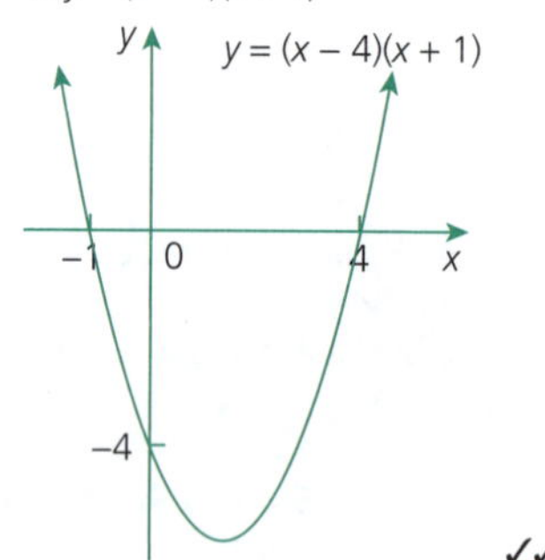

✓✓

2 a

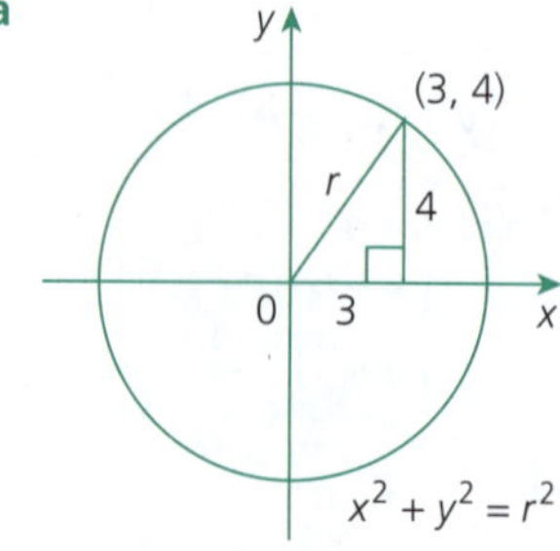

Using Pythagoras:

$r^2 = 3^2 + 4^2$

$= 25$

$r = 5$ ✓

$\therefore x^2 + y^2 = 25$ ✓

b Subs. (2, 9) in $y = a^x$

$9 = a^2$ ✓

$a = 3$ (note, from diagram, $a > 0$)

$\therefore y = 3^x$ ✓

c As curve has x-intercepts of 0 and 6, then equation of the form $y = ax(x - 6)$.

Subs. $(-2, -48)$ in $y = ax(x - 6)$ ✓

$-48 = a(-2)(-2 - 6)$

$-48 = 16a$

$a = -3$

$\therefore y = -3x(x - 6)$ ✓

d Subs. $\left(4, -\frac{1}{2}\right)$ in $y = \frac{c}{x}$

$-\frac{1}{2} = \frac{c}{4}$ ✓

$4 \times -\frac{1}{2} = 4 \times \frac{c}{4}$

$c = -2$

$\therefore y = -\frac{2}{x}$ ✓

3 a $x^2 - 5x + 1 = 2x - 5$

$x^2 - 7x + 6 = 0$

$(x - 6)(x - 1) = 0$

$x = 6, 1$ ✓

Subs. $x = 6$ in $y = 2x - 5$

$= 2(6) - 5$

$= 7$

Subs. $x = 1$ in $y = 2x - 5$

$= 2(1) - 5$

$= -3$

$\therefore$ (6, 7) and (1, −3) ✓

b $2x^2 + 3x + 4 = x^2 - x + 9$

$x^2 + 4x - 5 = 0$

$(x + 5)(x - 1) = 0$

$x = -5, 1$ ✓

Subs. $x = -5$ in $y = x^2 - x + 9$

$= (-5)^2 - (-5) + 9$

$= 39$

Subs. $x = 1$ in $y = x^2 - x + 9$

$= 1^2 - 1 + 9$

$= 9$

$\therefore$ (−5, 39) and (1, 9) ✓

Chapter 7

Rates and Proportion

Intermediate Test

p. 92

1 $0.4 \times 365 = 146$

$\therefore$ 146 mm/year = 14.6 cm/year ✓

2 a $10\,000 \text{ m}^2 = 1$ ha

$20 \times 10\,000 = 200\,000$

$\therefore$ 200 000 g/ha = 200 kg/ha ✓

b $100 \times 100 \times 100 = 1\,000\,000$
$\therefore 1\,000\,000\,\text{cm}^3 = 1\,\text{m}^3$
$0.08 \times 1\,000\,000 = 80\,000$
$\therefore 80\,000\ \text{c/m}^3 = \$800\ /\text{m}^3$ ✓

3 Bolt: 100 m/9.58 s
$100 \div 9.58 \times 60 \times 60$
$= 37\,578$ (nearest whole)
$\therefore 37\,578$ m/h = 37.578 km/h ✓
Cheetah: 100 m/5.95 s
$100 \div 5.95 \times 60 \times 60$
$= 60\,504$ (nearest whole)
$\therefore 60\,504$ m/h = 60.504 km/h ✓
Difference = 22.926
$\therefore$ Cheetah is faster by 23 km/h. ✓

4 a

h	0	1	2	3	4
D	0	90	180	270	360

✓

b $D = 90h$ ✓

c 90 km/h ✓

d 75 minutes $= 1\frac{1}{4}$ hours = 1.25 hours
Subs. $h = 1.25$ in $D = 90h$:
$D = 90 \times 1.25$
$= 112.5$
$\therefore$ 112.5 km ✓

e Second car: $D = 60h$

Distance (D)
420
360
300
240
180
120
60
0
1
2
3
4
Hours (h)
$D = 90h$
$D = 60h$

✓

f Subs. $h = 3.5$ in $D = 90h$
$D = 90 \times 3.5$
$= 315$
$\therefore$ first car travelled 315 km
Subs. $h = 3.5$ in $D = 60h$
$D = 60 \times 3.5$
$= 210$
$\therefore$ second car travelled 210 km
Difference = 105
$\therefore$ First car travelled 105 km more. ✓

g Subs. $D = 180$ in $D = 90h$
$180 = 90h$
$h = 2$
$\therefore$ first car takes 2 hours
Subs. $D = 180$ in $D = 60h$
$180 = 60h$
$h = 3$
$\therefore$ second car takes 3 hours
Difference = 1
$\therefore$ Second car takes 1 hour longer. ✓

5 a $C = kA$ ✓
Subs. $C = 235\,000$ and $A = 200$ in
$C = kA$
$235\,000 = 200k$
$k = 1175$
$\therefore C = 1175A$ ✓

b Subs. $A = 310$ in $C = 1175A$
$C = 1175 \times 310$
$= 364\,250$
$\therefore \$364\,250$ ✓

6 a $d = k\sqrt{h}$ ✓
Subs. $d = 4.16$ and $h = 1.69$ in $d = k\sqrt{h}$
$4.16 = k \times \sqrt{1.69}$
$4.16 = 1.3k$
$k = \frac{4.16}{1.3}$
$= 3.2$
$\therefore d = 3.2\sqrt{h}$ ✓

b Subs. $d = 8$ in $d = 3.2\sqrt{h}$
$8 = 3.2\sqrt{h}$
$\sqrt{h} = \frac{8}{3.2}$
$= 2.5$
$h = 6.25$
$\therefore$ 6.25 metres ✓

c Subs. $h = 18$ in $d = 3.2\sqrt{h}$
$d = 3.2\sqrt{18}$
$= 13.58$ (2 dec. pl.)
As $13.58 > 12$, then the person can see the island. ✓

Advanced Test

1 20 g/125 mL
$20 \div 125 \times 1000 = 160$
$\therefore$ 160 grams/1000 mL = 160 g/L ✓

2 $y = kx$
If $x = 1$, then $y = k$
If $x = 2$, then $y = 2k$
$\therefore$ If x is doubled, then y is doubled. ✓

3 $y = kx^2$
If $x = 1$, then $y = k$
If $x = 2$, then $y = 4k$
$\therefore$ If x is doubled, then y is multiplied by 4. ✓

4 a $y = k\sqrt{x}$ ✓
Subs. $x = 16$ and $y = 8$ in $y = k\sqrt{x}$
$8 = k \times \sqrt{16}$
$8 = 4k$
$k = 2$ ✓
$\therefore y = 2\sqrt{x}$

b Subs. $x = 4$ in $y = 2\sqrt{x}$
$y = 2\sqrt{4}$
$= 4$ ✓
Subs. $x = 49$ in $y = 2\sqrt{x}$
$y = 2\sqrt{49}$
$= 14$ ✓
Subs. $y = 24$ in $y = 2\sqrt{x}$
$24 = 2\sqrt{x}$
$\sqrt{x} = 12$
$x = 144$ ✓

x	4	16	49	144
y	4	8	14	24

5 a Truck A:
Gradient $= \frac{\text{Rise}}{\text{Run}}$
$= \frac{480}{6}$
$= 80$
$\therefore D = 80t$ ✓
Truck B:
Gradient $= \frac{600}{6}$
$= 100$
$\therefore D = 100t$ ✓

b Consider the difference in the two equations:
As $100t - 80t = 20t$, then to find differences between the two trucks use $D = 20t$.
Subs. $t = 4$ in $D = 20t$
$D = 20 \times 4$
$= 80$
$\therefore$ 80 km more ✓✓

c Subs. $D = 60$ in $D = 20t$
$60 = 20t$
$t = 3$
$\therefore$ After 3 hours. ✓✓

6 a $C = kn^2$ ✓
Subs. $n = 3$ and $C = 64.8$ in $C = kn^2$
$64.8 = k \times 3^2$
$64.8 = 9k$
$k = 7.2$
$\therefore C = 7.2n^2$ ✓

b Subs. $C = 45$ in $C = 7.2n^2$
$45 = 7.2n^2$
$n^2 = \frac{45}{7.2}$
$= 6.25$
$n = 2.5$
$\therefore$ 2.5 cm ✓

c Subs. $n = 4.3$ in $C = 7.2n^2$
$C = 7.2 \times (4.3)^2$
$= 133.13$ (2 dec. pl.)
$\therefore$ Costs \$133.13. ✓

7 Let $c = kd^2$
$6250 = k(25)^2$
$625k = 6250$
$k = 10$ ✓
$\therefore c = 10d^2$
$c = 10(40)^2$
$= 16\,000$
$\therefore$ Cost is \$160. ✓

Chapter 8

Pythagoras and Trigonometry

Intermediate Test

pp. 120–121

1 $PR^2 = PQ^2 + QR^2$ ✓

2 a $k^2 = 12^2 + 16^2$ ✓
$= 400$
$\therefore k = \sqrt{400}$
$= 20$ ✓

b $h^2 = 1.5^2 + 0.8^2$ ✓
$= 2.89$
$h = \sqrt{2.89}$
$= 1.7$ ✓

3 a $x^2 = 15^2 - 11^2$ ✓
$= 104$
$x = \sqrt{104}$ ✓

b $y^2 = 8^2 - 3^2$ ✓
$= 55$
$y = \sqrt{55}$ ✓

4 a $16^2 = 8^2 + 14^2$ ✓
$256 \neq 260$
Not right-angled. ✓

b $2.5^2 = 2^2 + 1.5^2$ ✓
$6.25 = 6.25$
Yes, right-angled. ✓

5

6, d, 6 ⇛ 6, d, 6

$d^2 = 6^2 + 6^2$ ✓
$= 72$
$d = \sqrt{72}$ ✓
$= 8.485\dot{2}$
$= 8.5$ (to 1 dec. pl.)
The diagonal is of length 8.5 cm. ✓

6 a 0.122 ✓

b 1.342 ✓

c 19.700 ✓

d 48.801 ✓

7 a 18° ✓

b 37° ✓

8 a $\frac{x}{17} = \sin 29°$ ✓
$\therefore x = 17 \sin 29°$
$= 8.2417\ldots$
$\therefore$ 8.2 cm (to 1 dec. pl.) ✓

b $\frac{m}{26.3} = \tan 42°$ ✓
$m = 26.3 \tan 42°$
$= 23.6806\ldots$
$= 23.7$ (to 1 dec. pl.) ✓

c $\frac{y}{23.2} = \cos 34°19'$ ✓
$\therefore y = 23.2 \cos 34°19'$
$= 19.1616\ldots$
$= 19.2$ (to 1 dec. pl.) ✓

d $\frac{17.2}{t} = \tan 39°20'$
$t = \frac{17.2}{\tan 39°20'}$ ✓
$= 20.98936\ldots$
$= 21.0$ (to 1 dec. pl.) ✓

9 a $\tan\theta = \frac{17}{8}$ ✓
$\therefore \theta = 65°$ ✓

b $\sin\theta = \frac{18}{21}$ ✓
$\theta = 59°$ ✓

c $\cos\theta = \frac{19}{42}$ ✓
$\theta = 63°$ ✓

10

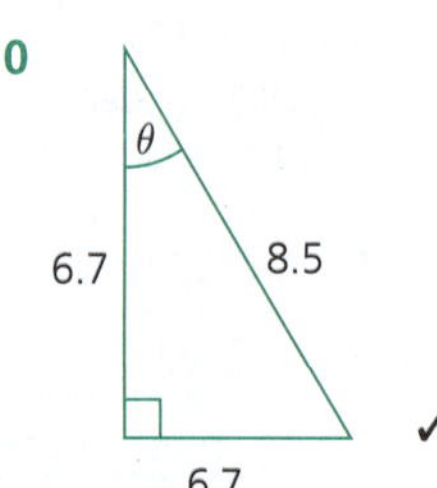

✓

$\cos\theta = \frac{6.7}{8.5}$
$\therefore \theta = 38°$ ✓
Ladder makes angle of 38° with wall.

11

h, 38°, 50

$h = 50 \tan 38°$ ✓
$= 39.064\ldots$
$= 39$ ✓
Banya tree is 39 metres high.

12 a

N, 175°

✓

b

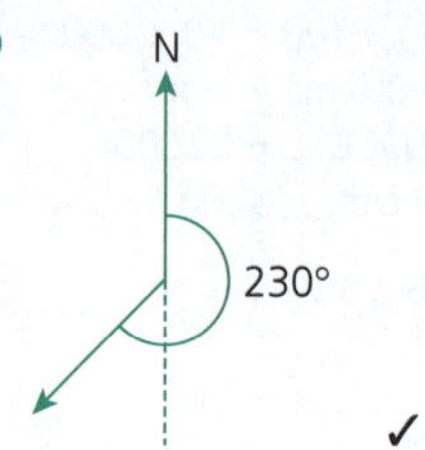

✓

13

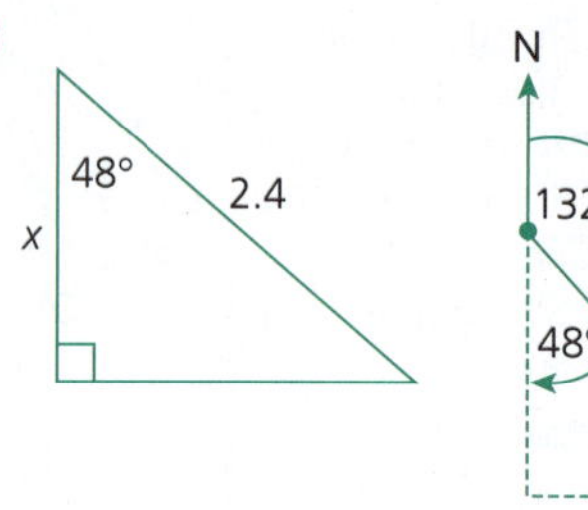

✓

$\frac{x}{10} = \cos 48°$ ✓
$x = 2.4 \cos 48°$
$= 1.6059\ldots$
$= 1.6$
Mary is 1.6 km south of her sheep. ✓

Advanced Test

pp. 122–123

1 a $y^2 = 5^2 + 8^2$ ✓
$= 89$
$y = \sqrt{89}$
$= 9.4$ (to 1 dec. pl.) ✓

b $t^2 = 15^2 - 14^2$ ✓
$= 29$
$t = \sqrt{29}$
$= 5.4$ (to 1 dec. pl.) ✓

2

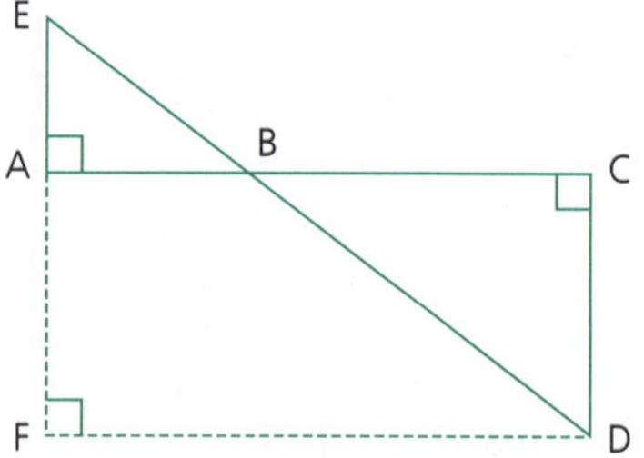

Complete rectangle ACDF, giving
DF = AC = 24 and EF = EA + AF (= CD)
$= 7 + 11$
$= 18$ ✓
Then, $DE^2 = EF^2 + DF^2$
$= 18^2 + 24^2$ ✓
$= 900$
$DE = \sqrt{900}$
$= 30$ cm ✓

3

11, 2, 5, 5, x, 9 ⇛ 2, 5, x

$x^2 = 2^2 + 5^2$
$= 29$ ✓
$\therefore x = \sqrt{29}$
$= 5.4$ (to 1 dec. pl.)
Perimeter $= 9 + 5 + 11 + 5.4$
$= 30.4$ cm ✓

4

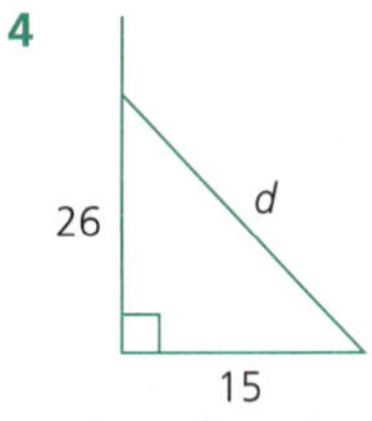

$d^2 = 26^2 + 15^2$
$= 901$
$d = \sqrt{901}$
$= 30.0166$ ✓
Four wires, so length
$= 4 \times 30.01666$
$= 120.066$
$= 120.1$ (to 1 dec. pl.)
Length $= 120.1$ metres ✓

5

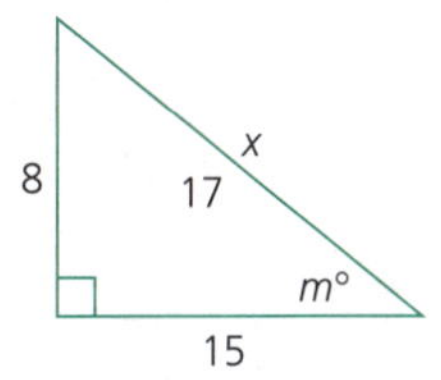

$x^2 = 8^2 + 15^2$
$= 289$
$x = \sqrt{289}$
$= 17$ ✓

a $\tan m° = \frac{8}{15}$ ✓

b $\sin m° = \frac{8}{17}$ ✓

c $\cos m° = \frac{15}{17}$ ✓

6 a 60°57′ ✓

b 77°42′ ✓

c 63°28′ ✓

d 54°52′ ✓

7

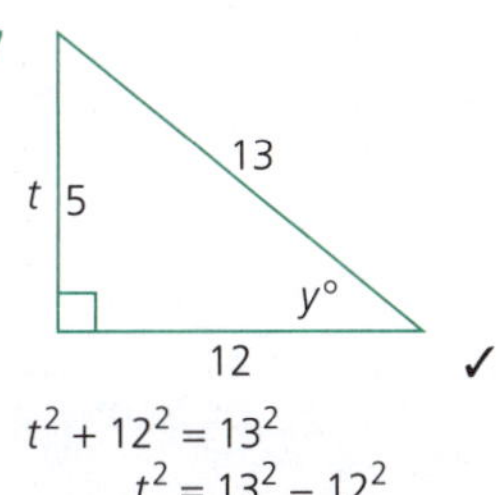

✓

$t^2 + 12^2 = 13^2$
$t^2 = 13^2 - 12^2$
$= 25$
$t = \sqrt{25} = 5$ ✓

$\therefore \tan y° = \frac{5}{12}$ ✓

8 a $\frac{x}{28} = \cos 46°17'$ ✓
$x = 28 \cos 46°17'$
$= 19.3505\ldots$
$= 19.4$ cm (to 1 dec. pl.) ✓

b $\frac{y}{4.8} = \tan 32°51'$ ✓
$y = 4.8 \tan 32°51'$
$= 3.09932\ldots$
$= 3.1$ cm (to 1 dec. pl.) ✓

c $\frac{t}{18.4} = \sin 53°8'$ ✓
$t = 18.4 \sin 53°8'$
$= 14.7206\ldots$
$= 14.7$ cm (to 1 dec. pl.) ✓

d $\frac{14.9}{y} = \cos 29°26'$ ✓
$y = \frac{14.9}{\cos 29°26'}$
$= 17.108\ldots$
$= 17.1$ cm (to 1 dec. pl.) ✓

9 a $\cos m° = \frac{19}{29}$ ✓
$m° = 49°4'$ ✓

b $\tan m° = \frac{29}{19}$ ✓
$m° = 56°46'$ ✓

10

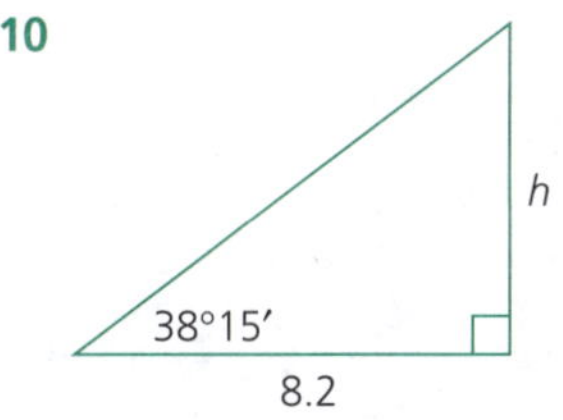

✓

$\frac{h}{8.2} = \tan 38°15'$ ✓
$h = 8.2 \tan 38°15'$
$= 6.464\ldots$
$= 6.5$ (to 1 dec. pl.)
Height of the flag pole is 6.5 metres. ✓

11

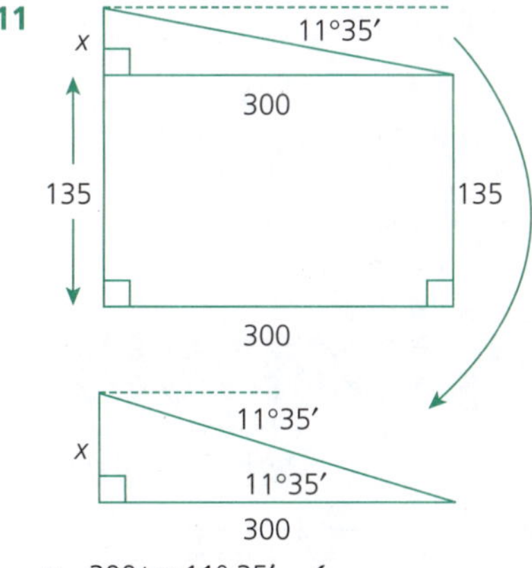

$x = 300 \tan 11° 35'$ ✓
$= 61.49\ldots$
$= 61.5$ (to 1 dec. pl.)
Height of LMC centre $= 135 + 61.5$ ✓
$= 196.5$
Height is 196.5 metres. ✓

12

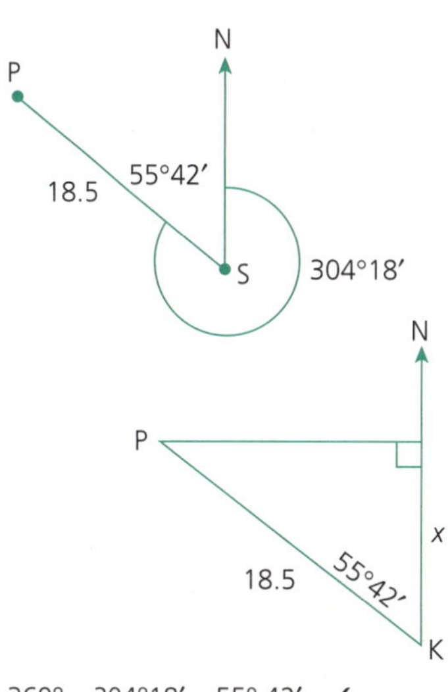

$360° - 304°18' = 55° 42'$ ✓
$\frac{x}{18.5} = \cos 55° 42'$
$x = 18.5 \cos 55° 42'$ ✓
$= 10.425\ldots$
$= 10.4$ (to 1 dec. pl.)
Johannes is 10.4 km North of Kotara. ✓

13

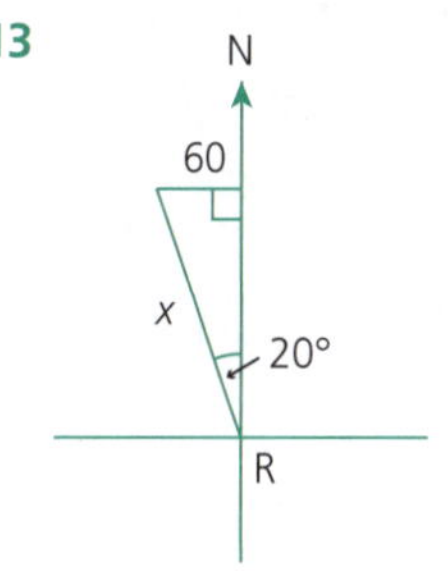

$\frac{60}{x} = \sin 20°$ ✓
$x = \frac{60}{\sin 20°}$
$= 175.4$ (1 dec. pl.) $\therefore$ 175.4 km ✓

Chapter 9

Measurement

Intermediate Test

pp. 135–136

1 a Perimeter $= 12 + \frac{1}{2} \times 2 \times \pi \times 6$ ✓
$= 30.84955592$
$= 30.85$ (to 2 dec. pl.)
$\therefore$ The perimeter is 30.85 cm. ✓

b Perimeter $= 18 + \frac{3}{4} \times 2 \times \pi \times 9$ ✓
$= 60.41150082$
$= 60.41$ (to 2 dec. pl.)
$\therefore$ The perimeter is 60.41 cm. ✓

2 Perimeter $= 36 + \frac{1}{2} \times 2 \times \pi \times 6$ ✓
$= 54.84955592$
$= 54.8$ (to 3 sig. fig.)
$\therefore$ The perimeter is 54.8 cm. ✓

3 a Area = 12×10
$= 120$
$\therefore$ The area is 120 cm^2. ✓

b Area = $\frac{1}{2} \times 12(8 + 20)$ ✓
$= 168$
$\therefore$ The area is 168 cm^2. ✓

4 Area = $\frac{1}{4} \times \pi \times 4^2$
$= 4\pi$
$\therefore$ The area is 4π cm^2. ✓✓

5 $SA = 2\pi r^2 + 2\pi rh$
$= 2 \times \pi \times 7^2 + 2 \times \pi \times 7 \times 20$ ✓
$= 1187.522023$
$= 1187.52$ (to 2 dec. pl.)
$\therefore$ The surface area of 1187.52 cm^2. ✓

6 $SA = 2\pi rh$ (curved surface area)
$= 2 \times \pi \times 0.1 \times 14$ ✓
$= 8.79645943$
$= 9$ (to nearest whole number)
$\therefore$ The curved surface area is 9 m^2. ✓

7 Area of trapezium = $\frac{1}{2} \times 4(6 + 9)$
$= 30$
$\therefore$ Sum of area
$= 2 \times 30 + 5 \times 3 + 6 \times 3 + 4 \times 3 + 9 \times 3$ ✓
$= 132$
$\therefore$ The surface area is 132 cm^2. ✓

8 a Using Pythagoras:

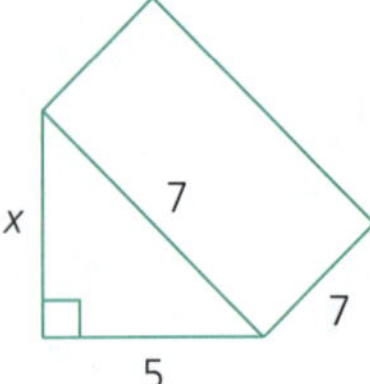

$x^2 = 7^2 - 5^2$
$= 24$
$x = \sqrt{24}$ ✓
$V = \frac{1}{2} \times 5 \times \sqrt{24} \times 7$
$= 85.732141\ldots$
$= 85.73$ (2 dec. pl.)
$\therefore$ Volume of 85.73 cm^3. ✓

b radius = 6 cm
$V = \frac{1}{2} \times \pi r^2 h$
$= \frac{1}{2} \times \pi \times 6^2 \times 16$ ✓
$= 904.7786842\ldots$
$= 904.78$ (2 dec. pl.)
$\therefore$ Volume of 904.78 cm^3. ✓

9 $V = \pi r^2 h$
$= \pi \times 4^2 \times 10$ ✓
$= 160\pi$
$\therefore$ Volume of 160π cm^3. ✓

10 $V = \pi r^2 h$
$= \pi \times 8^2 \times 23$
$= 4624.424386$ ✓
$= 4624$ (to nearest whole number)
$\therefore$ The volume is 4624 m^3.
$\therefore$ The capacity is 4624 kL. ✓

11

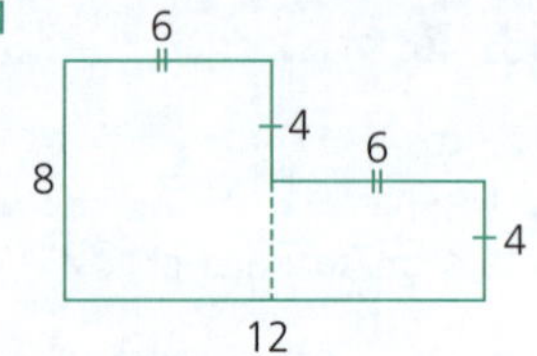

Area = $8 \times 6 + 6 \times 4$
$= 72$
$V = 72 \times 5$
$= 360$ ✓
$\therefore$ The volume is 360 cm^3.
$\therefore$ The capacity is 360 mL. ✓

12 Present capacity of container A
$= \frac{1}{2} \times 12 \times 5 \times 10$
$= 300$ ✓
New capacity of container B
$= 340 + 300$
$= 640$
$\therefore$ Container B contains 640 cm^3 ✓
$\therefore 640 = 8 \times 20 \times d$
$640 = 160d$
$160d = 640$
$d = 4$
$\therefore$ The depth of water is 4 cm. ✓

Advanced Test

pp. 137–138

1 Perimeter
$= 9 + 9 + \frac{1}{6} \times 2 \times \pi \times 9$ ✓
$= 27.42477796$
$= 27.42$ (to 2 dec. pl.)
$\therefore$ The perimeter is 27.42 cm. ✓

2 a Perimeter = $16 + 16 + \frac{1}{\cancel{4}_1} \times 2 \times \pi \times \cancel{16}^4$
$= 32 + 8\pi$
$\therefore$ Perimeter is $(8\pi + 32)$ cm. ✓✓

b

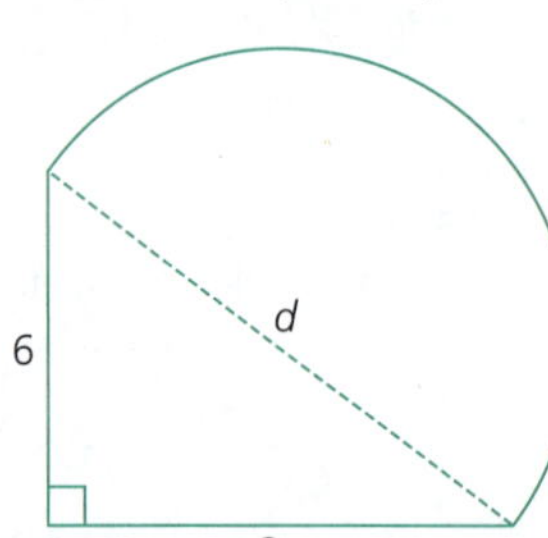

$d^2 = 6^2 + 8^2$
$= 36 + 64$
$= 100$
$d = \sqrt{100}$
$= 10$ ✓
$\therefore$ Perimeter = $6 + 8 + \frac{1}{\cancel{2}} \times \cancel{2} \times \pi \times 5$
$= 14 + 5\pi$
$\therefore$ The perimeter is $(5\pi + 14)$ cm. ✓

3

12

20

B

Area = $12 \times 8 + \pi \times 4^2$ ✓
$= 146.2654825$
$= 146$ (3 sig. fig.)
$\therefore$ The area is 146 cm^2. ✓

4

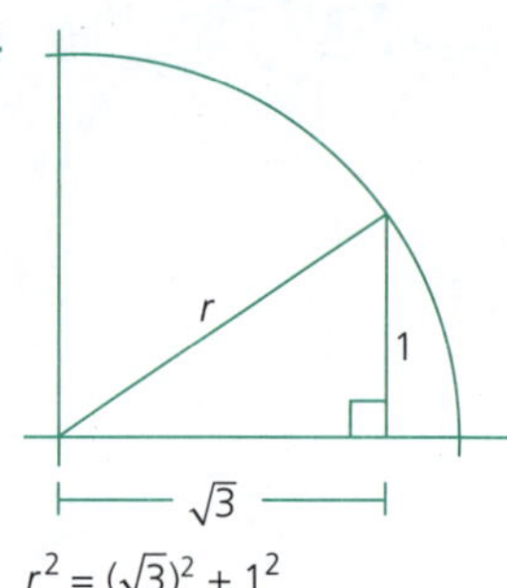

$r^2 = (\sqrt{3})^2 + 1^2$
$= 3 + 1$
$= 4$
$r = 2$
$\therefore$ Area = $\frac{60}{360} \times \pi \times 2^2$ ✓
$= \frac{1}{6} \times \pi \times 4$
$= \frac{2\pi}{3}$
$\therefore$ The area is $\frac{2\pi}{3}$ $units^2$. ✓

5 Area of kite = $\frac{1}{2}xy$
$= \frac{1}{2} \times 12 \times 9$
$= 54$
$\therefore$ The area is 54 cm^2. ✓✓

6 Curved surface area
$= 2\pi rh$
$= 2 \times \pi \times 6 \times 20$ ✓
$= 753.9822369$
$= 753.982$ (to 3 dec. pl.)
$\therefore$ The curved surface area is 753.982 cm^2. ✓

7 Area of parallelogram = 16×12
$= 192$
$\therefore$ Area is 192 cm^2. ✓
Volume = 192×5
$= 960$
$\therefore$ Volume is 960 cm^3. ✓

8

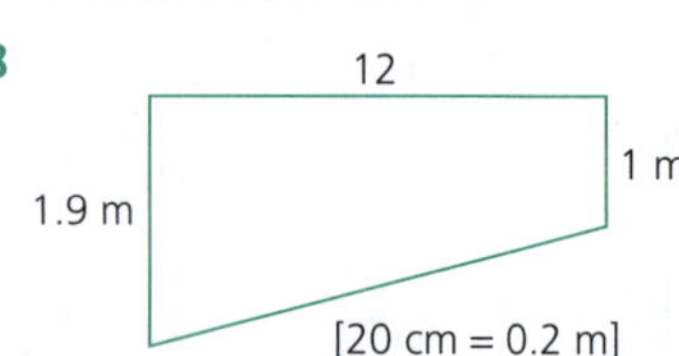

Area of trapezium = $\frac{1}{2} \times 12\,(1.9 + 1)$
$= 6 \times 2.9$
$= 17.4$ ✓
$\therefore$ Volume = 17.4×7
$= 121.8$
Capacity = 121.8 kL
$\therefore$ The capacity is 121.8 kL. ✓

9 Capacity = 4 L
$\therefore$ Volume = 4000 cm^3
$\therefore 4000 = \pi \times 10^2 \times h$ ✓
$4000 = 100\pi h$
$\therefore h = \frac{4000}{100\pi}$
$= \frac{40}{\pi}$
$= 12.732\,395\,45$
= 13 (to nearest whole number)
$\therefore$ The depth of the paint is 13 cm. ✓

10 Area of front surface
$= 10 \times 18 - 5 \times \pi \times 1^2$
$= 164.292\,0367$ ✓
$\therefore$ Volume $= 164.292\,0367 \times 6$
$= 985.752\,2204$
= 985.75 (to 2 dec. pl.)
$\therefore$ The volume is 985.75 cm^3. ✓

11 Area $= \frac{1}{2} \times 4a(2a + 1 + 5a + 6)$
$= 2a(7a + 7)$ ✓
$\therefore$ Volume $= 2a(7a + 7) \times 2a$
$= 4a^2(7a + 7)$
$= 28a^2(a + 1)$ ✓
$\therefore$ The volume is $28a^2(a + 1)$ cm^3.

12 Surface area of cube:
$6s^2 = 24$
$s^2 = 4$
$s = 2$
$\therefore$ Cube has side length of 2 metres ✓
$V = lbh$
$= 2 \times 2 \times 0.1$
$= 0.4$
As 1 m^3 = 1000 L,
$0.4 \times 1000 = 400$
$\therefore$ There are 400 litres in the cube. ✓

13 $V = \frac{3}{4} \times \pi r^2 h$
$= \frac{3}{4} \times \pi \times 3^2 \times 4$ ✓
$= 84.823\,001\,65\ldots$
= 84.82 (2 dec. pl.)
$\therefore$ Volume of 84.82 cm^3. ✓
$SA = \frac{3}{4} \times 2\pi rh + \frac{3}{4} \times 2\pi r^2 + 2 \times 3 \times 4$ ✓
$= \frac{3}{4} \times 2 \times \pi \times 3 \times 4 + \frac{3}{4} \times 2 \times \pi \times 3^2 + 2 \times 3 \times 4$
$= 122.960\,1866\ldots$
= 122.96 (2 dec. pl.)
$\therefore$ Surface area of 122.96 cm^2. ✓

Chapter 10

Geometry

Intermediate Test

pp. 160–161

1 a $x + 72 + 64 = 180$
$x = 44$
(angle sum of $\triangle$ is 180°) ✓

b $x + 52 = 90$ (complementary angles)
$x = 38$ ✓

c $x - 10 = 70$
$x = 80$
(vertically opposite angles) ✓

d $x + 90 + 90 + 51 = 360$
$x = 129$
(angle sum of a quadrilateral) ✓

e $x + 73 = 180$
$x = 107$
(co-interior angles) ✓

f $x + x + 90 = 180$
$x = 45$
(angles on a straight line) ✓

g $x + 90 + 70 + 80 = 360$
$x = 120$
(angles about a point) ✓

h $x = 60$
(angle in an equilateral triangle) ✓

i $x = 55$
(base angles of isosceles $\triangle$) ✓

2 D. In a parallelogram opposite sides are parallel. ✓

3 a Congruent ✓

b Similar ✓

4 C. In a rhombus the diagonals bisect each other at right angles. ✓

5 True ✓

6 $x = 50$ (corresponding angles in congruent $\triangle$s are equal)
$y = 4$ (corresponding sides in congruent $\triangle$s are equal) ✓✓

7 $m = 47$
(corresponding angles in similar triangles are equal)
$\frac{n}{6} = \frac{5}{10}$ $\therefore n = 3$ ✓✓
(corresponding sides in similar triangles are in the same ratio)

8 a Angle sum of polygon
$= (n - 2) \times 180°$
$= 13 \times 180°$
$= 2340°$ ✓✓

b Interior angle $= \frac{2340°}{15}$
$= 156°$ ✓

c The exterior angle is supplementary to the interior angle.
Therefore, exterior angle
$= 180° - 156°$
$= 24°$ ✓

9 Angle sum of polygon $= (n - 2) \times 180$
$\therefore 1080 = (n - 2) \times 180$
$1080 = 180n - 360$
$180n = 1440$
$n = 8$ ✓✓

10 a (SSS) ✓

b (AAS) ✓

c (SAS) ✓

11 $y = 59$ (corresponding angles in similar figures) ✓
$\frac{x}{12} = \frac{5}{15}$ (corresponding sides in similar figures) ✓
$\therefore x = 4$
$\frac{z}{7} = \frac{15}{5}$ (corresponding sides in similar figures) ✓
$\therefore z = 21$

12 a $y = 65$ (corresponding angles and AB || CD) ✓

b $x = 54$ (alternate angles AB || CD) ✓

c $z = 61$ (angle sum of $\triangle$ABC) ✓

13 a In $\triangle$s ABX and ZYX
$\angle XAB = \angle XZY$ (given)
$\angle BXA = \angle YXZ$ (vertically opposite)
$\therefore \triangle ABX \,|||\, \triangle ZYX$ (equiangular) ✓✓

b $\frac{x}{10} = \frac{3}{6}$ (corresponding sides in similar triangles in the same ratio)
$x = 5$ ✓

14 a In $\triangle$s ADC and CBA
AD = CB (S) given
DC = BA (S) given
AC = AC (S) common
$\therefore \triangle ADC \equiv \triangle CBA$ (SSS) ✓✓✓

b $\angle ADC = \angle CBA$ (corresponding angles of congruent triangles) ✓

Advanced Test

pp. 162–163

1 $3x + 2x + x = 180$ (angle sum of triangle ABC)
$6x = 180$
$x = 30$ ✓✓

2 Figure ABCD is a rhombus.
$x + 40 = 180$ (co-interior angles and AB || DC)
$x = 140$ ✓✓

3 Angle sum of exterior angles of polygon
$= 360$
$\therefore 5x = 360$
$x = 72$ ✓✓

4 $\angle ABC = \angle BCA = 75°$ (base angles of isosceles $\triangle$ABC)
$2x + 75 + 75 = 180$ (angle sum of $\triangle$ABC)
$2x = 30$
$\therefore \angle CAB = 2x° = 30°$ ✓✓

5 $70 + 60 + x = 180$ ($\angle$BCD is cointerior to $\angle$CDE and BC || ED)
$\therefore x = 50$ ✓✓

6 B is not true (diagonals of a rhombus are not equal, unless rhombus is a square). ✓

7 B. I and III are similar, since $\frac{6}{3} = \frac{4}{2}$. ✓

8 I and II are congruent (AAS test). ✓

9 $\frac{x}{4} = \frac{12}{2}$ (corresponding sides of similar △s are in the same ratio)
$\therefore x = 24$ ✓✓

10 a Triangle A is congruent to triangle C. ✓

b Triangle B is similar to triangle C ✓
The corresponding sides are in the same ratio = 1 : 1

c DE corresponds to NP. ✓

d ∠NMP corresponds to ∠DFE. ✓

e True. The area and perimeter of two congruent triangles is always the same. ✓

11 a True. ∠BCA = ∠YZX
(corresponding angles in similar △s are always equal) ✓

b False. AB ≠ XY
(corresponding sides in similar △s are not equal) ✓

c False.
(areas of similar △s are not equal) ✓

d True. $\frac{AB}{XY} = \frac{AC}{XZ}$
(corresponding sides in similar △s are in the same ratio) ✓

12 a Size of interior angle
$= \frac{(n-2) \times 180}{n}$
$= \frac{(11-2) \times 180}{11}$
$= \frac{1620}{11}$
$= 147\frac{3}{11}°$ ✓✓

b Angle sum of exterior angles
$= 360°$
$\therefore 12x = 360$
$x = 30$ ✓✓
Therefore, exterior angle of regular dodecagon = 30°.

c Let x = number of sides of the polygon.
$\therefore 9x = 360$
$x = 40$
Therefore, the polygon has 40 sides. ✓✓

13 a In △s ABC and QPC
∠ACB = ∠QCP = 90° (A)
(vertically opposite angles)
∠BAC = ∠PQC(A)
(alternate and AB ∥ PQ)
AC = QC = 10 cm (S) given
∴ △ABC ≡ △QPC (AAS) ✓✓✓

b $a = b$ (AB = PQ corresponding sides in congruent △s)
△ABC is right-angled using Pythagoras' Theorem
$b^2 = 6^2 + 10^2$
$b^2 = 36 + 100$
$b^2 = 136$
$b = \sqrt{136}$
$\therefore a = b = \sqrt{136}$
$c = 6$ (PC = BC corresponding sides in congruent △s) ✓✓✓

14 a In △s ABC and ADE
∠CAB = ∠EAD (common)
∠BCA = ∠DEA (corresponding and BC ∥ DE)
∴ △ABC ||| △ADE (equiangular) ✓✓

b Let DB = x cm.

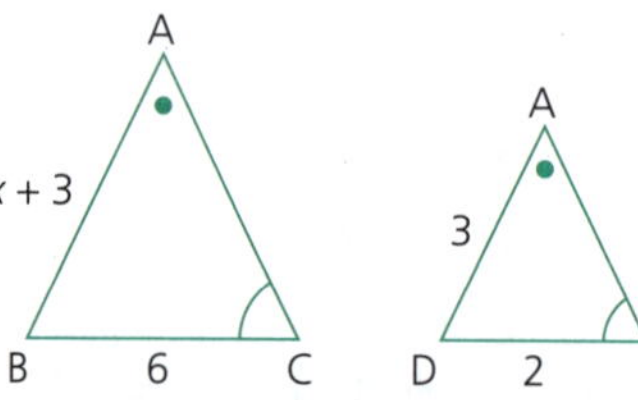

$\frac{x+3}{3} = \frac{6}{2}$ (corresponding sides in similar △s are in the same ratio)
$\frac{x+3}{3} = 3$
$x + 3 = 9$
$x = 6$
∴ The length DB = 6 cm. ✓✓

Chapter 11 Statistics

Intermediate Test

p. 183

1 C as it has two modes: 3 and 4 ✓

2 a Mean $= \frac{26 + 32 + 28 + 39 + 30}{5}$
$= 31$
∴ Mean number of points is 31. ✓

b No change means the team needs to score the mean
∴ Team needs to score 31 points. ✓

c Mean after 6 games $= \frac{31 \times 5 + 37}{6}$
$= 32$
∴ Mean after 6 games is 32. ✓

d Mean after 7 games = 30
Points scored in 7th game
$= 7 \times 30 - 6 \times 32$
$= 18$ ✓
Score was 28 – 18.
∴ Team lost by 10 points. ✓

3 a Mean = 4 ✓
Mode = 5 ✓
Range = 6 – 1
= 5 ✓

b 1, 2, 3, 3, 3, 4, 4, 4, 5, 5, 5, 5, 6, 6
Median = 4 ✓
Lower quartile = 3
Upper quartile = 5
∴ Interquartile range = 2 ✓

c

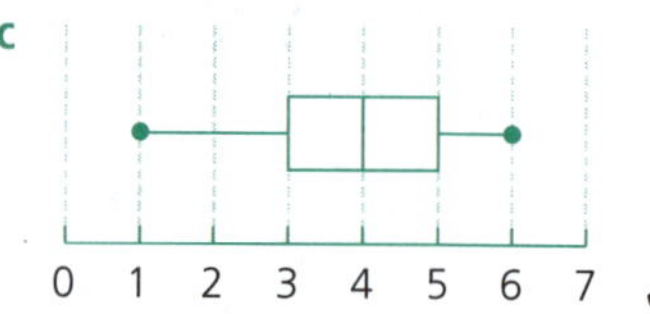

✓✓

4 a Highest was 24 out of 25; lowest was 3 out of 25. ✓✓

b **i** 50% ✓

ii 75% ✓

iii 88% is 22 out of 25
∴ 25% ✓

c 25% of 28 students = 7
∴ 7 students ✓

d 72% is 18 out of 25
∴ bottom half ✓

e Negatively skewed ✓

5 a **i** Wildfires: 51 – 8 = 43
Diamonds: 57 – 19 = 38
∴ Wildfires has bigger range. ✓

ii Wildfires: 20.5; Diamonds: 36
∴ Diamonds has bigger mode. ✓

iii Wildfires: 19; Diamonds: 41
∴ Diamonds has bigger median. ✓

b Wildfires is positively skewed; Diamonds is negatively skewed. ✓

6 a

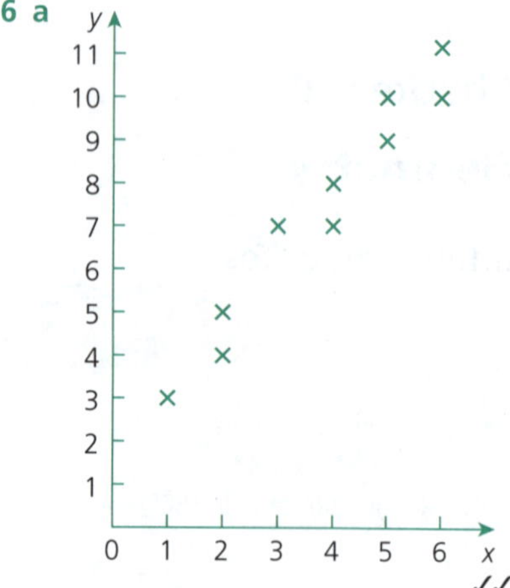

✓✓

b y ✓

c Positive ✓

d Strong ✓

Advanced Test

pp. 184–185

1 a Normally distributed ✓

b Bi-modal ✓

c Negatively skewed ✓

d Positively skewed ✓

2 a Mode = 10; Range = 35 ✓✓

b 18 ✓

c Lower quartile = 10;
Upper quartile = 31 ✓
∴ Interquartile range = 21 ✓

d

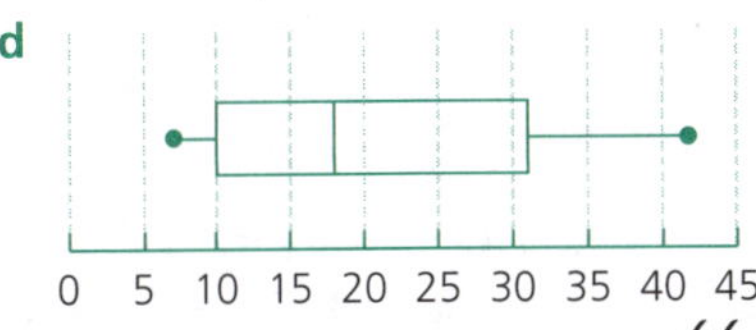

✓✓

3 a 30 ✓

b Negatively skewed ✓

c The class mean is $9\frac{1}{6}$. This means that this class's mean is higher than the year mean. ✓✓

4 a \$10 ✓

b 75% ✓

c 100 books ✓

d Shop A had IQR of 25, which is higher than Shop B's IQR of 15. ✓✓

5 a

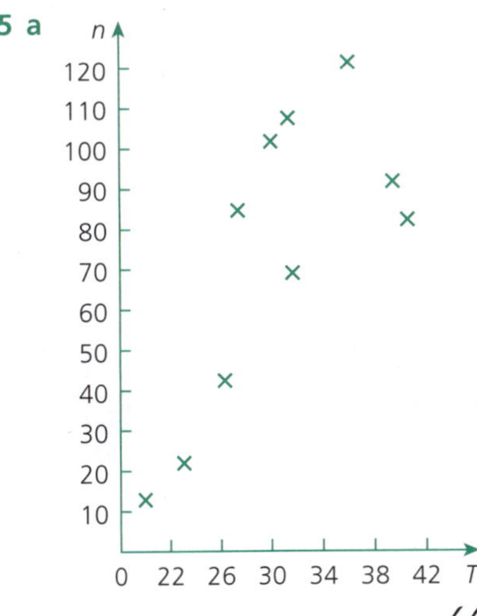

✓✓

b Positive ✓

c Strong relationship. The higher the temperature, the more people at the beach. ✓✓

6 Lower quartile = 7.4;
Upper quartile = 10.1
∴ Interquartile range = 10.1 – 7.4
= 2.7 ✓

7 a Weak ✓

b 50 ✓

c $a = 45$; $b = 40$ ✓✓

Chapter 12
Probability

Intermediate Test

pp. 200–201

1 $P(\text{head}) = \frac{1}{2}$

a $P(\text{HH}) = \frac{1}{2} \times \frac{1}{2}$
$= \frac{1}{4}$ ✓

b $P(\text{at least are head}) = 1 + P(\text{TT})$
$= 1 - \frac{1}{4}$
$= \frac{3}{4}$ ✓

2 a

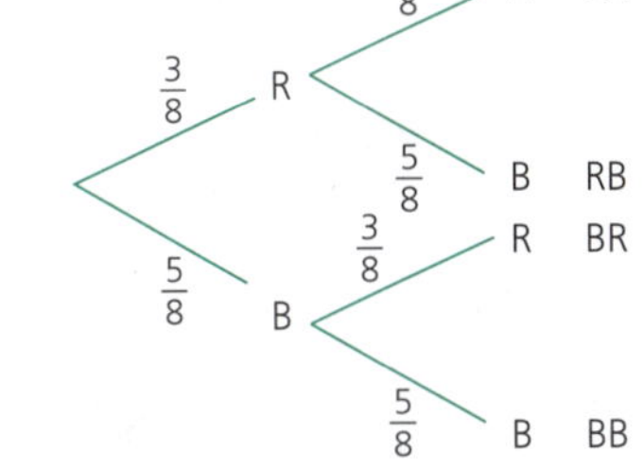

✓

b i $P(\text{BB}) = \frac{5}{8} \times \frac{5}{8}$
$= \frac{25}{64}$ ✓

ii $P(\text{RB}) + P(\text{BR}) = \frac{3}{8} \times \frac{5}{8} + \frac{5}{8} \times \frac{3}{8}$
$= \frac{15}{64} + \frac{15}{64}$
$= \frac{15}{32}$ ✓

3 a As $P(3) = \frac{1}{4}$, $P(\text{green}) = \frac{1}{3}$
∴ $P(3 \text{ and green}) = \frac{1}{4} \times \frac{1}{3}$
$= \frac{1}{12}$ ✓

b $P(\text{even}) = \frac{1}{2}$, $P(\text{not blue}) = \frac{2}{3}$
∴ $P(\text{even and not blue}) = \frac{1}{2} \times \frac{2}{3}$
$= \frac{1}{3}$ ✓

4 a As 100 males,
$P(\text{male smoker}) = \frac{27}{100}$ ✓

b As 100 females,
$P(\text{female non-smoker}) = \frac{64}{100}$
$= \frac{16}{25}$ ✓

c As 200 people,
$P(\text{non-smoker}) = \frac{137}{200}$ ✓

5 There are 80 commuters involved.

a $P(\text{bus}) = \frac{37}{80}$ ✓

b $P(\text{train only}) = \frac{43}{80}$ ✓

6 a $P(\text{head and 5}) = \frac{1}{12}$ ✓

b $P(\text{tail and odd}) = \frac{3}{12} = \frac{1}{4}$ ✓

c $P(\text{head and less than 3}) = \frac{2}{12} = \frac{1}{6}$ ✓

7 a

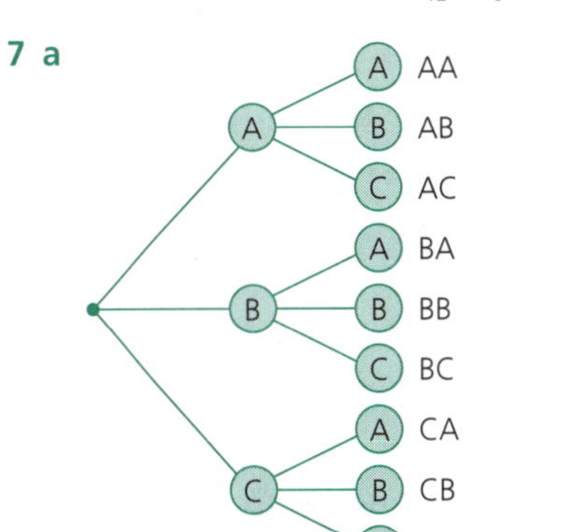

✓✓

b i $P(\text{BB}) = \frac{1}{9}$ ✓

ii $P(\text{BA}) = \frac{1}{9}$ ✓

iii $P(\text{at least 1 B}) = \frac{5}{9}$ ✓

iv $P(\text{AA, BB or CC}) = \frac{3}{9} = \frac{1}{3}$ ✓

v $P(\text{different letters}) = 1 - P(\text{same})$
$= 1 - \frac{1}{3}$
$= \frac{2}{3}$ ✓

8 a

4/7 G
3/6 G
3/6 W
3/7 W
4/6 G
2/6 W

✓✓

b i $P(\text{GW}) = \frac{4}{7} \times \frac{3}{6}$
$= \frac{2}{7}$ ✓

ii $P(\text{WG}) = \frac{3}{7} \times \frac{4}{6}$
$= \frac{2}{7}$ ✓

iii $P(\text{WW}) = \frac{3}{7} \times \frac{2}{6}$
$= \frac{1}{7}$ ✓

9 a $P(\text{less than } 6) = \frac{5}{20} = \frac{1}{4}$ ✓

b $P(\text{even, given that more than } 10)$
$= \frac{5}{10} = \frac{1}{2}$ ✓

c $P(\text{factor of } 24\text{, given that multiple of } 4)$
$= \frac{3}{5}$ ✓

Advanced Test

pp. 202–203

1 a

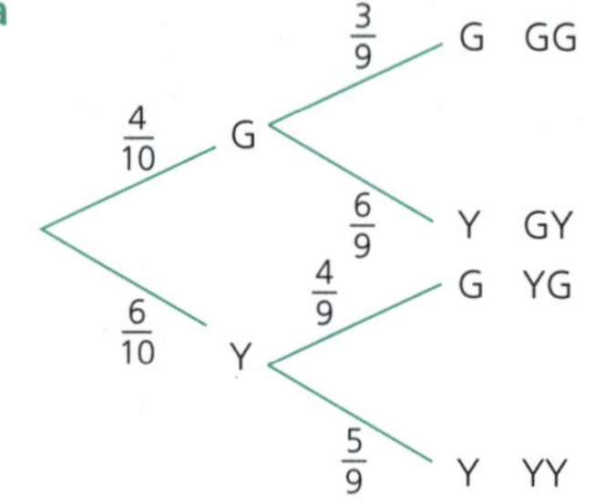

✓

b i $P(GY) + P(yG)$
$= \frac{4}{10} \times \frac{6}{9} + \frac{6}{10} \times \frac{4}{9}$
$= \frac{24}{90} + \frac{24}{90}$
$= \frac{48}{90}$
$= \frac{8}{15}$ ✓

ii $P(\text{at least 1 yellow})$
$= 1 - P(\text{no yellow})$
$= 1 - P(GG)$
$= 1 - \frac{4}{10} \times \frac{3}{9}$
$= 1 - \frac{12}{90}$
$= \frac{78}{90}$
$= \frac{13}{15}$ ✓

2 a 100 women surveyed
$P(\text{women voted Lib–NP}) = \frac{45}{100}$
$= 0.45$ ✓

b 200 people surveyed
85 people voted ALP
$\therefore P(\text{voted ALP}) = \frac{85}{200}$
$= 0.425$ ✓

3 a i As $18 + 16 + 12 + 15 = 61$
$\therefore P(\text{like PDHPE}) = \frac{61}{100}$
$= 0.61$ ✓

ii As $13 + 18 = 31$
$\therefore P(\text{like English, not Science})$
$= \frac{31}{100}$
$= 0.31$ ✓

b As $14 + 16 + 12 + 12 = 54$,
there are 54 students who like Science.

i As $14 + 16 = 30$
$\therefore P(\text{like English}) = \frac{30}{54}$
$= \frac{5}{9}$ ✓

ii $P(\text{like all 3}) = \frac{16}{54}$
$= \frac{8}{27}$ ✓

4 a

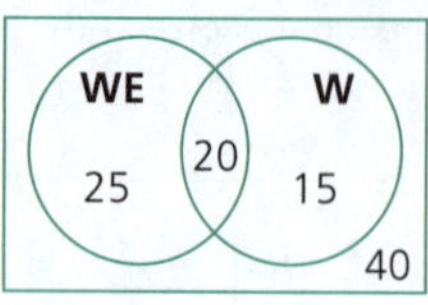

WE: Weekend sport
W: Week sport
✓

b $P(\text{both}) = \frac{20}{100} = \frac{1}{5}$ ✓

c $P(\text{week, not weekend}) = \frac{15}{55} = \frac{3}{11}$ ✓

5 a $P(\text{bat and bowler left hand})$
$= \frac{6}{30} = \frac{1}{5}$ ✓

b $P(\text{right-hand bowler, given left-hand batsman}) = \frac{4}{10} = \frac{2}{5}$ ✓

6 a

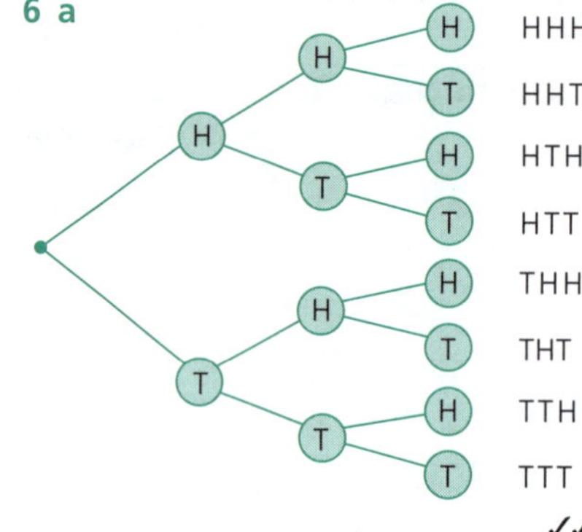

✓✓

b i $P(TTT) = \frac{1}{8}$ ✓

ii $P(HTT, THT, TTH) = \frac{3}{8}$ ✓

iii $P(\text{at least one H})$
$= 1 - P(\text{no heads})$
$= 1 - P(TTT)$
$= 1 - \frac{1}{8}$
$= \frac{7}{8}$ ✓

7 a

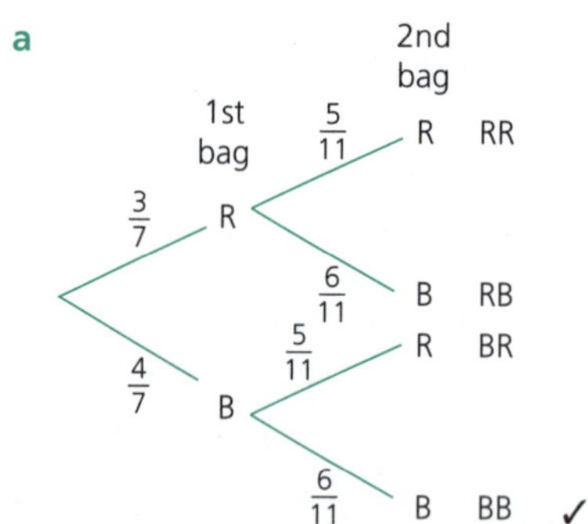

✓

b i $P(RR) = \frac{3}{7} \times \frac{5}{11}$
$= \frac{15}{77}$ ✓

ii $P(\text{RR or BB}) = \frac{3}{7} \times \frac{5}{11} + \frac{4}{7} \times \frac{6}{11}$
$= \frac{39}{77}$ ✓

iii $P(\text{RB or BR}) = 1 - P(\text{same})$
$= 1 - \frac{39}{77}$
$= \frac{38}{77}$ ✓

iv $P(\text{at least a blue}) = 1 - P(\text{no blue})$
$= 1 - P(RR)$
$= 1 - \frac{15}{77}$
$= \frac{62}{77}$ ✓

8 a

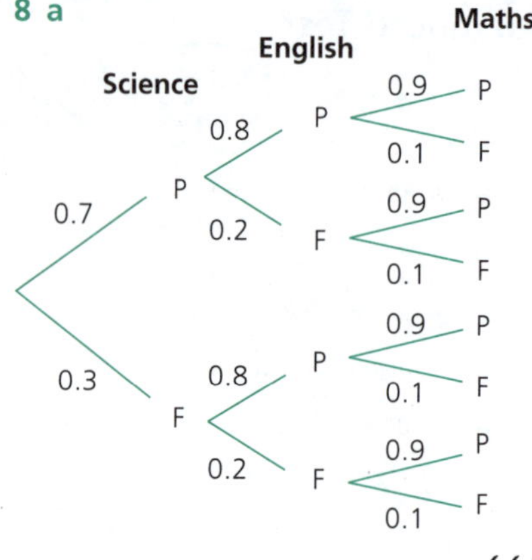

✓✓✓

b i $P(PPP) = 0.7 \times 0.8 \times 0.9$
$= 0.504$ ✓

ii $P(FFP) = 0.3 \times 0.2 \times 0.9$
$= 0.054$ ✓

iii $P(\text{FPP, PFP or PPF})$
$= 0.3 \times 0.8 \times 0.9 + 0.7 \times 0.2 \times 0.9 + 0.7 \times 0.8 \times 0.1$
$= 0.398$ ✓

Worked Solutions to Sample Examinations

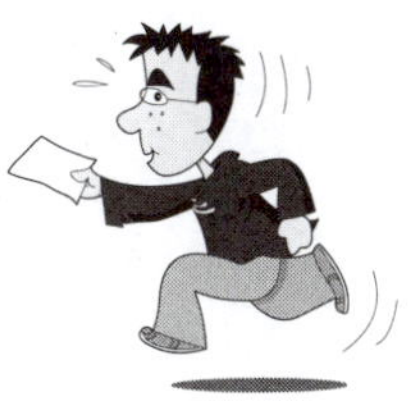

Examination Paper 1

pp. 204–208

1 a Total pay
$= \$13.40 \times 30 + \$13.40 \times 1.5 \times 6 + \$13.40 \times 2 \times 5$
$= \$656.60$
$\therefore$ Delta is paid \$656.60. ✓

b $I = Prn$ ✓
$= 6800 \times 0.0725 \times 8$
$= 3944$
$\therefore$ The interest is \$3944. ✓

c Tax $= 3572 + 0.325 \times 25\,420$ ✓
$= 11\,833.5$
$\therefore$ Rhett pays \$11 833.50 income tax. ✓

d **i** Cost $= 0.8 \times 80$
$= 64$
$\therefore$ Mitchell pays \$64. ✓

ii Laura's cost $= 0.9 \times 0.8 \times 80$
$= 57.60$
$\therefore$ Laura pays \$57.60
$\therefore$ Laura's discount is \$22.40. ✓

e Final price $= 76 \times 0.85 \times 0.8$
$= 51.68$
$\therefore$ The drill will cost \$51.68. ✓

f Value $= 680 \times 0.82^2$
$= 457.232$
$\therefore$ Value is \$457.23. ✓

g 6% p.a. = 0.5% per month
4 years = 48 months
Compounded amount
$= 25\,000 \times 1.005^{48}$
$= 31\,762.229\,03$ ✓
$= 31\,762.23$ (to 2 dec. pl.)
$\therefore$ Compounded amount is \$31 762.23.
$\therefore$ The interest is \$6762.23. ✓

2 a **i** $2 - (x + 3) = 2 - x - 3$
$= -1 - x$ ✓

ii $4(2x - 5) - 3(4x - 2)$
$= 8x - 20 - 12x + 6$ ✓
$= -4x - 14$ ✓

b **i** $12ab - 4a = 4a(3b - 1)$ ✓

ii $x^{12} + x^6 = x^6(x^6 + 1)$ ✓

c **i** $\frac{2x}{3} + \frac{x}{4} = \frac{8x + 3x}{12}$
$= \frac{11x}{12}$ ✓

ii $\frac{6x^2y}{3ab} \div \frac{xy^2}{2a} = \frac{{}^{2}\cancel{6}x^2y}{{}_{1}\cancel{3}ab} \times \frac{2\cancel{a}}{\cancel{x}y^2}$ ✓
$= \frac{4x}{by}$ ✓

d $4x^{-3} \div 2x^{-4} = 2x$ ✓

e $(3x - 4)(2x + 1) = 6x^2 + 3x - 8x - 4$
$= 6x^2 - 5x - 4$ ✓

3 a $4 + 2x \le 6$
$2x \le 6 - 4$
$2x \le 2$
$x \le 1$ ✓

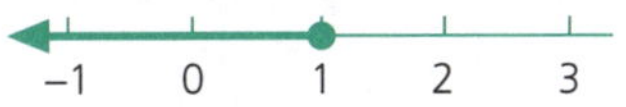

✓

b ${}^{4}\cancel{12}\left[\frac{4y - 5}{\cancel{3}_1}\right] = {}^{3}\cancel{12}\left[\frac{3y + 5}{\cancel{4}_1}\right]$
$16y - 20 = 9y + 15$ ✓
$16y - 9y = 15 + 20$
$\frac{7y}{7} = \frac{35}{7}$
$y = 5$ ✓

c $S = \frac{12}{2}[2 \times 4 + (12 - 1) \times -2]$
$= 6[8 - 22]$ ✓
$= 6 \times -14$
$= -84$ ✓

d $3y^2 - 75 = 0$
$3y^2 = 75$
$y^2 = 25$
$y = \pm 5$ ✓

e $x^2 - 3x + 2 = 0$
$(x - 2)(x - 1) = 0$
$x = 2, 1$ ✓

f Let b = cost of burger
d = cost of drink
$\therefore 3b + 2d = 1580$ (1)
$2b + d = 980$ (2) ✓
$2 \times (2)$ $4b + 2d = 1960$ (3)
$(3) - (1)$ $b = 380$
$\therefore$ The hamburger cost \$3.80. ✓

4 a **i** $d = \sqrt{(3 + 2)^2 + (-1 - 4)^2}$
$= \sqrt{25 + 25}$
$= \sqrt{50}$
$\therefore \sqrt{50}$ units ✓

ii $\text{MP} = \left(\frac{-2 + 3}{2}, \frac{4 - 1}{2}\right)$
$= \left(\frac{1}{2}, \frac{3}{2}\right)$
$= \left(\frac{1}{2}, 1\frac{1}{2}\right)$ ✓

iii $m = \frac{-1 - 4}{3 + 2}$
$= \frac{-5}{5}$
$= -1$ ✓

b $y = 2x - 3$

x	0	1	2
y	−3	−1	1

✓

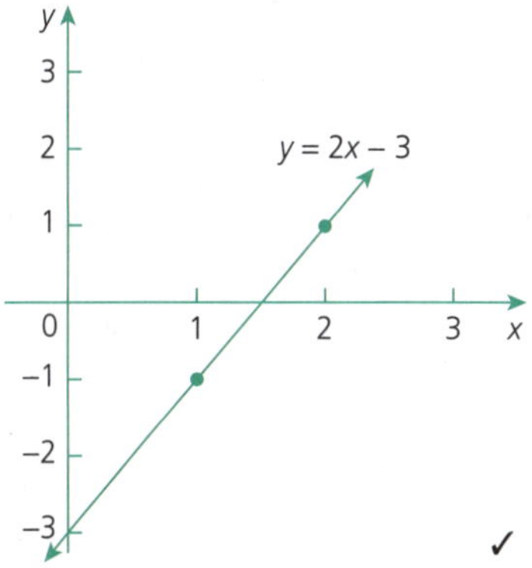

✓

c Subs. $(5, p)$ into $2x - y + 3 = 0$
$\therefore 2(5) - p + 3 = 0$ ✓
$10 - p + 3 = 0$
$13 = p$
$\therefore p = 13$ ✓

d $m = \frac{-4}{2}$ $b = 4$ ✓
$= -2$
$y = mx + b$
$\therefore y = -2x + 4$ ✓

e D. $y = 4 - x^2$
(Parabola, concave down, y-intercept of 4, x-intercept of ± 2) ✓

5 a **i** $30 \times 60 \times 60 = 108\,000$
$\therefore$ 108 000 m/h = 108 km/h ✓

b **i** $160 \div 8 = 20$
$\therefore$ 20 L/min ✓

ii $V = 20t$ ✓

iii

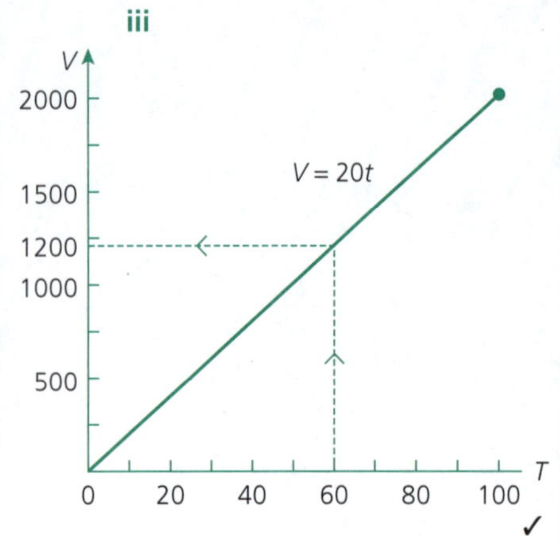

✓

iv From graph, 1200 L ✓

v Subs. $V = 6000$ in $V = 20t$

$6000 = 20t$

$t = 300$

∴ 300 minutes, or 5 hours. ✓

c **i** $y = kx$ ✓

ii Subs. $x = 9$ and $y = 36$ in $y = kx$

$36 = k \times 9$

$9k = 36$

$k = 4$

$\therefore y = 4x$ ✓

iii Subs. $x = 15$ in $y = 4x$

$y = 4 \times 15$

$y = 60$ ✓

d As directly proportional, then $W = kn$.

Subs. $n = 24$ and $W = 444$ in $W = kn$

$444 = k \times 24$

$24k = 444$

$k = 18.5$

$\therefore W = 18.5n$ ✓

6 a $x^2 = 12^2 - 10^2$

$= 144 - 100$

$= 44$

$x = \sqrt{44}$ ✓

b **i** $\frac{x}{50} = \sin 27°16'$

$x = 50° \sin 27°16'$ ✓

$= 22.906\,625\,02$

$= 22.91$ (to 2 dec. pl.) ✓

ii $\frac{24}{x} = \cos 37°$

$x = 24 \times \cos 37°$ ✓

$= 30.051\,255\,8$

$= 30.05$ (to 2 dec. pl.) ✓

c $\cos\theta = \frac{17.6}{23.8}$

$\theta = 42°$ ✓

d $\frac{h}{30} = \tan 28°$

$h = 30 \times \tan 28°$ ✓

$= 15.951\,2895$

$= 16$ (to the nearest whole number)

∴ The height is 16 m. ✓

e $\frac{x}{30} = \cos 40°$

$x = 30 \times \cos 40°$ ✓

$= 22.981\,333\,29$

$= 23.0$ (to 1 dec. pl.)

∴ The distance is 23.0 km. ✓

7 a Per $= 20 + \frac{1}{2} \times 2 \times \pi \times 10$ ✓

$= 51.415\,926\,54$

$= 51.42$ (to 2 dec. pl.)

∴ The perimeter is 51.42 cm. ✓

b

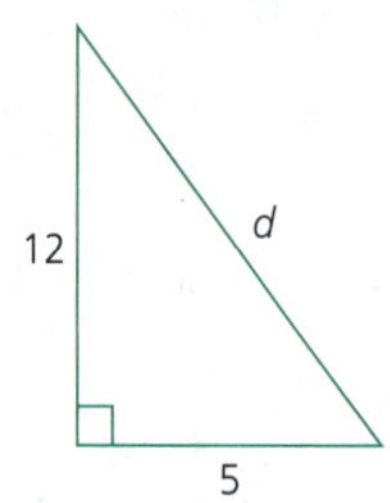

$d^2 = 12^2 + 5^2$

$= 144 + 25$

$= 169$

$d = \sqrt{169}$

$= 13$ ✓

$\therefore$ Area $= \frac{1}{2} \times 5 \times 12 + \frac{1}{2} \times \pi \times 6.5^2$

$= 96.366\,144\,81$

$= 96.37$ (to 2 dec. pl.)

∴ The area is 96.37 cm^2. ✓

c $SA = 2\pi r^2 + 2\pi rh$

$= 2 \times \pi \times 15^2 + 2 \times \pi \times 15 \times 40$ ✓

$= 5183.627\,878$

$= 5183.63$ (to 2 dec. pl.)

∴ Surface area is 5183.63 cm^2. ✓

d $V = \pi r^2 h$

$= \pi \times 6^2 \times 24$ ✓

$= 2714.336\,053\ldots$

$= 2714.34$ (2 dec. pl.)

∴ Volume of 2714.34 cm^3. ✓

e $V = \frac{1}{2}\pi r^2 h$

$= \frac{1}{2} \times \pi \times 0.8^2 \times 3$ ✓

$= 3.015\,928\,947\ldots$

$= 3.016$ (3 dec. pl.)

∴ volume of 3.016 m^3

∴ Capacity of 3016 L ✓

(as 1 m^3 = 1000 L).

8 a **i** ∠ sum of pentagon $= 3 \times 180°$

$= 540°$

∴ each ∠ = 108°

$\therefore x = 72$ ✓

ii Exterior ∠ of regular octagon

$= \frac{360}{8}$

$= 45°$

$\therefore x = 135$ ✓

(co-interior ∠s, parallel lines)

b Exterior ∠ of regular decagon

$= \frac{360}{10}$

$= 36$

∴ each ∠ is 36° ✓

c △ACB ≡ △EDF

(ASA test) ✓

d △ACE ||| △BCD

(equiangular or 2 ∠s equal) ✓

e **i** $\frac{x}{5} = \frac{4}{3}$ ✓

$x = \frac{20}{3}$

$= 6\frac{2}{3}$ ✓

ii $\frac{x}{x + 4} = \frac{6}{10}$ ✓

$10x = 6x + 24$

$4x = 24$

$x = 6$ ✓

f ∠DBC = 60° (alternative ∠s equal, AD || BC)

∠BCE = 30° (angle sum of △BCE) ✓

9 a **i** α 77 ✓

β 76 ✓

ii Lowest score = 47

Highest score = 96

Median = 76

Lower quartile = 60

Upper quartile = 83 ✓✓

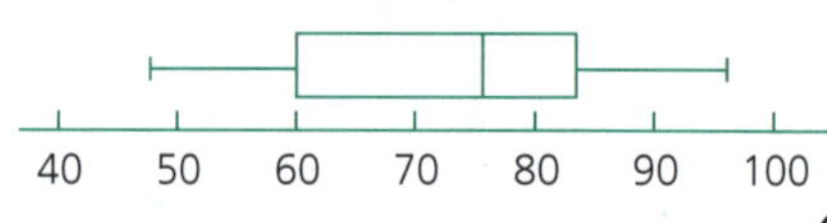

✓

b P(even, given a multiple of 3) $= \frac{2}{5}$ ✓

c See p. 190 for table.

i $P(1) = \frac{1}{36}$ ✓

ii $P(7) = \frac{0}{36}$

$= 0$ ✓

iii P(more than 20) $= \frac{6}{36}$

$= \frac{1}{6}$ ✓

d

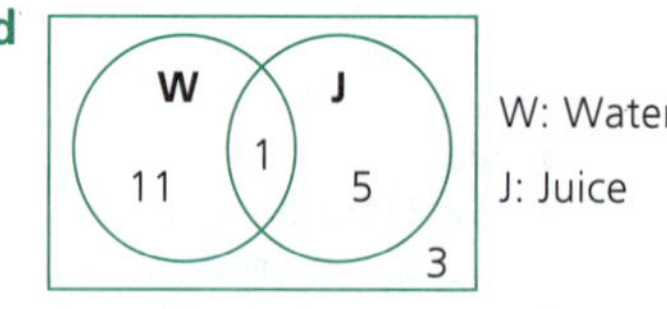

∴ One person ✓

Examination Paper 2

Part A .... pp. 209–216

1 $8.23 \div (2.7 \times 6.3)$

$= 0.483\,833\,039$

$= 0.48$ (to 2 dec. pl.)

∴ C ✓

2 Consider each alternative.
In D, subs. (2, 2) in $y = x^2 - x$:
$$2 = 2^2 - 2$$
$$= 2$$
∴ (2, 2) lies on the parabola
∴ D ✓

3 75% of scores between lower quartile and maximum mark.
As $0.75 \times 40 = 30$, then 30 students scored between 12 and 18.
∴ C ✓

4 Radius $= \sqrt{16} = 4$
∴ B ✓

5 Subs. $x = 3$ and $y = 6$ in $y = kx$:
$$6 = k \times 3$$
$$3k = 6$$
$$k = 2$$
∴ A ✓

6 $219\,000 = 2.19 \times 10^5$
∴ B ✓

7 $x < 2$
∴ A ✓

8 $12\frac{1}{2}\% = 0.125$
∴ B ✓

9 $2000 \div 375 = 5\frac{1}{3}$
∴ 5 full glasses
∴ D ✓

10 $\frac{5x}{4} + \frac{x}{3} = \frac{15x + 4x}{12}$
$$= \frac{19x}{12}$$
∴ C ✓

11 $d = \sqrt{(2+2)^2 + (4-1)^2}$
$$= \sqrt{16 + 9}$$
$$= \sqrt{25}$$
$$= 5$$
∴ 5 units
∴ A ✓

12 Mass $= 0.025 \times 63$
$$= 1.575$$
∴ B ✓

13 Pay $= 496 \div 40 \times 37$
$$= 458.8$$
∴ \$458.80
∴ A ✓

14 $V = \pi \times 2^2 \times 2$
$$= 8\pi$$
∴ Volume of 8π cm^2.
∴ B ✓

15 $BD^2 = 8^2 + 5^2$
$$= 64 + 25$$
$$= 89$$
$BD = \sqrt{89}$ NEW
$$= 9.433\,981\,132$$
= 9 (to the nearest whole number)
∴ D ✓

16 $y = 2x - 1$ has gradient 2. The perpendicular has gradient of $-\frac{1}{2}$.
Consider the alternatives, $y = -\frac{1}{2}x$ has gradient of $-\frac{1}{2}$, ∴ C. ✓

17 The four outcomes are HH, HT, TH, TT.
∴ $P(TT) = \frac{1}{4}$
∴ B ✓

18 $2a^2 = 2 \times 3^2$
$$= 18$$
∴ C ✓

19 $3a^4 \times 2a^2 \times a = 6a^7$
∴ A ✓

20 $V = Ah$
$$= 15 \times 6$$
$$= 90$$
∴ Volume is 90 cm^3.
∴ D ✓

21 4 6 8 (8) 12 16 20
∴ IQR = 16 − 6
= 10
∴ D ✓

22 $\sin\theta = \frac{12}{13}$
∴ C ✓

23 $3(a + b) - (a - b) = 3a + 3b - a + b$
$$= 2a + 4b$$
∴ C ✓

24 $(x - 3)(x + 1) = 0$
$$x = 3, -1$$
∴ B ✓

25 $x^2 - 16 = (x - 4)(x + 4)$
∴ A ✓

26 $\frac{x}{x + 12} = \frac{4}{20}$
$$20x = 4x + 48$$
$$20x - 4x = 48$$
$$\frac{16x}{16} = \frac{48}{16}$$
$$x = 3$$
∴ B ✓

27

7 8 9 10 11 12

∴ B ✓

28 $x = 180 - 2 \times 70$
$$= 180 - 140$$
$$= 40$$
∴ A ✓

29 As 2 h 15 min = 2.25 h,
∴ Dist. = 70×2.25
= 157.5
∴ closest to 158 km
∴ B ✓

30 Per $= 14 + \frac{1}{2} \times 2 \times \pi \times 7$
$$= 35.991\,148\,58$$
= 36 (nearest whole number)
∴ D ✓

31 Each of the 3 △s have 40°, 60°, 80°.
Now 9 cm is opposite 60° in all 3 △s
∴ all congruent
∴ C ✓

32 $(4x^{-3})^2 = 4x^{-3} \times 4x^{-3}$
$$= 16x^{-6}$$
$$= \frac{16}{x^6}$$
∴ B ✓

33 Games $= 2 + 3 + 5 + 4 + 2$
$$= 16$$
∴ 16 games
∴ D ✓

34 $P(\text{green or black}) = \frac{8}{23}$
∴ B ✓

35 Volume = 27 cm^3
∴ Side = 3 cm
∴ Area of face = 9 cm^2
∴ A ✓

36 $\frac{5bc}{3} \div \frac{25bc}{9} = \frac{{}^1\cancel{5}\,\cancel{bc}^1}{\cancel{3}_1} \times \frac{\cancel{9}^3}{{}_5\cancel{25}\,\cancel{bc}_1}$
$$= \frac{3}{5}$$
∴ D ✓

37

y, 4, O, 2, x

Area $= \frac{1}{2} \times 2 \times 4$
$$= 4$$
∴ Area is 4 units^2
∴ B ✓

38

y, 0, x

∴ D ✓

39 $\frac{x}{52} = \cos 41°$
$$x = 52 \times \cos 41°$$
$$= 39.244\,898\,17$$
= 39 (to nearest whole number)
∴ BC is about 39 cm
∴ D ✓

40 Value $= 25\,400 \times 0.83$
$$= 21\,082$$
∴ \$21 082
∴ B ✓

41 Area $= 5 \times 6 + \frac{1}{2} \times 6 \times 4$
$= 30 + 12$
$= 42$
∴ Area is 42 cm^2
∴ A ✓

42 13 scores
∴ 7th score is median
∴ 6 is median
∴ C ✓

43 $y = 2x^2 - 1$ is concave up with y-intercept of -1.
∴ C ✓

44 $\frac{1}{2}$ of $xy = \frac{xy}{2}$
∴ B ✓

45

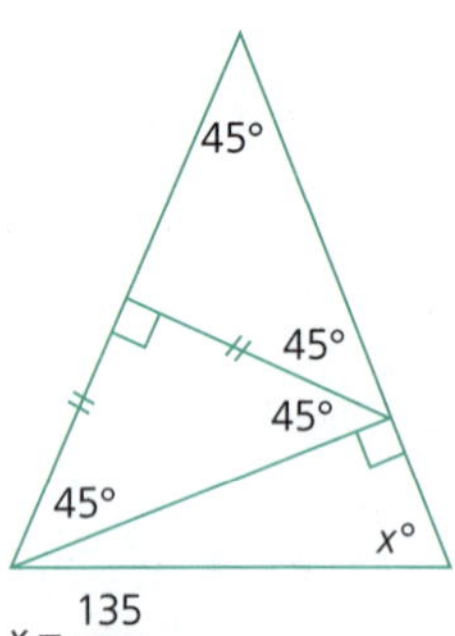

$x = \frac{135}{2}$
$= 67\frac{1}{2}$
∴ D ✓

Part B .... pp. 214–216

46 a i

2nd ball

1st ball

- R ($\frac{3}{5}$)
  - R ($\frac{2}{4}$): RR
  - B ($\frac{2}{4}$): RB
- B ($\frac{2}{5}$)
  - R ($\frac{3}{4}$): BR
  - B ($\frac{1}{4}$): BB ✓

ii (A) $P(RR) = \frac{3}{5} \times \frac{2}{4}$
$= \frac{3}{10}$ ✓

(B) $P(\text{RB or BR}) = \frac{3}{5} \times \frac{2}{4} + \frac{2}{5} \times \frac{3}{4}$
$= \frac{3}{5}$ ✓

b Wage $= 40 \times 18.42 + 14 \times 3.45$
$= 785.1$
∴ \$785.10 ✓

c Shaded area $= (20 \times 10) - (6 \times 6)$
$= 164$ cm^2 ✓

d $\tan C = \frac{17}{14}$
∴ $C = 51°$ ✓

e

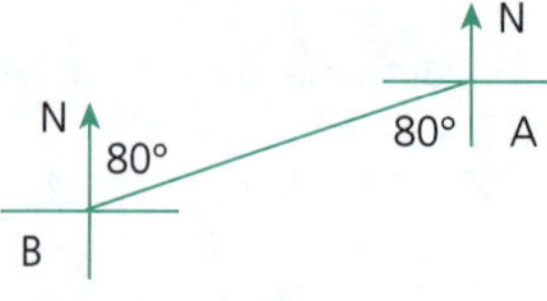

∴ Bearing of B from A is $180 + 80 = 260°$. ✓

f $V = \pi r^2 h$
$= \pi \times 6^2 \times 40$
$= 4524$ cm^3 (to the nearest cm^3) ✓

g Cash price = \$1489
Cost by instalments
= \$250 + \$70 × 24
= \$1930
Amount saved by paying cash
= \$1930 – \$1489
= \$441 ✓

h Tax = 54 547 + 0.45 × 8400
= 58 327
∴ Vance pays \$58 327 income tax. ✓

i Area to be painted
$= 96 \times 2$
$= 192$ m^2
Number of litres required
$= 192 \div 16$
$= 12$
Number of 4 litre cans
$= 3$
Cost of painting
= 3 × \$28.20
= \$84.60 ✓

j $I = Prn$
$= \frac{\$5400 \times 8.5 \times 4}{100}$
= \$1836 ✓

k Surface area of prism
$= 2(6 \times 4) + 2(4 \times 2) + 2(6 \times 2)$
$= (2 \times 24) + (2 \times 8) + (2 \times 12)$
$= 88$ cm^2 ✓

l 90 m/s
= 5400 m/min
= 324 000 m/h
= 324 km/h ✓

m Tax = $\frac{1}{3}$ of \$1500
= \$500
Remainder = \$1000
Bills = $\frac{1}{2}$ of \$1000
= \$500
∴ Peter banks \$1000 – \$500
= \$500 ✓

47 a Normal hourly rate
\$1004 ÷ 40
\$25.10 per hour ✓

b Amount earned for overtime
\$1229.90 – \$1004
= \$225.90
Number of hours paid for overtime
= \$225.90 ÷ \$25.10
= 9
Number of hours of overtime worked
$= 9 \div 1\frac{1}{2}$
$= 6$ ✓

c New hourly rate
= \$25.10 + 5% of \$25.10
= \$26.36 (to the nearest cent) ✓

48 a $c = 85 + \frac{75n}{4}$
When $n = 5$,
$c = 85 + \frac{75 \times 5}{4}$
$= 178.75$
∴ The electrician charges \$178.75 for 5 hours work. ✓

b If $c = 235$, find n.
$c = 85 + \frac{75n}{4}$
∴ $235 = 85 + \frac{75n}{4}$
$\frac{75n}{4} = 235 - 85$ ✓
$\frac{75n}{4} = 150$
$75n = 600$
$n = \frac{600}{75}$
$= 8$,
∴ The electrician worked for 8 hours. ✓

49 a Midpoint $= \left(\frac{x_1 + x_2}{2}, \frac{y_1 + y_2}{2}\right)$
$[A(\overset{x_1}{x}, \overset{y_1}{5}); B(\overset{x_2}{-1}, -\overset{y_2}{3})]$
$= \left(\frac{3 + -1}{2}, \frac{5 + -3}{2}\right)$
$= \left(\frac{2}{2}, \frac{2}{2}\right)$
$= (1, 1)$
∴ The midpoint of the interval AB is the point (1, 1). ✓

b

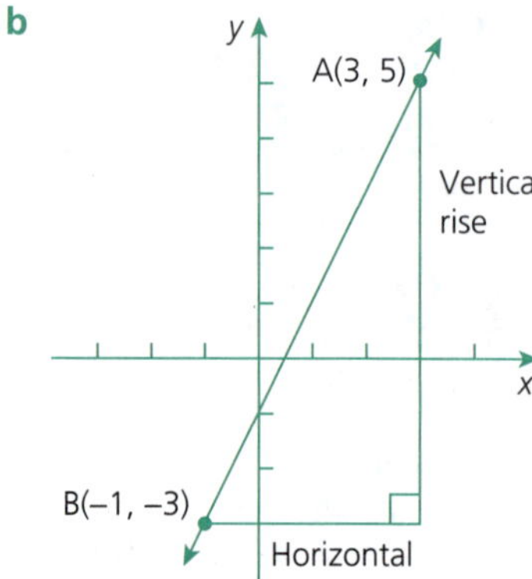

Vertical rise = 8
Horizontal run = 4

$$\text{Gradient} = \frac{\text{Verical rise}}{\text{Horizontal run}}$$
$$= \frac{8}{4}$$
$$= 2$$

Note: The gradient is positive.
∴ The gradient m of the line AB = 2. ✓

c The equation of any line is of the form $y = mx + b$, where m = gradient, and b = y-intercept.
For the line AB, $m = 2$ and $b = -1$,
∴ the equation of line AB is
$y = 2x - 1$. ✓

50 a P(movies and shopped) = $\frac{10}{60} = \frac{1}{6}$ ✓

b P(movie, given beach) = $\frac{11}{25}$ ✓

c P(shop, given beach or movies)
$= \frac{16}{48} = \frac{1}{3}$ ✓

51 a $y = 65$
[∠DCE is corresponding angle to ∠BAC and AB || CD] ✓

b $z + 50 + 65 = 180$
[∠ACE is a straight angle]
∴ $z = 65$ ✓

c $x + 65 + 65 = 180$
[angle sum of triangle ACB]
∴ $x = 50$
or
$x = 50$
[∠CBA is alternate to ∠BCD and AB || CD] ✓

Index

Index

Index

Index

NOTES

NOTES

NOTES